BIOLOGICAL AND MEDICAL PHYSICS, BIOMEDICAL ENGINEERING

For further volumes:
http://www.springer.com/series/3740

BIOLOGICAL AND MEDICAL PHYSICS, BIOMEDICAL ENGINEERING

The fields of biological and medical physics and biomedical engineering are broad, multidisciplinary and dynamic. They lie at the crossroads of frontier research in physics, biology, chemistry, and medicine. The Biological and Medical Physics, Biomedical Engineering Series is intended to be comprehensive, covering a broad range of topics important to the study of the physical, chemical and biological sciences. Its goal is to provide scientists and engineers with textbooks, monographs, and reference works to address the growing need for information.

Books in the series emphasize established and emergent areas of science including molecular, membrane, and mathematical biophysics; photosynthetic energy harvesting and conversion; information processing; physical principles of genetics; sensory communications; automata networks, neural networks, and cellular automata. Equally important will be coverage of applied aspects of biological and medical physics and biomedical engineering such as molecular electronic components and devices, biosensors, medicine, imaging, physical principles of renewable energy production, advanced prostheses, and environmental control and engineering.

Gustavo García Gómez-Tejedor
Martina Christina Fuss
Editors

Radiation Damage in Biomolecular Systems

With 137 Figures

Springer

Editors
Prof. Gustavo García Gómez-Tejedor
Instituto de Física Fundamental
Consejo Superior de Investigaciones
Científicas
Serrano 113-bis
28006 Madrid
Spain
g.garcia@iff.csic.es

Dr. Martina Christina Fuss
Instituto de Física Fundamental
Consejo Superior de Investigaciones
Científicas
Serrano 113-bis
28006 Madrid
Spain
mc.fuss@iff.csic.es

ISSN 1618-7210
ISBN 978-94-007-9582-2 ISBN 978-94-007-2564-5 (eBook)
DOI 10.1007/978-94-007-2564-5
Springer Dordrecht Heidelberg London New York

Printed on acid-free paper

Springer is part of Springer Science+Business Media (www.springer.com)

Contents

Preface

Since the discovery of x-rays and radioactivity, ionizing radiations have been widely applied in medicine both for diagnostic and therapy purposes. Risks associated to radiation exposure and handling led to a parallel development of the radiation protection area. For years, macroscopic magnitudes as the absorbed dose, or the energy deposited along radiation tracks, have been used in those applications to quantify radiation effects. Although this approach can be appropriate for high radiation fluxes it does not give information about low dose stochastic effects or radiation damage induced relatively distant from the irradiated regions. This motivated systematic studies of radiation damage at the molecular level. Pioneering experiments done by Sanche and co-workers in 2000 showed that low energy secondary electrons, which are abundantly generated along radiation tracks, are the main responsible for radiation damage through successive interactions with the molecular constituents of the medium. Apart from ionizing processes, which are customary related to radiation damage, below the ionization level low energy electrons can induce molecular fragmentation via dissociative processes as internal excitation and electron attachment. These ideas prompted collaborative projects between different research groups from European countries together with other specialists from Canada, USA and Australia.

This book is intended to summarize some of the advances achieved by these research groups after more than ten years of studies on radiation damage in biomolecular systems.

An extensive first part deals with recent experimental and theoretical findings on radiation induced damage on the molecular level. Satisfying their broad importance as secondary particles in radiation-matter interactions, it includes many contributions on electron and positron collisions with biologically relevant molecules (water, nucleobases and other DNA/RNA building blocks, amino acids, and other polyatomic molecules of biological interest). Following, X-ray and ion interactions are covered.

Part II addresses different approaches to radiation damage modelling. After a general overview of Monte Carlo methods, chapter contents feature some detailed accounts of mostly theoretical efforts to obtain the necessary input data on positron, electron, proton, and ion transport in water and DNA components and the resulting simulations. Closing this part, a sophisticated application in clinical context is presented.

In the third part, biomedical aspects of radiation effects are treated on different scales. After an introductory review on DNA damage repair, radiation-induced alterations in genes, chromosomes, cells and tissues, and, finally, the whole organism level (radiotherapy) are described. After the rather physics-oriented focus of the previous parts, the gradual transition to biology and medicine with growing size of the object studied is here illustrated.

Finally, Part IV is dedicated to current trends/novel techniques in radiation research and the applications hence arising. It includes new developments in radiotherapy and related cancer therapies as well as technical optimizations of accelerators and totally new equipment designs, giving a glimpse of the near future of radiation-based medical treatments.

We acknowledge the Spanish Ministerio de Ciencia e Innovación (Project FIS 2009-08246), Universidad Nacional de Educación a Distancia and Consejo Superior de Investigaciones Científicas for their financial support for meetings and discussions which motivated this book.

Madrid

Gustavo García Gómez-Tejedor
Martina Christina Fuss

Acronyms

3DCRT	Three-dimensional conformal radiation therapy
A	Adenine
AER	Auger-electron-emitting radiopharmaceuticals
ALARA	As low as reasonably achievable
BER	Base-excision repair
C	Cytosine
CA	Chromosome aberration
CC	Close-coupling
CL	Crosslinks
CNDO	Complete neglect of differential overlap
CRT	Chemoradiation therapy
CS	Cross section
CT	Computed tomography
CTV	Clinical target volume (demonstrated and/or suspected tumour)
DCS	Differential cross section(s)
DDR	DNA damage response
DEA	Dissociative electron attachment
DFT	Density functional theory
DNA	Deoxyribonucleic acid
DOSD	Dipole oscillator strength distribution
DSB	Double-strand break(s)
DWA	Dielectric wall accelerator
EDTA	Ethylenediaminetetraacetic acid
EF	Enhancement factor(s)
ELF	Energy loss function
EMMA	Electron Model for Many Applications
ESD	Electron stimulated desorption
ESRF	European Synchrotron Radiation Facility
EUROCARE	European Cancer Registry-Based Study of Survival and Care of Cancer Patients
FFAG	Fixed field alternating gradient

FISH	Fluorescence *in situ* hybridisation
FWHM	Full width at half maximum
G	Guanine
GFP	Green fluorescence protein
GNP	Gold nanoparticle(s)
GO	Gene ontology
GTV	Gross tumour volume (demonstrated tumour)
HDR	High dose-rate brachytherapy)
HGI	High gradient insulator
HPLC	High performance liquid chromatography
HR	Homologous recombination
IAM	Independent atom model
IARC	International Agency on Research of Cancer
ICRU	International Commission on Radiation Units and Measurements
IMPT	Intensity modulated proton therapy
IMRT	Intensity-Modulated Radiation Therapy
IR	Incident radiation
LEE	Low energy electron(s)
LET	Linear energy transfer
LNT	Linear non-threshold
MBRT	Minibeam radiation therapy
MC	Monte Carlo
ML	Monolayer(s)
MMR	Mismatch repair
MRS	Magnetic resonance spectroscopy
MRSI	Magnetic resonance spectroscopic imaging
MRT	Microbeam radiation therapy
MS	Mass spectrometry
NDC	Negative differential conductivity
NER	Nucleotide-excision repair
NHEJ	Nonhomologous end-joining
NLS	Nuclear localization sequence
PAMELA	Particle Accelerator for Medical Applications
PCC	Premature chromosome condensation
PCR	Polymerase chain reaction
PDT	Photodynamic therapy
PES	Potential energy surface
PET	Positron emission tomography
PI	Post-irradiation
PRRT	Peptide-receptor radionuclide therapy
Ps	Positronium, the bound-state of an electron and a positron
PTV	Planning target volume
PVDR	Peak-to-valley dose ratio
qRT-PCR	Quantitative real-time PCR
RAE	Resonant Auger electron

RBE	Relative biological effectiveness
RIBE	Radiation-induced bystander effect
RIDGE	Region of increased gene expression
RIT	Radio-immunotherapy
RNA	Ribonucleic acid
ROS	Reactive oxygen species
RPA	Random phase approximation
RT	Radiation therapy
SB	Strand break(s)
SCAR	Screening-corrected additivity rule
SE	Secondary electron(s)
SIB	Simultaneous integrated boost
SMC	Schwinger Multichannel
SPECT	Single photon emission coupled tomography
SSB	Single strand break(s)
SSRT	Stereotactic synchrotron radiation therapy
T	Thymine
TCS	Total cross section
TEPC	Tissue-equivalent proportional counter
THF	Tetrahydrofuran
TNF	Tumor necrosis factor
TNI	Transient negative ion(s)
TOF	Time-of-flight
TPS	Treatment-planning system
Tris	Tris (hydroxymethyl) aminomethane
TRK	Thomas-Reiche-Kuhn
U	Uracil
UHV	Ultra high vacuum
UV	Ultraviolet
WHO	World Health Association
XPS	X-ray photoemission spectroscopy

Part I
Radiation Induced Damage at the Molecular Level

Chapter 1
Nanoscale Dynamics of Radiosensitivity: Role of Low Energy Electrons

Léon Sanche

Abstract This chapter addresses the nanoscale dynamics involved in the sensitization of biological cells to ionizing radiation. More specifically, it describes the role of low energy electrons (LEE) in radiosensitization by gold nanoparticles and chemotherapeutic agents, as well as potential applications to radiotherapy. The basic mechanisms of action of the LEE generated within nanoscopic volumes by ionizing radiation are described in solid water ice and various forms of DNA. These latter include the subunits (i.e., a base, a sugar or the phosphate group), short single strands (i.e., oligonucleotides) and plasmid and linear DNA. By comparing the results from experiments with the different forms of the DNA molecule and theory, it is possible to determine fundamental mechanisms that are involved in the dissociation of the subunits, base release and the production of single, double-strand breaks and cross-links. Below 15 eV, LEE localize on DNA subunits to form transient negative ions. Such states can damage DNA by dissociating into a stable anion and radical fragment(s), via dissociative electron attachment, or by decaying into dissociative electronically excited states. LEE can also transfer from one DNA subunit to another, particularly from a base to the phosphate group, where they can induce cleavage of the C-O bond (i.e., break a strand). DNA damage and the corresponding nanoscale dynamics are found to be modified in the presence of other cellular constituents. For example, condensing on DNA the most abundant cellular molecule, H_2O, induces the formation of a new type of transient anion whose parent is a H_2O-DNA complex.

L. Sanche (✉)
Group in the Radiation Sciences, Faculty of Medicine, Université de Sherbrooke, Sherbrooke, QC Canada J1H 5N4
e-mail: Leon.Sanche@USherbrooke.ca

G.G. Gómez-Tejedor and M.C. Fuss (eds.), *Radiation Damage in Biomolecular Systems*, Biological and Medical Physics, Biomedical Engineering,
DOI 10.1007/978-94-007-2564-5_1, © Springer Science+Business Media B.V. 2012

1.1 Introduction

When high-energy radiation interacts with living tissue it produces a range of structural and chemical modifications that can affect biological function. These modifications occur via the production of intermediate species, which include excited atoms and molecules, radicals, ions, and secondary electrons (SE). Intermediate species are created in large quantities ($\sim 4 \times 10^4$ by a 1 MeV particle) within nanoscopic volumes along ionization tracks [1–3] and carry most of the energy of the initial fast particle. Most SE have low energies with a distribution that lies essentially below 70 eV and a most probable energy around 9–10 eV [3, 4] as shown in Fig. 1.1. This figure exhibits the SE density produced by high-energy proton and He^+ absorbed in water as function of SE energy. Obviously, electrons of low energy ($\leq$30 eV) are much more numerous than those at higher energies. Below about 300 eV, electrons have thermalization distances of the order of 10 nm [5], *which essentially define the initial volumes of energy deposition by high energy radiation.* In these nanoscopic volumes, usually called "spurs", the highly excited atomic, molecular and radical species, ions, and low-energy electrons (LEE) can induce non-thermal reactions within femtosecond times. A majority of the reactive species, which initiate further chemical reactions, are created by the SE. Thus, to fundamentally understand radiosensitivity of biological systems, we must investigate the nanoscale dynamics of these radiation energy deposition processes. In particular, the mechanisms of action of LEE with biomolecules must be known *in the condensed phase* and ultimately *within biological tissue and cells.*

The nanoscale dynamics of the pre-chemical stage of radiation damage can be illustrated by considering the simple example of the initial interaction of a fast charged particle with a molecular solid composed of organic molecules R-H. As the fast charged particle passes near the molecule R-H, that molecule is perturbed by the rapid change of electric field. Because this perturbation leaves the kinetic

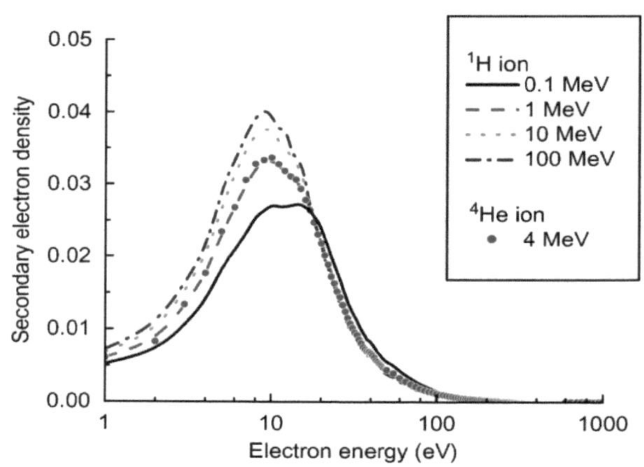

Fig. 1.1 Effect of primary proton and $^4He^+$ on the energy distribution of secondary electrons generated in water [4]. (Reprinted with permission from Reference 4. Copyright 2003 American Chemical Society)

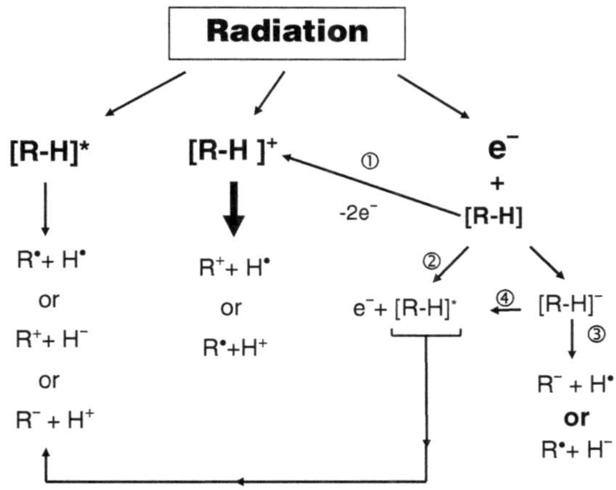

Fig. 1.2 Initial events induced by a fast charged particle that penetrates an organic or bioorganic solid composed of molecules R-H (H = hydrogen, R = rest of molecule). Events induced by secondary electrons are labeled 1 to 4

energy and momentum of the fast particle practically unchanged, the energy transfer can be described as an absorption of electromagnetic radiation by the molecules of the medium [6–8]. This absorption can lead to the formation of electronically excited species [R-H]*, and ionization (i.e., $[R\text{-}H]^+ + e^-$) as shown in Fig. 1.2, and multiple ionization ($[R\text{-}H]^{n+} + ne^-$) [6]. The most probable energy loss of fast primary charged particles to produce [R-H]* and ionization lies about 22 eV [3, 9]. Hence, most of the energy of high energy particles is deposited within irradiated systems by this emission of a succession of low energy quanta. From the values of the optical oscillator strengths for the dissociative electronic excited states of hydrocarbons [9] and a comparison with the normalized dipole oscillator strength distribution for DNA and liquid H_2O [1, 3], one can estimate that about 20% of the energy deposited by fast charged particles in organic matter, including biological and cellular material, leads to [R-H]* production, whereas the rest leads to ionization. The ionization energy is shared as the kinetic energy of SE and potential energy of the cation, with the largest portion of the energy going to SE [3]. The products of ionization and electronic excitation that lead to an hydrogen atom abstraction are shown in Fig. 1.2, as an example of possible fragmentation produced by ionizing radiation; for simplicity, products resulting from multiple ionizations are not shown, but the reaction paths are essentially the same as for single ionization.

A dissociative electronic state [R-H]* can produce two radicals by homologous bond scission or an ion pair (left vertical arrow in Fig. 1.2); however, when ionization occurs, the situation is more complex *due to the emission of at least one SE*. If the positive ion $[R\text{-}H]^+$ is created in a dissociative state, then a cation and a radical can be formed as shown by the larger vertical arrow in Fig. 1.2. The remaining reactions shown in Fig. 1.2 are due to the SE. By interacting with another nearby [R-H] molecule, the SE can produce [10], depending on its energy, further ionization (pathway 1) and/or dissociation (pathway 2), or it can temporarily attach to a nearby molecule to form a temporary transient anion state $[R\text{-}H]^-$, which can subsequently dissociate into the products $R^- + H^\bullet$ or $R^\bullet + H^-$, as shown by the pathway 3 on the right of Fig. 1.2. Alternatively, the electron temporarily captured by the molecule R-H can be reemitted with less energy leaving the molecule in an electronic excited state [R-H]*, which can dissociate as shown in pathway 4. If the temporary state $[R\text{-}H]^-$ is the ground state of the manifold of states of the anion and [R-H] has a positive electron affinity, then the captured electron may permanently stabilize on [R-H] forming a stable anion.

The "resonance" phenomenon which causes the formation of a transient anion [11] usually occurs below 15 eV and rarely above 30 eV. Thus, the electron-molecule interaction at low energies (0–30 eV) can be described in terms of resonant and non-resonant or direct scattering. The latter occurs at all energies above the energy threshold for the observed phenomenon, because the potential interaction is always present. *When the yield of damage to a molecule is measured as a function of electron energy (i.e., the yield function), direct scattering produces a smooth usually rising signal that does not exhibit any particular features.* However, resonance scattering occurs only when the incoming electron occupies a previously unfilled orbital of the molecule. Such an orbital exists at a precise energy [12, 13], and thus, resonance scattering occurs only at specific energy that corresponds to the formation of transient anions. At the resonance energy, the yield of damage or production of a dissociated species is usually enhanced, and a strong peak is observed in the yield function. *The dependence on incident electron energy of the formation of the products shown in Fig. 1.2 or other damage yields is, therefore expected to exhibit pronounced maxima superimposed on an increasing monotonic background that is the result of direct scattering.*

Electron resonances are well-described in the literature and many reviews contain information relevant to this scattering phenomenon [10–17]. There are two major types of resonances or transient anions [11]. If the additional electron occupies a previously unfilled orbital of the target molecule in its ground state, or of a basic subunit of a large biomolecule in its ground state, then the transitory state is referred to as a single-particle or "shape" resonance. The term "shape" resonance applies more specifically when temporary trapping of the electron is due to the shape of the electron-molecule potential, which retains the electron due to an angular momentum barrier. When the transitory anion is formed by two electrons occupying previously unfilled orbitals, the resonance is called "core-excited" and may be referred to as a two-particle, one-hole state. In this case, the electron is captured by the positive electron affinity of an electronically excited state of the molecule, or in the case of large biomolecules, by the electron affinity of a basic subunit of the molecule (e.g., in DNA a base, a sugar or phosphate group). If a momentum barrier in the electron-molecule (or electron-subunit) potential also contributes to retain the electron in the molecule (or in a subunit), the transient anion is referred to as a core-excited shape resonance. All electron resonance processes occur when an incident electron is captured into a usually unfilled orbital of a molecule for a time which is greater than the usual scattering time. If the transient anion state is dissociative and the resonance lifetime is greater than about half a vibration period of the anion, the latter dissociate. The process is called dissociative electron attachment (DEA).

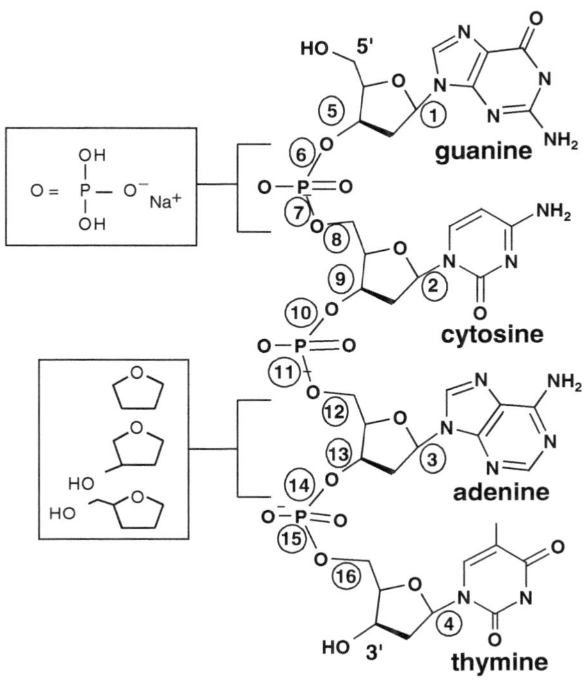

Fig. 1.3 Nomenclature of a four-base single strand of DNA (GCAT). Molecules representing the phosphate group, having Na^+ as the counter ion, and sugar units are shown in the windows on the left (top and bottom, respectively)

The present review article focuses on DNA damage induced by LEE. DNA damage is strongly linked to biological radiosensitivity, because of the essential role played by this molecule in maintaining cellular function. The DNA molecule consists [18] of two long polynucleotide anti-parallel strands composed of repeated sugar-phosphate units. Four bases are covalently linked to the sugar moiety of the backbone and hydrogen bonding between the bases holds the two strands together. The sugar-phosphate backbone of a short DNA strand supporting the four bases guanine (G), cytosine (C), adenine (A) and thymine (T) is shown in Fig. 1.3. According to the example given in Fig. 1.2, secondary species, generated along the radiation tracks close to or within DNA can react with this molecule and cause mutagenic, genotoxic, and other potentially lethal lesions [19–22]. Because LEE are the most abundant of the secondary species produced by the primary interaction, they are expected to play a crucial role in DNA damage.

Review articles on LEE-biomolecule interactions have appeared in the literature [23–25]. A complete account of the condensed-phase experimental results on LEE-induced damage to DNA and its constituents has been published recently by the author [26]. The article contains a description of the relevant experimental techniques. Quantum mechanical models developed to describe LEE scattering from DNA and its subunits have also been completely summarized recently [27]. The insertion of theoretical and experimental data into Monte Carlo calculations and their relationship to radiobiological effectiveness has been reviewed by Nikjoo and

Lindborg [5]. Recently, Zhang and Tan [28] incorporated the structure of DNA into Monte Carlo models of DNA damage induced by LEE. In the present article, we provide additional information on condensed-phase experiments with LEE, that have yet to be reviewed and summarize the earlier experimental data required to link the basic mechanisms of LEE-DNA interactions to radiosensitivity. The most pertinent experiments are described in Sect. 1.3. The main objective is to link the mechanisms responsible for the damage induced to DNA by LEE to radiosensitivity and discuss potential applications.

1.2 Role of LEE in radiosensitization and radiation therapy

The important role of LEE in radiosensitization can be understood from the curves appearing in Figs 1.1, 1.4 and 1.5. Fig. 1.4 represents the yield of DNA strand breaks (SB) induced by 3–100 eV electrons [29]. It shows that electrons of 10 eV induce single strand breaks (SSB) and double strand breaks (DSB) in DNA with a probability similar to that of 100 eV electrons, an energy at which electrons have the highest cross section to damage molecules, principally due to their high ionization yield [14, 30]. Other results obtained below 5 eV indicate that SSB between 0–5 eV are created with similar amplitudes [31]. Electrons below 15 eV are not quite as efficient to break DNA as 100-eV electrons, but they still possess a relatively high probability to induce damage. However, electrons of 100 eV and higher energies are not produced in very high numbers by high-energy ionizing radiation, as can be seen from the distribution in Fig. 1.1.

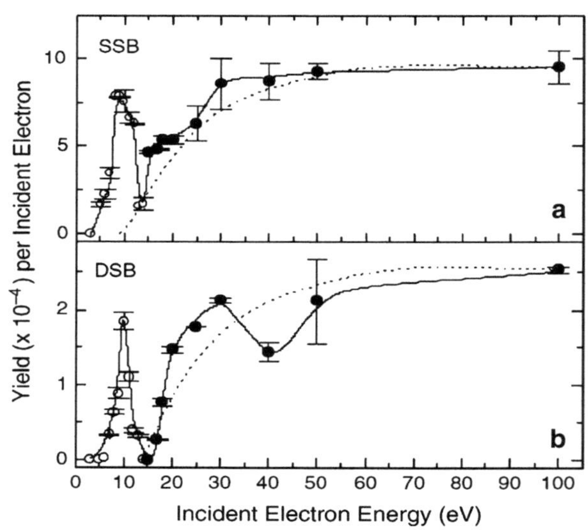

Fig. 1.4 Yields for the induction of (a) single and (b) double strand breaks in DNA films induced by 3–100 eV electron impact [29]. (Reprinted with permission from Reference 29. Copyright 2003 American Chemical Society)

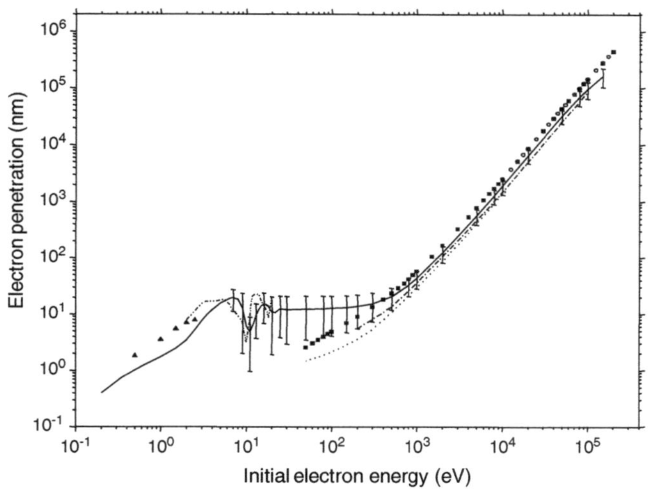

Fig. 1.5 Variation of the electron penetration range in liquid water at 25 °C as a function of initial electron energy between 0.2 eV and 150 keV [145]. (Reprinted with permission from Reference 145)

Another interesting aspect of LEE is their short range in biological matter [32], which is essentially composed of water. Fig. 1.5 shows the penetration range of electrons in water as a function of their energy. Electrons of energy 1–30 eV have a very short range in H_2O and also in DNA [33]. Hence, the damage they produce can be confined within a range of a few biomolecules (e.g., to the DNA of cancer cells and nearby water and proteins). Furthermore, they can easily produce clustered damage in large biomolecule; i.e. a type of lesions which is difficult to repair. In summary, electrons in the 1–30 eV range (i.e., LEE) have considerable efficiency to break DNA and produce multiply damage sites, they are created in much larger numbers than those of higher energies and they have the shortest possible range in biological tissue. They possess all the characteristics to relocate the distribution of radiation energy within nanoscopic volumes, thus increasing within such volumes the radiation dose by orders of magnitude. *In other words, controlling the local density of LEE and the reactions they induced should result in the control, within a short range, of a large amount of the energy deposited by high energy radiation in cells.*

Considering both that DNA is the most important molecular target in radiotherapy [34, 35] and the characteristics of LEE shown in Figs 1.1, 1.4, and 1.5, there are two ways to increase radiosensitization with LEE: *increase their number near DNA or make DNA more sensitive to LEE.* The former can be achieved by placing metallic particles near the DNA of cancer cells, such as gold nanoparticles (GNP), which absorb more radiation energy and hence produce locally a larger amount of LEE [36]. Fig. 1.6a shows the ratios between the photon absorption cross section (in $cm^2/mole$) of gold and water as a function of photon energy [37]. These ratios give a good indication of the higher probability of energy absorption by GNP

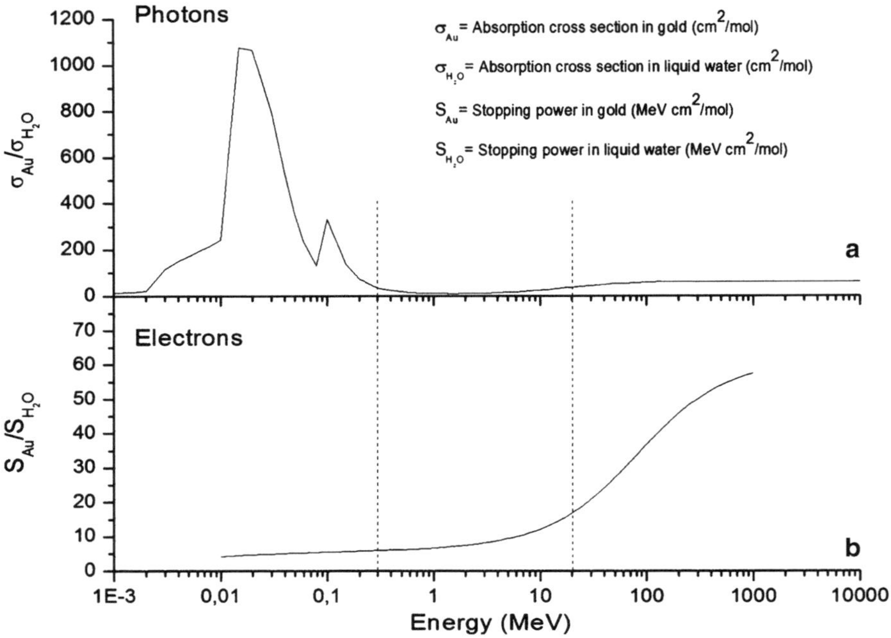

Fig. 1.6 a) Ratio between the photon absorption cross section of gold and water as a function of photon energy [37, 38]. Patients are usually treated with photons of energies lying between the limits set by the vertical dashed lines. b) Ratio between the electron stopping power of gold and that of water vs electron energy [38, 39]. In a and b, these ratios represent the enhancement of probability of energy absorption from primary photons or electrons by gold compared to water, which is the main constituent of cells. (Reprinted with permission from References 37, 38, 39. Copyright 1985 Elsevier)

[37, 38]. Patients are usually treated with primary photons of 0.3 to 20 MeV for which $\sigma_{Au}/\sigma_{H_2O}$ lies between 10 and 40. These ratios are low compared to those for energies between 10–100 keV in Fig. 1.6, but they are still substantial. Furthermore, as seen in Fig. 1.6b, the ratio of the stopping power for gold atoms [39, 40] to that for water molecules varies from 5–10 for the Compton and photoelectrons generated by these clinical photons. Interestingly, most photons inelastically scattered in the Compton effect have ratios $\sigma_{Au}/\sigma_{H_2O}$ of 10 to 1100, but many scatter away from the tumor volume. In other words, the primary photons in radiotherapy generate electrons and photons which themselves have a high probability of interacting with GNP. Another possibility to increase the amount of LEE in cancer cells or near the DNA of cancer cells is to develop carriers containing radionuclides that emit low-energy β particles and/or short-range Auger electrons as in targeted radionuclide therapy [41, 42].

A higher sensitivity to LEE can be achieved by binding a small molecule to DNA, which can amplify the DNA-LEE interaction. *If this small radiosensitizing molecule is also a chemotherapeutic agent then optimal conditions can be achieved in chemoradiation therapy (CRT)* (i.e., in the treatment of cancer with chemotherapy and high energy radiation). In addition to the benefit of radiation and chemotherapy, radiosensitivity is enhanced in this case. *However, for the chemotherapeutic agent to also serve as a radiosensitizer, the drug must be present near the DNA of the cancer cell at the time of irradiation (i.e., concomitant CRT).*

1.3 Experimental methods

This section focuses on experimental techniques to investigate in the condensed phase the damage induced to DNA and its subunits by LEE. Until recently, such studies had to be performed under ultra high vacuum (UHV) conditions to keep the surface of the biomolecular solid target free from impurities from the ambient atmosphere. In any case, vacuum conditions are always necessary to produce a LEE beam of a well-defined energy from an emitter or a monochromator. The biological samples are prepared as films whose thickness is sufficiently small ($\lesssim$10 nm) to prevent charging, which arise from thermalization of electrons within the target. Depending on the substance to be investigated, the thin films to be electron bombarded under UHV must be prepared by different techniques. Gases or liquids, of significant vapor pressure at room temperature, can be leaked into the UHV chamber from a point close to a cryogenically cooled metal substrate, onto which they condense. Substances that are solids at room temperature can be heated in an oven in front of the metal substrate, to produce a flux of molecules that condense onto the metal surface. However, if the molecule cannot be heated to sublimate or evaporate without decomposing, then it must be prepared outside the UHV vacuum system in a clean environment and afterward transferred to UHV [26].

A typical LEE irradiator [43] is shown in Fig. 1.7. The apparatus consists of two UHV chambers separated by a gate valve. The chamber on the right is a load-lock system in which the samples are introduced. It can be pumped to a base pressure in the 10^{-9} Torr range with an oil-free turbomolecular drag pump station. This chamber can contain various target holders or, as shown in Fig. 1.7, a resistively heated oven equipped with an activated shutter. The oven can be transferred into the main chamber on the left (pressure $\sim 10^{-10}$ Torr) for vacuum deposition of the solid compound onto a clean polycrystalline Pt substrate held at room or cryogenic temperature. Once loaded into a miniature oven and degassed, the compound is sublimated onto the Pt substrate. The latter is fixed to sample holder that can be rotated in front of the LEE gun and mass spectrometer. The integrity of the sublimated films can be verified *in situ* by X-ray photoelectron spectroscopy [44] and outside of the vacuum by chromatography. The average film thickness can be determined within 50% accuracy by measuring the mass of the condensed film with a quartz crystal microbalance.

For biomolecular compounds that might decompose upon sublimation, two different techniques have been developed to produce thin films on metal substrates. When dry multilayer films are required, a solution of the compound is made and a

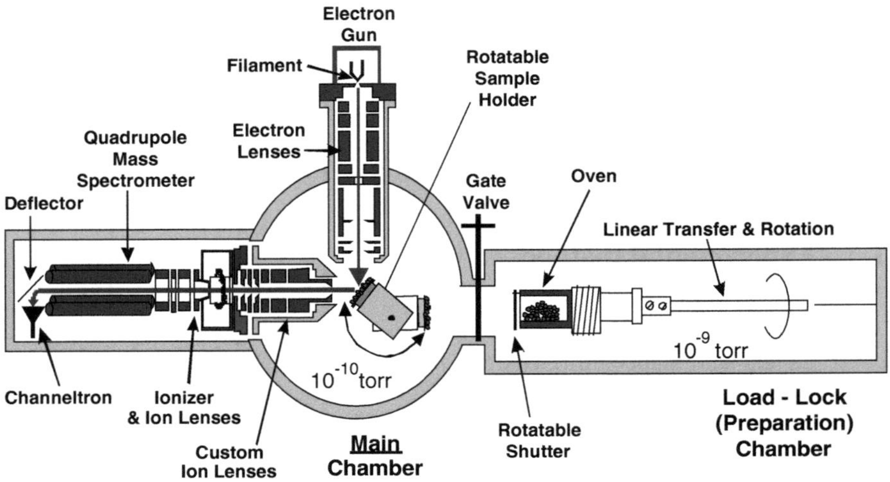

Fig. 1.7 Schematic overview of the type of apparatus used to investigate LEE-induced damage and the desorption of ions and neutral species induced by electron impact on thin molecular and bio-organic films [43]. The thin films can be formed by the condensation of molecules evaporated from an oven or leaked in front of a metal substrate fitted to the rotatable sample holder. Alternatively, the samples can be introduced from a controlled atmosphere into the main UHV chamber on the left via the load-lock on the right. (Reprinted with permission from reference 43. Copyright 2001 American Institute of Physics)

small aliquot of the solution is lyophilized on a tantalum substrate [33]. The sample preparation and manipulations are performed within a sealed glove box under a pure dry nitrogen atmosphere. Several samples are afterwards transferred from the glove box to a load-lock similar to that shown in Fig. 1.7. Then, from the chamber on the right, samples can be introduced into the main chamber, where they are placed on a rotary multi-sample holder. Each sample can be positioned in front of a LEE gun. Typical characteristics of focused electron beams in LEE irradiators are: $\approx$3 mm diameter spot, full-width at half-maximum energy distribution varying from 0.03 to about 0.3 eV and beam currents between 5–400 nA. The average film thickness of the film is usually estimated from the amount deposited and the density [33]. After irradiation the samples are returned to the glove box where they are dissolved in appropriate solvents for analysis.

When only a single layer of a relatively large biomolecule is needed, a uniform layer can be formed on a gold substrate by chemisorption if the molecule does not fragment on the metal. The technique is essentially the same as that utilized to prepare self-assembled monolayers [45, 46]. It can serve to prepare films of oligonucleotides of various lengths. However, compared to self-assembled monolayers, the large biomolecules are not necessarily well-ordered on the substrate. The gold substrate is usually prepared by vacuum evaporation of high-purity gold (99.9%) onto freshly cleaved preheated mica slides [46,47]. These slides are dipped for at least 24 h in an aqueous solution of highly purified oligonucleotides. With this procedure, one monolayer [46, 48] is chemically anchored to the gold substrate via a phosphotioate modification on each deoxycytosine nucleotide (i.e., substitution of the double-bounded oxygen atoms by double-bonded sulfur at the phosphorus). Depending on the site of the modification and their length and numbers, the oligos can be made to lie more or less parallel or perpendicular to the gold surface. After rinsing with a copious amount of nanopure water and dried under nitrogen flow, each slide is divided into smaller samples, which are introduced in a UHV preparation chamber (e.g., on the right in Fig. 1.7) for degassing. Afterwards, the samples are transferred from a load-lock chamber to the main LEE-irradiation chamber, via a gate valve, as shown in the schematic diagram of Fig. 1.7.

Some of the damage induced by LEE impact on biomolecular films can be assessed by monitoring the ions and neutral species that desorb in vacuum, while the film is being bombarded. Such measurements can be performed by placing the sample near a mass spectrometer, as shown in Fig. 1.7. A LEE beam, emanating from an electron monochromator or a focusing electron gun, impinges onto the sample. Neutral species desorbed from the films can be ionized by a laser beam close to the film's surface and focused onto the mass spectrometer [49]. Such measurements do not allow the determination of absolute yields. In order to determine the absolute desorption yields of neutral products, their formation must be related to a pressure rise within a relatively small volume. In this case, a mass spectrometer measures within a small UHV chamber the partial pressure increase due to the desorption of a specific fragment induced by LEE impact on a thin film [46,48,50]. Once the irradiated samples are extracted from vacuum, the damage and molecular fragments can, in principle, be identified by various standard methods of chemical analysis. In practice, however, the quantity of recovered material and fragments are so small that an efficient method of damage amplification is required to observe any type of fragmentation. One method of damage amplification consists of using a target film, in which a small modification at the molecular level can cause a large conformal change. SB in plasmid supercoiled DNA have so far been investigated with such a method. Owing to the supercoiled configuration, a single bond rupture in a plasmid of a few thousand base pairs can cause a conformational change in the geometry of the DNA, and hence can be detected efficiently by electrophoresis. The method separates and quantifies the following configurations: supercoiled, nicked circle, full-length linear, crosslink (CL) and short linear forms [51, 52]. When the initial configuration is highly pure SC and the damage is a linear function of dose, the nicked circle and linear forms can be associated to single strand breaks (SSB) and double strand breaks (DSB), respectively. The procedure is repeated for different electron energies and periods of bombardment. Unless otherwise stated, the SC plasmids used to produce the results reported in the present article were pGEM-3Zf(-) having 3197 base pairs.

The huge amplification factor obtained with plasmid DNA for SB and CL does not exist for other types of DNA damage, so that the quantity of fragments produced from a collimated electron beam is not sufficient for chemical analysis. To produce sufficient degraded material a new type of LEE irradiator was developed to bombard a hundred times more material [53]. The biomolecules are spin coated onto the inner

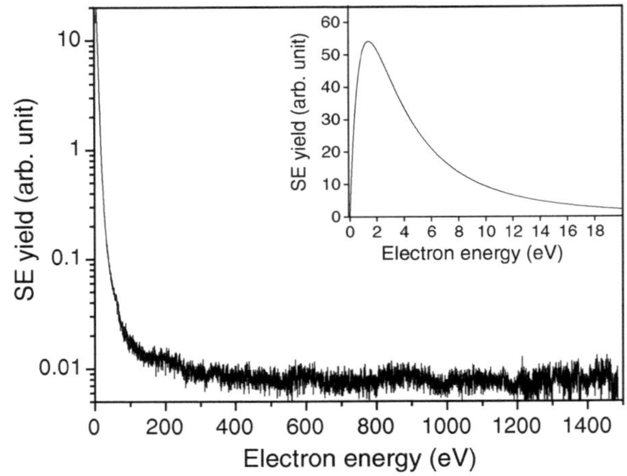

Fig. 1.8 Energy spectrum of secondary electron emission from a gold surface induced by incident 1.5-keV X-rays [55]. (Reprinted with permission from reference 55. Copyright 2009 American Chemical Society)

surface of tantalum cylinders. Up to ten cylinders may be placed on a rotary platform housed in an UHV system, where their inner walls are bombarded by a diverging LEE beam. This technique allows the total mixture of products resulting from LEE bombardment of DNA and its subunits to be analyzed by HPLC/UV, HPLC/MS and gas chromatography/MS [53, 54]. When analysis is performed only by HPLC, the identification of the products and their yields is determined by calibration with authentic reference compounds [54].

It is also possible to use as a source of LEE the emission of SE from a metal surface exposed to soft X-rays. As an example, the SE energy distribution emitted from a gold substrate, induced by 1.5 keV Al K_α X-rays, is shown in Fig. 1.8 [55]. The LEE distribution between 0 and 20 eV is shown on a linear scale in the insert of Fig. 1.8. The LEE distribution has a peak at 1.4 eV. Ninety-six percent of the SE have energies below 30 eV, and the average energy for these electrons is 5.9 eV. So essentially all electrons emitted from such a metal substrate have energies lower than 30 eV. Such a LEE source can be exploited [55] to investigate DNA damage under well-defined gaseous atmospheres and specific humidity levels. In this case, the DNA films are deposited on an insulator (glass substrate) and also on an electron-emitting metal surface. The damage produced on the glass substrate arises from energy absorption from X-rays, whereas that produced on the gold substrate arises from energy absorption from both the X-ray beam and SE emitted from the metal surface. The difference in damage yields recorded with the metal and glass substrates is therefore essentially due to the interaction of LEE with DNA.

The apparatus, developed by Alizadeh et al. [56] to perform such experiments is shown in Fig. 1.9. It is composed of a stainless steel chamber evacuated by a mechanical pump to pressure less than 5 mTorr, and connected to a pressure gage (i.e. baratron A) and an adjustable leak valve (B) connected to a nitrogen gas source. A negative potential of 3.4 kV applied to a concave aluminum cathode (C) through a

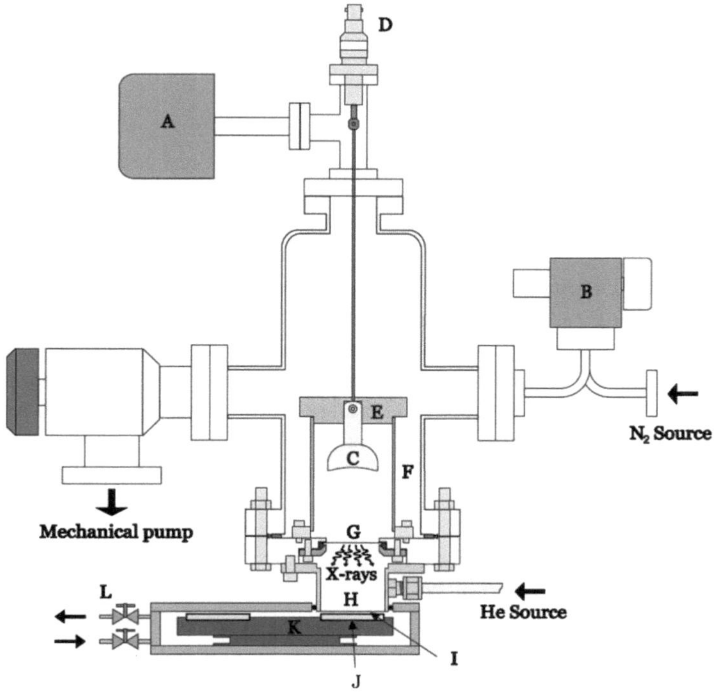

Fig. 1.9 Schematic view of the apparatus used to irradiate DNA samples with 1.5 keV Al K_α X-ray photons under a controlled atmosphere at atmospheric pressure and temperature [56]: (A) baratron, (B) adjustable leak valve, (C) concave aluminum cathode, (D) high voltage electric feedthrough, (E) glass-ceramic (Macor) support, (F) quartz tube, (G) aluminum foil target, (H) He-filled enclosed volume, (I) thin foil of Mylar, (J) aluminum plate as sample holder, (K) rotating disk, and (L) gas circulation valves. X-rays are generated by electrons from the gas discharge in F, which strike G. X-rays that pass through H and I, and not absorbed by the gaseous atmosphere, strike the DNA film deposited on a metal substrate attached to J

high-voltage electrical feedthrough (D) causes a discharge. The plasma current can be controlled and stabilized by the nitrogen gas pressure. The electrons from the discharge strike, a thin Al foil (G) that emits characteristic K_α X-rays with energy of about 1.5 keV towards a He-filled side enclosed volume (H). The produced X-rays cross the helium gas and then a thin foil of mylar (I) to enter a small chamber, where plasmid DNA films deposited on different substrates are supported by Al plates (J). These plates are set at different positions around a brass rotating disc (K) to allow irradiation of samples directly by X-rays, for different periods of time (i.e., various radiation doses) in presence of specific amounts of gases or vapours introduced by valves (L). The samples are positioned very close to the mylar foil to avoid excessive photon absorption by the surrounding atmosphere.

1.4 Damage to biomolecules induced by LEE

1.4.1 Amorphous solid water

Radiation damage to DNA within the nucleous of cells can generally be classified into two major groups, referred to as "direct" and "indirect". At least 50% of the damage induced by high energy radiation within cells is due to the "indirect" effect [20, 57, 58]. Whereas "direct" damage results from interaction of primary radiation and SE with the genome, "indirect" damage is attributed to the immediate species formed by ionizing radiation in the vicinity, but outside, the volume occupied by DNA. Owing to the large quantity of water in cells, most of the immediate species are H and OH radicals and solvated electrons formed by LEE interacting with water molecules surrounding DNA. These species can further react with water or diffuse to DNA, where they can trigger other reactions and damage the molecule. The production of species from LEE-interaction with condensed water molecules is therefore of considerable relevance to fully understand the indirect effect of radiation on DNA.

Fragmentation of condensed-phase H_2O by LEE impact has been investigated mainly with amorphous ice films. The yield function for desorption of H^- [59–61], H_2 [62, 63], $D(^2S)$, $O(^3P)$, and $O(^1D_2)$ [64, 65] were recorded in the range 5–30 eV. Most of these functions exhibit resonance structures below 15 eV, which are characteristic of transient anion formation. From anion yields, DEA to condensed H_2O was shown to result principally in the formation of H^- and the $OH^\cdot$ radical from dissociation of the $^2B_1$ state of H_2O^- located in the 7–9 eV region. Smaller contributions arise from the $^2A_1$ and $^2B_2$ anionic states, which are formed near 9 and 11 eV, respectively [59, 60]. At higher energies, nonresonant processes, such as dipolar dissociation (e.g., $[R\text{-}H]^* \rightarrow R^+ + H^-$ or $R^- + H^+$ in Fig. 1.2) lead to H_2O fragmentation with the assistance of a broad resonance that extends from 20 to 30 eV [60]. Kimmel et al. [64, 65] measured the ESD of $D_2\left(X^1\Sigma_g^+\right)$, $D(^2S)$, $O(^3P_{j=2,1,0})$, and $O(^1D_2)$ desorption yields from amorphous ice versus incident electron energy. An apparent threshold was found at ~6.5 eV with a steadily increasing intensity. Above this threshold, the $D(^2S)$ intensity also increases rapidly and exhibits a broad resonance for ~14–21 eV. Above 7 eV, direct electronic excitation of the $^{3,1}B$ states lead to $H^\cdot$ and $OH^\cdot$ formation. From 10 eV, ionization progressively takes over and dominates energy losses. The ensemble of these reactions leads to an abundant production of OH and H radicals and H_2 molecules.

The integral cross sections per scatterer (i.e. elastic collision, phonon excitations, vibrational excitations, electronic excitations and ionization) for 1–100 eV electron scattering in an amorphous film of ice condensed at a temperature of 14 K have been measured by Michaud et al. [66]. The integral cross sections were determined relative to the total from a two-stream multiple-scattering analysis of the electron energy distribution backscattered from the film. The magnitude of the electronic excitation and ionization cross sections and various features found in their energy

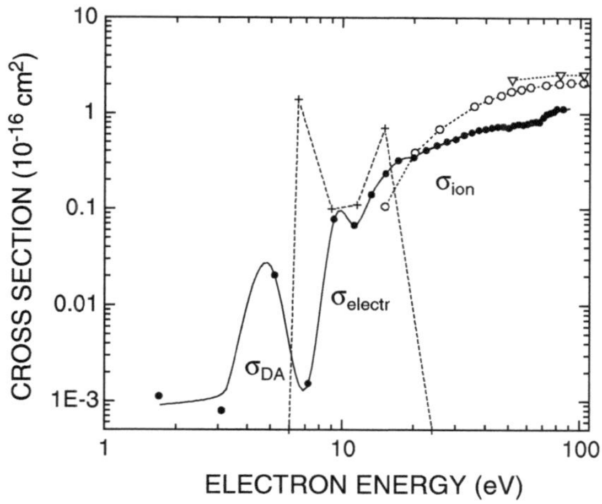

Fig. 1.10 Integral cross section ascribed to the sum of dissociative electron attachment (σ_{DA}), vibrational excitation (above 1 eV energy loss), electronic excitations (electr), as well as ionization (ion) processes in amorphous ice [66]. Electron scattering cross section reported for electronic excitations of water in the gas phase; $\cdots + \cdots$, from Ness and Robson [146]. Measured total electron impact ionization cross section for water in the gas phase; $\cdot\cdot$ O$\cdots\cdot$, Djuric *et al.* [147] $\cdots \nabla, \cdots$, Bolorizadeh and Rudd [148]. (Reprinted with permission from reference 66. Copyright 2003 Radiation Research)

dependence are shown by the solid line in Fig. 1.10. Within the energy range of LEE (0–30 eV), the cross sections for excitation of the electronic states leading to $OH^{\cdot}$, $H^{\cdot}$ and H_2 production varies from 10^{-19} to about 10^{-17} cm^2 in amorphous ice [66]. These values compare with similar cross sections of (1–3) $\times$ 10^{-18} cm^2 per nucleotide for inducing SB in DNA with LEE [67].

1.4.2 DNA basic components

In order to understand the basic mechanisms involved in LEE-induced damage in DNA both experimentalists and theoreticians have investigated LEE interactions with molecules of increasing complexity [24, 26, 27]. Gas and solid phase experiments have been performed with isolated components of DNA (the base, phosphate, sugar and water subunits) [24]. Experiments with DNA were performed with thin films of short strands or plasmids [25]. Owing to the size and complexity of the DNA molecule, details on the mechanisms involved in the bond breaking processes could usually only be inferred by comparing the results of LEE experiments with DNA to those obtained with the basic building blocks of the molecule. In the condensed phase, the bases were studied independently, whereas sodium dihydrophosphate and tetrahydrofuran with its derivatives were chosen as analogs of the phosphate group and the sugar ring of DNA, respectively [24, 26]. The nomenclature of the latter two

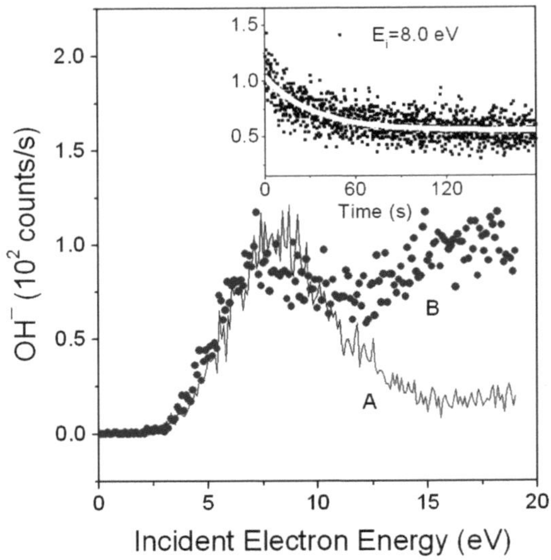

Fig. 1.11 OH^- yield function from (A) a NaH_2PO_4 film and (B) a self-assembled monolayer of 20 base-pair single-stranded DNA [68]. The inset shows the time dependence of the OH^- signal from the phosphate film recorded at an incident electron energy of 8 eV. In the inset, the solid line is an exponential fit to the data and the vertical scale is the same as that in the figure. (Reprinted with permission from reference 68. Copyright 2006 Elsevier)

appears in the rectangles on the left of Fig. 1.3 and the corresponding units in a DNA strand on the right. An example is provided in the case of sodium dihydrophosphate in this subsection.

To understand LEE-induced damage to the phosphate group in DNA, Pan and Sanche [68] measured the electron-stimulated desorption (ESD) of H^-, O^- and OH^- anions emitted from thin films of sodium dihydrogen phosphate under bombardment with 0–19 eV electrons. The yield functions exhibit a single broad peak with maxima at 8.8 ± 0.3 eV, 8.0 ± 0.3 eV, and 7.3 ± 0.3 eV, respectively, and a continuous rise above 15 eV. In each curve, the structure was attributed to DEA and the continuous rise to dipolar dissociation, both causing scission of the O-H, $P = O$ and P-O bonds, respectively. Rupture of these bonds was accompanied by the corresponding desorption of the stable anions H^-, O^- and OH^-. The incident electron energy dependence of the OH^- yields is shown in A of Fig. 1.11. The yield function exhibits a single broad peak with a maximum at 7.3 eV. Such a resonant peak is a typical signature of the DEA process [10, 69], which can be expressed as $e + NaH_2PO_4 \rightarrow [NaH_2PO_4]^- \rightarrow [NaHPO_3] + OH^-$. The inset shows the time dependence of the anion signal at an incident electron energy of 8.0 eV. The solid line in the inset is an exponential fit to the data. From measurements of this time dependence and those of H^- and O^-, the effective cross section to damage the molecule in the film was found to be very high, about 10^{-15} cm^2 [68].

Curve B of Fig. 1.11 exhibits the OH^- yield function obtained from LEE-bombardment of a 40-base oligonucleotide (i.e., a short single DNA strand) chemisorbed on a gold substrate [70] and having a thickness similar to that of the sodium dihydrogen-phosphate film ($\sim$5 nm). The similarity and coincidence in energy of the results obtained from the short DNA strands and that recorded from the NaH_2PO_4 film is a strong indication that in both cases the same transient anion (i.e., the same $[NaH_2PO_4]^-$ state) is involved. The signal producing the broad OH^- peak seen in curve A of Fig. 1.11, is expected to arise from P-OH bond cleavage, through the DEA pathway

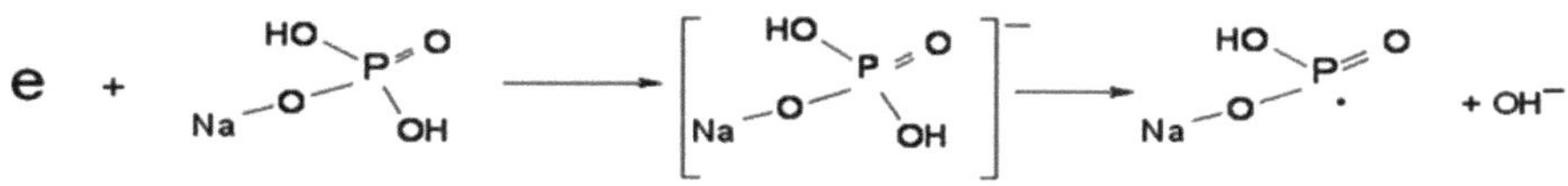

Within DNA, scission of this bond would produce a SB [71] but no OH^- would desorbed from the film. Furthermore, with Na^+ as the counter ion no OH group is present in the backbone. However, in the film of 40-base single strands of DNA, used to produce curve B in Fig. 1.11, the counter ion was H^+ [70]. When the Na^+ counter ion is replaced by H^+, an OH group is formed and cleavage of the P-O^- bond perpendicular to the chain can produce a OH^- desorption signal below 10 eV (i.e., curve B in Fig. 1.11) that matches the one seen from curve A. Thus, the OH^- signal producing the broad peak in curve B has been interpreted to arise from the decay of the same transient anion, as in NaH_2PO_4, but into the pathway

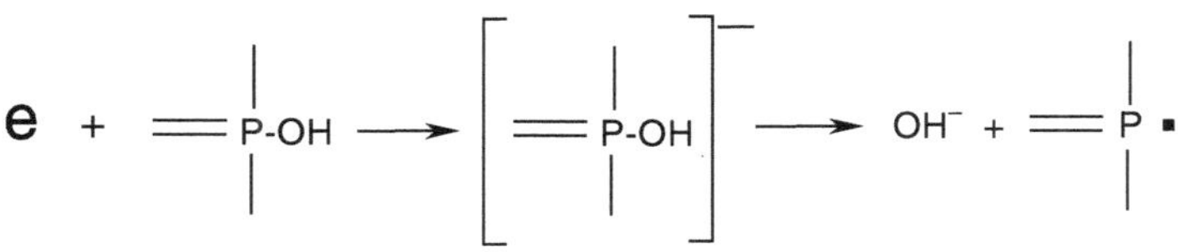

within the backbone of DNA. Similar comparisons with DNA were made with ESD signals from films composed of the other basic subunits of DNA [26].

1.4.3 Short single DNA strands

We have seen in curve B of Fig. 1.11 the desorption of OH^- from a short DNA strand stimulated by LEE impact. To obtain more details on the mechanisms of DNA damage, the products remaining in such films after LEE bombardment were analyzed by Zheng et al. [71]. They analyzed by HPLC the degradation products from the tetramer GCAT, its abasic forms and CGTA. These oligonucleotides, which constitute the simplest form of DNA containing the four bases (G, C, A and T), made the analysis of degradation products much easier than would be the case for longer single strand and double stranded configurations. Samples, prepared by spin coating inside tantalum cylinders, were irradiated by 10-eV electrons from the diverging beam LEE irradiator mentioned in Sect. 1.3. The HPLC analysis was first focused on SB and detachment of non-modified subunits of the tetramers CGTA and GCAT, which included monomeric components (nucleobases, nucleosides and mononucleotides), and oligonucleotide fragments (dinucleotides and trinucleotides)

[71]. The incident electron current and irradiation time were adjusted to give an exposure well within the linear regime of the dose response curves and an equal number of electrons to each sample. The nomenclature of GCAT with the potential sites of cleavage yielding non-modified fragments (i.e., 1 to 16) is shown in the right of Fig. 1.3.

The reaction of LEE with the tetramers led to the release of all four non-modified nucleobases with a bias for the release of nucleobases from terminal position. The release of nucleobases from tetramers was ascribed to N-glycosidic bond cleavage via DEA from initial electron capture by the base, as previously shown in the cleavage of thymidine to thymine in the condensed [54] and gas phase [72]. All major non-modified fragments were formed except for those corresponding to breakage at positions 7, 10, 11 in Fig. 1.3. Cleavage of the backbone gave fragments with and without a terminal phosphate, but the yield of fragments with a phosphate was much greater than that without a phosphate. This indicated that *LEE induce the cleavage of phosphodiester bonds to give non-modified fragments with a terminal phosphate rather than a terminal hydroxyl group*.

Based on this result and previous interpretations of SB in DNA [24], Zheng et al. [71] postulated that rupture of the phosphodiester bond was initiated by the formation of a dissociative transient anion on the phosphate group. There are two possible pathways leading to cleavage of the phosphodiester bond: (1) scission of the C-O bond corresponding to breaks at positions 5, 8, 9, 12, 13 and 16 in Fig. 1.3 and (2) cleavage of the P-O bond resulting in breaks at positions 6, 7, 10, 11, 14, 15 in Fig. 1.3. However, 95% of the products from the HPLC analysis corresponded to those resulting from breakage at positions 5, 8, 9, 12, 13 and 16. *Thus, Zheng et al. concluded that cleavage of the phosphodiester bond primarily takes place via C-O bond cleavage leading to the formation of a sugar radical and a terminal phosphate anion* [71]. The cleavage of C-O and P-O bonds was previously reported in electron spin resonance studies of argon ion and γ irradiated hydrated DNA [73–75]. These studies also showed that C-O bond cleavage was the dominant process.

In subsequent investigations, Zheng et al. measured the yields of the previously-mentioned products as a function of electron impact energy on GCAT [76]. From 4 to 15 eV, scission of the backbone gave similar non-modified fragments to those previously observed at 10 eV. This result indicated that *phosphodiester bond cleavage involves cleavage of the C-O bond rather than the P-O bond over the entire 4 to 15-eV range*. Many of yield functions of fragments exhibited a maximum near 6 eV, a large peak at 10–12 eV followed by a dip at 14 eV. The maxima were interpreted as due to the formation of transient anions (i.e., core-excited and core-excited shape resonances) leading to fragmentation. These resonances dominated bond dissociation. All four non-modified bases were released from the tetramer within the 4–15 eV range, by cleavage of the N-glycosidic bond [76]. Above 14 eV, the electron resonances did not dominate the yield functions, which were interpreted to arise from fragmentation via direct electronic excitation of dissociative states.

Afterwards, Zheng et al. [77] verified experimentally the theoretical hypothesis of electron transfer from a base to the phosphate group of DNA. According to calculations [78, 79] an electron captured by a base can transfer to the phosphate group forming at this position a transient anion that dissociates by breaking the C-O bond. Zheng et al. [77] analyzed the products induced by 4–15 eV electrons incident on two abasic forms of the tetramer GCAT, i.e., XCAT and GCXT, where X represents the base replaced by a hydrogen atom. With the exception of the missing base, the same fragments as those from irradiated GCAT were observed. Their results demonstrated that when a base is removed at a particular position in a small DNA strand, cleavage of the adjacent C-O bond of the backbone by LEE of 6 eV is inhibited, thus practically eliminating strand breaks in the chain at this position [77]. *Thus, the presence of a base is needed to produce C-O bond rupture next to that base (e.g. to make a break in GCAT at positions 12 and 13 in Fig. 1.3 adenine must be bound at position 3). It is difficult to explain this result without invoking electron capture by a base followed by electron transfer to the corresponding phosphate group.* This phenomenon was not observed at higher energies. Since electron transfer from a base π^* to a C-O σ^* orbital had been shown theoretically to occur at energies below 3 eV [78, 79], Zheng et al. [77] suggested that the incident 6-eV electron electronically excites a base before transferring to the C-O orbital. In so doing, the incident electron has energies below 3 eV after exciting the base. They founded their suggestion on the existence of electronically excited states of the DNA base within the 3.5 to 6 eV range as measured by electron-energy-loss spectroscopy [80]. This hypothesis implies a strong decay of core-excited resonances of the bases into electronically inelastic channels, a phenomenon that has been demonstrated theoretically by Winstead and McKoy [81].

Later, Li et al. undertook a systematic study of 11-eV electron induced damage to very small single-strand homo-oligonucleotides of increasing length [82]. The products arising from the reaction of LEE with dThd, pT, Tp, pTp, pTpT, TpTp, pTpTp and TpTpT, where pT or Tp represent thymidine phosphate, were analyzed by HPLC. Their results showed that the addition of a phosphate to the terminal positions of the monomers and dimers resulted in a considerable increase in total damage (i.e., terminal phosphate groups efficiently capture 11 eV electrons and this capture leads to considerable damage). This increase could be correlated to electron-beam experiments performed on thin films of NaH_2PO_4 [68] (see previous Sect. 1.4.2 and Fig. 1.11), and tetrahydrofuran [83]. The results of these experiments show that the phosphate group has a very large cross section (10^{-15} cm^2) [68] for 7 to 12 eV electron induced fragmentation which is much larger than that of the furyl ring [83, 84].

Although the addition of a terminal phosphate group increased total damage, Li et al. found that such addition decreased base release and phosphodiester cleavage. Thus, the initial capture of electrons by a phosphate group does not lead to formation of a transient anion that causes base release and phosphodiester bond cleavage. The work of Li *et al.* [82] therefore confirmed the hypothesis [78] that the electron must first attach to the base in order to break the *N*-glycosidic bond or transfer to the $P = O$ π^* orbital to break the phosphodiester C-O bond. Since continuous stretches of DNA do not have terminal phosphate groups, direct capture of LEE by the phosphate group followed by the formation of products from the phosphate should probably be considered as a minor process to produce SB in cellular DNA.

Other studies by Li et al. [84] included the effect of base sequences in a series of oligonucleotide trimers. These authors analyze the damage remaining in films of TXT, where X represents one of the four bases of DNA, after LEE irradiation [85]. Using HPLC-UV analysis, several known fragments were quantified from the release of nonmodified nucleobase (T and X) as well as from phosphodiester C-O bond cleavage (pT, pXT, Tp, and TXp). The total damage was estimated by the magnitude of the parent peaks in the chromatogram of nonirradiated and irradiated samples. When trimers were irradiated with 10 eV electrons, the total damage decreased 2-fold in the following order: TTT > TCT > TAT > TGT. The release of nonmodified nucleobase (giving from 17 to 24% of the total products) mainly occurred from the terminal sites of trimers (i.e., T), whereas the release of central nucleobase was minor (C) or not at all detected (A and G). In comparison, the formation of products arising from phosphodiester bond cleavage accounted for 9 to 20% of the total damage and it partitioned to the four possible sites of cleavage. These results indicated that the initial LEE capture and subsequent bond breaking within the intermediate anion depend on the sequence and electron affinity of the base, with the most damage attributed to the most electronegative base, T. This result agrees well with the recent calculations by Gu et al. [86], which indicate that electron capture by pyrimidines may be most efficient pathway to SB and base release.

The ensemble of the results summarized in this section led to the proposition of a model for LEE-induced SB and base release in DNA. The model is shown schematically in Fig. 1.12. It illustrates the pathways leading to base release and C-O bond cleavage following initial electron capture by a nucleobase. The incoming electron first forms either a shape, core-excited or core-excited shape resonance of the nucleobase. However, about 0.5 eV below the energy of the first electronic excitation threshold of the base, only shape resonances can be formed. The lifetime of these resonances is usually sufficiently long below 3 eV to lead to molecular dissociation via DEA [12, 13]. At higher energies, this is not the case and core-excited types of resonances are usually those, which lead to dissociation because of their longer lifetimes [12, 13]. According to the scheme of Fig. 1.12, the transient anion on a base can decay into three channels: (1) the elastic channel on the left where the electron is re-emitted with the same energy (E_o); (2) the direct DEA channel leading to fragmentation of the nucleobase or base release; and (3) the electronically inelastic channel, which can lead to the electronic excitation of a base and the release of a very low energy electron (e.g., $E_o = 0$–3 eV). This latter channel is only possible with core-excited types of resonances. In case 1 and 3, the electron can be re-emitted into the continuum (e_c^-) or transfer (e_t^-) within DNA. When the extra electron transfers to and localizes on the phosphate group (i.e., a transient anion of the phosphate moiety is formed), the C-O σ bond has a high probability to break via DEA. According to the previously mentioned investigations and recent theoretical calculation on electron diffraction [87], it is reasonable to assume that breakage of the C-O bond within DNA occurs principally via pathway 3 for electron energies above about 5 eV. Thus, LEE-induced damage in the range 5–15 eV can be

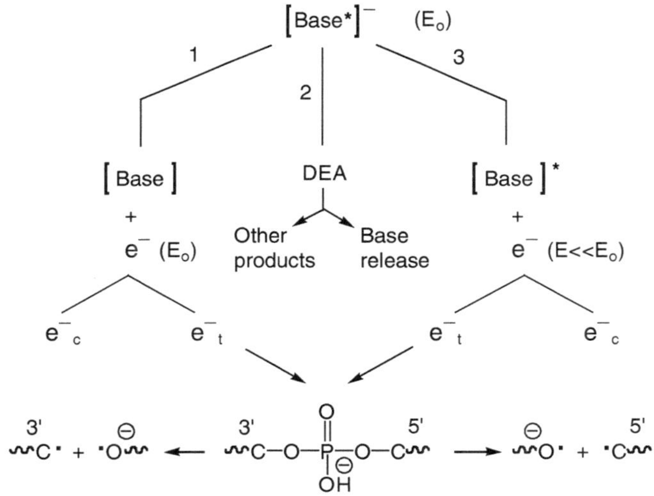

Fig. 1.12 Decay channels of transient negative ions of DNA bases formed at an initial electron energy of E_o. Pathways 1–3 represent the elastic ($E = E_o$), dissociative electron attachment (DEA), and electronically inelastic ($E \ll E_o$) channels, respectively. e_c^- represents the electron re-emitted into the continuum. e_t^- represents the transfer of an electron from the base to the phosphate unit. The transient anion formed on the phosphate unit by electron transfer dissociates, leading to C-O bond cleavage

discussed in terms of theoretical models [27, 79] for electrons of lower energies, since after electronic excitation in pathway 3 the electron can have energies within 0–3 eV.

Further evidence of the occurrence of reaction 3 was later provided by studies of the reactions with oligonucleotide trimers 5′-TpXpT-3′ where X represents thymine (T) or 5-bromouracil (5BrU) [88]. Analysis by HPLC-UV of films made of these trimers, which were irradiated with 10-eV electrons, led to the quantification of seven known fragments: those which arise from the release of non-modified nucleobase (T and 5BrU) and those arising from C-O bond cleavage (pT, Tp, pXT, TXp), as well as TUT, resulting from debromination of T5BrUT. When T5BrUT was irradiated with LEE, TUT was the major product. This indicates that the presence of BrU in DNA shifts the reaction pathway of LEE toward the predominant formation of U, likely by way of DEA, giving the uracilyl radical and bromide anion. In addition, a relatively large percentage of fragments arising from N-C and C-O bond cleavage contained uracil rather than 5-bromouracil (i.e., uracil, pUT, and TUp), indicating that a single 10 eV electron induces double events, i.e., Br-C and C-N cleavage or Br-C and C-O cleavage. The latter two cleavages produced with a single electron correspond to pathway 3 in Fig. 1.12. An incoming electron is first captured by the positive electron affinity of an excited state of BrU (a core-excited resonance). Then it leaves BrU in an electronically excited state, while transferring on the phosphate group, where it forms a transient anion that dissociates by cleaving the C-O bond. The remaining electronically excited BrU* dissociates forming the uracil-yl radical, which transforms to uracil, resulting in the production of pUT or TUp.

The model proposed in Fig. 1.12 does not specify which base has the largest probability for LEE capture. This probability may depend on sequence and electron energy, but the capture cross section is difficult to determine experimentally. If we assume that the yield of detached nucleobases or C-O bound cleavage is directly proportional to electron capture, we find from the result of Zheng et al. [76] that thymine has the largest capture cross section in GCAT. At 10 eV and below the order is T > G > A > C, but at 12 eV it changes to T > G > C > A. However, for such a comparison to be valid, the increased capture by a terminal base would have to be eliminated (i.e., the end effect). Thus, this result only serves to indicate the effect of electron energy on capture cross section. The TXT oligonucleotides used in the previously discussed experiments of Li et al. represent a more valid model [85]. Their results show essentially undetectable base release from the middle base (X = G, C, A or T). Whereas the lowest total damage was found with X = G, the same oligonucleotide gave the largest *percentage of damage* into C-O bond cleavage. Solomun et al. [89] studied DNA damage as a function of G content in DNA. Their experiments with 33 bases oligonucleotides, containing thymine and between 1 and 4 guanine bases, showed that the total damage induced by 1-eV electron correlates linearly with the number of G in the sequence. Their finding clearly showed that, at least within a T-oligonucleotide, G could have a large cross section for electron capture leading to damage. These results are consistent with the stability calculations of Gu et al. [90] of guanine nucleotides in aqueous solution.

In a different set of experiments, Ray et al. [91] measured LEE (<2 eV) transmission through self-assembled monolayers of short DNA oligomers made of 15 adenine bases which were substituted by different quantities of guanine bases. The electrons that were not transmitted were captured by the layer. Hence, the transmission signal reflected the efficiency of the layer towards capturing electrons. The authors found that the capturing probability scaled with the number of G bases in the single-stranded oligomers and depended on their clustering level. From two-photon photoelectron spectroscopy experiments, they found that once captured, the electrons do not reside on the bases. In fact, according to Fig. 1.12, these electrons should return to the continuum or be transferred to the phosphate group. The measurements of Ray et al. also indicate that guanine has a high cross section for capturing non-thermal electrons when substituted in a homogeneous sequence. Similar experiments with the other bases would be necessary, however, to reach a conclusion on the ability of G to capture electrons compared to that of the other bases.

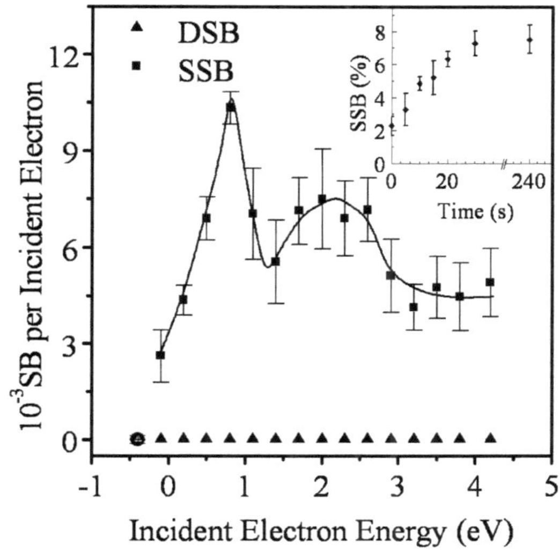

Fig. 1.13 Yields of SSB and DSB induced by 0–4.2 eV electron impact on supercoiled plasmid DNA films [31]. The inset shows the dependence of the percentage of circular DNA (i.e. SSB) on irradiation time for a beam of 0.6 eV electrons of 2 nA. (Reprinted with permission from reference 31. Copyright 2004 American Physical Society)

1.4.4 Plasmid and linear DNA

The damage induced to plasmid and linear DNA films by LEE has been investigated by (1) measuring ESD of anion radicals, (2) imaging the breaks by atomic force and scanning tunneling microscopies and (3) analyzing, after bombardment, the change of topology by gel electrophoresis. The results obtained are well described in the literature and summarized in a previous review article [26]. In this section, we only summarize the results of analysis of DNA damage by gel electrophoresis which are the most relevant to radiosensitization.

LEE-induced damage to thin films of plasmid DNA below 4 eV was investigated in UHV by Martin et al [31]. Exposure response curves were obtained at different energies for SSB and DSB. The inset in Fig. 1.13 shows the measured dependence of the percentage yields of circular (i.e., SSB) DNA on irradiation time for 0.6 eV incident electrons. The other curves exhibit two peaks in the yield function of SSB from plasmid DNA, with maxima of $(1.0 \pm 0.1) \times 10^{-2}$ and $(7.5 \pm 1.5) \times 10^{-3}$ SSB per incident electron, at electron energies of 0.8 and 2.2 eV, respectively [31]. Since at such low energies electronic excitation is energetically impossible, these peaks provide unequivocal evidence for the role of shape resonances in DNA damage.

Martin et al. [31] compared the results in Fig. 1.13 with those obtained from electron scattering from gaseous DNA base and sugar analogs [92]. The solid curve in Fig. 1.13, which reproduces in magnitude and line shape the SSB yield function, is similar to that obtained by a model that simulates the electron capture cross-section as it might appear in DNA owing to the π^* single-particle anion states of the bases. In this simulation, the attachment energies were taken from electron transmission measurements [92] and the peak magnitudes were scaled to reflect the inverse energy dependence of the electron capture cross-sections. Assuming an equal numbers of each base in DNA, the contributions from each base were simply added. The relationship between the resonances in the base and SSB in DNA offers support for the charge transfer mechanism proposed by Barrios et al. [78]; i.e., pathway 1 in Fig. 1.12.

As shown in Fig. 1.4, the incident electron energy dependence of the yields of SSB and DSB induced by 5–100 eV electron impact on plasmid DNA [29, 51, 93] also exhibits strong resonance features. The data were recorded in the linear regime of the dose-response curves and hence represent SSB and DSB produced by a single electron interaction. Whereas the DSB yield begins near 6 eV, the apparent SSB yield threshold near 4–5 eV is due to the cut-off of the electron beam at low energies in these experiments. Both yield functions have a peak around 10 eV, a pronounced minimum near 14–15 eV followed by an increase between 15 and 30 eV, and a roughly constant yield up to 100 eV. Below 15 eV, the strong peaks around 10 eV clearly show that the damage occurs via the formation of transient anions. Above 15 eV, as shown by the dotted line, the yield increase monotonically and saturates above 50 eV. The form of these yield functions were later understood from the results of fragmentation induced by LEE to the various subunits of the DNA molecule, including its structural water [26]. The strong energy dependence of DNA SB below 15 eV in Fig. 1.4 can be attributed to the initial formation of core-excited transient anions of specific DNA subunits decaying into the DEA and/or dissociative electronic excitation channels [24, 26]. To date this phenomenon has been confirmed by a large number of experiments with LEE impinging on films of DNA of various topologies [25, 26]. As explained in the previous section, pathway 3 of Fig. 1.12 is expected to contribute significantly to the formation of SSB at 10 eV.

If interstrand electron transfer also occurs, it may explain how a single electron interaction within DNA can produce a DSB with only 7–10 eV of energy as seen from Fig. 1.4. For example, a core-excited resonance on a phosphate group could decay by electron transfer to the other strand while leaving behind a dissociative electronic state breaking the C-O bond. The electron transferred to the opposite strand could attack the sugar-phosphate backbone to cause another SB via DEA, a process which according to Fig. 1.10 requires little energy. Such breaks on adjacent strands would be very close to each other and hence more difficult to repair by the cell than more distant DSB.

1.4.5 Single- and double-stranded DNA in the presence of cellular constituents

Although vacuum experiments with dry films of pure DNA and its basic constituents were essential to unveil fundamental mechanisms of LEE-induced damage, experimental conditions did not correspond to those found in cells. It is now well established that the processes induced by LEE impact on a molecule are dependent on its environment [26, 94, 95]. It is therefore crucial, if we are to apply knowledge of LEE-biomolecule interactions to practical problems in radiation protection and therapy, to show how the fundamental mechanisms, derived from gaseous and thin film studies, are affected and modified within the environment of the cell. We therefore expect that, as the field evolves, the complexity of targets investigated under LEE bombardment will further increase, up to a point, where the action of LEE in cells can be deduced with appreciable certainty. The next step toward a systematic comprehension of cellular damage has been the investigation of DNA in the presence of specific cell constituents, which lie close to or are bonded to DNA (e.g., water, O_2, ions and DNA-binding proteins).

Ptasińska and Sanche [96] measured the yields of the anions H^-, O^- and OH^- desorbed by 3–20 eV electron impact on GCAT films under hydrated conditions. Three monolayers (ML) of water were deposited on GCAT films; an amount corresponding, on average, to 5.25 H_2O molecules per nucleotide at the surface of the oligomer film [97]. Assuming a uniform water distribution, such two-component films represent DNA with the addition of 60% of the first hydration shell. Below 15 eV, the desorbed anions were found to emanate principally from a new type of dissociative core-excited transient anions formed via electron capture by a DNA-H_2O complex. The formation of transient anions from GCAT-H_2O complexes was most obvious in the O^- and OH^- yield functions. The 9-eV resonance in the yield function for O^- desorption from GCAT was replaced by a new one peaking near 11–12 eV and having a slightly reduced width. Similar results were obtained with the OH^- signal. Isotope experiments with ^{18}O-labeled water and deuterium indicated that O^- and OH^- at 11–12 eV emanate from the complex. The binding site of H_2O was found to be located at the negatively charged oxygen of the phosphate group, as expected from the experiment of Falk et al. [98, 99] on the binding energy of water to DNA. The investigations of Ptasińska and Sanche [96] further indicated that H^- desorbs upon LEE impact by dissociation of a transient anion of the complex, with bond cleavage on the H_2O portion at this site.

Shortly after Orlando et al. [100] investigated theoretically LEE diffraction of 5–30 eV electrons in hydrated B-DNA 5′-CCGGCGCCGG-3′ and A-DNA 5′-CGCGAATTCGCG-3′ sequences. The theoretical results were compared with experiments [100] on the damage induced by 5–30 eV electrons in hydrated plasmid DNA. Variations in the calculated electron density for the sequences occurred in energy regimes associated with DEA, direct electronic excitation, and dissociative ionization. Since their diffraction intensities were essentially localized on structural water, they postulated that compound H_2O:DNA states may contribute to the energy dependence of the LEE induced SSB and DSB. More specifically, the energy of the resonances they observed in those yields were shifted to higher energies by a value similar to that observed by Ptasińska and Sanche in ESD from hydrated GCAT films [96].

Ions are also an important constituent of cells; they play a role in the stabilization of the B-DNA conformation in vitro by their interaction with the major and minor grooves, as well as the negatively charged phosphate. Dumont et al. [101]

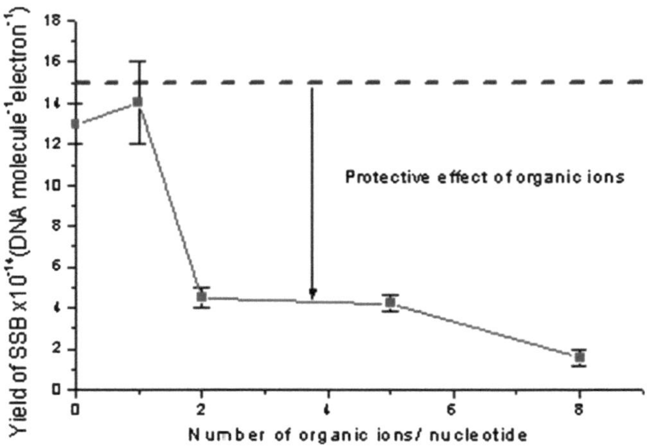

Fig. 1.14 Experimental (■) and projected (- - -) yields of SSB at constant DNA film thickness, as a function of the number of organic ions/nucleotide [101]. The dashed line is the expected yield if the interaction between DNA and organic ions had no effect. Each data points correspond to the mean value of 6 samples irradiated 15s (i.e., in the linear portion of the dose-response curve) ± standard deviation. (Reprinted with permission from reference 101. Copyright 2011 American Institute of Physics)

investigated the influence of the organic cation Tris and anion EDTA on the mechanisms of SSB induction in plasmid DNA by 10 eV electrons. Tris and EDTA were chosen for two reasons. First, 10 mM Tris with 1mM EDTA constitute the standard buffer solution for many in vitro experiments with DNA. Furthermore, the Tris molecule possesses an amine group (NH_2), while EDTA molecule contains four carboxylic acid groups (COOH). Because of their functional groups, these organic ions can mimic some aspects of the protein structure. In addition, the histone proteins, which are intimately associated with the DNA of eukaryotic cells, are rich in basic amino acids such as lysine and arginine which have positively charged amino groups. Experiments with these molecules can therefore provide some insight into the effect of proteins on LEE-induced DNA damage.

In their experiments, Dumont et al. [101] incorporated Tris and EDTA (with ratio 10 Tris: 1 EDTA) at various concentrations within plasmid DNA films of different thicknesses. The yield of SSB was found to decrease dramatically as a function of the number of organic ions per nucleotide. Fig. 1.14 shows the experimental yield of SSB (per incident electron, per molecule) measured at 5 ML thickness as a function of the number of ion per nucleotide in LEE-bombarded plasmid DNA films. The presence of only 2 organic ions/nucleotide is sufficient to decrease the yield of SSB by approximately 70%, while the presence of 8 organic ions/nucleotide decrease the yield of SSB by 88%.

Later, the experiments of Dumont et al. [101] were extended to the energy range 1 eV to 60 keV [102]; both SSB and DSB induced by electron impact on thin films of plasmid DNA under vacuum were measured [102]. Two characteristic salt concentrations were bombarded: one organic ion/nucleotide, which is necessary to neutralize the charge on the phosphate group of the backbone and hence corresponds to minimal salt concentration; and six organic ions/nucleotide, which corresponded to a high level of protection at 10 eV in the experiment of Dumont et al. [101]. Electron energies (i.e., 1, 10, 100 and 60,000 eV) characteristic of specific phenomena were chosen. The mechanisms occurring at 1 and 10 eV are shown in Fig. 1.12. Around 100 eV, electron scattering within DNA films produces maximum ionization and hence the largest density of LEE within the film. Finally, 60 keV electrons can be taken as representative of the fast electrons, such as Compton electrons, created during patient treatment with photons of 1–20 MeV [103].

The interaction of organic cations with DNA was found to protect the molecule from SSB and DSB induced by electrons of 1 eV to 60 keV [102] in similar proportions as those found by Dumont et al. [101] for SSB. It was concluded that the cation-DNA interaction may decrease the capture cross section and lifetime of transient anions formed by 1 and 10 eV electrons on base and phosphate groups of DNA. Electron transfer from transient anions to Tris cations or proton transfer from Tris to anions could contribute in reducing the lifetime of resonance formed on such DNA units. Since bond dissociation is highly sensitive to lifetime [12, 13], such a modification could reduce DNA damage. At higher energies, a multitude of direct electron scattering processes could also be modified by the presence of organic cations within the groove of DNA. However, at 100 eV and 60 keV, the DNA molecule was found to be less protected against SSB than at the lower energies, indicating that non resonant mechanisms are less affected by organic salts (i.e., transient anions formed within DNA are more sensitive to environmental changes than other mechanism of LEE induced damage). The reverse trend was observed for DSB induced by 100 eV electrons. The fact that at 100 eV DSB may arise from the high density of LEE may account for this different trend. These electrons would produce a higher density of radicals and ions around DNA owing to their higher ionization and electronic excitation cross sections [30], thus increasing the probably of breaking two strands with a single primary event.

Ptasińska et al. [104] investigated the role of two selected amino acids, i.e. glycine and arginine, on the tetramer GCAT, during exposure to 1 eV electrons, in order to provide insight on the effect of protein constituents on pathway 1 of Fig. 1.12. Surprisingly, at low ratios (R) of amino acid:GCAT (i.e., $R < 1$), particularly in the case of glycine, an increase in the total fragmentation yield of oligonucleotides was observed. At higher ratios ($R = 1$–4) of amino acid:GCAT, either glycine or arginine protected DNA from the interaction of electrons. Hence, the amino acid probably diminished electron capture by GCAT and/or the lifetime of the transient anion undergoing dissociation (i.e., DEA).

The effect of protein binding to DNA on total damage induced by 3-eV electrons was investigated by Solomon and Skalicky [105]. HPLC purified thiolated oligonucleotides with the structure 5′-SH-$(CH_2)_6$-$(dT)_{25}$-3′ were deposited onto gold-coated glass chips resulting in immobilized single-strand DNA monolayers about 4 nm thick. A fluorescent dye-marked complementary strand, Cy5-dA_{25} was used as a probe to monitor DNA damage and added in a solution containing the bombarded glass chip. When hybridization occurred, the $(dT)_{25}$ was considered to be intact, so that by monitoring the fluorescence intensity from the probe on the chip following LEE bombardment, the authors were able to measure the yield of damage as a function of fluence. Such exposure curves were obtained for the oligonucleotide with and without binding of the single-strand binding *E. coli* protein. The cross sections for total damage were found to be $1.6 \times 10^{-17}\,cm^2$/nucleotide and $3.8 \times 10^{-18}\,cm^2$/nucleotide in the case of pure DNA and the protein-DNA complex, respectively. Since at 3 eV the only mechanism that leads to bond breaking in DNA is DEA, the protein binding interaction was postulated to decrease LEE induced damage via modification of the resonance parameters of the transient anion involved in the process.

Another way to get closer to cellular conditions is to develop techniques to investigate LEE-induced DNA damage under atmospheric conditions at different levels of hydration. Such conditions are quite different from those of UHV, where DNA retains only structural H_2O (i.e., 2.5 water molecules per nucleotide) within its structure and thus adopts the compressed A-form. Cellular DNA is mainly in the B-form and contains about 20 bound water molecules per nucleotide (Γ) in the first ($\Gamma < 9$) and second ($9 < \Gamma < 20$) layers of hydration [97, 106]. The first layer consists of contiguous surface water, while the second layer is composed of amorphous water.

Brun et al. [55] were first to measure DNA damage induced by LEE under atmospheric conditions. They used the method explained at the end of Sect. 1.3 with an apparatus similar to that described in Fig. 1.9. Five ML films of plasmid DNA (3197 base pairs) deposited on glass and gold substrates were irradiated with 1.5 keV X-rays in UHV and under room atmosphere containing 65% humidity. The total damage was analyzed by agarose gel electrophoresis. Fig. 1.15a shows the percentage loss of supercoiled DNA as a function of incident photon fluence (i.e., photons $\cdot$ cm^{-2}) in UHV for 5 ML DNA films on glass and gold substrates. This damage arises essentially from SSB. Within experimental error, the loss of the supercoiled form is a linear function of the photon fluence. The percentage yields derived from the slope of these curves are given on the left of the first horizontal line of Table 1.1. The enhancement factors (EF) derived from these values appear in the middle column. The EF is calculated by dividing the yield obtained with the DNA on the gold substrate by that measured with DNA films deposited on glass. As expected, *the gold substrate enhances DNA damage owing to photoemission of LEE*. Fig. 1.15b presents the dose-response curves for the same substrates under atmospheric conditions at about 65% humidity. The loss of supercoiled DNA is a linear function of fluence in both cases. The percentage yields derived from the slope of these curves are given in the second horizontal line of Table 1.1. The EF increases to 1.9 under this atmosphere. As shown in the bottom line of Table 1.1, the damage caused by X-rays is increased by a factor of 2 in the presence of water whereas that caused by X-rays and LEE is increased by a factor of 2.6.

From estimates of the energy deposited in DNA by X-rays and LEE, the SE distribution emitted from gold, (see Fig. 1.8) and from the slopes of the curves of

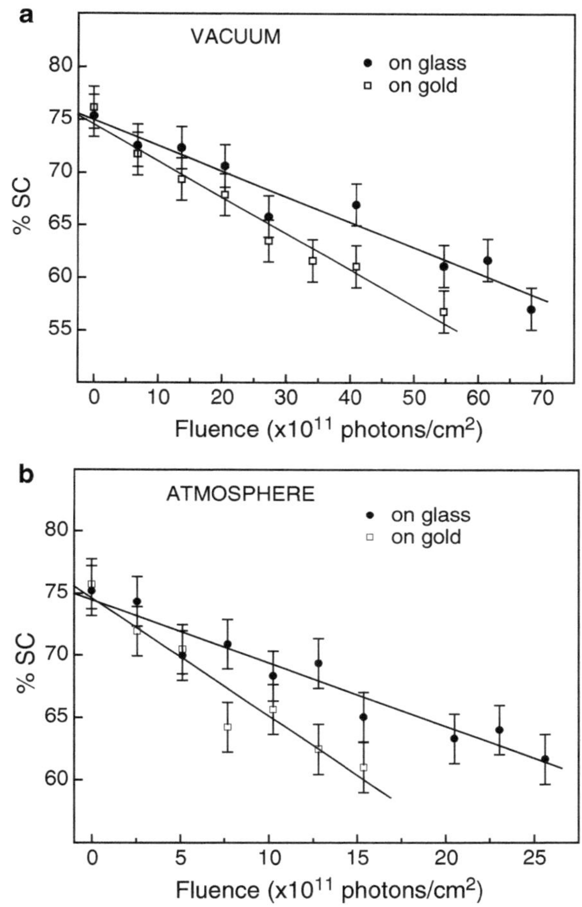

Fig. 1.15 X-ray exposure curves for the loss of the supercoiled DNA, **a**) in vacuum on glass and gold substrates and **b**) under atmospheric conditions on glass and gold substrates [55]. The points represent the means of three independent experiments and the error bars represent standard deviation of the means. (Reprinted with permission from reference 55. Copyright 2009 American Chemical Society)

Table 1.1 Percentage yield per 10^{12} photons · cm^{-2} for the loss of supercoiled DNA in 5 ML films deposited on glass and gold substrates and held under vacuum or atmospheric conditions. EF is the enhancement factor of the yield in going from glass to gold. G_L and G_X are the G values (number of damages/100 eV) for LEE and X-rays, respectively

	Substrate				
Environment	Glass	Gold	EF	G_L	G_X
vacuum (V)	2.4 ± 0.2	3.4 ± 0.3	1.4	4 ± 2	0.42 ± 0.04
atmosphere (A)	4.7 ± 0.5	8.7 ± 0.9	1.9	6 ± 2	0.40 ± 0.04
A/V	2.0	2.6			

Fig. 1.15a, Brun et al. [55] estimated the amount of damage produced per units of energy deposited, usually referred to as the G values. *G* values of 44 ± 6 nmol/J (0.42 D/100 eV) were found for 1.5 keV X-rays in vacuum and 400 ± 200 nmol/J (4 D/100 eV) for LEE, where D represents one damaged DNA molecule. In the case

of the atmospheric experiments, taking into account additional energy absorption by water molecules, they estimated from the slopes of Fig. 1.15b, G values of 42 ± 6 nmol/J (0.40 D/100 eV) for 1.5 keV X-rays and 600 ± 200 nmol/J (6 D/100 eV) for LEE. Thus, in the presence of water, the G value of LEE further increases by 50% whereas that of X-rays remains the same within instrumental error. As in the similar work of Cai et al. [107] performed only under UHV, the results of Brun et al. illustrate that SE are much more efficient than X-ray photons to induce DNA damage. In other words, when the same amount of energy is deposited in DNA by photons or low-energy SE, the latter produce much more damage.

There are a number of mechanisms that can account for the differences between G values for X-rays and LEE, although the details cannot be determined exactly at this stage. Whereas X-rays and fast electrons produce electronic and vibrational excitation in a similar proportion [9, 30, 108] the relative abundance of electronic excitations may be much larger for LEE due to the formation of transient anions. Such a difference results in a higher number of dissociative states produced per unit energy. Also, owing to charge polarization and the formation of transient anions, it takes less energy to break a chemical bond with a LEE than with a photon. The threshold energy for breaking a DNA strand with a photon lies around 7 eV [109], whereas a LEE can induce a SSB in DNA by DEA via shape resonances located at 0.8 and 2.3 eV [31] as shown in Fig. 1.13. Core-excited resonances induce SSB and DSB [33, 51, 110] from 5 to 15 eV with an efficiency as large as at 100 eV (see Fig. 1.4). The cross section to produce a SSB in DNA via DEA is on the order of 2×10^{-18} cm^2 per nucleotide and similar at 0.8 eV and around 10 eV [67], suggesting that low-lying shape resonances and core-excited resonances in DNA are quite efficient for breaking the phosphodiester backbone. Thus, SE emitted from gold with an energy distribution peaking around 1.4 eV (see Fig. 1.8) can cause considerable DNA damage via DEA. In other words, the dissociative processes are triggered with much less energy with LEE giving higher G values.

According to the results of the first column of numbers in Table 1.1, adding the hydration layers around DNA, while keeping all other experimental parameters the same, increases radiation damage by about a factor of 2.6 on gold compared to 2.0 on glass. The gold substrate increases considerably the density of LEE in the DNA film and the interaction of these latter with H_2O produces the species described in subsection 1.1. According to the EF on gold these species are highly efficient in damaging DNA. In dilute solution of DNA, the hydroxyl radical ($OH^\bullet$) is considered to be the secondary species formed by water radiolysis that produces SB in DNA. Thus, the added damage on gold was considered to be due essentially to $OH^\cdot$ radicals produced by LEE [55], but the change of DNA conformation to the B form could also play a role.

The increase in DNA damage in the presence of H_2O is in good agreement with recent estimates of the relative contribution to DNA damage in cells via the direct and indirect effects of radiation. Von Sonntag has estimated that the direct effects contribute about 40% to cellular DNA damage, while the effects of water radicals amount to about 60% [20]. The work of Krisch et al. on the production of strand breaks in DNA initiated by $HO^\bullet$ radical attack sets the direct effects contribution at 50% [57]. From the experiments of de Lara et al., one can estimate that 40–50% of the lesions induced in DNA by low linear energy transfer radiation arise from direct energy deposition events, many of which ionize DNA [58]. In other words, if we assume that the direct effects of radiation in cells corresponds to the damage measured by Brun et al. in vacuum with X-rays, we deduce from Table 1.1 a 49% and 61% contribution of the indirect effect (i.e., that caused by the presence of water molecules).

More recently, using the apparatus shown in Fig. 1.9, Alizadeh et al. [56] measured DNA damage induced by LEE and soft X-rays under dry nitrogen and oxygen at atmospheric pressure and temperature. Five-ML plasmid DNA films deposited on tantalum and glass substrates were exposed to Al K_α X-rays of 1.5 keV in the two different environments. From the yields, G values were extracted for X-rays and LEE. In the gaseous N_2 environment, the G values were found to be 65 ± 6 nmol/J for X-rays and 227 ± 15 nmol/J for LEE. These values agreed with those obtained for SSB in similar experiments under vacuum [107] and show similarity between vacuum and N_2 environments. Thus, it appears that the excited species produced by 1.5 keV photons and LEE in the N_2 atmosphere surrounding DNA did not largely contribute to the damage. Under the O_2 atmosphere, G values were 124 ± 9 and 415 ± 15 nmol/J for X-rays and LEE, respectively, i.e., in an oxygenated atmosphere the G values were 1.9 and 1.8 higher than the corresponding values in a nitrogenated environment, respectively. The results of Alizadeh et al. [56] also indicated a higher ($\sim$3-fold) effectiveness for LEE relative to 1.5 keV X-rays in causing SSB in both environments. These authors discussed their results by considering reactions with DNA of the species created by the radiation in gaseous N_2 and O_2 and oxygen fixation of primary damage. From these considerations, they suggested that the oxygen fixation mechanism, which is highly effective in increasing radiobiological effectiveness, under aerobic conditions, is operative on the type of damage created at the early stage of DNA radiolysis (i.e., at times $t < 10^{-12}$ s) by LEE.

1.5 Low energy electrons in concomitant chemoradiation therapy (CRT)

In many clinical studies with agents such as 5′ fluorouracil and cisplatin, it has been demonstrated that treatments with a chemotherapeutic drug delivered with concurrent radiation increased the survival of patients as compared to non-synchronal treatments [35, 111]. This observation has been attributed to a superadditive effect on tumor regression, due to a synergistic interaction between the radiation and the drug. Superadditive effects in concomitant CRT are expected to depend on the ways different drugs can increase the effects of radiation. The case of the chemotherapeutic agent cisplatin [112] is discussed in this section. Essentially two mechanisms have been proposed to explain the observed superadditive effect of

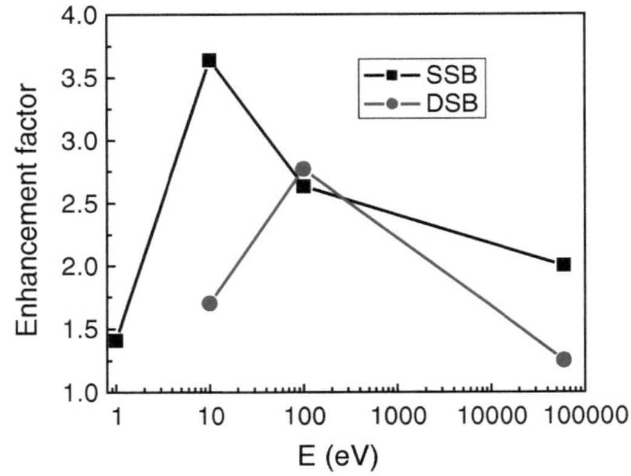

Fig. 1.16 Enhancement of damage arising from the bonding of two cisplatin molecules to DNA as a function of electron energy [116]. The enhancement factor is expressed as the yield of SSB and DSB with cisplatin bound to DNA divided by the yields without bound cisplatin, respectively. (Reprinted with permission from L. Sanche. Chem. Phys. Lett. 474, 1–6 (2009))

cisplatin on tumor regression. Studies in tissue cultures [113] and tumor-bearing mice [114] suggest inhibition of repair of radiation-induced damage to DNA, but an increase in the immediate species created by the primary radiation in cells can cause additional damage when cisplatin is covalently bound to DNA [115].

Zheng et al. [116] measured SSB and DSB induced by the impact of 1, 10, 100 and 60,000 eV electrons on thin solid films of DNA with and without cisplatin chemically bound to the molecule. In their experiments, they deposited lyophilized films of the pure plasmid and of plasmid-cisplatin complexes onto a clean tantalum foil. Under their conditions, 90% of the plasmid-cisplatin complexes consisted of a cisplatin molecule bound to DNA preferentially at the N7 atom of guanine [112, 117] producing an intrastrand adduct. The SSB and DSB were detected by measuring the transformation induced by electron impact, from the initial supercoiled configuration of the molecule to the circular or linear forms. Exposure response curves were obtained at each energy for DNA and cisplatin-DNA complexes of different ratios R of cisplatin molecules to that of plasmid DNA. The yields expressed as the percentage of SSBs and DSBs per electron and molecule were obtained from the initial slope of the exposure response curves. These yields saturated as a function of R [116]: beyond eight cisplatin molecules per plasmid, no increase in SB was observed within instrumental errors up to $R = 20$. The ratio of the yields with and without the presence of cisplatin, i.e., the EF, is plotted as a function of electron energy in Fig. 1.16 for $R = 2$. It maximizes at 10 eV for SSB and at 100 eV for DSB. *Only two cisplatin molecules bound to a plasmid composed of 3197 base pairs increase the number of SSB at 10 eV and DSB at 100 eV by factors of 3.7 and 2.7, respectively.*

Details on the mechanism of LEE-induced damage to DNA by 1 eV electrons have been discussed in subsections 1.3 and 1.4. According to the discussion, the yields at 1 eV arise exclusively via DEA and SB are induced by pathway 1 shown in Fig. 1.12. Hence, *the results of Fig. 1.16 at 1 eV unambiguously show that cisplatin can enhance DNA damage by increasing the magnitude of the DEA process within DNA*. Although the EF is only 1.4 at 1 eV, it represents a huge rise in capture cross section, if we attribute the increase damage to electron capture by cisplatin. Since only two cisplatin molecules per 3197 base pairs increase damage by a factor of 1.4, the capture cross section must be orders of magnitude higher at the site of cisplatin. The cisplatin molecule has a shape resonance near zero eV which leads to DEA, but the magnitude of the process is not known [118]. Close to zero eV, DEA cross sections often reach huge values of the order of 10^{-13}–10^{-15} cm^2 owing to the 1/k momentum factor, which enters into the expression of the captured cross section [119–121]. Furthermore, in DNA this cross section could be increased by base to base electron transfer along the chain, which would act to draw additional electrons to the site of cisplatin. This idea that a modified molecular site may act as a sink for non-thermal electrons has previously been tested by doping the surface of solid saturated hydrocarbon films with a simple molecule having a shape resonance at low energy [122, 123]. In the case of an n-hexane surface doped by O_2, electron capture by O_2 was increased by four orders of magnitude [122]. Another possible explanation for the large EF obtained with only two cisplatin molecules may be related to modifications induced in DNA by cisplatin. Cisplatin modifies the topology of DNA, which could weaken certain bonds and promote DEA within the backbone [117].

From the results shown in Fig. 1.4 and the explanations in subsection 1.4, the large EF plotted in Fig. 1.16 at 10 eV can be interpreted as due to an increase in the formation of SSB and DSB caused by the formation of core-excited resonances. At 10 eV, the exact process which is enhanced by the presence of cisplatin may *a priori* arise from pathways 1 and 3 in Fig. 1.12. Since pathway 1 is not very effective above 3 eV, it is probably the mechanism of pathway 3 that is enhanced by the presence of cisplatin. Replacing pathway 1 by pathway 3, the rest of the explanation for the 10 eV enhancement process becomes similar to that for 1-eV electrons.

As seen from Fig. 1.4, above 15 eV both the SSB and DSB yields rise monotonically from an apparent threshold and reach a plateau near 30 eV. Above 30 eV, many nonresonant mechanisms exist that can contribute to the observed DNA damage, such as transitions to excited states of the neutral molecule or its cations. The possible reactions are shown for a molecule RH by pathways 1 and 2 in Fig. 1.2. Hence, there exists a plethora of reactions that could be enhanced by the presence of cisplatin in DNA and would contribute to the EF shown in Fig. 1.4 at 100 eV. It is impossible, in this case, to determine which reactions are increased by binding cisplatin to DNA. However, according to the EF in Fig. 1.16, cisplatin increases more resonance processes at 10 eV leading to SSB than those occurring at 100 eV.

Contrary to the behavior of the EF for SSB, the EF for DSB increases from 10 to 100 eV. This behavior may seem strange since the probability to cut each individual strand must decrease according to the data for SSB. However, owing to the large ionization cross section at 100 eV [30], a single 100 eV electron can induce considerable multiple ionizations. With LEE thermalization distances of the order of the film thickness [5, 124], the probability of breaks occurring on adjacent chain sites by one or more SE created from ionization by the initial 100 eV electron or a SE is high and probably accounts for the higher EF for DSB at 100 eV. Moreover, the increase in ionization cross section due to the presence of a Pt atom in cisplatin increases the quantity of LEE in the film.

In radiotherapy, cancer is usually treated with 0.2–20 MeV photons from X-ray tubes, cobalt sources and electron accelerators. These primary photons transfer most of their energy to cancer cells by Compton scattering [103]. From conservation of momentum, electrons with a wide energy distribution are created which includes those of 60 keV. Almost all Compton electrons have energies sufficiently high such that description of their energy deposition into biological matter in terms of the same fundamental interaction can be calculated using the Born approximation [121]. Thus, results obtained with 60 keV electrons may be taken to represent the direct effects of radiation on DNA caused by clinical photons. Assuming that the EF in Fig. 1.16 are essentially produced by a modification of the action of LEE, then the sizable EF retained at 60 keV may be largely attributed to the effect of cisplatin on the action of secondary LEE produced by the high energy electrons along their path. The lower EF at 60 keV compared to that at 10 eV and 100 eV possibly accounts for other less sensitive processes and electrons produced with higher energies, both of which may contribute less significantly to the enhancement. In conclusion, the results of this section show that the formation of transient negative ions plays an important role, not only in radiobiological damage, but also in the enhancement of this damage with a chemotherapeutic agent.

1.6 Gold nanoparticles as radiosensitizers

Recently, there has been considerable interest in the potential use of GNP as radiosensitizers in the treatment of cancer with ionizing radiation [125–127]. For such treatments, GNP have two interesting properties: they increase the absorption of radiation energy and they accumulate preferentially in cancer cells [128, 129]. The first experimental evidence of GNP radiosensitization was provided by Hainfeld et al. [130]. Intravenous injection of 1.9-nm diameter GNP into mammary tumour-bearing mice combined to 250 kVp X-rays resulted in a one-year survival of 86% versus 20% with X-rays alone. More recently, Chang et al. [131] showed that, in a melanoma tumour-bearing mice model, 13-nm diameter GNP in conjunction with a single dose of 25 Gy from a 6 MeV electron beam led to a more pronounced reduction of the tumour volume than in the control groups [131]. A two-fold increase of apoptotic cells in animals treated with GNP and irradiation was achieved compared to irradiation alone. This result shows that it may not be necessary to irradiate tumors loaded with GNP with low energy photons having a high $\sigma_{Au}/\sigma_{H_2O}$ ratio (see Fig. 1.6) to obtain a reasonable EF. As explained in Sect. 1.2 many other factors are involved in the physical process of radiosensitization by GNP.

These pioneering results opened the exciting perspective in oncology of a more efficient and specific radiotherapy treatment. However, the amount of GNP administered to mice in these experiments was much too large for application in the radiotherapy of human cancer. This amount had to be decreased considerably, while maintaining as large as possible the radiosensitizing effect. Thus, efforts have been made to identify the mechanisms responsible for radiosensitization by performing X-rays irradiation experiments with specific cell components including protein [132] and supercoiled DNA in solution [133–135] and high-energy electron irradiations in UHV with plasmid DNA films [36]. The cellular sensitivity to radiation was expected to be considerably increased by the binding of GNP to DNA, since the proximity of GNP would cause an increase in the production of short-range (see Fig. 1.5) secondary electrons capable of depositing a considerable portion of their energy into the DNA molecule.

In their efforts to understand DNA damage in the presence of GNP, Foley et al. [133] investigated the indirect effect of radiation (i.e., essentially DNA damage induced by OH radicals). A solution of supercoiled DNA, bound to GNP in a GNP:DNA ratio of 100:1 was irradiated by 100 kVp X-rays. The maximum enhancement in damage was about a factor of two for a dose of 0.5 to 2 Gy, in good agreement with the dose calculations of Cho [136]. Brun et al. [137] and Butterworth et al. [135] also investigated damage induced to DNA by OH radicals. They reported the results of an investigation of four key-parameters governing GNP radiosensitization of DNA in solution, namely, nature of buffer, DNA:GNP molar ratio, GNP diameter and incident X-ray energy. Brun et al. performed irradiations with a clinical source and tested concentration ratios up to 1:1, five sizes of GNP from 8 to 92 nm and six effective X-ray energies from 14.8 to 70 keV. The most efficient parameters were large-sized GNP, high molar concentration and 50-keV photons, which could potentially result in a dose EF of 6. However, the EF decreased according to the scavenging capacity of the buffers [135].

Carter et al. [134] developed a Monte Carlo method to investigate the generation and transport of electrons in plasmid DNA dissolved in water. They chose to model the small gold nanoparticles (3-nm gold nanoparticles) chemisorbed on DNA. In the simulation, X-rays from a 100 kVp tungsten source interacted directly with water and gold. They also probed experimentally the nanoscale spatial profile of energy deposition created by GNP in aqueous solution with the inclusion of radical scavengers in the solution to reduce the diffusion distance of hydroxyl radicals, all the way down to a few nanometers or less from DNA. The resolution afforded by the scavengers made it possible to characterize the spatial profile of electron energy deposition of the GNP. The average calculated enhancement of DNA damage due to GNP caused by hydroxyl radicals was found to be $\sim$20%, which was much lower than their experimental observation (150%).

The direct effect of radiation (i.e., damage induced directly on DNA by SE and other secondary products formed from the initial radiation interaction) was investigated by Zheng et al. [36]. Relatively thick ($\sim$0.3 and 2.9 μm) films of plasmid DNA were bombarded by 60-keV electrons with and without electrostatically bound GNP. SSB and DSB were measured by agarose gel electrophoresis. The probabilities for formation of SSB and DSB from the exposure of 1:1 and 2:1 gold nanoparticle:plasmid mixtures to fast electrons increased by a factor of about 2.5 compared to neat DNA samples. For monolayer DNA adsorbed on a thick gold substrate, the damage per DNA molecule increased by an order of magnitude relative to that observed in the 2.9 μm thick pure DNA films. Most of the energy of the 60 keV electrons was absorbed by the thick film, whereas for 1-ML DNA essentially all the energy was absorbed by the gold substrate. Furthermore, as seen from Fig. 1.5, LEE have an effective range of about 10 nm; so, in a 2.9 μm thick film, the damage due to any secondary LEE emitted from the gold substrate would appear in about 0.4% of the total sample. Thus, the increase in damage per DNA molecule in going from thick to thin films deposited on a gold substrate could be interpreted as arising from SE whose distribution is shown in Fig. 1.8.

However, the interaction of a 60 keV electrons with a GNP does not necessarily produce the same SE energy distribution as that expected from the surface of a thick gold foil (Fig. 1.8), which can absorb all of the primary beam [138, 139]. The GNP do not act as a solid for electrons, and the consideration for bulk and thick foils cannot be applied, unless the GNP have a large diameter. SE have a probability to be ejected from GNP given by the relativistic Möller cross section [138], which provides the number of collision within the nanoparticle. Furthermore, any fast charged particle passing near a GNP generates virtual photons that produce a distribution of SE, consisting mainly of LEE, as explained in the introduction. Thus, the SE distribution from GNP is expected to be composed essentially of LEE, whose maximum in the distribution lies at higher energy than that shown in Fig. 1.8. In the case of 60 keV electrons, it must also be considered that a single gold atom has a mass absorption coefficient approximately seven times larger than that of biological material [40] as shown in Fig. 1.6.

Thus, the results of both GNP-DNA and pure DNA film experiments suggest that the enhancement of the direct effect of radiation by GNP is principally due to the production of additional LEE caused by the increased absorption of ionizing radiation energy by gold, in the form of GNP or a thick gold substrate. Zheng et al. [36] discussed their results in terms of the basic interaction mentioned in Sect. 1.2. Since LEE have a range of about five times the diameter of the DNA helix and they are created in large numbers by any kind of ionizing radiation, the authors concluded that *the radiosensitizing properties of GNP should be universal and should exist for any type of high-energy radiation, including the 0.3–20 MeV photon beams commonly used in radiotherapy* as shown in the animal studies by Chang et al. [131]. Furthermore, since on average only one GNP per DNA molecule was needed to obtain the sizable increase in damage, the results of Zheng et al. [36] indicated that by targeting GNP to the DNA of cancer cells, these nanoparticles should be applicable to patients and may thus offer a novel approach to radiotherapy treatments. In recent cell experiments, GNP were targeted to the DNA in the nucleus, by linking coded peptides to the gold surface. Interestingly, a chemotherapeutic effect [140] was observed in agreement with further LEE-impact data on DNA-GNP complexes [141] and recent animal studies [142] Such vectorized GNP thus hold promise as chemoradiotherapeutic agents.

1.7 Gold-nanoparticle enhancement of DNA damage induced by anti-cancer drugs and radiation

It was shown in Sect. 1.5 that cisplatin increases LEE interactions with DNA and damage to the molecule, primarily via modification of electron resonances below 15 eV. Since, according to the previous Sect. 1.4, GNP increase the local density of LEE, binding GNP to a cisplatin-DNA complex should boost the effect of cisplatin and further increase DNA damage. This hypothesis was tested by binding electrostatically GNP to a cisplatin-DNA complex. Dry films of pure plasmid DNA and DNA-cisplatin, DNA-GNP and DNA-cisplatin-GNP complexes were bombarded by 60 keV electrons [143]. The yields of SSB and DSB were measured as described in Sect. 1.4 in the experiments of Zheng et al. [36] For ratios of GNP to DNA of 1:1 and cisplatin to DNA of 2:1, the EF of SSB lied between 2–2.5 and increased to 3 with cisplatin-GNP-plasmid in ratio of 2:1:1. This modest increase could be only additive and not related to the interaction of the additional LEE on cisplatin. *For DSB, however, the binding of both GNP and cisplatin to DNA created a spectacular increase in the EF; i.e., DSB were increased by a factor of 7.5 with respect to pure DNA*. From the results of the previous section and these EF values, it appears quite obvious that the additional DSB to the cisplatin-DNA-GNP complex arises from the generation of additional SE from the GNP. Since the diameter of DNA is 2 nm, SE from GNP of energy lower than 300 eV, which have an effective range of about 10 nm [5, 33], (see Fig. 1.5) could reach up to 10 DNA molecules in the films. These SE can therefore interact very efficiently with DNA-cisplatin-GNP complexes.

This synergy observed between GNP and cisplatin in the case of DSB could arise from a number of basic phenomena. First, we can consider the possibility of two-event processes triggered by the interaction of a single 60 keV electron with a GNP. As mentioned in Sect. 1.4, the yield of DSB is expected to be highly dependent on two-event processes, such as the damage created by two LEE within the range of the distance between 10 base pairs ($\sim$4 nm). According to Figs 1.5 and 1.6, within 10 nm of its site, a single gold atom increases the density of LEE by an average factor of about 7 in the case of 60 keV electron irradiation. Hence, we expect a GNP which contains thousands of gold atoms to produce a dramatic increase in this density. This phenomenon should increase considerably DSB formation by two LEE interactions on opposite strands within a distance of 10 base pairs. The much higher EF obtained by combining cisplatin and GNP compared to GNP alone may also arise in part from the energy requirement to produce a DSB. Since cisplatin locally modifies the topology of DNA (117,144), the different topology could potentiate DSB formation. Combined with the electron affinity and chemical reactivity of cisplatin, these modifications could appreciably lower the energy required to break two adjacent strands.

Further research is needed to assess the clinical potential of cisplatin-GNP-DNA complexes and more generally the combination of platinum chemotherapeutic agents with GNP. Biochemical carriers of such complexes to the DNA of cancer cells still need to be developed and tested in cell uptake experiments. If successful, these studies should be followed by treatment of tumor-bearing animals to determine the efficiency of the drug with radiation exposure. We hope that the present review article will motivate research in this direction.

Acknowledgments This work was supported by the Canadian Institutes of Health Research and the Marie Curie international incoming fellowship program of the European Commission. Thanks are extended to Drs E. Alizadeh, A.D. Bass, M.A. Huels, D. Hunting, M. Michaud, L.G. Caron and R. Wagner for helpful comments and corrections. The efficient collaboration of Ms Francine Lussier in the elaboration of this article is gratefully acknowledged.

References

1. V. Cobut, Y. Frongillo, J.P. Patau, T. Goulet, M.-J. Fraser, J.-P. Jay-Gerin, Radiat. Phys. Chem. **51**, 229 (1998)
2. I. Abril, R. Garcia-Molina, C.D. Denton, I, Kyriakou, D. Emfietzoglou, Radiat. Res. **175**, 247 (2011)
3. J.A. LaVerne, S.M. Pimblott, Radiat. Res. **141**, 208 (1995)
4. S.M. Pimblott, J.A. LaVerne, Rad. Phys. Chem. **76**, 1244 (2007)
5. H. Nikjoo, L. Linkbord, Phys. Med. Biol. **55**, R65 (2010)
6. M. Inokuti, Radiat. Eff. Def. Sol. **117**, 143 (1991)
7. N.F. Mott, H.S.W. Massey, *The Theory of Atomic Collisions* (Clarendon, Oxford, 1965)
8. K. Takayanagi, Prog. Theor. Phys. Suppl. Jpn **40**, 216 (1967)
9. D. Srdoc, M. Inokuti, I. Krajcar-Bronic, E. Waibel, Y. Hatano, I. Kaplan, in *Atomic and Molecular Data for Radiotherapy*, edited by M. Inokuti (IAEA, CRP, 1995) p.547
10. L. Sanche, Scanning Microscopy **9**, 619 (1995)
11. G.J. Schulz, Rev. Mod. Phys. 45, 378 (1973); *ibid* **45**, 423 (1973)
12. M. Allan, J. Electr. Spectr. Rel. Phenom. **48**, 219 (1989)
13. L. Sanche, in *Excess Electrons in Dielectric Media*, edited by J.-P. Jay-Gerin, C. Ferradini (CRC Press, Boca Raton, 1991)
14. L.G. Christophorou, *Electron-Molecule Interactions and their Applications* (Academic Press, Orlando, 1984)
15. H.S.W. Massey, *Negative Ions* (University Press, London, 1976)
16. R.E. Palmer, P. Rous, Rev. Mod. Phys. **64**, 383 (1992)
17. L. Sanche, Surface Science **451**, 82 (2000)
18. R.L.P. Adams, J.T. Knowler, D.P. Leader, *The Biochemistry of the Nucleic Acids* (Chapman and Hall, New York, 1981)
19. A.F. Fuciarelli, J.D. Zimbrick, *Radiation Damage in DNA: Structure/Function Relationships at early Times (*Batelle, Columbus, 1995)
20. C. von Sonntag, *The Chemical Basis for Radiation biology* (Taylor and Francis, London, 1987)
21. O. Yamamoto, *Aging Carcinogenesis and Radiation Biology* (Plenum, New York, 1976)
22. J.F. Ward, *Advances in Radiation Biology 5* (Academic Press, New York, 1977)
23. L. Sanche, Phys. Scripta **68**, C108 (2003)
24. L. Sanche, Eur. Phys. J. D – Atom. Mol. Opt. Phys. **35**, 367 (2005)
25. L. Sanche, in *Radiation induced molecular phenomena in nucleic acid: A comprehensive theoretical and experimental analysis*, ed. by Manoj K. Shukla, J. Leszczynski (Springer, Netherlands, 2008)

26. L. Sanche, in *Radicals in Nucleic Acids*, edited by M. Greenberg (Wiley, 2009)
27. L. Caron, L. Sanche, in *Radical and radical ion reactivity in nucleic acid chemistry*, edited by P. Čársky, R. Čurik (Taylor & Francis, 2010) (submitted)
28. L. Zhang, Z. Tan, Radiat. Environ. Biophys. **49**, 15 (2010)
29. M.A. Huels, B. Boudaïffa, P. Cloutier, D. Hunting, L. Sanche, J. Am. Chem. Soc. **125**, 4467 (2003)
30. M. Inokuti, *Atomic and Molecular Data Needed for Radiotherapy and Radiation Research* TECDOC-799 (IAEA Press, Vienna, 1995)
31. F. Martin, P.D. Burrow, Z. Cai, P. Cloutier, D.J. Hunting, L. Sanche, Phys. Rev. Lett. **93**, 068101(2004)
32. J. Meesungnoen, J,-P. Jay-Gerin, A. Filali-Mouhim, S. Mankhetkorn, Radiat. Res. **158**, 657 (2002)
33. B. Boudaïffa, P. Cloutier, D. Hunting, M.A. Huels, L. Sanche, Radiat. Res. **157**, 227 (2002)
34. V.T. Devita Jr, S. Hellman, S.A. Resenberg, *Cancer: Principles and Practice of Oncology* (Lippincott Williams and Wilkins, New York, 2001)
35. M.W. Saif, R.B. Diaso, in *Chemoradiation in Cancer Therapy*, edited by Hak Choy (Humana Press, Totowa, 2003) Chap. 2
36. Y. Zheng, D.J. Hunting, P. Ayotte, L Sanche, Radiat. Res. **169**, 19 (2008)
37. National institute of standards and technology database: http://physics.nist.gov/PhysRefData/Xcom/Text/XCOM.html
38. National institute of standards and technology database: http://physics.nist.gov/PhysRefData/Star/Text/contents.html
39. Stopping powers for electrons and positions, ICRU reports 37 (1984)
40. Stopping Power and Range Tables for Electrons. Available online at http://physics.nist.gov/PhysRefData/Star/Text/ESTAR.html
41. R.W. Howell, Med. Phys. **19**, 1371 (1992)
42. J.L. Humm, R.W. Howell, D.V. Rao, Med. Phys. **21**, 1901 (1994)
43. M.A. Herve du Penhoat, M.A. Huels, P. Cloutier, J.P. Jay-Gerin, L. Sanche, J. Chem. Phys. **114**, 5755 (2001)
44. D.V. Klyachko, M.A. Huels, L. Sanche, Radiat. Res. **151**, 177 (1999)
45. M.D. Porter, T.B. Bright, D.L. Allara, C.E.D. Chidsey, J. Am. Chem. Soc. **109**, 3559 (1987)
46. P. Dugal, M.A. Huels, L. Sanche, Radiat. Res. **151**, 325 (1999)
47. N. Mirsaleh-Kohan, A.D. Bass, L. Sanche, Langmuir **26**, 6508 (2010)
48. P. Dugal, H. Abdoul-Carime, L. Sanche, J. Phys. Chem. B **104**, 5610 (2000)
49. M. Bazin, S. Ptasińska, A.D. Bass, L. Sanche, E. Burean, P. Swiderek, J. Phys. Condens. Matter **22**, 084003 (2010)
50. H. Abdoul-Carime, P.C. Dugal, L. Sanche, Radiat. Res. **153**, 23 (2000)
51. B. Boudaiffa, P. Cloutier, D. Hunting, M.A. Huels, L. Sanche, Science **287**, 1658 (2000)
52. S. Cecchini, S. Girouard, M.A. Huels, L. Sanche, D.J. Hunting, Biochemistry **44**, 1932 (2005)
53. Y. Zheng, P. Cloutier, J.R. Wagner, L. Sanche, Rev. Sci. Instrum. **75**, 4534 (2004)
54. Y. Zheng, P. Cloutier, D.J. Hunting, J.R. Wagner, L. Sanche, J. Am. Chem. Soc. **126**, 1002 (2004)
55. E. Brun, P. Cloutier, C. Sicard-Roselli, M. Fromm, L. Sanche, J. Phys. Chem. B **113**, 10008 (2009)
56. E. Alizadeh, P. Cloutier, D. Hunting, L. Sanche, J. Phys. Chem. B **115**, 4523 (2011)
57. R.E. Krisch, M.B. Flick, C.N. Trumbore, Radiat. Res. **126**, 251 (1991)
58. C.M. deLara, T.J. Jenner, K.M.S. Townsend, S.J. Marsden, P. O'Neill, Radiat. Res. **144**, 43 (1995)
59. P. Rowntree, L. Parenteau, L. Sanche, J. Chem. Phys. **94**, 8570 (1991)
60. W.C. Simpson, M.T. Sieger, T.M. Orlando, L. Parenteau, K. Nagesha, L. Sanche, J. Chem. Phys. **107**, 8668 (1997)
61. W.C. Simpson, T.M. Orlando, L. Parenteau, K. Nagesha, L. Sanche, J. Chem. Phys. **108**, 5027 (1998)
62. G.A. Kimmel, T.M. Orlando, Phys. Rev. Lett. **77**, 3983 (1996)
63. G.A. Kimmel, T.M. Orlando, C. Vezina, L. Sanche, J. Chem. Phys. **101**, 3282 (1994)

64. G.A. Kimmel, T.M. Orlando, Phys. Rev. Lett. **75**, 2606 (1995)
65. G.A. Kimmel, T.M. Orlando, P. Cloutier, L. Sanche, J. Phys. Chem. B **101**, 6301 (1997)
66. M. Michaud, A. Wen, L. Sanche, Radiat. Res. **159**, 3 (2003)
67. R. Panajotovic, F. Martin, P. Cloutier, D. Hunting, L. Sanche, Radiat. Res. **165**, 452 (2006)
68. X. Pan, L. Sanche, Chem. Phys. Lett. **421**, 404 (2006)
69. L. Sanche, Phys. Rev. Lett. **53**, 1638 (1984)
70. X. Pan, L. Sanche, Phys. Rev. Lett. **94**,198104 (2005)
71. Y. Zheng, P. Cloutier, D. Hunting, L. Sanche, J.R. Wagner, Am. Chem. Soc. **127**, 16592 (2005)
72. H. Abdoul-Carime, S. Gohlke, E. Fischbach, J. Scheike, E. Illenberger, Chem. Phys. Lett. **387**, 267 (2004)
73. D. Becker, A. Bryant-Friedrich, C. Trzasko, M.D. Sevilla, Radiat. Res. **160**, 174 (2003)
74. D. Becker, Y. Razskazovskii, M. Callaghan, M.D. Sevilla, Radiat. Res. **146**, 361 (1996)
75. L. Shukla, R. Pazdro, D. Becker, M.D. Sevilla, Radiat. Res. **163**, 591 (2005)
76. Y. Zheng, P. Cloutier, D.J. Hunting, J.R. Wagner, L. Sanche, J. Chem. Phys. **124**, 64710(2006)
77. Y. Zheng, R. Wagner, L. Sanche, Phys. Rev. Lett. **96**, 208101 (2006)
78. R. Barrios, P. Skurski, J. Simons, J. Phys. Chem. B **106**, 7991 (2002)
79. J. Berdys, I. Anusiewicz, P. Skurski, J. Simons, J. Am. Chem. Soc. **126**, 6441 (2004) and references therein
80. P.L. Lévesque, M. Michaud, W. Cho, L. Sanche, J. Chem. Phys. **122**, 224704 (2005)
81. C. Winstead, V. McKoy, Phys. Rev. Lett. **98**, 113201 (2007)
82. Z. Li, Y. Zheng, P. Cloutier, L. Sanche, J.R. Wagner, J. Am. Chem. Soc. **130**, 5612 (2008)
83. Y.S. Park, H. Cho, L. Parenteau, A.D. Bass, L. Sanche, J. Chem. Phys. **125**, 074714 (2006)
84. C. Jäggle, P. Swiderek, S.-Ph. Breton, M. Michaud, L. Sanche, J. Phys. Chem. B **110**, 12512 (2006)
85. Z. Li, P. Cloutier, L. Sanche, J.R. Wagner, J. Am. Chem. Soc. **132**, 5422 (2010)
86. J. Gu, J. Wang, J. Leszczynski, Nucl. Acids Res. **38**, 5280 (2010)
87. L.G. Caron, S. Tonzani, C.H. Greene, L. Sanche, Phys. Rev. A **78**, 042710 (2008)
88. Li et al., submitted
89. T. Solomun, H. Seitz, H. Sturm, J. Phys. Chem. **113**, 11557 (2009)
90. J. Gu, Y. Xie, H.F. Schaefer, J. Phys. Chem. B **114**, 1221 (2010)
91. S.G. Ray, S.S. Daube, R. Naaman, Proc. Natl. Acad. Sci. USA **102**, 15–19 (2005)
92. K. Aflatooni, G.A. Gallup, P.D. Burrow, J. Phys. Chem. A **102**, 6205 (1998)
93. B. Boudaiffa, P. Cloutier, D. Hunting, M.A. Huels, L. Sanche, Méd. Sci. **16**, 1281 (2000)
94. A.D. Bass, L. Sanche, Radiat. Environ. Biophys. **37**, 243 (1998)
95. N. Mirsaleh-Kohan, A.D. Bass, L. Sanche, J. Chem. Phys. **134**, 015102 (2011)
96. S. Ptasińska, L. Sanche, Phys. Rev. E **75**, 031915 (2007)
97. N.J. Tao, S.M. Lindsay, A. Rupprecht, Biopolymers **28**, 1019 (1989)
98. M. Falk, K.A. Hartman, R.C. Lord, J. Am. Chem. Soc. **85**, 387 (1963)
99. M. Falk, Can. J. Chem. **43**, 314 (1965)
100. T.M. Orlando, D. Oh, Y. Chen, A.B. Aleksandrov, J. Chem. Phys. **128**, 195102 (2008)
101. A. Dumont, Y. Zheng, D. Hunting, L. Sanche, J. Chem. Phys. **132**, 045102 (2010)
102. Y. Zheng, L. Sanche, J. Chem. Phys. **133**, 155102 (2010)
103. H.E. Johns, J.R. Cunningham, *The physics of Radiology*, (C.C. Thomas Publisher, Springfield, Illinois, 1983)
104. S. Ptasińska, Z. Li, N.J. Mason, L. Sanche, Phys. Chem. Chem. Phys. **12**, 9367 (2010)
105. T. Solomun, T. Skalicky, Chem. Phys. Lett. **453**, 101 (2008)
106. S. Swarts, M. Sevilla, D. Becker, C. Tokar, K. Wheeler, Radiat. Res. **129**, 333 (1992)
107. Z. Cai, P. Cloutier, L. Sanche, D. Hunting, Radiat. Res. **164**, 173 (2005)
108. Shimamura J, Takayanagi K, *Electron-Molecule Collision. Plenum*, (New York, 1984)
109. M. Folkard, K.M. Prise, B. Vojnovic, B. Brocklehurst, B.D. Michael, Int. J. Radiat. Biol. **76**, 763 (2000)
110. X. Pan, P. Cloutier, D. Hunting, L. Sanche, Phys. Rev. Lett. **90**, 208102 (2003)
111. J. Landry, W. Blackstock, R.M. McRae, G. Yang, H. Choy, in *Chemoradiation in Cancer Therapy*, edited by H. Choy (Humana Press, Totowa, NJ, 2003)

112. V. Brabec, in *Platinum-based Drugs in cancer therapy*, edited by L.R. Kelland, N. Farrell (Humana Press Inc, Totowa, NJ, 2000)
113. J.D. Zimbrick, A. Sukrochana, R.C. Richmond, Int. J. Radiat. Oncol. Biol. Phys. **5**, 1351 (1979)
114. P. Lelieveld, M.A. Scoles, J.M. Brown, R.F. Kallman, Int. J. Radiat. Oncol. Biol. Phys. **11**, 111 (1985)
115. L. Dewit, Int. J. Radiat. Oncol. Biol. Phys. **13**, 403 (1987)
116. Y. Zheng, D.J. Hunting, P. Ayotte, L. Sanche, Phys. Rev. Let. **100**. 198101 (2008)
117. R.N. Bose, Mini-Rev. Med. Chem. **2**, 103 (2002)
118. Q.-B. Lu, S. Kalantari, C.-R. Wang, Mol. Pharmaceut. **4**, 624 (2007)
119. N.F. Mott, H.S.W. Massey, *The Theory of Atomic Collisions* (Clarendon, Oxford, 1965)
120. A.D. Bass, J. Gamache, P. Ayotte an L. Sanche, J. Chem. Phys. **104**, 4258 (1996)
121. H.S.W. Massey, E.H.S. Burhop, H.B. Gilbody, *Electron Collisions with Molecules and Photoionization* (Clarendon Press, Oxford, 1969)
122. K. Nagesha, L. Sanche, Phys. Rev. Lett. **81**, 5892 (1998)
123. E.T. Jensen, L. Sanche, J. Chem. Phys. **129**, 074703 (2008)
124. R. Naaman, L. Sanche, Chem. Rev. **107**, 1553 (2007)
125. D.M. Herold, I.J. Das, C.C. Stobbe, R.V. Iyer, J.D. Chapman, Int. J. Radiat. Biol. **76**, 1357 (2000)
126. R. Whyman, Gold Bull. **29**, 11 (1996)
127. W. Chen, J. Zhang, J. Nanosci. Nanotechnol. **6**, 1159 (2006)
128. A. Anshup, J.S. Venkataraman, C. Subramaniam, R.R. Kumar, S. Priya, T.R.S. Kumar, R.V. Omkumar, A. John, T. Pradeep, Langmuir **21**, 11562 (2005)
129. T. Niidome, K. Nakashima, H. Takahashi, Y. Niidome, Chem. Commun. 1978 (2004)
130. J.J. Hainfeld, D.N. Slatkin, H.M. Smilowitz, Phys. Med. Biol. **49**, N309 (2004)
131. M. Chang, A. Shiau, Y. Chen, C. Chang, H. Chen, C. Wu, Cancer Sci. **99**, 1479 (2008)
132. E. Brun, P. Duchambon, Y Blouquit, G. Keller, L. Sanche, C. Sicard-Roselli, Rad. Phys. Chem. **78**, 177 (2009)
133. E.A. Foley, J.D. Carter, F. Shan, T. Guo, Chem. Commun. **25**, 3192 (2005)
134. J.D. Carter, N.N. Cheng, Y. Qu, G.D. Suarez, T. Guo, J. Phys. Chem. B **111**, 11622 (2007)
135. K.T. Butterworth, J.A.Wyer, M. Brennan-Fournet, C.J. Latimer, M.B. Shah, F.J. Currell, D.G. Hirst, Radiat. Res. **170**, 381 (2008)
136. S.H. Cho, Phys. Med. Biol. **50**, N163 (2005)
137. E. Brun, L. Sanche, C. Sicard-Roselli, Colloids Surf. B **72**, 128 (2009)
138. L. Reimers, H. Drescher, J. Phys. D **10**, 805 (1977); J. Goldstein II, D.E. Newbury, P. Echlin, D.C. Joy, A.D. Romig Jr., C.E. Lyma, *Scanning electron microscopy and X-ray microanalysis* (Plenum Press, New York, 1992) 2nd ed
139. J. Schou, in *Physical processes of the interaction of fusion plasmas with solids*, edited by W. Hofer, J. Roth (Academic Press, New York, 1993)
140. B. Kang, M.A. Mackey, M.A. El-Sayed, J. Am. Chem. Soc. **132**, 1517 (2010)
141. Y. Zheng, P. Cloutier, D.J. Hunting, L. Sanche, J. Biomed. Nanotechnol. **4**, 469 (2008)
142. E. Hébert, P.-J. Debouttière, M. Lepage, L. Sanche, D.J. Hunting. Int. J. Radiat. Biol. **86**, 692–700 (2010)
143. Y. Zheng, L. Sanche, Radiat. Res. **172**, 114 (2009)
144. J. Malinge, M. Giraud-Panis, M. Leng, J. Inorg. Biochem. **77**, 23 (1999)
145. J. Meesungnoen, J.-P. Jay-Gerin, private communication
146. K. F. Ness, R. E. Robson, Phys. Rev. A **38**, 1446–1456 (1988)
147. N. L. Djuric, I. M. Cadez, M. V. Kupera, Int. J. Mass Spectrom. Ion Proc. **83**, R7–R10 (1988)
148. M. A. Bolorizadeh, M. E. Rudd, Phys. Rev. A **23**, 882–887 (1986)

Chapter 2
The Role of Secondary Electrons in Radiation Damage

Stephan Denifl, Tilmann D. Märk, and Paul Scheier

Abstract In the present contribution we will show that low energy electrons (<10 eV), which are formed as secondary species in the interaction of ionizing radiation with biological matter, can effectively damage molecules of biological matter. The underlying mechanism is dissociative electron attachment, where the transient negative ion (TNI) state decays by dissociation into a fragment anion and one or more neutral fragments. We investigated this process with building blocks of DNA (nucleobases) and proteins (amino acids). We studied these compounds under isolated conditions in the gas phase and also when they were embedded in cold helium droplets. The latter experiments allow the study of clusters of biomolecules and moreover, the effect of environment on the electron capture process and the decay of the TNI formed can be elucidated. This bears important consequences for drawing conclusions from the gas phase results to the damage of secondary electrons in a cellular environment.

2.1 Introduction

About one decade ago, the potential of low energy electrons in damaging DNA has been realized. Sanche and co-workers showed that single and double strand breaks can be observed in plasmid DNA after irradiation with electrons with an energy below 20 eV [1, 2]. This observation gains highly on importance when the efficient production of secondary electrons by the interaction of energetic radiation (α, β, γ, and ions) with biological matter is considered. The known reference value is about 10^5 electrons per MeV deposited primary radiation in matter [3].

S. Denifl (✉) • T.D. Märk • P. Scheier
Institut für Ionenphysik und Angewandte Physik and Center of Biomolecular Sciences Innsbruck (CMBI), Universität Innsbruck, A-6020 Innsbruck, Austria
e-mail: stephan.denifl@uibk.ac.at; tilmann.maerk@uibk.ac.at; Paul.Scheier@uibk.ac.at

G.G. Gómez-Tejedor and M.C. Fuss (eds.), *Radiation Damage in Biomolecular Systems*, Biological and Medical Physics, Biomedical Engineering,
DOI 10.1007/978-94-007-2564-5_2, © Springer Science+Business Media B.V. 2012

Thus, secondary electrons are the most abundant of all secondary species (electrons, ions, radicals) produced by this interaction. Another remarkable fact is their kinetic energy distribution, which shows that the vast majority is formed in the energy region below 100 eV with its maximum pointing towards zero eV [4, 5]. Another additional source for low energy electrons may be intermolecular Coulombic decay which is a kind of non-local autoionization process [6]. Thereby a single hole in an inner shell (produced by the initial high energy radiation) is replaced by two vacancies in the outer valence shells of two adjacent molecules in a molecular framework and an additional free electron with low kinetic energy. Independent of the way of formation, it is then the action of such low energy electrons in driving severe damage to the DNA which can lead to mutagenic, genotoxic and other potential DNA lesions. It was supposed that this direct DNA damage leads to about 1/3 of the overall damage of the genome of a living cell, while 2/3 is ascribed to indirect damage from free radicals produced by energy deposited in water molecules (a cell consists of 70%–80% of water) and other biomolecules surrounding DNA [7]. However, recent results from experiments with prehydrated electrons (formed by the radiolysis of water) challenged the convential notation of indirect DNA damage ascribed mainly to action of the OH radical [8, 9]. Prehydrated electrons are weaker bound than hydrated electrons and have much higher quantum yield than hydrated electrons and OH radicals [8].

Placing emphasis now on direct damage, laboratory experiments with plasmid DNA like the experiments of the Sanche group are therefore crucial for determining quantum yields for cell damage by low energy electrons. Another important aspect turned out to be the electron energy dependence of DNA strand breaks which give a hint on the underlying molecular mechanism. The measurements showed a resonant behaviour instead of a monotonic increase above the threshold energy which would be characteristic for ionization. Thus rather dissociative electron attachment has to be considered as the decisive step because such resonant behaviour is characteristic for this process. Thus it was proposed that DNA damage starts with electron attachment to a site of DNA with subsequent decay of the transient negative ion formed [1, 2].

2.2 Electron attachment to biomolecules in the gas phase

Motivated by the electron irradiation experiments with plasmid DNA a large number of elastic and inelastic electron scattering experiments with simple biomolecules have been carried out. While other chapters of this book cover the theoretical description of electron induced radiation damage (see also for example Refs. [10–12]), electron scattering experiments with thin biomolecular films of different complexity [3, 13], or experiments with other projectiles like for example atomic multiply charged ions [14], the focus of the present book chapter lies on electron attachment experiments with isolated biomolecules and biomolecularclusters. In this case high vacuum conditions provide well-defined conditions for electron scattering experiments; moreover, if charged products are formed they can be detected instantaneously by mass spectrometry. The electron attachment processes discussed here are reactions like:

$$M + e^- \rightarrow M^{*-} \rightarrow \ldots . \qquad (2.1)$$

Thereby a free electron attaches to a molecule M which leads to the formation of a transient negative ion (TNI) state M^{*-} in about 10^{-16} s. This state can be electronically or/and vibrationally excited and decays depending on the molecular properties either by electron detachment or dissociation. The formation of a stable molecular anion via radiative stabilization is much less likely. However, for some systems with a positive electron affinity the molecular anion becomes also mass spectrometrically observable by storage of the deposited excess energy (which is equally to the initial electron energy plus the electron affinity) in the vibrational degrees of freedom of the polyatomic molecule like for example in the case of SF_6^-/SF_6 [15]. However, what is shown below, for most biomolecules no long-lived molecular anion is formed under isolated conditions and in this case only anions formed by dissociative electron attachment are observable by mass spectrometry. As mentioned before, reaction (2.1) is a resonant process, i.e. the electron energy must fit for capture into a quasibound state of the molecule. Thus, experimentally the major interest lies in the determination of resonance energies for certain anionic products which can be determined from the corresponding anion efficiency curves.

A typical experimental apparatus used for electron attachment studies to biomolecules in the gas phase is shown in Fig. 2.1a. The setup is a crossed neutral/electron beam apparatus which consists of a neutral source, an electron source and a mass spectrometer for mass analysis of the charged reaction products [16]. Mass analyzed ions are detected using a channeltron type secondary electron multiplier. Since most biomolecules are non-volatile compounds, their vapour-pressure has to be increased in order to form a sufficient neutral density in the reaction zone with the electron beam. For this often thermal heating is applied which is at least for small biomolecules an easy and efficient method. However, for a few less stable compounds it may be a severe problem, since in this case also thermally decomposed products interact with the electron beam and any resulting ion yield will contaminate the ion yields resulting from intact molecules. Possible ways to differentiate the ion yields are then temperature dependency measurements or reference to pyrolysis studies, where thermal decomposition products are determined. Alternative sources for intact neutral biomolecules like laser desorption or laser induced acoustic desorption may be promising but have been hardly used so far [17]. However, these methods may be fruitful for production of complex neutral targets in the gas phase like (oligo-)nucleotides which thermally degrade already at moderate temperatures.

Anions are formed in the crossing region of the neutral beam with the electron beam. For the exact determination of resonance energies a monochromatized electron beam is of high benefit since the measured ion yield is a convolution of the true cross section and the electron energy distribution. In the present studies a

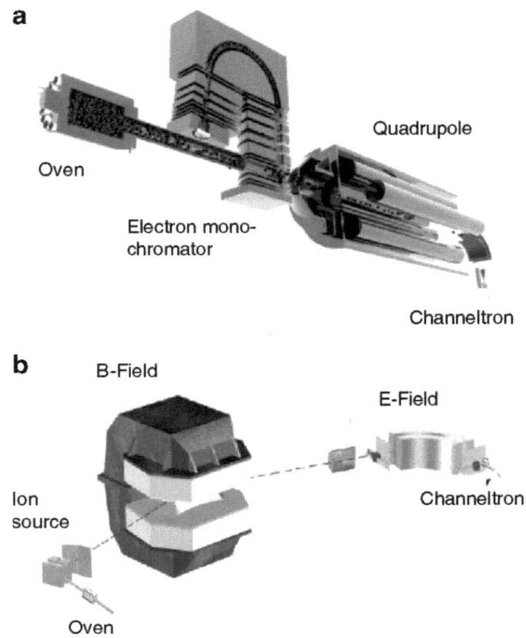

Fig. 2.1 Schematic view of two experimental setups used for the present investigations, (**a**) a hemispherical electron monochromator in combination with a quadruople mass spectrometer and (**b**) a two sector field mass spectrometer equipped with a conventional Nier-type ion source (see text)

hemispherical electron monochromator is used as monochromatizing element for the electron energy beam generated by a hairpin filament. Moreover, before and after the hemispheres a dedicated lens system focuses the electron beam into the analyzer and the crossing zone with the neutral beam, respectively. Typical electron energy resolutions achievable are between 50–100 meV with beam currents of up to a few tens of nA. These beam currents are high enough to measure most of anions formed upon electron attachment to biomolecules. However, for the measurement of anions with very low intensity higher sensitivity can be gained by using a standard ion source without a monochromatizing element as shown for the second setup in Fig. 2.1b. Thereby the pure beam of thermally emitted electrons from a filment is used which amounts easily to μA or mA. However, in this case the electron energy resolution is typically around 1 eV.

For mass analysis of anions formed in electron attachment studies different kind of mass spectrometers can be used. Often quadrupole mass spectrometers are used like in the monochromator setup of Fig. 2.1a, which provide a mass resolution $m/\Delta m$ of about a few hundred. This limitation in mass resolution may be not disturbing since one can assume that due to the thermal heating procedure anyway only rather small biomolecules may be measured. However, in DEA often isobaric fragment anions are formed (e.g., O^-/NH_2^-) which appear on the same nominal mass and cannot be separated by a quadrupole mass spectrometer. Thus our group also uses a double focusing sector field mass spectrometer as a second setup (see Fig. 2.1b) [18] which provides mass resolutions of a few 1000. Even when no isobaric anions are present, mass peaks can be calibrated then with high precision, which allows an exact determination of the composition of fragment anions formed in DEA. Moreover, the combination of a magnetic sector and an electric sector allows different scan techniques to study the decay of metastable anions in an unimolecular decomposition process. One possibility is the so called mass analyzed kinetic energy scan (MIKE) technique where a decay of a mass analyzed anion in the field free region between the magnetic and electric sector is monitored by scanning the electric sector field voltage. A decay product with mass m_2 formed in the decay of a mass analyzed anion with m_1 (which passes the electric sector at V_1) will only pass the electric sector at the reduced voltage $(m_2/m_1)V_1$. Moreover, this technique can be also used to study collision induced decay of mass selected anions which can be used for structure analysis. This is especially important to distinguish isomeric anions, which may show a different collision induced fragmentation pattern.

In a third experimental setup we also studied clusters of biomolecules to elucidate the response of larger neutral biomolecular complexes upon low energy electron exposure. Since biomolecules have low vapour pressure, we embedded thereby simple building blocks into cold helium droplets [19], where they immediately cool down and form agglomerates, i.e. for example nucleobase pairs, hydrated biomolecules, etc., can be formed. Although the low temperature of the droplets (0.38 K) [20] is far from true biological conditions, these experiments allow to study the aspects of inelastic electron scattering from biomolecular complexes on the molecular and nanoscopic level. Moreover, they perfectly allow the study of effects on the decay of biomolecular transient negative ions embedded in the droplets. The experimental technique to generate such doped droplets is to form pure helium droplets in a free jet expansion and let them pass a differentially pumped pick-up cell which is filled with vapour of biomolecules. For this the biomolecular samples are heated in a small oven connected to the pick-up cell. The average size of the cluster can be varied by changing the temperature of the oven which will result in a different partial pressure in the pick-up cell. After electron capture by the doped droplets, anions formed are mass analyzed with a standard double focusing mass spectrometer.

2.3 DEA studies with building blocks of DNA/RNA and proteins

Since the pioneering experiments by the Sanche group, where double and single strand breaks in electron irradiated plasmid DNA films were observed, a major impetus arose to study electron attachment to the DNA (and RNA) building blocks. Amino acids are other important targets, when biological tissue is exposed to ionizing radiation. DNA is wrapped in chromosomes in a packed way around proteins. Thus also amino acids, the building blocks of proteins, are studied intensively to elucidate the radiation damage. In the following the most important aspects of DEA will be discussed first for nucleobases and then for amino acids.

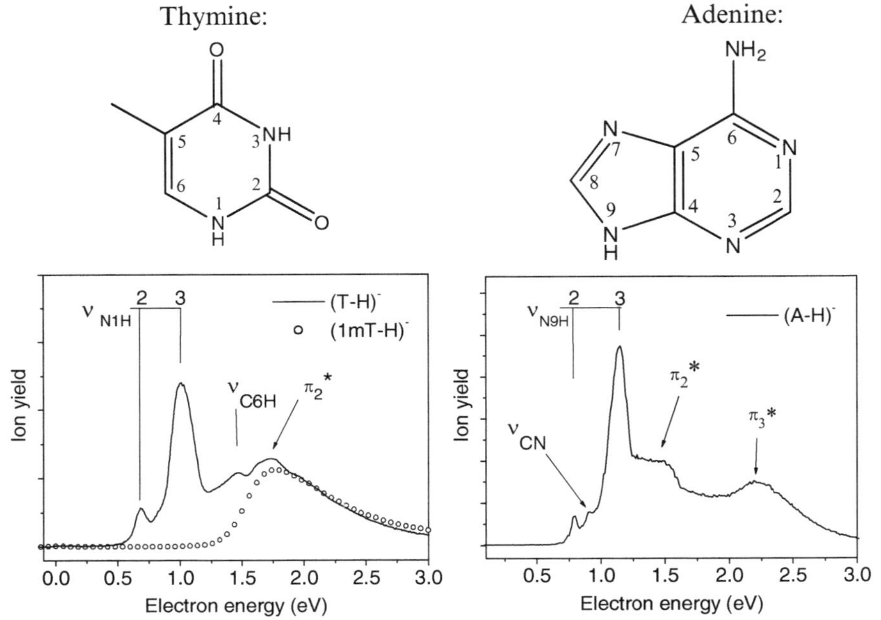

Fig. 2.2 Ion yield as the function of the initial electron energy of the dehydrogenated parent anion $(T\text{-}H)^-$ (left panel) and $(A\text{-}H)^-$ (right panel) formed upon DEA to the nucleobases thymine and adenine, respectively [21, 22]. The corresponding molecular structures are shown on top. Also included in the left panel is the ion yield for the dehydrogenated parent anion of 1-methylthymine $(1mT\text{-}H)^-$ (circles) [27]. Moreover, energies of involved vibrational excitation modes and attachment energies for the relevant π^* orbitals are indicated (see text)

2.3.1 Nucleobases

Nucleobases consist of either one aromatic ring (pyrimidines like uracil and thymine) or two aromatic rings (purines like adenine and guanine); see also Fig. 2.2. Attachment of a free electron to an isolated nucleobase does not lead to a molecular anion [16, 18, 21, 22] which is observable on mass spectrometric timescales. Instead it will undergo autodetachment or dissociation. The most abundant fragment anion thereby produced is the dehydrogenated anion $(M\text{-}H)^-$ which is shown in Fig. 2.2 for thymine [21] and adenine [22]. For both molecules the ion yield consists of a combination of sharp structures and at least one broad peak. It is then the question, how these resonances are formed. The lowest lying valence orbitals available for electron capture are π^* orbitals which were observed in previous electron transmission spectroscopy experiments [23]. However, it should be noted that nucleobases possess rather high dipole moments, which are above the critical value of 2.5D to form a dipole bound anion [24]. The existence of dipole bound states for nucleobases has been proven in photoelectron spectroscopy experiments with nucleobases formed in a microplasma [25] and moreover, in Rydberg electron transfer experiments [26]. In free electron attachment to an isolated nucleobase a dipole bound anion formed is vibrationally excited and thus unstable towards autodetachment. However, a possible doorway exists by coupling of the dipole bound state with the dissociative σ^*(N1-H) state, i.e. by tunnelling of the N1 hydrogen to form a $(M–H)^-$ [27]. Thus the sharp peaks (at 0.7 eV, 1.1 eV and 1.47 eV for thymine; at 0.8 eV, 0.92 eV and 1.15 eV for adenine, respectively) shown in Fig. 2.2 represent vibrational progressions of various stretching modes as assigned in the figure (vibrational Feshbach resonances [28]). The rather broad resonances (at 1.8 eV for thymine and 1.44 eV and 2.21 eV for adenine, respectively) may arise from vibronic coupling of a non-dissociative and a dissociative state but in this case a valence state is involved as intermediate (for example the π^*_2 resonance coupling with the σ^*(N3-H) state in the case of thymine) [27].

In line with the experimental results another important aspect of this assignment is that the cleavage of bonds can be selectively chosen by the initial electron energy. All gas phase experiments with partially deuterated or methylated nucleobases clearly showed that no hydrogen loss from the carbon positions of the molecules occurs below 4 eV [22, 29, 30]. Instead only hydrogens from nitrogen positions are involved and the loss from N1 and N3 site could be distinguished (see Fig. 2.2). This site selectivity is even more dramatic in the H^- channel. The corresponding ion yield is shown in Fig. 2.3 and consists of a series of partially overlapping resonances at about 5.5 eV, 6.8 eV, 8.5 eV and 10 eV [31, 32]. Due to the low electron affinity of H^- ($\sim$0.75 eV) the endothermicity of H^- formation is higher than compared to $(M\text{-}H)^-$ and all sites become only thermodynamically accessible above 4–5 eV [31]. By measurements with partially deuterated or methylated nucleobases one can show that each of the 4 resonances is formed by hydrogen loss from different position of the thymine molecule as assigned in Fig. 2.3. Thus we have here the case of a true bond and site selective process, which is not only a consequence of different threshold energies for the various isomers. Instead it might be related to particular electronic structures of the associated transient negative ions formed at the different resonance energies [31].

In addition to DEA arising from the cleavage of a single N–H or C–H bond, DEA to nucleobases show many more fragmentation channels which are also accompanied by dissociation of the ring structure. These processes mainly occur above 5 eV and as shown in [18] they considerably contribute to the total cross section for negative ion formation. For example, CNO^- is the dominant fragment anion for thymine at these energies [33].

From these experiments three main conclusions for the description of radiation damage of biological matter can be deduced:

(i) The sharp peaks of the dehydrogenated parent anion may not be likely present in the ion yield for complex biomolecules since they are formed by hydrogen loss from the N1 position. In the DNA framework the 1N–H bond is replaced by the glycosidic bond to the sugar molecule. The experimental proof of this prediction has been given by DEA to the nucleoside thymidine [34] and moreover, thymine methylated at the 1N position, where the $(M\text{-}H)^-$ ion

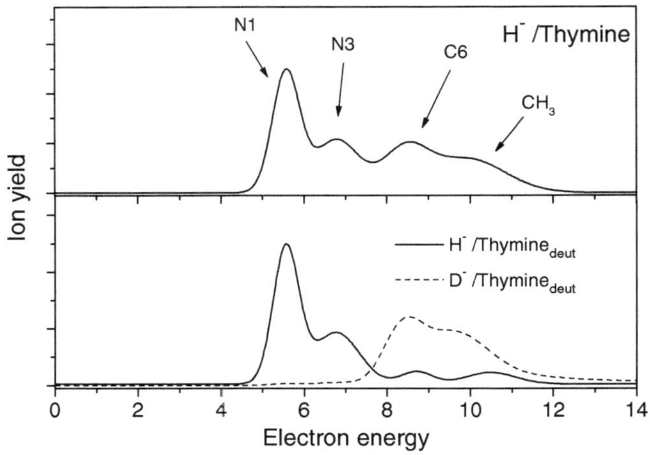

Fig. 2.3 Ion yield of H^- upon DEA to thymine (upper panel). Each peak corresponds to H^- formation from a different site, which can be deduced from experiments with partially deuterated or methylated thymine [31]. For example, the lower panel shows the H^- and D^- formed upon DEA to thymine deuterated at the carbon positions [32]

yield indeed just showed the broad resonance at about 1.8 eV [29]. Moreover, photoelectron spectroscopy with solvated uracil showed a dominance of valence states compared to dipole states [25].

(ii) The situation is different for the H^- channel, where the carbon sites are available also for nucleobases embedded in the DNA-strand. Indeed, the ion yield of H^- resulting from hydrogen loss from the carbon positions resembled in a remarkably way the quantum yield for strand breaks [1]. Thus one may speculate that hydrogen abstraction from carbon positions may initiate double strand breaks in DNA in addition to other pathways proposed recently [35], which involves for example direct electron attachment to the phosphate group at low energies [36].

Moreover, electron capture by an isolated nucleobase always leads to dissociation or alternatively the TNI decays by electron emission. The question is then, if this will be also the case if the nucleobase is surrounded by other molecules, or if excess energy will be transferred to the neighbours by intermolecular vibrational redistribution. Experiments with nucleobases embedded in helium droplets can answer this question [37]. Indeed electron attachment experiments with nucleobases embedded in helium droplets show a pronounced change of the anion abundances. In contrast to the gas phase we can observe a parent anion on mass spectrometric timescales (few μs), and the fragmentation pattern changes in favour of the dehydrogenated parent anion. While for the isolated molecule $(M\text{-}H)^-$ is formed in considerable amounts just below 4 eV, the abundance in the droplets is strongly increased above this energy [37] as shown in Fig. 2.4 for adenine embedded in helium droplets. In contrast other fragment anions, which are formed by cleavage of the ring, are suppressed in the

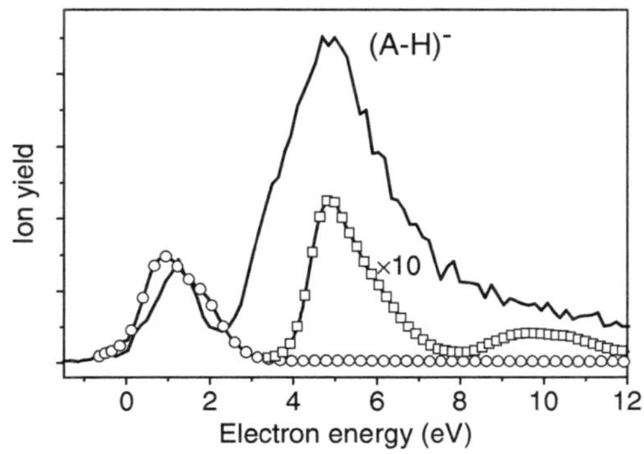

Fig. 2.4 Ion yield of the dehydrogenated parent anion $(A\text{-}H)^-$ of adenine resulting from electron attachment to helium droplets doped with adenine (solid line), $(A\text{-}H)^-$ formed upon DEA in the gas phase (line with open circles) and the sum of all other product anions observed upon DEA to isolated adenine (multiplied with a factor of ten; line with open squares). The electron energy scale of the droplet yield is shifted downwards by 1.1 eV to account for the energy required for the electrons to penetrate the droplet (see [38])

cluster. Similar effects we also observed recently for another aromatic molecule, the explosive TNT [39], where fragmentation was completely quenched. The effect observed for nucleobases results from the different dissociation mechanism for H-loss and ring cleavage: while the former undergoes direct dissociation along a purely repulsive potential energy curve, the latter implies cleavage of multiple bonds and possibly considerable rearrangement of the molecule. These processes are time-consuming and can be effectively quenched by the environment. We note that, however, at least for nucleobases still hydrogen loss occurs, which is therefore an intermediate fragmentation product which became frozen in helium droplets by energy relaxation.

2.3.2 *Amino acids*

Fig. 2.5 shows the molecular structure of the simple amino acids glycine and alanine. For both molecules we have studied DEA in the gas phase in recent years [40, 41]. The molecules are characterized by the carboxylgroup COOH, amino group NH_2 and different side groups. Like for nucleobases also DEA to amino acids shows some characteristic features for this class of molecules. No intact parent ion on mass spectrometric time scales is observable after capture of a free electron and the most abundant fragment anion is $(M\text{-}H)^-$ [41–46]. The resonant fragmentation pattern often shows also peaks for loss of the carboxyl group and amino group and moreover, fragments formed upon complex rearrangement. As shown in Fig. 2.5 for glycine and alanine, the dehydrogenated parent anion $(M\text{-}H)^-$ is formed in a threshold peak at about 1.2 eV (indicated by a steep onset with vibrational structure

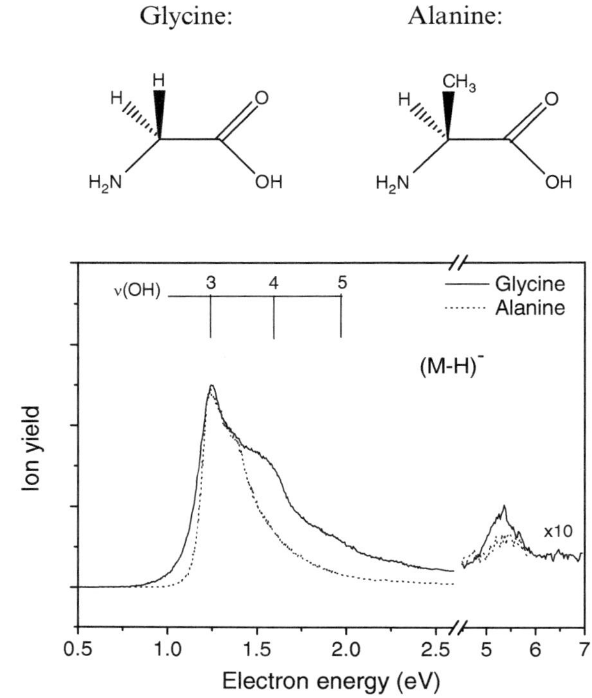

Fig. 2.5 Ion yield as a function of the initial electron energy of the dehydrogenated parent anion (M-H)$^-$ formed upon DEA to glycine (solid line) and alanine (dotted line) (see also [40, 41]). The corresponding molecular structures are shown on top. Also indicated are thresholds for vibrational excitation in competition to DEA (see text). The ion yield above 4.5 eV was multiplied with a factor of ten

on the higher energy side) and a second peak at about 5.3 eV [40, 41]. Early DEA studies ascribed the structure to initial electron attachment to the π^* orbital of the COOH group which is coupled to the repulsive σ^* (O–H) orbital [40]. However, recent calculations questioned this DEA mechanism and instead a direct electron capture into the purely repulsive σ^* (O–H) orbital was proposed, which is a very broad resonance of more than 5 eV width [47]. The vibrational structure can be assigned to the excitation of the $\upsilon(\mathrm{OH}) = 3$–5 strechting modes (see Fig. 2.5) in competition to DEA. The observed dips were interpreted as cusps [48]; however, they may also be interpreted as vibrational Feshbach resonances arising from dipole bound states as suggested recently for glycine [49]. The latter may be possible for some conformers of simple amino acids formed in the vaporisation by the thermal heating (the energetically lowest structure has a subcritical dipole moment). Experimentally no confirmation for or against this prediction was achieved so far; however, site selectivity studies with methylated esters of amino acids confirmed that the structure below 5 eV can be ascribed indeed to hydrogen loss from the carboxyl group [45]. In contrast, hydrogen loss at the 5.3 eV resonance can be ascribed to the amino group [50]. In the course of our DEA experiments with the amino acid valine, we also investigated the unimolecular und collision induced decomposition of (M-H)$^-$ on μs timescales [50]. Thereby it turned out that the (M-H)$^-$ anions formed at about 1.2 eV and at about 5.3 eV show a different fragmentation pattern which is another confirmation that the resonances have a

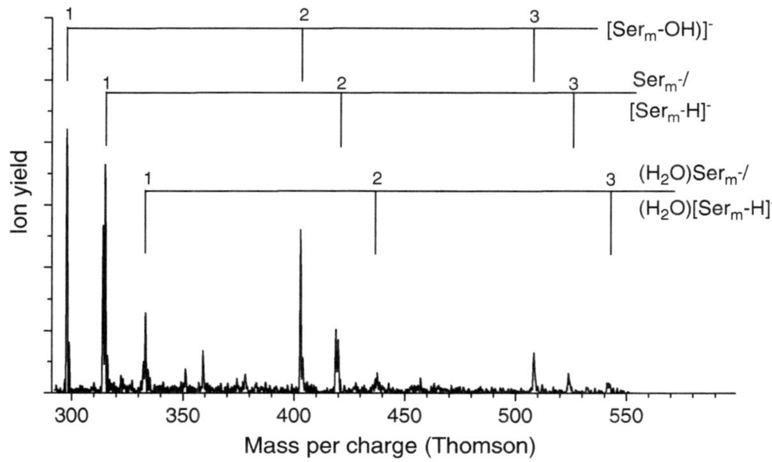

Fig. 2.6 Negative ion mass spectrum of serine cluster ions formed upon electron attachment to serine clusters embedded in helium droplets (see also [51]). The electron energy was about 2.5 eV

different nature, i.e. two different isomers are involved. Moreover, the metastable fragmentation pattern turned out to be the same, when $(M-H)^-$ is formed upon DEA or by deprotonation in matrix-assisted laser desorption ionization [50]. This indicates that the anion has lost its memory on the initial formation process when the dissociation starts.

We also studied recently the formation of amino acid clusters in helium droplets for glycine, alanine and serine [51]. The observed electron induced chemistry turned out to be significantly different to that for the isolated molecule. In the latter case the $(M-H)^-$ channel is dominant, while for the clusters the parent cluster ions become a dominant channel. However, fragmentation is not fully quenched like in the case for nucleobases: some fragmentation channels are still present such as OH loss and O loss. As shown in Fig. 2.6, OH loss becomes strongly significant for serine and exceeds the abundance of the corresponding parent cluster ions [51]. This strong abundance can be explained by the molecular composition of serine, because in this molecule the side chain has a facile-leaving OH group. It should be also noted that the abundance of $(M-OH)^-$ for serine clusters becomes stronger when the cluster size is increased. This might be related to a previous spectroscopic observation where a progressive stabilization of the zwitterionic structures in serine clusters was observed when the cluster size is increased from dimer up to the pentamer [52].

2.4 Conclusions and outlook in the future

Low energy electrons can effectively damage biomolecules by means of dissociative electron attachment in the electron energy range below 10 eV. When the molecule is isolated, the fragmentation above the electron energy of 3–4 eV is often severe;

for example, in the case of nucleobases cleavage of the aromatic ring may happen. At low electron energies abundant formation of the dehydrogenated parent anion occurs. This hydrogen loss turned out to be highly bond and site selective, i.e. by setting the electron energy to a certain resonance a specific bond in the molecule is cleaved. When biomolecules are embedded in the cluster, the surrounding can prevent dissociation after TNI formation by fast energy relaxation and caging effects. However, this effect is confined to a particular slow dissociation process like ring cleavage while a fast reaction, i.e. direct electronic dissociation along a purely repulsive potential energy surface, is still operative. This result is in line with the suggestion that severe damage of biological matter can be caused by low energy secondary electrons.

Acknowledgements This work has been supported by the FWF (P14900, P18052, P18804, P19073, and P22665), Vienna, the Austrian Academy of Sciences, Vienna, and the European Commission, Brussels. S.D. would like to thank Dr. F. Ferreira da Silva for the support in preparing two of the graphic illustrations.

References

1. Boudaiffa, B., Cloutier, P., Hunting, D., Huels, M.A., Sanche, L., Science **287**, 1658–1660 (2000)
2. Martin, F., Burrow, P.D., Cai, Z., Cloutier, P., Hunting, D., Sanche, L., Phys. Rev. Lett. **93**, 068101-1-4 (2004)
3. Sanche, L., Eur. Phys. J. D **35**, 367–390 (2005)
4. Cobut, V., Frongillo, Y., Patau, J.P., Goulet, T., Fraser, M.J., Jay-Gerin, J.P., Rad. Phys. Chem. **51**, 229–243 (1998)
5. Surdutovich, E., Obolensky, O.I., Scifoni, E., Pshenichnov, I., Mishustin, I., Solov'yov, A.V., Greiner, W., Eur. Phys. J. D **51**, 63–71 (2009)
6. Mucke, M., Braune, M., Barth, S., Forstel, M., Lischke, T., Ulrich, V., Arion, T., Becker, U., Bradshaw, A., Hergenhahn, U., Nature Physics **6**, 143–146 (2010)
7. Michael, B.D., O'Neill P., Science **287**, 1603–1604 (2000)
8. Wang, C.R., Nguyen, J., Lu, Q.B., J. Am. Chem. Soc. **131**, 11320–11322 (2009)
9. Sanche, L., Nature **461**, 358–359 (2009)
10. Gianturco, F.A., Sebastianelli, F., Lucchese, R.R., Baccarelli, I., Sanna, N., J. Chem. Phys. **128**, 174302-1-8 (2008)
11. Baccarelli, I., Gianturco, F.A., Grandi, A., Sanna, N., Lucchese, R.R., Bald, I., Kopyra, J., Illenberger, E., J. Am. Chem. Soc. **129**, 6269–6277 (2007)
12. Bouchiha, D., Gorfinkiel, J.D., Caron, L.G., Sanche, L., J. Phys. B: At. Mol. Opt. Phys. **39**, 975–986 (2006)
13. Sanche, L., Mass Spect. Rev. **21**, 349–369 (2002)
14. Bari, S., Sobocinski, P., Postma, J., Alvarado, F., Hoekstra, R., Bernigaud, V., Manil, B., Rangama, J., Huber, B., Schlathölter, T., J. Chem. Phys. **128**, 074306-1-6 (2008)
15. Graupner, K., Field, T.A., Mauracher, A., Scheier, P., Bacher, A., Denifl, S., Zappa, F., Märk, T.D., J. Chem. Phys. **128**, 104304-1-8 (2008)
16. Hanel, G., Denifl, S., Scheier, P., Probst, M., Farizon, B., Farizon, M., Illenberger, E., Märk, T.D., Phys. Rev. Lett. **90**, 188104-1-4 (2003)
17. Bald, I., Dabkowska, I., Illenberger, E., Angew. Chem. In. Ed. **47**, 8518–8520 (2008)

18. Huber, D., Beikircher, M., Denifl, S., Zappa, F., Matejcik, S., Bacher, A., Grill, V., Märk, T. D., Scheier, P., J. Chem. Phys. **125**, 084304-1-7 (2007)
19. Zappa, F., Denifl, S, Maehr, I., Bacher, A., Echt, O., Märk, T.D., Scheier, P., J. Am. Chem. Soc. **130**, 5573–5578 (2008)
20. Stienkemeier, F., Lehmann, K.K., J. Phys. B: At. Mol. Opt. Phys. **39**, R127–R166 (2006)
21. Denifl, S., Ptasinska, S., Cingel, M., Matejcik, S., Scheier, P., Märk, T.D., Chem. Phys. Lett. **377**, 74–80 (2003)
22. Denifl, S., Sulzer, P., Huber, D., Zappa, F., Probst, M., Märk, T.D., Scheier, P., Injan, N., Limtrakul, J., Abouaf, R., Dunet, H., Angew. Chem. In. Ed. **46**, 5238–5241 (2007)
23. Aflatooni, K., Gallup, G.A., Burrow, P.D., J. Phys. Chem. A **102**, 6205–6207 (1998)
24. Desfrancois, C., AbdoulCarime, H., Schermann, J.P., Int. J. Mod. Phys. B. **10**, 1339–1395 (1996)
25. Hendricks, J.H., Lyapustina, S.A., de Clercq, H.L., Bowen, K.H., J. Chem. Phys. **108**, 8–11 (1998)
26. Desfrancois, C., AbdoulCarime, H., Schermann, J.P., J. Chem. Phys. **104**, 7792–7794 (1996)
27. Burrow, P.D., Gallup, G.A., Scheer, A.M., Denifl, S., Ptasinska, S., Märk, T.D., Scheier, P., J. Chem. Phys. **124**, 124310-1-7 (2006)
28. Gallup, G.A., Fabrikant, I.I., Phys. Rev. A **83**, 012706-1-7 (2011)
29. Ptasinska, S., Denifl, S., Scheier, P., Illenberger, E., Märk, T.D., Angew. Chem. Int. Ed. **44**, 6941–6943 (2005)
30. Abdoul-Carime, H., Gohlke, S., Illenberger, E., Phys. Rev. Lett. **92**, 168103-1-4 (2004)
31. Ptasinska, S., Denifl, S., Grill, V., Märk, T.D., Illenberger, E., Scheier, P., Phys. Rev. Lett. **95**, 093201-1-4 (2005)
32. Ptasinska, S., Denifl, S., Grill, V., Märk, T.D., Scheier, P., Gohlke, S., Huels, M.A., Illenberger, E., Angew. Chem. Int. Ed **44**, 1647–1650 (2005)
33. Denifl, S., Ptasinska, S., Probst, M., Hrusak, J., Scheier, P., Märk, T.D., J. Phys. Chem. A **108**, 6562–6569 (2004)
34. Ptasinska, S., Denifl, S., Gohlke, S., Scheier, P., Illenberger, E., Märk, T.D., Angew. Chem. Int. Ed. **45**, 1893–1896 (2006)
35. Simons, J., Acc. Chem. Res. **39**, 772–779 (2006)
36. König, C., Kopyra, J., Bald, I., Illenberger, E., Phys. Rev. Lett. **97**, 018105-1-4 (2006)
37. Denifl, S.; Zappa, F.; Mahr, I.; Lecointre, J.; Probst, M.; Märk, T. D.; Scheier, P., Phys. Rev. Lett. **97**, 043201-1-4 (2006)
38. Denifl, S., Zappa, F., Mauracher, A., Ferreira da Silva, F., Bacher, A., Echt, O., Märk, T.D., Bohme, D.K., Scheier, P., Chem. Phys. Chem. **9**, 1387–1389 (2008)
39. Mauracher, A., Schöbel, H., Ferreira da Silva, F., Edtbauer, A., Mitterdorfer, C., Denifl, S., Märk, T.D., Illenberger, E., Scheier, P.,. Phys. Chem. Chem. Phys. **11**, 8240–8243 (2009)
40. Ptasinska, S., Denifl, S., Abedi, A., Scheier, P., Märk, T.D., Anal. Bioanal. Chem. **377**, 1115–1119 (2003)
41. Ptasinska, S., Denifl, S., Candori, P., Matejcik, S., Scheier, P., Märk, T.D., Chem. Phys. Lett. **403**, 107–112 (2005)
42. Sulzer, P., Alizadeh, E., Mauracher, A., Märk, T.D., Scheier, P., Int. J. Mass Spect. **277**, 274–278 (2008)
43. Papp, P., Urban, J., Matejcik, S., Stano, M., Ingolfsson, O., J. Chem. Phys. **125**, 204301-1-8 (2006)
44. Kocisek, J., Papp, P., Mach, P., Vasil'ev, Y.V., Deinzer, M.L., Matejcik, S., J. Phys. Chem. A **114**, 1677–1683 (2010)
45. Vasil'ev, Y.V., Figard, B.J., Voinov, V.G., Barofsky, D.F., Deinzer, M.L., J. Am. Chem. Soc. **128**, 5506–5515 (2006)
46. Abdoul-Carime, H., König-Lehmann, C., Kopyra, J., Farizon, B., Farizon, M., Illenberger, E., Chem. Phys. Lett. **477**, 245–248 (2009)
47. Gallup, G.A., Burrow, P.D., Fabrikant, I.I., Phys. Rev. A **79**, 042701-1-7 (2009)
48. Abouaf, R., Chem. Phys. Lett. **451**, 25–30 (2008)

49. Vasil'ev, Y. V., Figard, B. J., Barofsky, D. F., and Deinzer, M. L., Int. J. Mass. Spect. **268**, 106–121 (2007)
50. Flosadottir, H.D., Denifl, S., Zappa, F., Wendt, N., Mauracher, A., Bacher, A., Jonsson, H., Märk, T.D., Scheier, P., Ingolfsson, O., Angew. Chem. Int. Ed. **46**, 8057–8059 (2007)
51. Ferreira da Silva, F., Denifl, S., Märk, T.D., Ellis, A.M., Scheier, P., J. Chem. Phys. **132**, 214306-1-6 (2010)
52. Kong, X.L., Tsai, I.A., Sabu, S., Han, C.C., Lee, Y.T., Chang, H.C., Tu, S.Y., Kung, A.H., Wu, C.C., Angew. Chem. Int. Ed. **45**, 4130–4134 (2006)

Chapter 3
Electron Transfer-Induced Fragmentation in (Bio)Molecules by Atom-Molecule Collisions

Negative Ion Formation

Paulo Limão-Vieira, Filipe Ferreira da Silva, and Gustavo García Gómez-Tejedor

Abstract Ion-pair formation to gas phase molecules induced by electron transfer has been studied by investigating the products of collisions between fast potassium atoms and target molecules using a crossed molecular-beam technique. The negative ions formed in such collisions are TOF mass analysed. As far as (bio)molecules are concerned, TOF mass spectra at different collision energies reveal interesting anionic patterns with reduced fragmentation at lower impact energies. In the unimolecular decomposition of the temporary negative ion (TNI), complex internal rearrangement may involve the cleavage and formation of new bonds. In this chapter we report some of the recent achievements in negative ion formation of some polyatomic molecules with the special attention to biological relevant targets.

3.1 Introduction

It is now well established that low energy electrons provide an effective method for inducing chemical processes in a wide variety of media, including thin films, plasmas, aerosols, planetary and biological. Unlike photon induced dissociation,

P. Limão-Vieira (✉) • F. Ferreira da Silva
Laboratório de Colisões Atómicas e Moleculares, CEFITEC, Departamento de Física, FCT, Universidade Nova de Lisboa, 2829-516 Caparica, Portugal
e-mail: plimaovieira@fct.unl.pt; f.ferreiradasilva@fct.unl.pt

G.G. Gómez-Tejedor
Instituto de Física Fundamental, Consejo Superior de Investigaciones Científicas, 28006 Madrid, Spain

Departamento de Física de los Materiales, UNED, 28040 Madrid, Spain
e-mail: g.garcia@iff.csic.es

G.G. Gómez-Tejedor and M.C. Fuss (eds.), *Radiation Damage in Biomolecular Systems*, Biological and Medical Physics, Biomedical Engineering,
DOI 10.1007/978-94-007-2564-5_3,

the energy of the incident electron may be significantly below the energy required to excite the molecule to a dissociative electronic state, indeed electron induced dissociation may occur using electrons with meV energies. The mechanism for such molecular fragmentation has been ascribed to a process known as Dissociative Electron Attachment (DEA), in which a transition from the continuum ($e^- + ABC$) to a discrete state of the molecular anion $(ABC^{\#})^-$, represents a resonant process. Therefore the incoming electron is 'captured' before decaying, producing a negative fragment accompanied by one or more associated neutral counterpart(s) e.g. $e^- + ABC \rightarrow (ABC^{\#})^- \rightarrow B^- + AC$ or $B^- + A + C$. It is these neutral, often radical products that subsequently may initiate new chemistry within the medium. Such DEA or 'electron induced chemistry' is prevalent in many natural and industrial processes including: a) the formation of organic molecules within ice mantles on dusty grains in the interstellar medium, b) the control of fluorocarbon plasmas used to produce silicon chips, (c) the chemical modification of absorbates using electron patterning and scanning tunnel microscopy and d) mutagenic and other potentially lethal DNA lesions, such as single- and double-strand breaks (SSBs and DSBs, respectively).

However, it is the recent work on the role of DEA in biophysics and its possible explanation of DNA damage by ionising radiation that has attracted the most attention in the wider academic community. Boudaiffa et al. [1] (updated by Huels et al. [2]) revealed that the probability of DNA damage (for both SSBs and DSBs) induced by low energy electrons is enhanced at specific incident electron energies, these 'resonant' regions coinciding with those energies where DEA to the nucleotide bases of DNA leads to molecular dissociation. Furthermore, recent experiments have shown that DEA is both bond (C-H versus C-N) and site selective (N_1-H versus N_3-H) in the DNA bases, see e.g. [3]. Thus a new low energy mechanism for DNA damage has been established which may be described at a basic molecular level.

In this mechanism low energy electrons produced along the primary irradiation (X-rays, alpha particles, gamma rays) induce DEA at specific sites in the DNA helix forming parent anions that subsequently decay leading to rupture of the DNA helix. The rate defining step in such a mechanism is the formation of the parent anion, thus the pattern of strand breaks should follow the DEA cross section of the target molecules. The observed correlation between the patterns of DNA damage and the DEA cross sections of the nucleotide bases supports this model. This 'model' also provides a coherent explanation of the observed correlation between electron attachment rates to biomolecules and their carcinogenicity and may be used to suggest new compounds to be adopted in radiation therapy as treatment enhancing sensitizers. For example, the radiosensitization properties of halouracils (e.g. 5-XU, X = F, Cl, Br, I) have been known for several decades with irradiation of cells in which some DNA thymines have been replaced by halogenated uracils being shown to increase the frequency of both single and double strand breaks and chromosomal alterations (e.g., BrU) [4–6]. The DEA model provides an explanation for such effects with the introduction of a strongly electrophilic atom into the DNA (e.g. a halogen) leading to an enhancement in the DEA probability and thence increased probability for DNA destruction in cells containing such compounds

[7–9]. However many elementary collisional processes are not direct electron impact but rather depend upon electron transfer. It is generally recognized that a DNA molecule in equilibrium state does not have any free charge carriers so it would be useful to probe alternative routes by which electrons might form anions without the initial electron being 'free' since such a process may be a more realistic analogue for DNA damage under physiological conditions.

Alkali atoms have very low ionisation energies and are therefore excellent electron donors. Accordingly in our laboratory a new experiment has been developed to study for the first time, electron transfer processes with constituent molecules of DNA using potassium - molecule collisions [10]. In these experiments a positive ion K^+ and a molecular anion are formed, allowing access to parent molecular states which are not accessible by free electron attachment experiments [11, 12]. In particular states with a positive electron affinity can be formed, and the role of vibrational excitation of the parent neutral molecule can be studied by the collision dynamics [12]. Even if the free negative molecular ion is unstable in the gas phase, in the atom-molecule collision complex it can be stabilized [13]. The negative molecular ion lifetime will depend upon both the collision time (order of several tenths of fs) and the autodetachment time (order of a few 10^{-14} s). If the lifetime of the parent negative ion is longer than the fragmentation time, energy can be distributed over the available vibrational degrees of freedom and so change the fragmentation pathways. If the collision time is shorter than the dissociation time, collision induced dissociation is likely to take place and produce fragment ions with finite kinetic energies. Comparison of the fragmentation patterns by electron transfer in atom-molecule experiments and free electron molecule interactions will therefore allow us to develop a more detailed model of electron induced damage in DNA.

3.2 Fundamental aspects of electron transfer processes

Electron transfer in low energy atom molecule collisions is usually mediated by the crossing of the potential energy surfaces $M + AB$ and $M^+ + AB^-$, where M the electron donor is an alkali atom and AB an electron acceptor molecule. Due to the fact that at large atom-molecule distances the ionic potential energy surface lies above the covalent, due to the Coulomb potential there is a crossing point at small distances at which both potential energy surfaces have the same value [11,14]. The crossing distance R_c is inversely proportional to the difference $I_M - EA_{AB}$, where I_M is the ionisation potential of M and EA_{AB} the electron affinity of AB. As a consequence, in a collisional ion-pair formation, roughly speaking, only the lowest ionic state will be, involved. On the other hand, it should be noted that in these processes the electron transfer is only possible between configurations of the same symmetry and multiplicity [14]. This leads to the formation of a positive ion and a molecular anion allowing access to parent molecular states which are not accessible in free electron attachment experiments. In the latter processes, only anionic states above the vibrational ground state of the neutral parent can be probed [15, 16].

Even if the free negative molecular ion is unstable with respect to autodetachment, in the collision complex it can be stabilized at distances shorter than the crossing point between the two potential energy curves involved in the electronic transition, due to the attractive interaction with the positive ion [13]. The product anion lifetime will therefore depend upon both the collision time and the autodetachment or dissociation times. The electron transfer process is however, in general, endoergic i.e. requires a minimum collisional energy usually of the order of a few eV, and hence the incident neutral electron transferring species must have a high kinetic energy. Finally, it should be noted that since the collision energy can be chosen larger than the threshold for electron transfer, the negative molecular ion can be formed with an excess of internal energy which even might result in fragmentation. If the lifetime of the metastable ion is long, intra-molecular energy redistribution may occur competing therefore with direct dissociation [10, 17].

3.3 Experimental set-up

The crossed molecular beam set-up has been described elsewhere [18] and is mainly composed of two interconnected high-vacuum chambers equipped with a home-made Time-of-Flight (TOF) mass spectrometer. Briefly, a primary beam of fast neutral potassium (K) atoms is produced in a charge-exchange source similar to that referred by Aten and Los [19]. A commercial potassium ion source (operating at ~1100 K) produces K^+ ions that are accelerated to a given collision energy, presently limited to 30–100 eV, before entering a chamber where they resonantly charge-exchange with neutral potassium atoms, yielding a hyperthermal neutral potassium atom beam. Any residual K^+ ions that do not charge exchange are removed by electrostatic deflecting plates outside the oven. The resulting neutral potassium beam then enters a high-vacuum chamber where its relative intensity can be monitored (normally in the order of tens of pA and an estimated energy resolution of ~0.5 eV) using an iridium surface ionisation detector of the Langmuir - Taylor type. This detector samples the beam intensity but does not interfere with the beam passing to the collision region. It operates in a temperature regime that only allows detection of the fast beam. Although the beam intensity is only monitored prior and after the TOF mass spectra collection, its intensity is on average kept constant.

The molecular target is produced by simple effusion through a 1 mm capillary of a liquid or gas whereas solid samples are brought into the gas phase with an increasing vapour pressure through a lamp-heated oven.

The collision volume where both beams cross is located between two mutually spaced parallel plates with a separation of 1.2 cm. The anions produced in the collision region are extracted by a 250 V cm^{-1} pulsed electrostatic field. The pulse width and frequency are adjusted depending on the total TOF mass range. The same applies to the K^+ beam, which is pulsed with a 2 μs width and an adjusted frequency depending on the collision energy (typical values of 222 kHz are applied for 100 eV collision energy). The negative ions extraction pulse is properly delayed in respect to the K^+/K° beam, allowing the latter to travel to the interaction region prior to the anions extraction. The base pressure in the collision chamber is $\sim 2 \times 10^{-7}$ mbar and the working pressure does not rise upon heating the solid samples. Nevertheless, this pressure is more than enough to guarantee a binary collision process. The recorded anionic yields are obtained by subtracting the background signal from the sample signal. Mass spectra are calibrated through the well known anionic species formed in collisions of potassium atoms with nitromethane molecules [13].

The negative ion mass spectra presented here were obtained with a 1.4 m linear TOF mass spectrometer with an estimated resolution $m/\Delta m = \sim 150$. After extraction, ions are accelerated and guided into the TOF and detected with a channel electron multiplier from Dr. Sjuts Optotechnik GmbH operated in a pulse counting mode. Mass spectra were taken at different collision energies and for all anions, identified as products of dissociative electron attachment.

The liquid sample of nitromethane was purchased from Sigma-Aldrich with a minimum purity of $\geq 96\%$. All samples were degassed by a repeated freeze–pump–thaw cycle before admission to the collision chamber.

The solid samples of thymine and uracil were obtained from Sigma-Aldrich with a minimum stated purity of $\geq 96\%$ and 99%, respectively, and used as delivered. Samples were heated typically at 473 K and temperature controlled through a PID (a Proportional–Integral–Derivative) unit. In order to identify possible oven temperature effects on the present data, anion mass spectra were recorded for 100 eV potassium impact upon gas phase uracil at 473K and 458 K. No differences were observed in the relative peak intensities for the two measurements. The vacuum chamber as well as the extraction region and TOF electrodes, were kept heated in order to avoid any condensation of the solid samples.

3.4 Results

The topic of this chapter is a part of a broader experimental program concerning the analysis of the time-of-flight (TOF) mass experiments on electron transfer in collisions of nitromethane and pyrimidine molecules (thymine and uracil) with fast potassium atoms. These results have been published elsewhere [10, 13]. As typical experimental results, Fig. 3.1 shows a section of the TOF mass spectrum of the anion fragments formed in the collision system potassium-CH_3NO_2, whereas Figs. 3.2 and 3.3 are related to thymine and uracil, respectively. A background spectrum has been collected for each collision energy studied and subtracted to the molecule's TOF mass spectrum.

The three sets of molecules revisited here, nitromethane, thymine and uracil, have very large dipole moments (>3.4 D), and therefore can support a stable dipole-bound state (DBS) and so, under appropriate conditions, a weakly bound electron can be transferred to form a dipole-bound anion which may serve as a "doorway" to valence states [20].

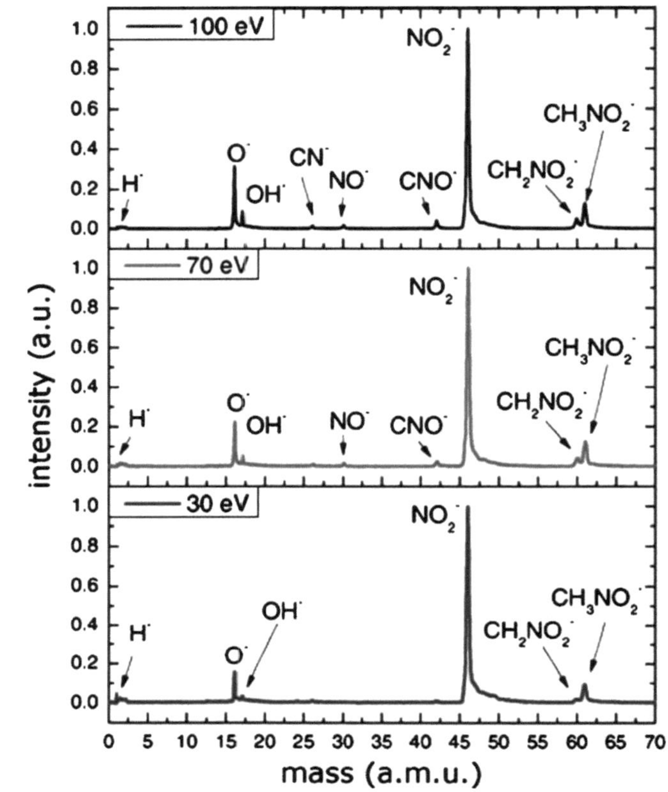

Fig. 3.1 Negative ion time-of-flight mass spectra of K + CH_3NO_2 at 100, 70 and 30 eV collision energy

3.4.1 Nitromethane, CH_3NO_2

Nitromethane has a slightly positive electron affinity, 0.44 eV, and may accommodate an extra electron to form a stable parent anion in a third body interaction. The negative ion TOF mass spectra of the anions formed in the collision of potassium atoms with nitromethane molecules, recorded at incident energy of 30, 70 and 100 eV, normalised to the most intense peak (NO_2^-), produce the parent ion $CH_3NO_2^-$, the dehydrogenated closed-shell anion $(CH_2NO_2)^-$, NO_2^-, CNO^-, OH^-, O^- and some other minor intensity fragment ions [13]. From the TOF mass spectra, we observe that the two major anionic fragments are NO_2^- and O^-, anions with masses 17, 26, 30, 42 and 60 having less than 10% of the total yield. At 30 eV collision energy, the O^- intensity is about 15% of NO_2^- peak, while at 100 eV O^- is about one-third of the dominant NO_2^- feature. This strongly suggests that the energy of the resonant state resulting in O^- is higher than that yielding NO_2^-

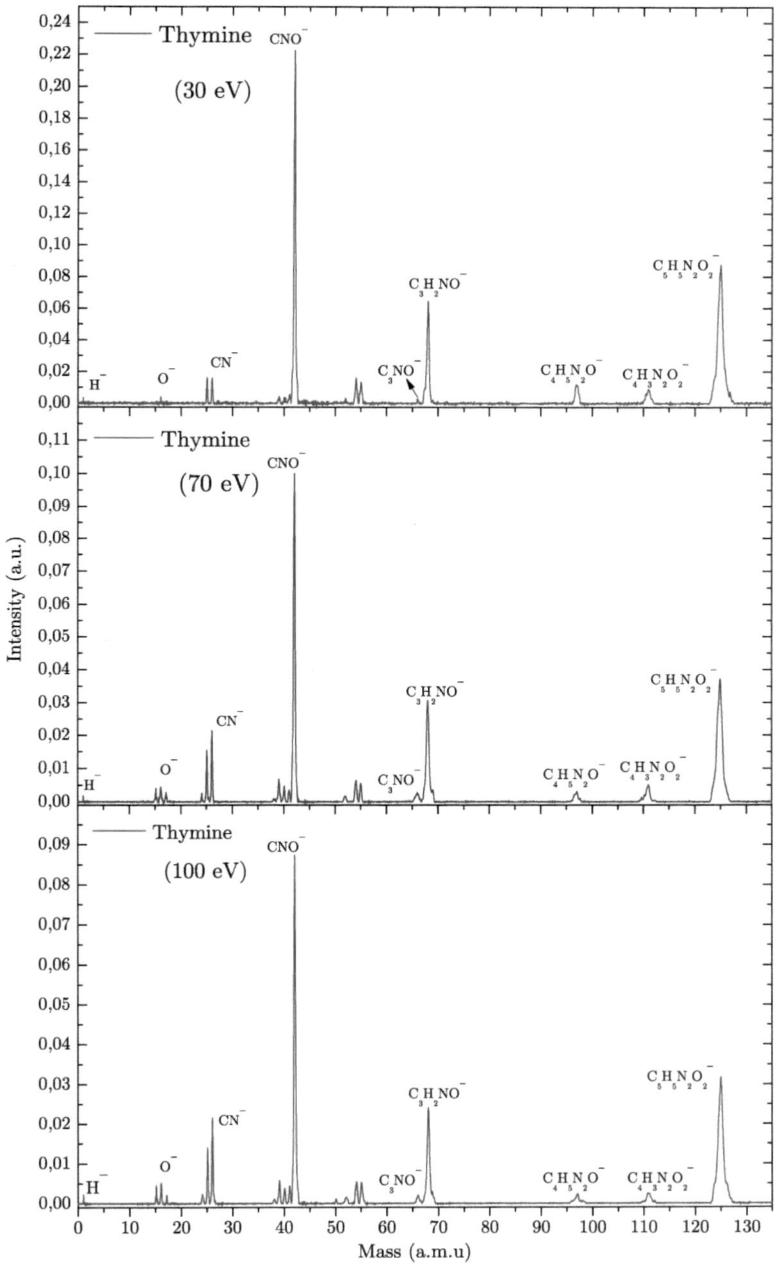

Fig. 3.2 Negative ion time-of-flight mass spectra of $K + C_5H_6N_2O_2$ at 30, 70 and 100eV collision energy

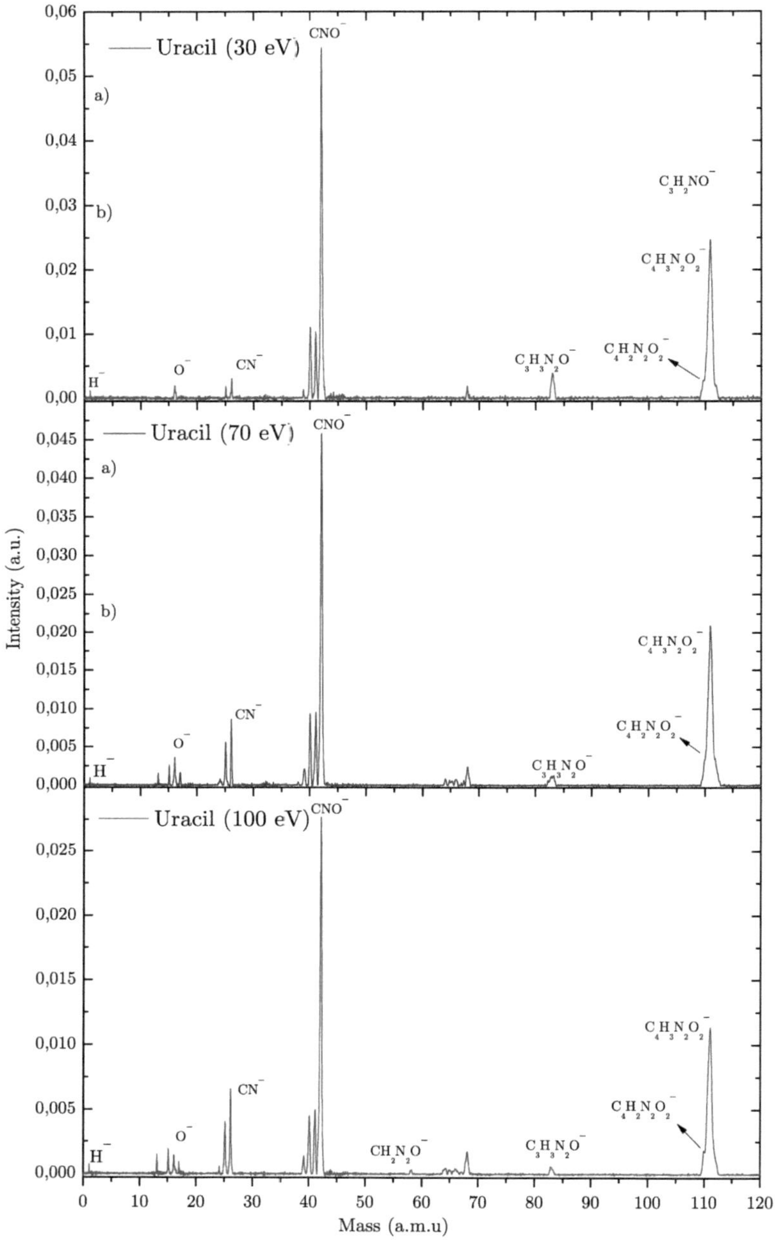

Fig. 3.3 Negative ion time-of-flight mass spectra of $K + C_4H_4N_2O_2$ at 30, 70 and 100eV collision energy

and is in agreement with the recent electron attachment spectroscopy results [21] and the previous DEA studies [22]. However while NO_2^- and O^- may result from single bond cleavage in the unimolecular decomposition of the transient negative ion, CNO^-, OH^- and CN^- formation may result from more complex reactions involving multi-bond cleavage and internal energy rearrangement. Theoretical calculations for electron attachment to nitromethane [23] revealed that, on a mass spectrometric timescale, the dipole bound anion serves as an efficient doorway to a valence state, which is quickly populated and may be stabilised by collisions with a third body partner.

Electron attachment to form a transient negative valence anion can occur via a nuclear excited Feshbach resonance as well as via dipole-supported shape and vibrational Feshbach resonances in the continuum as suggested in [23, 24]. The system may then be rationalised in terms of dipole states acting as doorways to the eventual formation of stable valence anions, as long as they exist in the continuum. In the case of Rydberg electron transfer experiments, Compton and co-workers [20] suggested that the ratio of the time spent in the pure dipole- (D^-) to valence-state (V^-) is given by the ratio of the density of states, ρ_{D-}/ρ_{V-} [20]. They found that this ratio is negligible; thus, they expected the system to spend most of the time in the valence configuration. If the anion is vibrationally excited, hot bands, and this internal energy is not removed by stabilisation with a third body, autodetachment becomes relevant. This is in agreement with the discussion on the electron attachment studies of Walker and Fluendy [25] and may also explain the observations of the present experiments, where the intensity of the $CH_3NO_2^-$ anion is never the strongest, suggesting that autodetachment competes with dissociation and dictates the intensity of the parent ion formation [13].

3.4.2 *Thymine, $C_5H_6N_2O_2$ and Uracil, $C_4H_4N_2O_2$*

Electron transfer from potassium atoms to thymine/uracil molecules produces the dehydrogenated closed-shell anion (T/U-H)$^-$, CNO^- and fragment ions, in intensities that mainly depend on the collision energy. Figs. 3.2 and 3.3 show the TOF mass spectrum of all the anions formed, but for the present discussion we will only focus our attention to the most intense negative ions, i.e. CNO^- and (T-H)$^-$. We, therefore restrict ourselves to the TOF mass spectrum at 100 eV collision energy, although we have performed measurements at 30 and 70 eV. Increasing energy in potassium-thymine/uracil collisions will produce vibrationally excited states of the low-lying anion state, which may lie in the continuum of the neutral molecule leading to autodetachment. The two most intense fragments formed are always CNO^- and (T/U-H)$^-$, which relative intensities increase with decreasing energy and are observed in the proportion of $\sim 3:1$, respectively (this fraction is kept for all collision energies studied). Therefore, this may indicate that the energy of the resonant states resulting in CNO^- and (T/U-H)$^-$ are somehow similar. This is consistent with the electron attachment results near 4 eV, see e.g. [26, 27].

Though, CNO^- must not be formed through direct electron capture but rather due to the transient $(T\text{-}H)^{-*}$ anion, which is in clear agreement with the fact that the dehydrogenated parent anion is an intermediate product for the formation of this anion. This picture can actually be seen from an interesting point of view regarding the strong coupling between the dipole bound state and the temporary anion state related to the occupation of valence σ^* orbitals, that actually play a significant role in electron capture by thymine molecules.

The LUMOs in thymine are the $\pi_1{}^*$, $\pi_2{}^*$ and $\pi_3{}^*$ at approximately 0.29, 1.71 and 3.83 eV [28], respectively. As the asymptotic limit of the $(T/U\text{-}H)^- - H$ molecular system is located at ~1.4 eV [26], the lowest resonance cannot lead to a dissociative attachment, whereas the second and third will give rise to the emission of $(T/U\text{-}H)^-$ in electron attachment experiments [26]. The electron can also be weakly bound due to the high value of the dipole moment of thymine ($\mu > 4.0$ D) yielding the existence of a stable dipole-bound state (DBS). Schiedt and co-workers [29] showed that this DBS lies very close in energy to the neutral molecule (62 ± 8 meV). Though, Burrow and co-workers have shown that this DBS serve as doorway to valence-bound states [30]. The three $\sigma_1{}^*$, $\sigma_2{}^*$ and $\sigma_3{}^*$ resonances in thymine have been calculated to lie at 2.4, 3.42 and 3.75 eV, respectively, with $\sigma_2{}^*$ primarily antibonding between C_6-H, methyl CH_2-H and N_3-H and $\sigma_3{}^*$ antibonding mainly at N_3-H. Moreover, the calculated wave functions for the DBS show these are placed on the side of the molecule that contains the C_6 and N_1, which resembles the orientation of the $\sigma_1{}^*$ [30]. Though, there is an excellent spatial overlap between the wave function of the DBS and the lowest valence $\sigma_1{}^*$ orbital, resulting in an avoided crossing of these anionic states when plotted as a function of the N_1-H internuclear distance. The H can actually tunnel through the barrier created by the avoided crossing. The calculations also suggested that the threshold for loss of H from coupling of the DBS with $\sigma_2{}^*$ at C_6H and C–H of the methyl group sites are too high in energy, and are not expected to play a significant role. Since the $\pi_1{}^*$ anion state is not accessible due to energy reasons for the production of $(T/U\text{-}H)^-$, removal of an hydrogen from the N_3-H site is not favourable through the $\sigma_1{}^*$ anionic state due to the small wave function amplitude at this position. Thus, the coupling is most likely to occur between the $\sigma_2{}^*$ and $\pi_2{}^*$ orbitals [30] or through vibronic coupling between these two temporary anions as reported by Allan and co-workers on planar unsaturated hydrocarbons substituted with halogens [31].

3.5 Conclusions

The dissociation of metastable anions has been demonstrated to cause DNA DSBs [1], recognised as a key source of radiation damage in living systems. The aim of the proposed experiments in the Lisbon laboratory is to observe the fragmentation patterns of nucleobase anions, notably $(\text{thymine/uracil}^-)^{\#}$, produced in collisions with accelerated potassium atoms. Electron transfer-induced processes in neutral low energy atom-molecule collisions are presently a relatively unexplored field, particularly in respect to biological relevant molecules. As the outermost electron in potassium is weakly bound, the proposed experiments may provide an analogy for electron transfer from electronically excited secondary neutrals in biological material.

Electron transfer mechanism yielding ion-pair formation in gas phase nitromethane, thymine and uracil through collisions with a hyperthermal neutral potassium beam, have been investigated using a crossed molecular-beam technique. The negative ions formed in the collision region are TOF mass analysed. Therefore the role of the isolated negative ion state seems to be relevant since it may stabilise by the presence of the K^+ ion, being the collision time estimated to be of the order of a few 10^{-14} s, during which the distance between two fragments have been increased by ~1–2 Å.

In atom-molecule collisions, the K^+ ion can strongly interact with the transient molecular anion, and if bond stretching is allowed in the target molecule during the electron transfer, strong vibronic coupling may occur. Since the collision energies are larger than the threshold for electron transfer (~4 eV), the negative molecular ion can be formed with an excess of internal energy which even might result in fragmentation dictating the nature of the negative fragment ions formed.

The results will be compared with recent "free" electron attachment studies, in order to achieve a more complete understanding of biomolecular anion fragmentation pathways in diverse electron attachment processes.

Acknowledgments PLV acknowledges the Portuguese Foundation for Science and Technology (FCT-MCTES) for the research grant POCI/FIS/58845/2004 & PPCDT/FIS/58845/2004, and together with GG acknowledges the Spanish-Portuguese Project HP2006-0042; Ministerio de Ciencia e Innovación (project FIS2009-10245), Spain is also acknowledge. FFS acknowledges FCT-MCTES for the SFRH/BPD/68979/2010 financial support. Some of this work forms part of the EU/ESF COST Actions: Electron Controlled Chemical Lithography (ECCL) CM0601, The Chemical Cosmos CM0805 and the Nano-scale Insights in Ion Beam Cancer Therapy (Nano-IBCT) MP1002.

References

1. Boudaiffa, B., Cloutier, P., Hunting, D., Huels, M.A., Sanche, L., Science. **287**, 1658–1660 (2000)
2. Huels, M.A., Boudaiffa, B., Cloutier, P., Hunting, D., and Sanche, L., J. Am. Chem. Soc. **125**, 4467–4477 (2003)
3. Ptasinska, S., Denifl, S., Grill, V., Märk, T.D., Illenberger, E., and Scheier, P., Phys. Rev. Lett. **95**, 093201 (2005)
4. Abdoul-Carime, H., Limão-Vieira, P., Petrushko, I., Mason, N.J., Gohlke, S., and Illenberger, E., Chem. Phys. Lett. **393**, 442–447 (2004)
5. Zamenhof, S., Degiovanni, R., Greer, S., Nature. **181**, 827–289 (1958)
6. Lawrence, T.S., Davis, M.A., Maybaum, J., Stetson, P.L., Ensminger, W.D., Radiat. Res. **123**, 192–198 (1990)
7. Abdoul-Carime, H., Huels, M.A., Bruning, F., Illenberger, E., Sanche, L., J. Chem. Phys. **113**, 2517–2521 (2000)

8. Abdoul-Carime, H., Huels, M.A., Illenberger, E., Sanche, L., J. Am. Chem. Soc. **123**, 5354–5355 (2001)
9. Denifl, S., Matejcik, S., Ptasinska, S., Gstir, B., Probst, M., Scheier, P., Illenberger, E., Märk, T.D., J. Chem. Phys. **120**, 704–709 (2004)
10. Almeida, D., Antunes, R., Martins, G., Eden, S., Ferreira da Silva, F., Nunes, Y., Garcia, G., and Limão-Vieira, P., Phys. Chem. Chem. Phys. **13**, 15657–15665 (2011)
11. Kleyn, A.W., Moutinho, A.M.C., J. Phys. B., **34**, R1–R44 (2001)
12. Limão-Vieira, P., Moutinho, A.M.C., and Los, J., J. Chem. Phys. **124**, 054306 (2006)
13. Antunes, R., Almeida, D., Martins, G., Mason, N.J., Garcia, G., Maneira, M.J.P., Nunes, Y., and Limão-Vieira, P., Phys. Chem. Chem. Phys. **12**, 12513–12519 (2010)
14. Kleyn, A.W., Los, J., and Gislason, E.A., Phys. Rep., **90**, 1–71 (1982)
15. Fenzlaff, H.P., Illenberger, E., Int. J. Mass Spect. Ion Proc. **59**, 185–202 (1984)
16. Christophorou L.G., McCorkle D.L., and Christodoulides, A.A. (1984) Electron-Molecule Interactions and Their Applications, vol. 2. Academic Press, Inc., NY
17. Limão-Vieira, P., Moutinho, A.M.C., Los, J., J. Chem. Phys. **124**, 054306 (2006)
18. Limão-Vieira, P., private communication
19. Aten, J.A., and Los, J., J. Phys. E. Sci. Instr. **8**, 408–410 (1973)
20. Compton, R.N., Carman Jr, H.S., Desfrancois, C., Abdoul-Carime, H., Schermann, J.P., Hendricks, J.H., Lyapustina, S.A., and Bowen, K.H.,. J. Chem. Phys. **105**, 3472–3478 (1996)
21. Alizadeh, E., Ferreira da Silva, F., Zappa, F., Mauracher, A., Probst, M., Denifl, S., Bacher, A., Märk, T.D., Limão-Vieira, P., Scheier, P., Int. J. Mass Spectrom. **271**, 15–21 (2008)
22. Sailer, W., Pelc, A., Matejcik, S., Illenberger, E., Scheier, P., and Märk, T.D.,. J. Chem. Phys. **117**, 7989–7994 (2002)
23. Sommerfeld, T., Phys. Chem. Chem. Phys. **4**, 2511–2516 (2002)
24. Gustev, G.L., and Bartlett, R.J., J. Chem. Phys. **105**, 8785–8792 (1996)
25. Walker, I.C., Fluendy, M.A.D., Int. J. Mass Spectrom. **205**, 171–182 (2001)
26. Denifl, S., Ptasinska, Probst, M., Hrusak, J., Scheier, P., and Märk, T.D., J. Phys. Chem. A **108**, 6562–6569 (2004)
27. Denifl, S., Ptasinska, S., Hanel, G., Gstir, B., Probst, M., Scheier, P., and Märk, T.D., J. Chem. Phys. **120**, 6557–6565 (2004)
28. Aflatooni, K., Gallup, G.A., Burrow, P.D., J. Phys. Chem. A **102**, 6205–6207 (1998)
29. Schiedt, J., Weinkauf, R., Neumark, D.M., Schlag, E.W., Chem. Phys. **239**, 511–524 (1998)
30. Burrow, P., Gallup, G., Scheer, A., Denifl, S., Ptasinska, S., Märk, T.D., and Scheier, P., J. Chem. Phys. **124**, 124310 (2006)
31. Skalicky, T., Chollet, C., Pasquier, N., Allan, M., Phys. Chem. Chem. Phys. **4**, 3583–3590 (2002)

Chapter 4
Following Resonant Compound States after Electron Attachment

A Quantum Modelling of the Dynamical Evolution in Molecular Anions

Ana G. Sanz, Francesco Sebastianelli, and Francesco A. Gianturco

Abstract The evolution pathways which can follow temporary, metastable electron attachment processes involving polyatomic molecules in the gas phase are analysed using a quantum method for the calculation of such transient anionic states as a function of molecular deformations. The method is specifically applied, as an example, to the two-dimensional deformations of bond geometries for the HCN molecule and shown to describe well the interplay between electronic and nuclear degrees of freedom.

4.1 Introduction

It is by now an accepted result that a great deal of the consequences, at the molecular level, of radiation exposures of biosystems and living tissues are produced by the secondary electrons coming from medium ionization [1, 2]. Such electrons are generally found to be energetically rather slow ($E_t \leq 15$ eV) and can therefore effectively interact with any of the DNA components, or with the larger DNA fragments, thereby inducing successive fragmentations and bond ruptures by depositing their excess energy into the molecular networks of the bonds [3, 4].

A.G. Sanz
Instituto de Física Fundamental, Consejo Superior de Investigaciones Científicas, 28006 Madrid, Spain
e-mail: ana@iff.csic.es

F. Sebastianelli • F.A. Gianturco (✉)
Department of Chemistry and CNISM, The University of Rome 'Sapienza', 00185 Rome, Italy
e-mail: f.sebastianelli@caspur.it; fa.gianturco@caspur.it

G.G. Gómez-Tejedor and M.C. Fuss (eds.), *Radiation Damage in Biomolecular Systems*, Biological and Medical Physics, Biomedical Engineering,
DOI 10.1007/978-94-007-2564-5_4, © Springer Science+Business Media B.V. 2012

From the theoretical standpoint, it therefore becomes important to be able to describe such initial electron–molecule interactions, to obtain information at the molecular level on the dynamical couplings between impinging electrons and molecular components (bound electrons and nuclei), and to ultimately follow the evolutions of the initial metastable states (the transient negative ions (TNIs)) into either stabilizing a bound molecular anion or into breaking selectively some of the bonds to yield neutral and anionic fragmentation products.

In the present work we therefore describe a quantum mechanical treatment of the nanoscopic scattering event between a single impinging electron and a single molecule in its initial equilibrium geometry (i.e. into its electronic and rotovibrational ground state). We will show, in fact, that the scattering process can lead to the occurrence of resonant collisions whereby the electron is temporarily attached to the target molecule and can evolve along different final channels that depend on the specific resonant compound state, on the vibrational energy content of the formed anion and on the selective dynamical couplings between the attached electron and the molecular nuclear motions along specific bonds.

Besides describing in some detail the computational method employed, we will carry out a detailed analysis of such evolutionary paths in the case of a fairly simple, but very important, linear polyatomic target like the H–C≡N molecular system.

We know, in fact, that besides its ubiquitous presence as a component of several systems of biological interest, hydrogen cyanide is also considered interesting as a species that can play an important role in many areas of gas-phase molecular physics: from the studies of interstellar formation of HCN and CNH molecules [5] to its postulated role as an initiator of complex organic molecules formation in the Interstellar Medium (ISM) [6] and in Titan's atmosphere [7]. One of the crucial features in many of such studies has been to establish and verify for this molecule the role played by electron-induced mechanisms which cause fragmentation via resonant electron attachment [8,9].

In particular, Burrow et al. [10] measured its electron transmission spectrum and found a fairly structureless, broad resonance centred at 2.26 eV, which was attributed to the occurrence of a π^*-type transient negative ion (TNI) formation. Earlier experiments by Tronc's group [11] had reported relative vibrational excitation cross sections and observed resonant features centred at 2.3 and 6.7 eV, assigned respectively to π^* and σ^* shape resonances. The most recent experiments by May, Kubala and Allan [12] reported absolute partial cross sections for the formation of CN^- in dissociative electron attachment (DEA) to HCN and DCN. They found a steep onset of those cross sections above the threshold for H + CN^- formation (1.5 eV), peaking at 1.85 eV and presenting broad structures due to the formation of vibrationally excited CN^-.

The earlier calculations of resonant cross sections by Jain and Norcross [13, 14] located the π^*-type TNI formation around 2.71 eV with a width of 1.9 eV at the equilibrium geometry of HCN, while finding that the stretching of the C–H and C≡N bonds, as well as the bending of the molecule, led to the appearance of a σ^* broad resonance as suggested by the earlier experiments [11], where the authors had observed strong stretching activity, also accompanied by bending vibrations.

More recent calculations that employed the R-matrix method for the HCN and the CNH molecules [15] found a $^2\Pi$ TNI resonance for the former species and for the latter, while only the HCN molecule exhibited around 8 eV a much smaller resonance of $^2\Sigma$ symmetry.

A very complete study on the DEA process for HCN/HNC molecules has also been carried out recently [16]. The dissociation mechanisms were analysed for both molecules and for their deuterated variants DCN/DNC: a curve crossing between $^2\Sigma$ and $^2\Pi$ complex potential energy surfaces of the metastable anions was computationally located, leading to a barrier for the bent structure and a suggested tunnelling mechanism for DEA processes leading to CN^- formation from HCN.

The methodology that we shall describe in the next Section allows one to follow precisely this behaviour of shape resonances, i.e. their energy changes and width shifts as the molecular geometries are modified along specific bonds. It will therefore be a test of our modelling to follow the behaviour of the postulated resonances of π^* and σ^* symmetry for HCN as they evolve into metastable fragments and localize their resonant wavefunctions into specific molecular regions as stretching and bending deformations are taking place. In other words, we wish to test out model quantum scattering calculations on the features of the established TNIs of HCN^- and follow them upon bond-deforming vibrational excitations of the molecular fragments. We shall show that the present quantum modelling is able to relate the changing lifetimes of such metastable states to the appearance of CN^- fragments and/or to the detachment of fragments. Our calculations are therefore expected to shed more light on the features of the resonant processes at low scattering energies and to link our findings with existing experiments and earlier calculations.

The next section 4.2 describes our computational method while section 4.3 reports our calculations, comparing them with the existing data. Our conclusions are finally given in section 4.4.

4.2 The quantum scattering model

4.2.1 Single-center expansion and static exchange (SE) forces

Both the target one-electron bound orbitals (MOs) and the impinging electron wavefunction are expanded as linear combinations of symmetry-adapted spherical harmonics [17], where the indexes label the relevant Irreducible Representation (IR) by p and its component by μ, while l labels the angular momenta and its component h.

$$\Phi_i^{p\mu}(r,\vartheta,\phi) = \frac{1}{r}\sum_{lh} u_{lh}^{i,p\mu}(r) X_{lh}^{p\mu}(\vartheta,\phi) \tag{4.1}$$

The initial assumption that the target molecule is adequately described by its ground-state Slater determinant (static exchange approximation) leads to the following simplified form for the scattering equations at each chosen (positive) energy ϵ

$$\left[\frac{1}{2}\nabla^2 + (E-\epsilon)\right] F(\mathbf{r}) = \int V(\mathbf{r},\mathbf{r}')F(\mathbf{r}')d\mathbf{r}' \tag{4.2}$$

that can be solved (once defined the correct potential for our system) using the Schwinger variational method as described in our earlier publications (e.g. see ref. [18]), and subject to the standard boundary conditions of the scattering problem [19] on the continuum solutions.

4.2.2 Correlation-polarization potential with model exchange forces

The static exchange approximation does not include the response of the target to the impinging electron, i.e. the correlation and polarization effects acting at short and at large electron–target distances respectively. That part of the overall interaction can be modelled by writing

$$V^{\mathrm{cp}}(\mathbf{r}) = \begin{cases} V^{\mathrm{corr}}(\mathbf{r}) & \text{for } \mathbf{r} \le \mathbf{r_{match}} \\ V^{\mathrm{pol}}(\mathbf{r}) & \text{for } \mathbf{r} > \mathbf{r_{match}} \end{cases} \tag{4.3}$$

both in the past and in the present work we have used the Lee-Yang-Parr form for $V^{\mathrm{corr}}(\mathbf{r})$ [20] and further employed the $V^{\mathrm{pol}}(\mathbf{r})$ as a function of the polarizability tensor, to describe successfully via the V^{cp} model of eq. (4.3) this additional part of the overall interaction potential [21] of eq. (4.2), as discussed below.

In order to perform scattering calculations at many energies and for fairly large molecular targets, we replaced an exact nonlocal bound–continuum exchange interaction of eq. (4.2) with a local energy-dependent exchange model potential. We have chosen the Free-Electron-Gas-Exchange ($FEGE$) potential proposed long ago [21,22]:

$$V_{FEGE} = \frac{2}{\pi}k_F(\mathbf{r})\left(\frac{1}{2} + \frac{1-\eta^2}{4\eta}ln\left|\frac{1+\eta}{1-\eta}\right|\right) \tag{4.4}$$

where k_F is the Fermi momentum and η is a parameter related to the molecule's ionization potential [22]. We have already found before that eq. (4.4) yielded a reasonably realistic description of such forces in a great variety of molecular systems [23–25] and therefore decided to employ the above modelling also in the present study.

4.2.3 Cross section calculations: solution of Volterra equations

Once the full interaction potential is defined, another computational option is to rewrite the scattering equations in homogeneous form to obtain the static-model-exchange-correlation-polarization (SMECP) approximation for the scattering

$$\left[-\frac{1}{2}\nabla^2 - \frac{1}{2}k^2 + \hat{V}^{\mathrm{st}} + \hat{V}^{\mathrm{cp}} + \hat{V}^{\mathrm{FEGE}}\right] F(\mathbf{r}) = 0 \tag{4.5}$$

Indicating by $V(\mathbf{r})$ the sum of the three local potentials $V(\mathbf{r}) = V^{st}(\mathbf{r}) + V^{cp}(\mathbf{r}) + V^{FEGE}(\mathbf{r})$, the local formulation of the scattering equation can be cast in terms of the usual partial wave expansion [19, 26]. After integrating over the angular variables eq. (4.5) takes the form

$$\left[\frac{d^2}{dr^2} - \frac{l(l+1)}{r^2} + k^2\right] F_{lh}^{p\mu}(r) = 2\sum_{l'h'} V_{lh,l'h'}^{p\mu}(r) F_{l'h'}^{p\mu}(r), \tag{4.6}$$

and the potential coupling elements are given as

$$\begin{aligned} V_{lh,l'h'}(r) &= \langle X_{lh}^{p\mu}(\hat{r}) | V(\mathbf{r}) | X_{l'h'}^{p\mu}(\hat{r}) \rangle = \\ &= \int d\hat{r} X_{lh}^{p\mu}(\hat{r}) V(\mathbf{r}) X_{l'h'}^{p\mu}(\hat{r}). \end{aligned} \tag{4.7}$$

The standard Green's function technique allows us to rewrite the previous differential equations in an integral form [26]

$$F_{lh}^{p\mu}(r) = \delta_{l,l'}\, j_l(kr) + \sum_{l'h'} \int_0^r dr'\, g_l(r,r') V_{lh,l'h'}(r') F_{l'h'}^{p\mu}(r'). \tag{4.8}$$

The upper limit of the integral on the right-hand side of the previous equation is a variable (integral equations with this property are called Volterra equations) and the integral vanishes for $r' > r$. The partial wave expansion form of the Green's function in eq. (4.8) is given in [27], and is known to vanish for $r > r'$

$$g_l(r,r') = \frac{1}{k}\left[j_l(kr)n_l(kr') - n_l(kr)j_l(kr')\right] \tag{4.9}$$

where $j_l(kr)$ and $n_l(kr)$ are Riccati-Bessel and Riccati-Neumann functions. The required K-matrix elements are obtained after the Volterra integration and provide in turn the S-matrix elements that finally give us the set of partialwave phaseshifts $\delta_l(k)$ for each coupled channel of the partial wave expansion.

$$S_l = S_l(k) = e^{2i\delta_l(k)}. \tag{4.10}$$

4.2.4 S-matrix, Q-matrix and eigenphase sums

For a scattering process characterized by many, overlapping resonances, the usual Breit-Wigner formula describing an isolated resonance [28] is no longer efficient. Thus, it is possible to extract the positions and widths of closely spaced resonant states by using the Q-matrix formalism [29]:

$$Q(E) = i\hbar S \frac{dS^{\dagger}}{dE} = -i\hbar \frac{dS}{dE} S^{\dagger} = Q^{\dagger}(E). \tag{4.11}$$

As proved by Smith in the 60's [30], eigenvalues of the Q matrix are equal to the delay time of the outgoing wavepacket. For a multichannel scattering process the trace of the Q matrix is related to the eigenphase sum δ_{sum}:

$$2\hbar \frac{d\delta_{\text{sum}}}{dE} = Tr\, Q(E) \tag{4.12}$$

where the eigenphase sum is defined as:

$$\delta_{\text{sum}}(E) = \sum_{l=1}^{\infty} \delta_l(E). \tag{4.13}$$

It can be shown that the first derivative of the eigenphase sum can be recast as a sum of Lorentzian functions, each associated with a resonance feature characterized by an energy position $\Re\{E\}$ and a width $\Im\{E\}$ [28]: this is the procedure we have followed here to analyse the resonant regions.

4.2.5 Modelling the DEA processes from computed resonances

The basic process we are envisaging to model is one where the excess resonant electron energy gets transferred into the internal degrees of freedom of the molecular bonds, thereby causing, in some instances, the breakup of the target molecule and the detachment of a stable anionic fragment (DEA). Whenever a single specific bond is responsible for the breakup of the metastable anion, one then observes a direct dissociation event and corresponding shifts in the computed resonance position and width while that bond is stretched, with a further dissociative behaviour of the one-dimensional (1D) potential curve associated to the (N+1)-electron molecular anion formed during the resonant process. This is essentially a 1D picture for the Dissociative Electron Attachment (DEA) mechanism whereby one is searching for a possible, dominant fragmentation via the breakup of one particular bond. It is, however, an approximate picture in the sense that the quantum dynamical coupling between the pseudo-bound electron and the bound nuclei is not explicitly included but only adiabatically taken into account through the rearrangement energy of the complex.

Several calculations can therefore be carried out by using different geometrical deformations of the system under study. By stretching one or more bonds where the resonant electron density maps showed, at the molecular equilibrium geometries, the presence of nodal planes (and therefore the occurrence of an antibonding localization along them of the excess resonant electron density) one can locate those special deformations which may lead to bond-breaking within the TNI. One then computes the potential energy curves associated with both the neutral N-electron target and the metastable (N+1)-electron TNI, adiabatically following in each case a selected bond or angle deformation. This is achieved by writing the energy balance in the familiar way (e.g. see ref. [25] and references therein)

$$E_{tot}(\mathbf{R}) = E_{res}(\mathbf{R}) + E_N(\mathbf{R}) - E_N(\mathbf{R}_{eq}) \qquad (4.14)$$

where $E_N(\mathbf{R})$ is the computed electronic energy (at the same level of accuracy of the calculations at the equilibrium geometry) for the neutral molecule at a set of geometries identified by $\mathbf{R}$ and E_{res} is the energy location of the resonant electron over the same range of molecular geometries. The corresponding widths of the resonances associated with the (N+1) states are also given at each computed molecular geometry [30] and could be represented, as we shall show below, on the same data grid as that for the real energies of eq. (4.14) since they correspond to the imaginary component $\Gamma_{res}(R)$ of the complex energy expression:

$$E = E_{res} + i\Gamma_{res} \qquad (4.15)$$

It therefore follows that from the overall computational behaviour of the (N+1)-electron energies, E_{tot}, and of their widths, Γ_{res}, as a function of the geometry $\mathbf{R}$, one can have some idea of the effects on the molecular anion's stability caused by the metastable excess electron attached to the whole system.

4.3 Results for a test case and general discussion

The chief aim of the present work consists in showing that, using rigorous scattering theory, albeit with a simplified modelling of interaction forces in order to be able to handle larger systems as those of biological interest (e.g. DNA components, aminoacids, etc.), it becomes possible to follow the evolution of specific TNIs associated to individual resonances and to do so via selective "cuts" of the potential energy surfaces (PESs) which describe that evolution in terms of nuclear geometry modifications with respect to the equilibrium configurations.

Hence, the computational indicators which are employed by our model to select the pseudo-1D paths along which resonances are analysed are provided by the spatial representation of the excess, resonant electron wavefunction over the grid of points given by the PESs of the TNI complex: the presence of nodal planes and the different features of this antibonding behaviour provide clues to which are the more likely "cuts" to follow either DEA or anionic stabilization. Furthermore, the imaginary component of the resonant energy provides indications on the possible competition of autodetachment processes during the resonance evolution.

The approach of our work on DNA damage, therefore, hinges on the analysis of the complex potentials on which the motion of the nuclei will take place during the attachment of the excess electron. In order to make the computations feasible for large systems, however, we will analyse the processes as adiabatic evolutions of the TNIs, without the explicit nuclear dynamics on a multidimensional PES: our previous experience on various systems tells us, in fact, that the qualitative picture of the fragmentation paths is still preserved [31–33].

In order to provide an illustrative example of the method, we shall report below our results for a fairly simple molecular partner, but yet rather instructive in terms of the processes involved: the HCN molecule, taken initially in its linear ground state.

4.3.1 The HCN resonances at fixed geometries

As discussed in 4.1, the HCN molecule has been assigned two distinct, low-energy resonances: one in the region of about 2.3 eV (of π^* character) and another in the region of about 6.7 eV, of σ^* character [11, 12]. The ensuing evolution of the system involves its fragmentation into the CN^- ($^1\Sigma, \nu'$), vibrationally excited anion, and H (^{2}S) atom, as extensively discussed in several papers [11–16]. In the present, illustrative calculation we have selected this system, with its three normal modes for nuclear deformations, to see how the initial resonant state formation can evolve into the above mentioned fragments and therefore how the fragmentation path could be identified from our adiabatic modelling of the complex PES for the relevant metastable anions.

The calculations were carried out following the multichannel scattering method described in section 4.2. The initial target wavefunction was generated by an SCF procedure at the Hartree-Fock (HF) level of the following quality: HF/6-311++G(3df,3pd). The bound and scattering orbitals were expanded at the center-of-mass of the target by using up to $L_{max} = 60$ components, while the potential was expanded up to $\lambda_{max} = 120$ multipolar terms. The scattering at short-range was carried out up to $R_{max} = 11.5$ Å and the discrete (r, θ, ϕ) grid of the scattered wavefunction involved a total of 1464x84x324 points. The computed dipole moment from our wavefunction of the ground state turned to be 3.26 D, with an experimental value of 2.98 [34]. The computed variations of the dipole moment was used to describe the long-range potential, although the resonances were analysed within the inner box region, the latter being varied till stability of the final resonance parameters was within 10%. The data in Fig. 4.1 report the features of the computed π^* resonant electron: on the right panel the real part of the scattering wavefunction is given, while on the left panel the corresponding LUMO orbital from the neutral calculations is shown (the H atom is on the left in the linear axis): there are certain

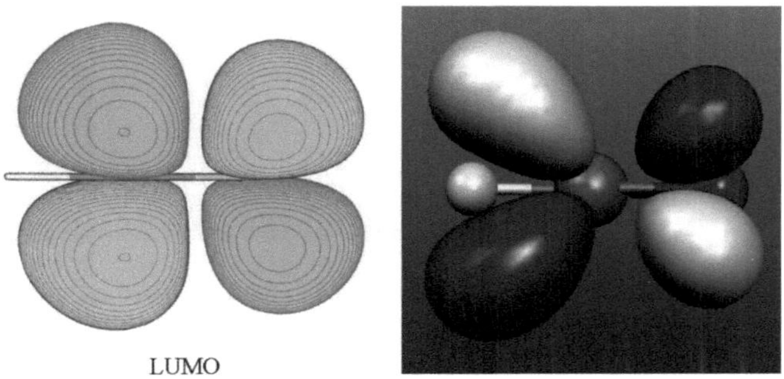

Fig. 4.1 Computed features of the π^* resonant orbital for HCN at its equilibrium geometry (CH=1.06Å, CN=1.11Å). Resonance energy E_{res} = 4.094eV and its width $\Gamma_{res}(R)$ = 0.596 eV. See main text for further details

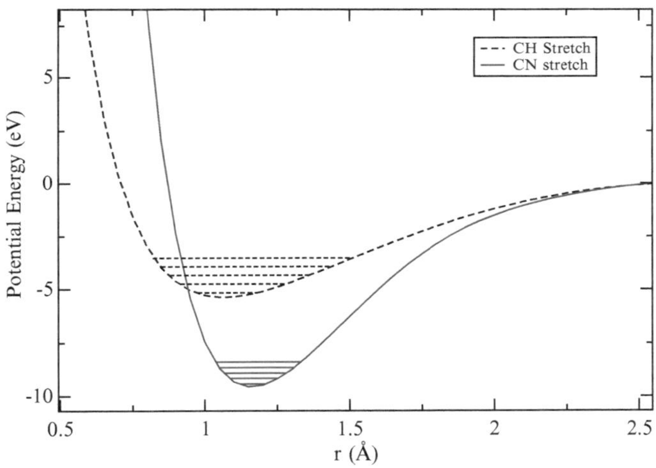

Fig. 4.2 Computed "diatomic" potential energy curves for the CH and CN fragments. The asymptotes correspond to either H or N atom dissociation and are artificially scaled together

similarities between the two spatial features since both the virtual and the scattering wavefunctions appear to be chiefly located on the two bond regions. However, the computed resonance is found at ~4.0 eV (as expected with our model, at about 1 eV higher than experiments [11]) and with similar width as that observed [12], while the LUMO eigenvalue is 0.94 eV. The experimental observation on vibrational excitation effects [11, 12] found strong C≡N and bending excitations during the resonant attachment into the π^* orbital and the appearance of a σ^* resonance at higher energy, with extensive C–H and C≡N stretching activity. In Fig. 4.2 we report the behaviour of the "diatomic" potential curves for the C–H and the C≡N

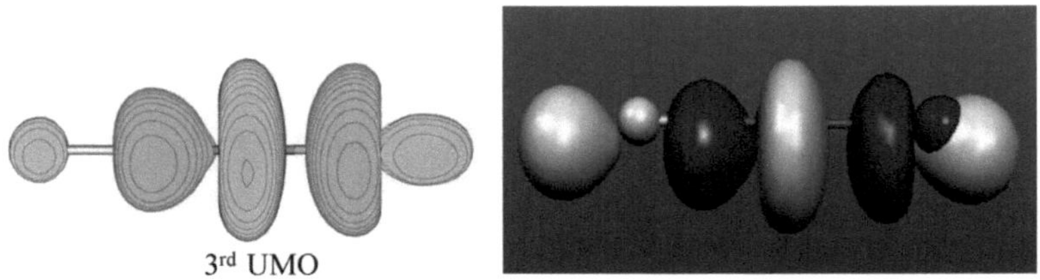

Fig. 4.3 Computed features of the $^2\Sigma^*$ resonance found for a stretched geometry (CH = 1.6Å, CN = 1.7Å) of linear HCN. Resonance energy E_{res} = 7.714 eV and its width $\Gamma_{res}(R)$ = 0.53 eV. See main text for further details

fragments, along the linear stretching motion of either of them, computed with the same accuracy of our HCN wavefunction. One clearly sees the greater strength of the CN bond and its reduced extension compared with CH when supporting an equal number of vibrational levels (five each in the picture). With this information we can now analyse more accurately the effects on resonant features upon stretching (and, later, bending) of these two bonds. One result presented by the data of Fig. 4.3 is given by the appearance of a σ^* resonance around 7 eV when both CH and CN bonds are stretched: the features of the scattering wavefunction are again similar to those of the 3rd virtual orbital from the neutral calculations, although the corresponding eigenvalue is 0.53 eV. The additional electron density due to the trapped particle is extended at this point over the whole molecular space. The resonance position is found from our calculations to be higher than experiments by nearly 1 eV, as it also occurred with the π^* resonance.

To summarize, our model calculations find the two resonances seen by experiments, albeit each at an higher energy. We also find that the $^2\Sigma$ resonant state only appears after stretching both bonds in a linear fashion.

4.3.2 *Stretching the C–H and C≡N bonds*

The previous computational studies of the resonances of the present example [14,16] indicated that the DEA process which produced the $CN^- + H$ fragmentation originated from the couplings between $^2\Pi$ and $^2\Sigma$ states during bending deformations which undergo avoided crossing between adiabatic states and show decay along the $^2A'(\sigma^*$-like) potential energy curve [16]. On the other hand, the CN stretching would lead to $HC^-(^3\Sigma^-)$ and $N(^4S)$ fragments coming now directly via the $^2\Sigma$ resonance [13]. It is therefore useful to see how the present modelling could be employed to follow the resonances' behaviour upon stretching both bonds, thereby creating a 2-dimensional PES and also examining the spatial modifications of the associated excess electron wavefunctions upon bond deformation.

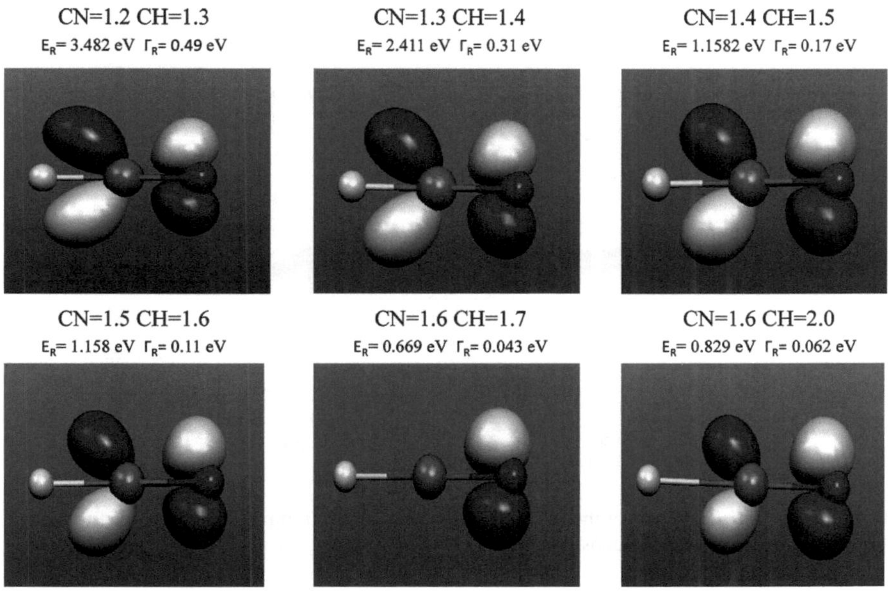

Fig. 4.4 Computed resonance features and spatial resonant wavefunctions (real part) for the $^2\Pi$ state as a function of the stretching of both C–H and C≡N bonds. All distances are in Å

Figure 4.4 reports the changes on the spatial features of the resonant wavefunction of the π^* symmetry when both the CN and the CH bonds are stretched for a range of values corresponding to bringing both bonds to excited vibrational levels (see Fig. 4.2). One clearly sees there two interesting features: (i) the excess electron localizes on the $(CN)^-$ spatial region, leaving the H atom and the CH bond , and (ii) the resonance position moves down in energy, while also becoming narrower, i.e. longer lived. Both features thus appear as the signature of CN^- formation as the linear molecule is stretched.

An interesting two-dimensional presentation of the PES for this metastable state (real component) of (N+1) electrons could be viewed from the data of Fig. 4.5. Following eq. (4.14), the potential values are scaled by the (N electron) energy at the equilibrium geometry. One clearly sees that the associated potential is initially decreasing upon bond stretching but rapidly produces a broad repulsive region as the bonds are stretched: the excess charge localizes on the (CN) region but the nuclei do not dissociate upon bond stretching, thus confirming the findings of [16]: the π^* resonance is not directly causing dissociation into $CN^- + H$.

When a similar procedure is applied to the $^2\Sigma$, as indicated in the panels of Fig. 4.6, we see that the location of the resonance moves to lower energies on stretching both bonds, also becoming narrower, but the corresponding charge of the resonant electron remains distributed over the whole molecule: the CH^- formation is therefore now a distinct possibility during bond stretching.

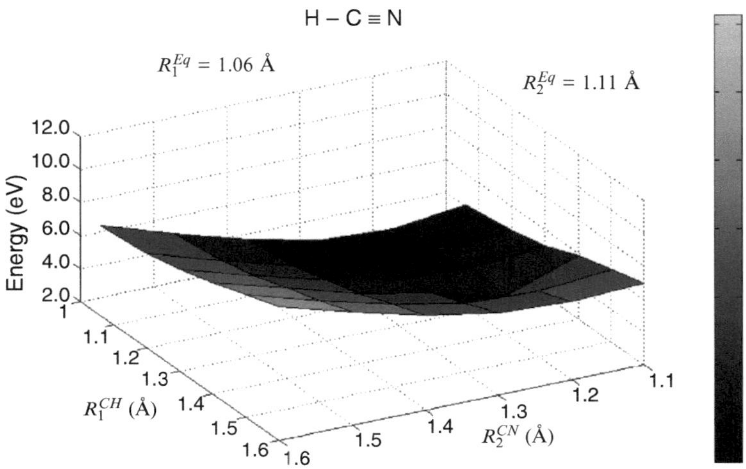

Fig. 4.5 Computed real part of the PES for the $^2\Pi$ (N+1) electron resonant state as a function of the linear stretching of both bonds. See text for details

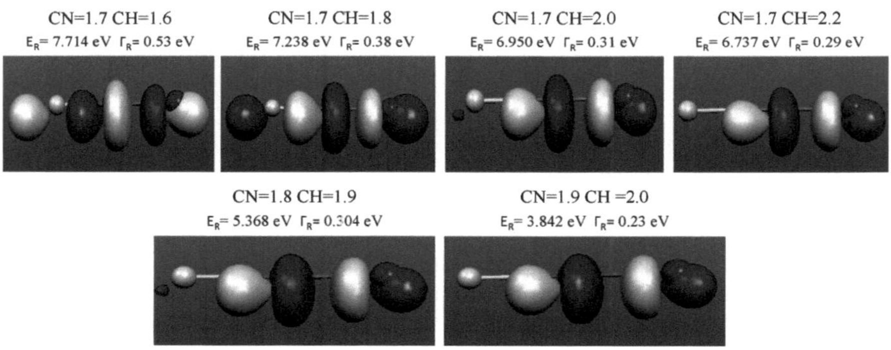

Fig. 4.6 Computed resonance locations and widths, together with excess electron maps, for the $^2\Sigma$ metastable state as a function of the stretching of both C–H and C≡N bonds. All distances are in Å

The corresponding behaviour of the resonance energy (real part) is shown by the 2D surface reported in Fig. 4.7. One clearly sees that the resonance decreases in energy when *both* bonds are stretched and the extra electron stabilizes on the molecule as both R_{CH} and R_{CN} increase in length. The corresponding adiabatic PES for the (N+1) electron system, referred to the equilibrium geometry electronic energy for N-electrons, is shown by the data reported in Fig. 4.8: this time we see that the $^2\Sigma$ potential surface decreases as both bonds increase, doing so more rapidly for the stretched R_{CN} values. In other words, this metastable anion exhibits dissociative behaviour along the CH coordinate provided that the CN partner is also vibrationally excited: it will cross the $^2\Pi$ curve of Fig. 4.5 in regions outside the

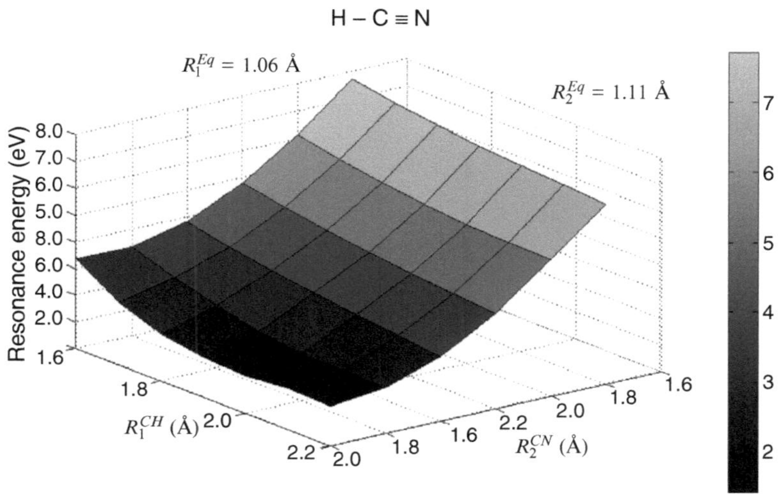

Fig. 4.7 Two-dimensional view of the $^2\Sigma$ resonance as a function of the stretching of both C–H and C≡N bonds

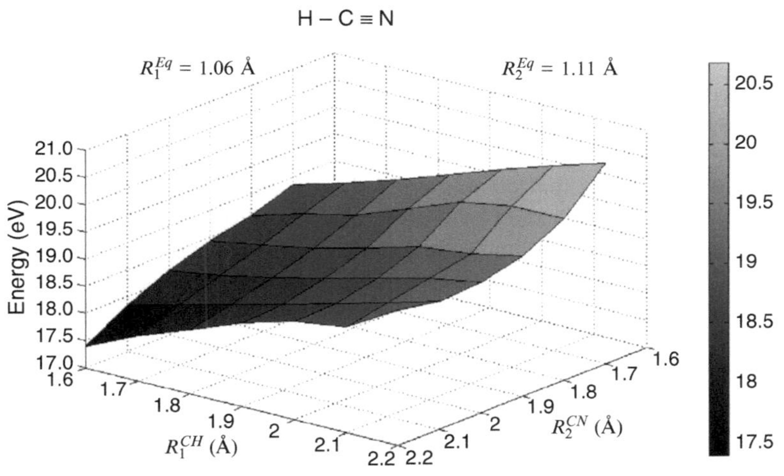

Fig. 4.8 Computed real part of the PES for the $^2\Sigma$ (N+1) electron resonant state as a function of the linear stretching of both bonds. See text for details

areas shown (for $R_{CH} \geq 2$ Å and $R_{CN} \sim 1.8$ Å, qualitatively speaking). This means, therefore, that the $^2\Pi$ resonance that leads to (CN^-) formation creates a PES for its nuclei that supports bound states, while a dissociative PES is formed via the $^2\Sigma$ resonance, which however indicates the excess charge to be distributed on both molecular fragments.

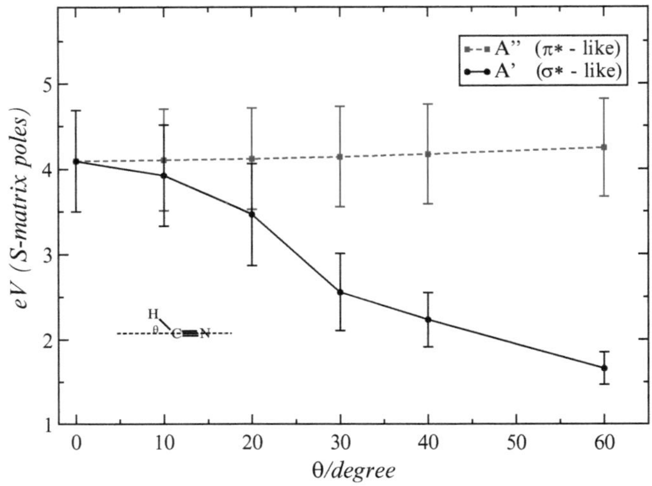

Fig. 4.9 Computed changes of the π^* resonance in linear geometry ($\theta = 0°$) as the angles are increased by bending the molecule. See text for details

4.3.3 Bending effects on the π^* resonance

It is already known that, by moving the HCN molecule away from linearity, the corresponding reduction of symmetry will remove the degeneracy of the two components for the $^2\Pi$ metastable state and provide an A' component of the same symmetry as that coming from the $^2\Sigma$ metastable state [13–15]. The ensuing crossing between the two initial ($^2\Sigma, ^2\Pi$) PESs will become avoided (a conical intersection in multidimensions) and the initial resonant compound can now undergo H-tunneling to dissociation into (CN^-) + H [16], producing a vibrationally excited CN^- fragment [12].

To show such a behaviour within our modelling of the process, we report in Fig. 4.9 the changes of the π^* resonance as the molecule is bent from linearity, keeping however $R_{CH} = 1.06$ Å and $R_{CN} = 1.7$ Å. One sees that the A" component mantains its original π-character and therefore changes little as the molecule moves away from linearity. On the other hand, the σ^*-like component is instead rapidly decreasing in energy while also becoming a narrower resonance. The excess electron is now stabilized and evolves into becoming a state close to threshold energy. We therefore can qualitatively expect that it will cross the neutral, N-electron potential and produce the bound fragments experimentally observed. The corresponding resonant electron wavefunction (real part) at the $\theta = 60°$ angle of Fig. 4.9, is shown for the A' resonant state in Fig. 4.10. We clearly see that the excess electron is now chiefly localizing onto the CN part of the molecule, indicating the above process to be the doorway state to CN^- stabilization upon bond stretching processes.

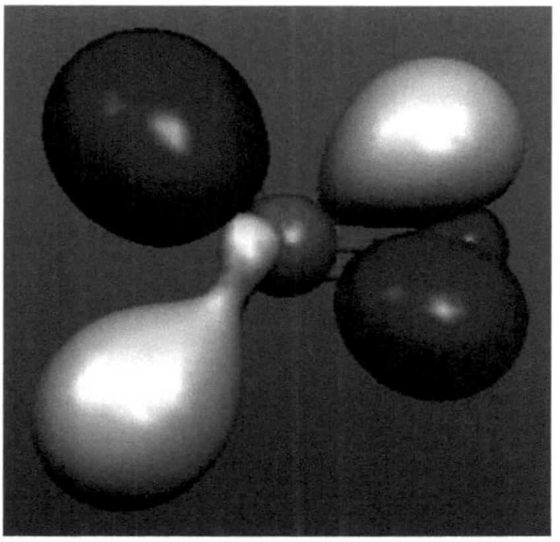

Fig. 4.10 Computed scattering wavefunction (real part) at $\theta = 60°$ geometry of Fig. 4.9. See text for further details

4.4 Present conclusions

In this work we have presented a possible quantum modelling of the fragmentation dynamics in polyatomic systems that takes place after an initial resonant electron attachment collision. The aim has been to assemble a computational tool which would allow us to try to follow, via a simpler adiabatic approach, the evolution of the energy location and of the associated width of any of the initially formed resonances in a DNA component or in a larger DNA fragment. The actual calculation of a complex PES for the (N+1) electron compound states created by the resonance thus becomes an instrument for predicting, or for confirming, the expected paths which would lead to the stabilization of a part of the initial molecule, hence the rupture of one of the DNA bonds in its helix or within one of its components. In spite of the simple nature of the interaction model we have employed, and without carrying out specific nuclear dynamics on the computed PES, we expect to be able to qualitatively discover the dominant evolution paths and those bonds which are most likely to undergo rupture during the energy rearrangement processes following electron attachment.

As an instructive example, and as a test of the method, we have selected a linear triatomic molecule, HCN, already extensively studied both theoretically and experimentally (see references in 4.1). The reasons for this choice stem from the relative simplicity of the system, its polyatomic nature that suggests 2D potentials to be realistic descriptions of the resonance evolutions, and the existence of several earlier studies which can provide information on both experiments and calculations.

The features and behaviour of our calculations for both stretching and bending deformations are analysed in some detail and indeed demonstrate the capabilities of this simple model to view the dissociation dynamics in a small polyatomic target and to be rather directly connected with existing experiments on the test system reported here.

Acknowledgements We thank the CASPUR Consortium for the computational help. One of us (A.G.S.) acknowledges the support of the Departamento de Posgrado y Especialización, Consejo Superior de Investigaciones Científicas (CSIC) for her stay at the University of Rome during the completion of this work.

References

1. C. Von Sonntag, *The chemical basis of Radiation Biology* (Taylor and Francis, London, 1987)
2. B. Boudaiffa, P. Cloutier, D. Hunting, M.A. Huels, and L. Sanche, Science **287**, 1658 (2000)
3. B.D. Michael and P. O'Neill, Science **287**, 1603 (2000)
4. A. Grandi, F.A. Gianturco, and N. Sanna, Phys. Rev. Lett. **93**, 048103 (2004)
5. M.P. Conrad and H.F. Schaefer III, Nature **274**, 456 (1978)
6. R.V. Yelle, Astrophys. J. **383**, 380 (1991)
7. V. Vuitton, P. Lavvas, R.V. Yelle, M. Galand, A. Wellbrock, G.R. Lewis, A.J. Coates, and J.-E. Wahlund, Planet Space Sci. **57**, 1558 (2009)
8. M. Inoue, J. Chem. Phys. **63**, 1061 (1966)
9. S.K. Srivastava, H. Tanaka, and A. Chutjian, J. Chem. Phys. **69**, 1493 (1978)
10. P.D. Burrow, A.E. Howard, A.R. Johnston, and K.D. Jordan, J. Phys. Chem. **96**, 7570 (1992)
11. F. Edard, A.P. Hitchcock, and M. Tronc, J. Phys. Chem. **94**, 2768(1990)
12. O. May, D. Kubala, and M. Allan, Phys. Rev. A **82**, 010701 (2010)
13. A. Jain and D.W. Norcross, Phys. Rev. A **32**, 134 (1985)
14. A. Jain and D.W. Norcross, J. Chem. Phys. **84**, 739 (1986)
15. H.N. Varambhia and J. Tennyson, J. Phys. B-At. Mol. Opt. Phys. **40**, 1211 (2007)
16. S.T. Chourou and A.E. Orel, Phys. Rev. A **80**, 032709 (2009)
17. e.g. see: F.A. Gianturco, D.G. Thompson, A. Jain, *Computational Methods for Electron-Molecule Collisions* (Plenum, New York, 1995)
18. F.A. Gianturco in: *Electron Collisions with Molecules, Clusters and Surfaces*, ed. by H. Ehrhardt and L.A. Morgan (Plenum Press, New York, 1994)
19. I. Baccarelli, F.A. Gianturco, A. Grandi, R.R. Lucchese, and N. Sanna, Adv. Quantum Chem. **52**, 189 (2007)
20. C. Lee, W. Wang, and R.G. Parr, Phys. Rev. B **37**, 785 (1998)
21. S.J. Hara, J. Phys. Soc. Jpn **22**, 710 (1967)
22. S. Salvini and D.G. Thompson, J. Phys. B **14**, 3797 (1981)
23. I. Baccarelli, F.A. Gianturco, A. Grandi, and N. Sanna, Int. J. Quant. Chem. **108**, 1878 (2008)
24. F.A. Gianturco and R.R. Lucchese, J. Phys. Chem. A **108**, 7056 (2004)
25. F. Carelli, F. Sebastianelli, I. Baccarelli, and F.A. Gianturco, Int. J. Mass Spectrom. **277**, 155 (2008)
26. R.R. Lucchese and F.A. Gianturco, Int. Rev. Phys. Chem. **15**, 429 (1996)
27. T.N. Rescigno and A.E. Orel, Phys. Rev. A **24**, 1267 (1981)
28. J.R. Taylor, *Scattering theory* (Wiley, New York, 1972)
29. A. Igarashi, I. Shimamura, Phys. Rev. A **70**, 012706 (2004)
30. F.T. Smith, Phys. Rev. **118**, 349 (1960)
31. T.P.M. Goumans, F.A. Gianturco, F. Sebastianelli, I. Baccarelli and J.L. Rivail, J. Chem. Theory Comput. **5**, 217 (2009)
32. F.A. Gianturco, F. Sebastianelli, R.R. Lucchese, I. Baccarelli, and N. Sanna, J. Chem. Phys. **128**, 174302 (2008)
33. C. Panosetti, I. Baccarelli, F. Sebastianelli, and F.A. Gianturco, Eur. Phys. J. D **60**, 21 (2010)
34. R.L. DeLeon and J.S. Muenter, J. Chem. Phys. **80**, 3992 (1984)

Chapter 5
Electron–Biomolecule Collision Studies Using the Schwinger Multichannel Method

Carl Winstead and Vincent McKoy

Abstract We review applications of the Schwinger multichannel method to low-energy electron collisions with polyatomic molecules of biological interest. After briefly describing the method, its implementation, and its strengths and limitations, we turn to a discussion of specific molecular systems, with an emphasis on studies related to radiation damage to DNA mediated by secondary electrons. Throughout, we situate our results in the context of calculated and experimental data on electron scattering, dissociative attachment, and other relevant processes.

5.1 Introduction

Radiation physicists and chemists have long understood that most of the damage done to living cells by ionizing radiation is indirect: through repeated collisions, mostly with water molecules, the primary radiation generates large numbers of secondary electrons, ions, and radicals, and these in turn undergo collisions of their own to generate still more secondary particles [1, 2]. Some of this vast number of reactive secondary particles generated along the radiation track may collide with, or, after solvation, diffuse into the proximity of important cellular molecules such as DNA. There they may cause damage such as strand breaking, base excision, and base dimerization. In modeling radiation damage, it was once widely assumed that slow secondary electrons—those that no longer had sufficient energy to

C. Winstead • V. McKoy (✉)
Noyes Laboratory of Chemical Physics, California Institute of Technology, Pasadena, California 91125, USA
e-mail: carl@schwinger.caltech.edu; mckoy@caltech.edu

G.G. Gómez-Tejedor and M.C. Fuss (eds.), *Radiation Damage in Biomolecular Systems*, Biological and Medical Physics, Biomedical Engineering,
DOI 10.1007/978-94-007-2564-5_5,

electronically excite or ionize water—could be neglected; they would not generate further reactive species, and would themselves quickly be solvated. Thus it was quite startling when, beginning in 2000, the group of Léon Sanche at the University of Sherbrooke demonstrated that even very slow electrons, with kinetic energies on the order of 1 eV, could induce single-strand breaks in plasmid DNA, while somewhat faster but still sub-ionization electrons could cause double-strand breaks [3–5]. These pioneering experiments stimulated a significant, and still ongoing, effort by research groups around the world to elucidate the mechanisms underlying slow-electron damage to DNA through studies of smaller molecules. That effort has largely been experimental, and while some additional work has been done on DNA and oligonucleotides (e.g., [6, 7]), most of the experiments have involved dissociative electron attachment to (or, in the condensed phase, electron-stimulated desorption of anions from) the fundamental subunits of DNA and RNA or molecular models of those subunits, including nucleobases and substituted nucleobases [8–37], molecules related to the backbone sugars [31,38–44], phosphate-containing compounds [31,45,46], and the nucleoside thymidine [47,48].

At the same time, however, there is clearly much that can be learned from theoretical and computational studies. Bound-state electronic structure calculations can be, and have been, of considerable use in identifying metastable anionic states that may trap slow electrons and in suggesting mechanisms by which those anionic states may couple to nuclear motion to promote dissociative attachment (e.g., [49, 50]). Ideally, though, one wishes to treat the electron–molecule interaction as a collision problem, subject to the appropriate (scattering) boundary conditions, in order to obtain detailed information, including not only the energies of such metastable states (scattering resonances) but also their lifetimes (resonance widths) and the probabilities of forming them (scattering cross sections). Indeed, at collision energies of more than 1 or 2 eV, where resonances tend to be broad and merged into the background of nonresonant scattering, and where coupling to electron-impact excitation channels can be important, bound-state methods become less and less applicable and a proper scattering treatment, correspondingly, more and more essential.

For sufficiently high electron energies, simplifying approximations render the electron–molecule scattering problem quite tractable. Thus, for example, at energies on the order of 10 keV and higher, the incident electron can be treated as a plane wave $\exp(i\vec{k}_0 \cdot \vec{r})$, where $\vec{k}_0$ is the wave vector related to the incident momentum $\vec{p}_0$ by $\vec{p} = \hbar\vec{k}_0$, that is scattered to produce outgoing plane-wave final states $\exp(i\vec{k}_f \cdot \vec{r})$ (first Born approximation). The molecule, meanwhile, can to good approximation be treated as a collection of independent scattering centers, because the internuclear distances are large compared to the projectile electron's de Broglie wavelength; thus the molecular scattering amplitude is just the coherent superposition of first-Born atomic scattering amplitudes that have been calculated and tabulated once and for all. At lower collision energies, though, the problem becomes increasingly complex. As the projectile electron's de Broglie wavelength increases, the independent-atom picture breaks down; as its energy becomes more comparable to the strength of the electron–molecule interaction potential, the Born approximation breaks down; at energies on the order of 100 eV and below, the cross sections of excitation and ionization channels can be large relative to the elastic cross section, and electron exchange becomes an important component of the interaction potential; below a few tens of eV, phenomena that depend on the detailed molecular electronic structure, such as shape resonances, core-excited resonances, and the opening of new electronic-excitation channels, come to dominate the scattering; and at impact energies on the order of 10 eV and less, where the kinetic energies of the valence electrons are comparable to that of the projectile, the molecular charge density can no longer be considered as fixed throughout the collision. In the low-energy range, then, one must consider the full electron–molecule scattering problem, in which the projectile and the N electrons of the target molecule are jointly described by a properly antisymmetrized $(N + 1)$-electron wavefunction that is computed in the field defined by the nuclei and that is subject to proper asymptotic boundary conditions. That is, one faces all of the challenges normally faced in computational quantum chemistry, including the rapid scaling of effort with molecular size that makes accurate quantum-chemical calculations on large molecules difficult, with the added complication of scattering boundary conditions that render the usual energy-minimizing variational principle of bound-state quantum chemistry inapplicable.

In this chapter, we describe a computational method, the Schwinger multichannel (SMC) method, whose design and implementation are intended to cope with the challenges just described. After outlining the SMC method, we will review recent studies in which we have applied it to explore low-energy electron interactions with RNA and DNA. In those studies, we have looked at electron scattering by the individual nucleobases, the backbone sugar deoxyribose, and the phosphate group. Moreover, as a step toward understanding how features of the scattering will change when those subunits are assembled into nucleic-acid chains, we have also looked at related molecules, both smaller (e.g., tetrahydrofuran and 3-hydroxytetrahydrofuran, as models of deoxyribose) and larger (e.g., methyl esters of phosphoric acid as models of the phosphate–sugar linkage), and we have studied larger assemblies, including nucleosides and the nucleotide deoxyadenosine 5′-monophosphate. Our principal aim in these studies has been to characterize the resonances (temporary anions) that dominate the low-energy elastic scattering and that may play a role in dissociative attachment and dissociative excitation. In so doing, we hope to provide useful cross-checks and occasional guidance to experimental studies, as well as, in some cases, to obtain information that would be difficult to extract from measurements. Of course, a large gap remains between calculations on electrons encountering isolated DNA fragments in vacuo and electron interactions with condensed-phase, solvated DNA. Accordingly, we conclude by briefly describing the prospects for bridging, or at least usefully narrowing, that gap.

We note at the outset that we do not aim to provide a comprehensive review of small-molecule studies related to low-energy electron-induced damage to biomolecules, or even of the computational portion of such studies. Calculations using bound-state methods have made many valuable contributions to understanding of electron-induced DNA damage, but we will not often have occasion to refer to that work; for some recent perspectives, see [51–53]. Our focus is much narrower: electron collision studies, primarily elastic, of constituents of RNA and DNA or close analogues of those constituents. Though we mostly review our own results, we attempt to situate them alongside complementary work by other researchers, experimental and computational, and to provide sufficient references to guide the reader to the relevant literature. We apologize in advance for the inevitable oversights and errors.

5.2 The Schwinger Multichannel (SMC) Method

As mentioned in Sect. 5.1, the many-electron scattering problem, though fully as difficult as the bound-state problem of quantum chemistry, cannot be treated by the energy-minimization procedure applicable to the latter. Instead, one typically applies some other optimization procedure that produces accurate results efficiently. In the time-independent picture, scattering is described by a wavefunction that behaves asymptotically as

$$\Psi^{(+)}(\vec{r}_1,\ldots,\vec{r}_{N+1}) \rightarrow \mathcal{A}\Bigg[\phi_0(\vec{r}_1,\ldots,\vec{r}_N)\exp(\mathrm{i}\vec{k}_0\cdot\vec{r}_{N+1}) + \sum_j f(\vec{k}_0,\vec{k}_j)\phi_j(\vec{r}_1,\ldots,\vec{r}_N)\frac{\exp(\mathrm{i}k_j r_{N+1})}{r}\Bigg], \quad (5.1)$$

where the wave vector $\vec{k}_0$ describes the direction and energy of incidence and likewise $\vec{k}_j$ describes the direction and energy of the outgoing scattered electron. The N-particle wavefunctions ϕ_0 and ϕ_j are the initial and final states of the target molecule, and $\mathcal{A}$ is an antisymmetrizer between the projectile electron and those of the target. Energy conservation, though not explicitly indicated, is assumed: that is, $E_0 + k_0^2/2 = E_j + k_j^2/2$, where E_0 is the energy of ϕ_0, E_j the energy of ϕ_j, and $k_{0,j}^2/2$ are the electron kinetic energies (Hartree atomic units, $\hbar = c = e = 1$, are henceforward asssumed). The summation over j includes all open electronic-excitation channels (we neglect ionization). The $(+)$ superscript on Ψ denotes that only outgoing spherical waves are included in the second term on the right, which describes the scattered flux. The information about every possible scattering process $\vec{k}_0 \rightarrow \vec{k}_j$ is contained in the coefficient $f(\vec{k}_0,\vec{k}_j)$, known as the scattering amplitude, that modulates those outgoing waves. The central role of the scattering amplitude naturally suggests developing variational principles in which it is the stationary quantity.

The SMC method was developed by Takatsuka and McKoy [54, 55] as an adaptation of the Schwinger variational procedure [56] to the specific situation of low-energy scattering involving multiple indistinguishable particles. For the present purposes, details of the derivation are not relevant, and we simply write down the variational expression for the scattering amplitude:

$$-2\pi f(\vec{k}_0,\vec{k}_j) = \left\langle \psi_j(\vec{k}_j) \mid V \mid \Psi^{(+)}(\vec{k}_0)\right\rangle + \left\langle \Psi^{(-)}(\vec{k}_j) \mid V \mid \psi_0(\vec{k}_0)\right\rangle - \left\langle \Psi^{(-)}(\vec{k}_j) \mid A^{(+)} \mid \Psi^{(+)}(\vec{k}_0)\right\rangle, \tag{5.2}$$

where $\psi_{0,j}(\vec{k}_{0,j})$ represent the antisymmetrized products of target states ϕ and plane waves appearing in Eq. 5.1, V is the electron–molecule scattering potential defined in terms of the N-electron and $(N+1)$-electron Hamiltonians via $V = H_{N+1} - H_0$, with $H_0 = H_N + T$ and T the kinetic energy operator of a free electron. $A^{(+)}$ is the operator

$$A^{(+)} = VP + \left(\frac{1}{N+1} - P\right)(E - H) - VG_P^{(+)}V, \tag{5.3}$$

with P a projection operator onto open, i.e., energetically accessible, scattering channels j, and $G_P^{(+)}$ the projected interaction-free Green's function subject to outgoing-wave boundary conditions,

$$G_P^{(+)} = \lim_{\varepsilon \to 0^+} \frac{P}{E - H_0 + \mathrm{i}\varepsilon}. \tag{5.4}$$

A key property that the SMC expression shares with the original Schwinger principle is that, despite the appearance of a Hamiltonian term in Eq. 5.3, all required matrix elements are independent of the asymptotic form of $\Psi^{(\pm)}$. Therefore, one may use square-integrable wavefunctions to approximate $\Psi^{(\pm)}$; in particular, they may be approximated in the same form as is used in conventional bound-state quantum chemistry, that is, by linear expansions in terms of configuration state functions that are, in turn, represented in terms of molecular orbitals constructed from Gaussian-type atomic orbitals. This convenient property allows us to make use of computational machinery developed for the bound-state problem to handle much of the work. Writing those expansions explicitly as

$$\Psi^{(+)}(\vec{k}_0) = \sum_m x_m(\vec{k}_0)\chi_m(\vec{r}_1,\ldots,\vec{r}_{N+1})$$

$$\Psi^{(-)}(\vec{k}_j) = \sum_m y_m(\vec{k}_j)\chi_m(\vec{r}_1,\ldots,\vec{r}_{N+1}) \tag{5.5}$$

with χ_m a configuration state function and unknown coefficients x_m and y_m, we proceed in the usual manner by inserting the expansions of Eq. 5.5 into Eq. 5.2 and requiring that f be stationary:

$$\frac{\partial f}{\partial x_m} = \frac{\partial f}{\partial y_m} = 0, \tag{5.6}$$

leading to a system of linear equations

$$\mathbf{A}\mathbf{x} = \mathbf{b}^{(0)}, \tag{5.7}$$

where elements of the square matrix $\mathbf{A}$ and column vector $\mathbf{b}^{(0)}$ are given by

$$\begin{aligned} A_{mn} &= \langle \chi_m \mid A^{(+)} \mid \chi_n \rangle \\ b_m^{(0)} &= \left\langle \chi_m \mid V \mid \psi_0(\vec{k}_0) \right\rangle. \end{aligned} \tag{5.8}$$

Having solved this linear system, one obtains the scattering amplitude from

$$-2\pi f(\vec{k}_0, \vec{k}_j) = \mathbf{b}^{(j)\dagger} \cdot \mathbf{x} \tag{5.9}$$

where † indicates complex-conjugate-transpose and $\mathbf{b}^{(j)}$ is defined analogously to $\mathbf{b}^{(0)}$ but for the outgoing wave:

$$b_m^{(j)} = \left\langle \chi_m \mid V \mid \psi_j(\vec{k}_j) \right\rangle. \tag{5.10}$$

For nonspherical targets such as molecules, one is of course interested in more than one direction of incidence $\vec{k}_0$ as well as more than one scattering direction $\vec{k}_j$; this simply means that $\mathbf{b}^{(0,j)}$ become rectangular matrices, with each column corresponding to a different direction, and the scattering amplitude in Eq. 5.9 correspondingly becomes a matrix.

In many ways, then, the SMC procedure is not too different from bound-state quantum chemistry: one proceeds from Gaussian atomic orbitals to molecular orbitals to configuration state functions comprising a linear space in which to apply a variational principle, with only the slight difference that the working equations take the form of a linear system rather than an eigenvalue problem. The major complication arises when we work back down to the level of one-electron functions in order to evaluate the matrix elements of $\mathbf{A}$ and $\mathbf{b}$. Because the functions $\psi_{0,j}$ occurring in $\mathbf{b}^{(0,j)}$ are built not only from Gaussians but from plane waves $\exp(i\vec{k}_{0,j} \cdot \vec{r})$, their evaluation requires electron–nucleus attraction integrals containing a Gaussian and a plane wave, as well as two-electron Coulomb matrix elements involving three Gaussians and a plane wave. The same types of mixed integrals arise when the Green's-function term in $A^{(+)}$, Eq. 5.3, is evaluated by introducing a spectral representation,

$$G_P^{(+)} = \frac{1}{(2\pi)^3} \lim_{\varepsilon \to 0+} \sum_{j \in \text{open}} \int d^3k' \frac{|\psi_j(\vec{k}')\rangle\langle\psi_j(\vec{k}')|}{E - (E_j + k'^2/2) + i\varepsilon}. \tag{5.11}$$

Fortunately, the requisite one- and two-electron mixed integrals can be evaluated accurately and efficiently using numerical procedures analogous to those used for integrals involving Gaussians only [57, 58]. The difficulty lies rather in the sheer number of such integrals: for every magnitude and direction of $\vec{k}_{0,j}$ required to evaluate the desired scattering amplitudes, and for every quadrature point $\vec{k}'$ necessary to evaluate the spectral representation in Eq. 5.11, there are $\mathcal{O}(N_g^3)$ such two-electron integrals to evaluate, where N_g is the number of contracted Gaussians. Moreover, each such set of these "raw" integrals must be reduced to $(N+1)$-electron matrix elements of **A** and $\mathbf{b}^{(0,j)}$, which requires a 3-index transformation analogous to the 4-index transformation required in post-Hartree–Fock bound-state quantum chemistry.

Given these computational demands, the only practical route to addressing larger molecules, including biomolecules, is to exploit massively parallel computing. Here we briefly describe the approach that we have followed in parallelizing the SMC code [59]; additional details can be found in [60, 61]. Our parallel SMC code is designed to run efficiently on both tightly-integrated multiprocessor computers and workstation clusters. It uses a loosely synchronous, single-program, multiple-data model; that is, each processor or core runs the same code but operates on a different subset of the data, and processors mostly operate independently of each other, synchronizing only when they need to communicate. Communication is handled through explicit message passing. Though not as straightforward to implement as other approaches (e.g., parallelization through compiler directives), this design has proven robust, portable, and flexible, adapting to many different architectures, communication networks, and message-passing libraries. To construct matrix elements from mixed Gaussian/plane-wave integrals, a set of integrals is (conceptually) arranged into a matrix, and each processor independently evaluates a different batch of integrals comprising a submatrix of that global integral matrix. Likewise, each processor evaluates its submatrix of a global transformation matrix that encodes both the three-index transformation and the transformation from molecular-orbital integrals to final matrix elements. All processors then cooperate in a distributed matrix multiplication that effects the transformation. Interprocessor communication is thus localized in a procedure, multiplication of large distributed matrices, that has a favorable computation to communication ratio, helping to achieve high efficiency and scalability.

Although the parallel SMC method has proven highly successful in treating low-energy collisions with larger polyatomics, some of its current limitations bear mentioning. As originally implemented, it was applicable only to single-configuration, closed-shell singlet representations of the target molecule's ground-state wavefunction and likewise to single-configuration singlet or triplet representations of its excited states. Limited, special-purpose generalizations to doublet and triplet targets have been made, but applications have been few. Restriction to singlet ground states is not a serious limitation in studies of biomolecules, but restriction to single-configuration representations of the target states is, particularly if one wishes to examine electron collisions at nuclear geometries well away from equilibrium, as in studies of dissociative attachment. We are currently completing a major overhaul that includes allowing multiconfigurational representations of the target states, as well as relaxation of the restriction on target spins, in order to address these limitations.

5.3 Applications

In this section, we review applications of the SMC method outlined above to biological molecules, and in particular to molecules related to the nucleic acids. Several mechanisms have been suggested by which DNA might temporarily trap slow electrons and by which those trapped electrons might promote dissociation events such as base excision or strand breaking. Our calculations address the trapping mechanism by identifying low-energy shape resonances that are associated with specific base, sugar, or phosphate subunits, and by examining how the energies of those resonances change when the fundamental subunits are incorporated into larger assemblies. As will be seen, though many open questions remain, our results to date support a significant role for the nucleobases in trapping electrons, in agreement with models suggested by others (e.g., [16, 50, 51, 62–64]) but do not identify any strong, narrow resonances associated with the phosphate or sugar moieties.

5.3.1 The Nucleobases

When considering where slow electrons might attach to DNA, the nucleobases are obvious candidates. From a chemical point of view, each base is an aromatic ring system that can be expected to have empty π^* valence orbitals available for temporary trapping of electrons, thus giving rise, in scattering terms, to strong low-energy shape resonances in the elastic cross section. In contrast, the backbone sugar is a saturated molecule in which one expects only higher-energy and broader (shorter-lived) resonances associated with σ^* orbitals and is therefore a less obvious site of attachment, while the possibilities for attachment to the phosphate are also unclear. Thus we begin by discussing the nucleobases and molecules related to them, though we will return later to the constituents of the backbone.

5.3.1.1 The Pyrimidines

As the smallest and simplest nucleobase, the RNA base uracil (Fig. 5.1) is a natural place to begin an examination of electron interactions with the nucleic acids, and it has, accordingly, received more experimental and theoretical attention to date than any other subunit except, perhaps, the furanose ring. Uracil has also been the subject of some controversy. Given that direct data on uracil's elastic electron cross section are extremely limited [23], vital evidence on the energies and widths of resonances, both in uracil and the DNA nucleobases, comes from experiments by Burrow and coworkers [16, 65], who measured the derivative of the transmitted current as a function of the energy of an electron beam passed through a gas-phase sample. These experiments identified three resonances in each of the bases.

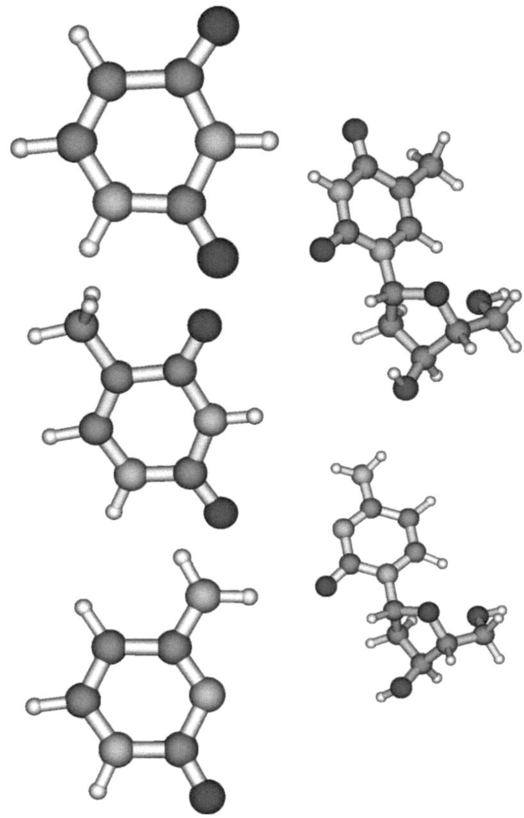

Fig. 5.1 The pyrimidine nucleobases uracil, thymine and cytosine (top to bottom, left) and the nucleosides deoxythymidine (top right) and deoxycytidine (bottom right). The molecules are not drawn to the same relative scale

In uracil, those resonances fell at 0.29, 1.58, and 3.83 eV and were assigned, on the basis of comparison with empirically scaled virtual-orbital energies from Hartree–Fock calculations, as the three expected π^* shape resonances [16, 65]. However, the first calculation on low-energy electron–uracil scattering [66] produced a different assignment. The lowest π^* resonance, π_1^*, was computed to lie at 2.27 eV and identified with the observed resonance at 1.58 eV, implying that the lowest-energy resonance, at 0.17 eV, must have some other origin. Subsequent calculations by a different group [67] likewise placed the π_1^* resonance at 2.2 eV. However, those calculations, which employed local-potential approximations to exchange and polarization components of the electron–molecule interaction, had been found to place the well-known π^* shape resonance of CO_2 about 2 eV too high [68], so similar errors for uracil might be expected. Moreover, significant arguments were advanced [69] against the resonance assignments proposed in [66].

Dissociative attachment to uracil has been extensively studied [15, 16, 20, 21, 23–25, 28–30]. The principal ion produced is found to be $(U{-}H)^-$, that is, the parent uracil molecule with a single hydrogen missing, while studies in which uracil is deuterated or methylated establish that the hydrogen is lost from the nitrogen atoms, and principally from that labeled N_1 in the standard ring numbering (the nitrogen that is *not* in between the keto carbons). The most prominent feature in the $(U{-}H)^-$ production cross section is a sharp peak at 1.0 eV. Neither this feature nor a weaker one at 0.69 eV appears to correlate with any of the measured or calculated elastic resonance positions just mentioned; rather, these are thought [16, 70] to be vibrational Feshbach resonances associated with temporary trapping of the projectile electron in the potential well created by the stong dipole moment of uracil. Because the electronic wavefunction of such a dipole-bound state is diffuse, it unclear whether such a resonance mechanism is relevant to nucleic acids in the condensed phase. However, there is also a broader peak near 1.8 eV in the dissociative attachment signal that may plausibly be associated with the resonance observed in transmission at 1.58 eV. As described in the preceding paragraph, that resonance has been variously assigned as π_1^* and π_2^*, both of which are compact valence states; without clarity on the correct assignment, though, it is difficult to be confident in any model of the dissociation mechanism.

In applying the SMC method to uracil, therefore, we employed different one-electron basis sets and levels of approximation in order to develop confidence in the consistency and convergence of our calculations [71]. The final results included an extensive treatment of polarization effects, represented as virtual excitations of the target molecule, and placed the three π^* shape resonances at 0.32, 1.91, and 5.08 eV. For comparison, with polarization omitted (thus treating the target charge density as frozen, the so-called static–exchange approximation), these energies shift upward to 2.1, 4.2, and 8.2 eV, reflecting the attractive nature of the polarization effect. The principal conclusion to be drawn from our results is that they strongly support the original assignment of the resonances observed in transmission [16, 65] by placing π_1^* near 0.3 eV rather than at 2.2 eV, while also placing π_2^* and π_3^* reasonably close to the second and third transmission features.

In computations, unlike gas-phase experiments, it is easy to take advantage of the symmetry properties of the molecule, both to speed the calculations and to assist in the analysis of results. In the case of uracil, we carried out separate scattering calculations for wavefunctions of A' and A'' symmetry, that is, even and odd, respectively, with respect to the molecular plane. The π^* resonances just discussed of course occur in the A'' component. In A', there is a broad peak near 8.5 eV similar to that seen in the 7–10 eV range in many other molecules that is likely associated with overlapping, short-lived σ^* resonances. However, there is also a sharp resonance peak at 1.45 eV, which in our experience is unusually low for a σ^* resonance. As discussed immediately below, this peak may be artifactual.

Subsequent studies of the π^* resonances of uracil have tended to confirm the picture outlined above and therefore to support the original resonance assignments [16, 65]. An expanded calculation by the group that originally placed π_1^* at 2.2 eV was interpreted as indicating a fourth π^* resonance (one more than expected on chemical grounds), now assigned as π_1^*, at 0.33 eV [72]. That interpretation was questioned [73], and the low-energy feature was subsequently found to be a computational artifact [74], leaving π_1^* at 1.7 eV, near the original position calculated by the same group. In contrast, a recent, extensive R-matrix calculation [75] places the three π^* resonances at 0.13, 1.94, and 4.95 eV, in reasonably close agreement with both experiment and, especially, our SMC results. Given this consistency and the fact that the SMC and R-matrix studies are the only all-electron calculations to date, the locations of the π^* resonances appear well established. However, for future reference, we note that the comparatively large disagreement between the calculated and measured positions of π_3^* is unexpected because the principal source of error, inadequate representation of polarization effects, should be less important for higher collision energies and shorter resonance lifetimes. On the other hand, the R-matrix calculation shows no indication of the apparent A' resonance that we saw at 1.45 eV, raising the possibility that it was a computational artifact.

Given their close chemical relationship to uracil, it is no surprise that the pyrimidine bases of DNA, thymine and cytosine (Fig. 5.1), should exhibit very similar scattering behavior. Our SMC calculations for these molecules, carried out in collaboration with Sergio Sanchez [76], yielded π^* resonance energies of 0.3, 1.9, and 5.7 eV for thymine, to be compared to experimental values [65] of 0.29, 1.71, and 4.05 eV; and of 0.5, 2.4, and 6.3 eV for cytosine, vs. experimental values of 0.32, 1.53, and 4.5 eV. Thus, as for uracil, we find agreement that, while far from perfect, confirms the experimental assignments, notably by locating the first π^* resonance well below 1 eV rather than near 2 eV as in more approximate calculations [67]. Once again, however, we note that the agreement between the SMC and experimental resonance positions is much poorer for π_3^* that for π_1^* and π_2^*, in contrast to expectation.

So far we have considered the electron-trapping properties of the pyrimidine bases only in isolation. An important question is how the metastable π^* anion states seen in the gas phase are affected by incorporation of the bases into DNA. Unfortunately, with present techniques and computational resources, accurate calculations on even the isolated bases are already challenging. One way to gain information about larger assemblies, however, is to carry out calculations at lower level on those larger molecules and compare to results obtained at both high and low level for the isolated bases. Thus our study of thymine and cytosine [76] also included calculations at the static–exchange level (where polarization is neglected) for thymine, cytosine, and the associated nucleosides, deoxythymidine and deoxycytidine (Fig. 5.1). Although static–exchange calculations cannot be expected to produce quantitative results for the positions or widths of low-energy resonances, the *relative* positions of the resonances in a base and its associated nucleoside may indicate the importance of environmental effects. Moreover, the shift due to polarization determined by comparing the full results for a given base to those obtained at the static–exchange level may be used to estimate the absolute positions of the resonances in the corresponding nucleotide. Our results indicate only slight upward shifts in energy for the π^* resonances in the nucleosides relative to the isolated bases. This is an encouraging outcome insofar as it indicates that calculations on the bases already capture the important physics, but of course it is not definitive: other sorts of environmental effects besides linking the base to the sugar, notably base pairing and base stacking, may be significant and remain to be explored. It is also consistent with gas-phase dissociative attachment experiments on thymidine [47, 48]: the yield of the anion in which thymidine has lost a single hydrogen shows a peak near 1.8 eV that is thought to arise from π_2^*, and the location of this peak is shifted little if any from that of the corresponding peak in thymine [26, 27] or 1-methylthymine [28].

We have mentioned above that, given the agreement between the SMC results and experimental data for the energies of the first two π^* resonances in each of the pyrimidine bases, our calculated energies for π_3^* are unexpectedly poor, in light of the general rule that polarization effects should be less, not more, important when the target has less time to respond to the passage of the projectile. This rule, however, is rooted in a specific physical picture of polarization, wherein the electric field of the approaching projectile perturbs the electronic wavefunction of the target molecule, inducing a dipole moment that makes the electron–molecule potential more attractive. In that picture, electric-dipole selection rules apply, and thus for a singlet ground state, only singlet-coupled virtual excitations should be needed to capture the effect. There is of course also a response when the projectile electron is within the extent of the molecular electron density. In order to capture both effects, local-potential methods typically join a long-range quadrupolar potential representing the charge–induced-dipole interaction onto a short-range correlation potential that is a function of the local electron density. In the SMC method and other all-electron approaches, the short-range relaxation effects are once again represented in terms of virtual excitations of the target. For a shape resonance in which the projectile is temporarily bound in a specific virtual valence orbital, the appropriate physical picture would seem to involve relaxation of the remaining orbitals, equivalent to performing a self-consistent-field calculation on the metastable anion. An equivalent and computationally more tractable procedure is to represent such orbital relaxation by including terms in the wavefunction that couple a "resonance orbital" to single excitations of the target; indeed, this picture is the basis of procedures developed for efficient representation of polarization effects on resonances [77–79]. Once again, though, the picture suggests that only singlet-coupled excitations of the target are important. Thus in designing our calculations on the pyrimidines, we emphasized singlet-coupled excitations, though we did include some triplet-coupled terms for uracil.

A different physical picture of short-range polarization suggests that triplet-coupled excitations may sometimes be quite important. Considering the perturbative response of the target to the projectile in terms of the actual electronic states of the target, rather than in terms of hole–particle configurations produced by promoting a single electron from an occupied to an arbitrarily-constructed virtual orbital, suggests that the low-energy elastic cross sections of molecules having low-lying triplet excited states may be strongly influenced by virtual (or real) channel coupling to those states. The pyrimidine nucleobases are prime examples of such systems: $\pi \rightarrow \pi^*$ excitations produce triplet excited states at energies as low as 3.5 eV in cytosine [80], 3.6 eV in thymine [81], and 3.7 eV in uracil [82]. Indeed, low-lying triplet states are generic in unsaturated and aromatic molecules, and Nenner and Schulz [83], who studied resonances in benzene (C_6H_6) and the nitrogen-substituted benzenes with formulas $C_4N_2H_4$ and $C_3N_3H_3$, had already suggested in 1975 that

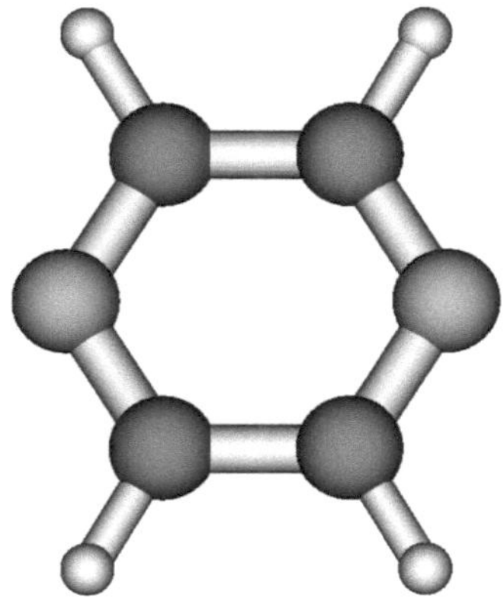

Fig. 5.2 Pyrazine, a high-symmetry model of the pyrimidine nucleobases

the highest π^* shape resonances in those molecules might be mixed with one or more of the triplet excited states. Moreover, such mixing was directly observed by Allan [84] in the form of the decay of the highest π^* resonance of benzene into triplet excited states.

As a model system for exploring the possible role of triplet coupling in the bases, we chose pyrazine (Fig. 5.2). Like the pyrimidine bases, it is a dinitrogen heterocycle, but its simplicity and high symmetry, together with the availability of experimental data on the resonance positions [83], make it a better test case. Calculations both with and without triplet-coupled virtual excitations demonstrated that their inclusion did indeed produce a large shift in the π_3^* resonance position, bringing it into much closer agreement with experiment, and that $^3(\pi \rightarrow \pi^*)$ excitations, specifically, account for a major fraction of the shift [85, 86]. This result provides strong evidence that resonant channel coupling between the elastic and $^3(\pi \rightarrow \pi^*)$ channels is likewise important in the nucleobases, and that omission of that coupling accounts for the relatively poor π_3^* resonance positions that we found in our SMC calculations on those bases.

To summarize the results of our studies of the pyrimidine bases, they support the locations of the the π^* shape resonances determined from electron transmission measurements [16, 65]. Given those locations, π_1^* appears to lie too low in energy to contribute to the electron-induced dissociation dynamics. As suggested by Burrow and coworkers [16], dissociative attachment at and below 1 eV instead appears to be explained by vibrational Feshbach resonances associated with the dipole-bound anion. On the other hand, π_2^*, lying between 1 and 2 eV, may well contribute to dissociative attachment in that energy range. It is important to note, moreover, that the mechanisms governing gas-phase dissociative attachment and those governing DNA damage in the condensed phase may well be different. Martin et al. [87] have studied electron-induced strand breaks in thin-film DNA and noted that the peak structure in the single-strand-break yield, though superficially similar to that seen in dissociative attachment to the pyrimidine nucleobases (e.g., [13, 20]), cannot be reconciled with the dissociative attachment spectra. They further point out that the

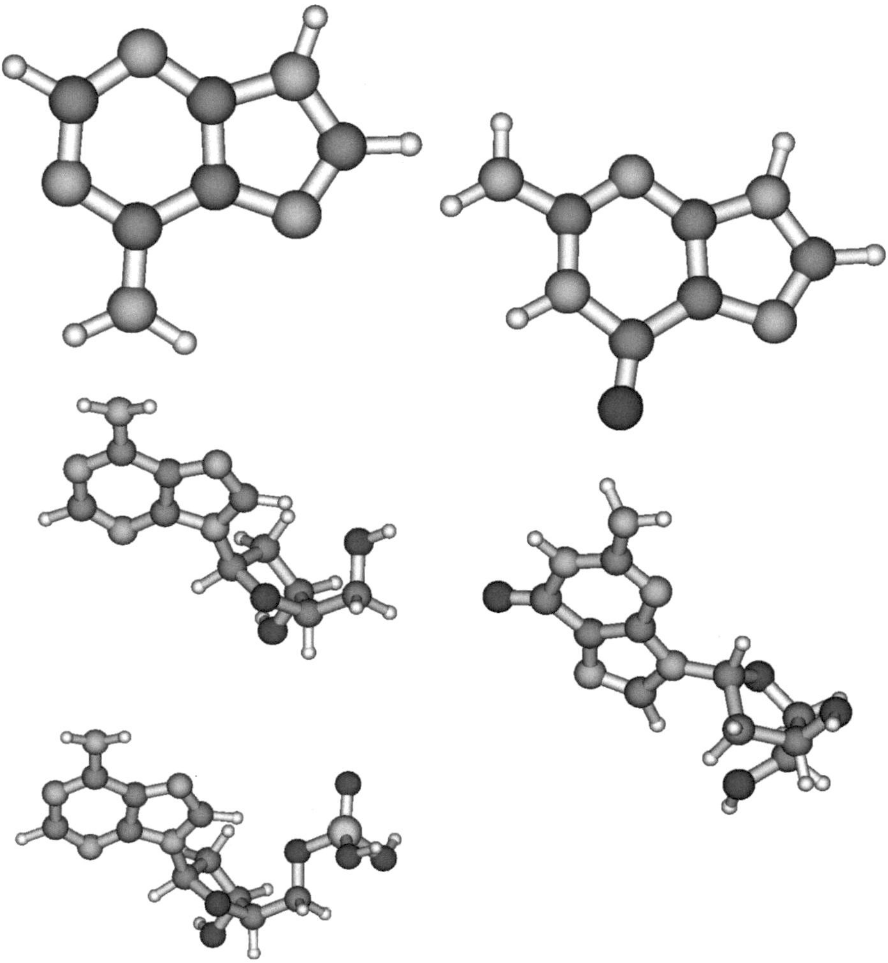

Fig. 5.3 The purine nucleobases adenine and guanine (top left and right), the nucleosides deoxyadenosine and deoxyguanosine (middle left and right), and the nucleotide deoxyadenosine-5′-monophosphate. The molecules are not drawn to the same relative scale

diffuse dipole-bound states required for the vibrational Feshbach resonance model may not exist in the condensed phase, and they argue that the principal strand-break mechanism is more likely to involve the nucleobase π^* resonances.

5.3.1.2 The Purines

The purine bases, adenine and guanine (Fig. 5.3), contain fused 5- and 6-membered rings, rather than the single 6-membered ring of the pyrimidines. This increase in size already makes computations much more difficult, but the purines moreover contain NH_2 groups that render them slightly nonplanar. Loss of reflection symmetry both adds to the computational cost and complicates the analysis of the results. In the pyrimidines, we could carry out separate calculations for the A' and A'' representations of the C_s point group, and resonances could be classified as σ or π according to which representation they occurred in, with the π^* resonances standing out clearly in A'', where background scattering is weak. In principle, scattering from the purines should make no use of symmetry, with both the σ^* and π^* resonances to be determined from a single calculation, against a large nonresonant background. To render our study of the purine bases [88] more tractable, we first determined the effect of enforcing planarity by carrying out static–exchange calculations on guanine in both planar and nonplanar geometries. The results indicated an upward shift of about 0.2 eV in the π^* resonance energies of the planar form. With this information in hand, we then carried out full calculations, including polarization, for the A'' representation of both planar guanine and planar adenine in order to obtain more accurate π^* resonance positions. As with the pyrimidines, we then proceeded to study the effect of incorporating the bases into larger moieties, namely the nucleosides deoxyguanosine and deoxyadenosine as well as the nucleotide 2′-deoxyadenosine-5′-monophosphate or dAMP (Fig. 5.3), at the static–exchange level of approximation.

Our calculations including polarization effects placed the three lowest π^* resonances of adenine at 1.1, 1.8, and 4.1 eV, compared to experimental values [65] of 0.54, 1.36, and 2.17 eV. Agreement is thus quite good for the first two resonances—though not as good as for the pyrimidines, even after allowing for a downward shift of 0.2 eV to compensate for imposing planar geometry—but much poorer for π_3^*, likely indicating that in this case also it mixes with core-excited resonances built on triplet excited states. The picture is somewhat different for guanine: the SMC calculation places the first three π^* resonances at approximately 1.55, 2.4, and 3.75 eV, whereas the electron transmission measurement places them at 0.46, 1.37, and 2.36 eV [65]. At first sight this comparison indicates much larger errors in the calculated positions for π_1^* and π_2^* than we saw in adenine or any of the pyrimidines, and a smaller error for π_3^*. However, as Aflatooni and coworkers already noted [65], the dominant guanine isomer in the gas phase is believed to be an enol form [89], rather than the keto isomer found in DNA, for which we did the calculation. In contrast, gas-phase adenine is mostly the keto form that we assumed [90]. Here we have, therefore, an example where theory provides information not directly available from experiment: assuming the same shift of about 0.5 eV needed to bring our calculated positions for π_1^* and π_2^* of adenine into agreement with the measurement, we can predict that the same resonances in the keto tautomer of guanine should lie at about 1.0 and 1.9 eV.

Comparison with the calculations of Tonzani and Greene [67] generally follows the pattern seen in the pyrimidines, with reasonable qualitative agreement but with their resonance positions shifted to higher energies compared to the SMC results: thus they place π_1^* at 2.4 eV in both adenine and guanine.

The results for deoxyguanosine, deoxyadenosine, and dAMP did not reveal any dramatic changes in the scattering behavior. However, comparison of the results for the nucleosides with the static–exchange results for adenine and guanine did show a slight upward shift, by 0.1 to 0.3 eV, in the positions of the three low-energy π^* resonances, with negligible further shifts upon attaching a phosphate to form the nucleotide. Thus, as in the pyrimidines, it appears that the π^* shape-resonant behavior is already modeled well at the level at the level of the isolated nucleobases.

5.3.2 *The DNA Backbone*

Components of the DNA backbone have already appeared in the calculations on nucleosides and dAMP described above; however, the focus in that work remained on the π^* resonances of the bases. It is natural to consider, though, whether strand-breaking might involve direct attachment of the electron to the backbone, as proposed early on by Li, Sevilla, and Sanche [49], rather than or in addition to attachment to a base followed by intramolecular electron transfer to the backbone, as in the model of Simons and coworkers [50, 51, 62–64]. Thus it is of interest to explore electron scattering by molecular models of the backbone constituents, deoxyribose and phosphate. As with the bases, the focus of our SMC calculations has been on identifying and characterizing low-energy resonances in the elastic cross section that might trap electrons long enough to promote dissociative attachment.

5.3.2.1 Deoxyribose

The DNA backbone consists of phosphate groups alternating with substituted furanose rings. If the phosphates connected to a given furanose are replaced by hydroxyls, and likewise the base bonded to C_1, the resulting molecule is of course 2-deoxyribose. However, there are many other ways to extract a plausible model of the sugar moiety. The simplest is to replace all ring linkages with H, resulting in tetrahydrofuran (THF). Retaining the oxygen at C_3 results in 3-hydroxytetrahydrofuran (3HTHF). These three structures are illustrated in Fig. 5.4. Obviously other models, such as tetrahydrofurfuryl alcohol [91–93], are possible, but in fact the majority of attention to date has been given to THF.

THF is a saturated system, and its ring is puckered rather than planar. Which particular puckered geometry is the absolute minimum of energy has been a topic of past debate, but what it is clear is that the various minima are of nearly equal energy and are separated by very small barriers [94]. Thus, although the geometry of deoxyribose in DNA is constrained by its phosphate and nucleobase linkages, the conformation of an isolated THF molecule at room temperature is constantly changing. One issue that arises, therefore, is whether the electron scattering cross section of THF is sensitive to its conformation; if so, gas-phase measurements may not comment directly on the conformation relevant to DNA, and comparison

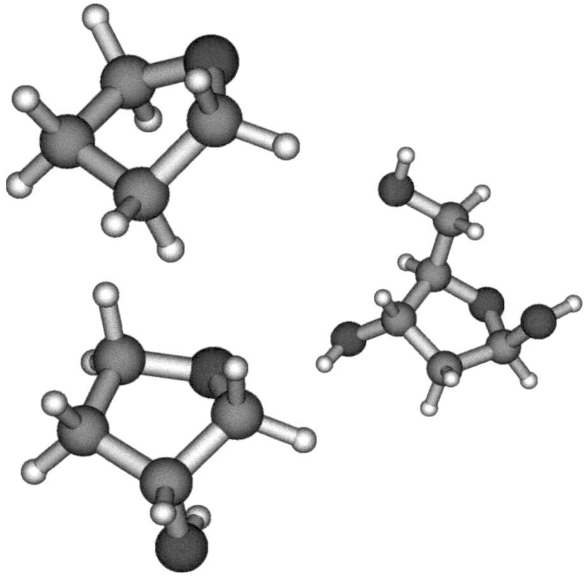

Fig. 5.4 Models of the sugar moiety in the nucleic acid backbone: tetrahydrofuran (top left), 3-hydroxytetrahydrofuran (bottom left), and 2-deoxyribose (right). The molecules are not drawn to the same relative scale

of calculated results to gas-phase experiments would require a conformational average. Accordingly, we began our study of THF [95] with preliminary SMC calculations at the static–exchange level for two different puckered geometries, one having C_2 symmetry and one having C_s symmetry, as well as for the C_{2v} (planar ring) geometry that had been imposed, for computational efficiency, in previous calculations [96,97]. The SMC calculations produced two broad resonance peaks whose energies were quite similar in the two puckered geometries; however, in C_{2v} the splitting between the peaks increases by about 3 eV, indicating that forcing the ring to be planar introduces significant error. The full calculation with polarization effects included was therefore carried out on the C_2 structure. The resulting integral cross section, Fig. 5.5, exhibits a broad resonance maximum at 8.3 eV and a still broader peak or shoulder centered between 13 and 14 eV. Each peak contains both ^2A and ^2B symmetry components, indicating that it arises from more than one resonance. Notably absent, however, are any narrow resonances, or indeed any evident resonances below those at about 8 eV, that could temporarily trap electrons and promote dissociative attachment. Indeed, experiment shows only very weak dissociative-attachment peaks, one near 1.25 and the other near 7.5 eV [42]. Several other calculations [96–98] and measurements [99–101] of low-energy elastic scattering by THF have been carried out, and these are in general agreement with our results. As an example, Fig. 5.6 shows differential cross sections (that is, scattering as a function of the angle between the incident beam and the direction of detection) at selected energies as determined by various groups. Although there are differences in detail, overall agreement is satisfactory.

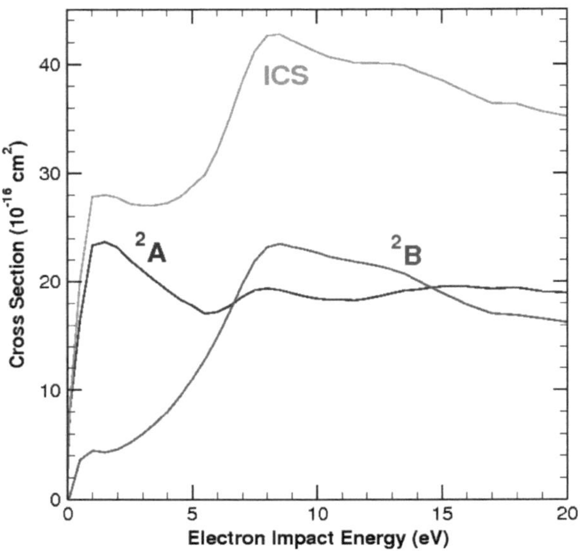

Fig. 5.5 Integral cross section (ICS) calculated by the SMC method [95] for elastic scattering of low-energy electrons by tetrahydrofuran in its C_2 conformation, together with the contributions from wavefunctions belonging to the ^{2}A and ^{2}B representations

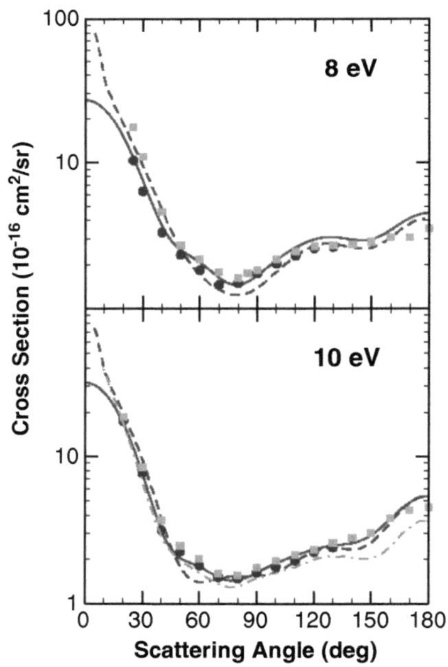

Fig. 5.6 Differential cross sections at 8 and 10 eV for elastic electron scattering by tetrahydrofuran. The solid curves are the SMC results [95], the dashed curves Kohn variational results of Trevisan and coworkers [96], the circles measurements by Colyer and coworkers [99], the squares measurements by Dampc and coworkers [101], and the chained curve measurements by Allan [100]

The apparent absence of long-lived, low-energy resonances in THF appears to argue that direct electron attachment to the sugar moiety of the backbone is not significant in promoting DNA damage. A counter-argument has been made that, because sugars including ribose [43, 102], fructose [42], and deoxyribose itself [40] are in fact extensively dissociated by slow electrons, THF "cannot be viewed as a model for the deoxyribose ring in DNA" [42]. Just the opposite conclusion was reached, though, from a comparative study of absolute dissociative-attachment cross sections: the presence of the hydroxyl group was found to enhance dissociative attachment in 3HTHF relative to THF by a factor of 30, leading the authors to argue that "a compound such as deoxyribose (2-deoxyribose), which contains 3 OH groups, . . . is not a suitable model for the response of the sugar ring in DNA, which does not possess those groups" [31]. Moreover, the pyranose rather than the furanose forms of ribose and deoxyribose predominate in the gas phase [103], and the same is true of fructose [104, 105]. To address these concerns, Illenberger and coworkers looked at dissociative attachment to 1,2,3,5-tetraacetylfuranose [44], intending simultaneously to avoid hydroxyl groups, to preserve the furanose structure in the gas phase, and to provide, via the acetyl groups, reasonable models of both the phosphate and the nucleobase. As in sugars, they observed extensive dissociation even at electron energies near 0 eV, although neither the products (apart from acetate anion) nor the mechanisms were clear. At any rate, the overall evidence demonstrates that sensitivity to electron-induced damage is controlled by the substituents attached to the furanose ring—in this last case the acetyls, in 3HTHF and sugars the hydroxyls, and in DNA the nucleobases and/or phosphates—rather than by the ring itself.

Before moving on to consider the possible role of the phosphate group, we briefly mention results from our SMC studies of some larger model compounds related to the sugar moiety. In collaboration with the experimental groups of Stephen Buckman and Michael Brunger, we studied elastic scattering by 3HTHF [106]. Overall, its electron scattering behavior proved very similar to that of THF, with two broad resonant maxima in the integral cross section at about 8.0 and 12.3 eV, and agreement between experiment and calculation is good. However, a puzzling and as yet unexplained result is that the measured differential elastic scattering cross sections of 3HTHF are generally slightly smaller than those of THF, while the calculated cross sections are generally slightly larger. Although one naively expects the elastic cross section to increase with the size and electron count of the target molecule, independent measurements of the THF [107] and 3HTHF [108] elastic cross section at 40 eV and above also indicate that scattering by 3HTHF is slightly weaker.

While experimental studies of gas-phase deoxyribose are, as noted above, complicated by isomerization to the pyranose form, in computational work one is free to choose the biologically relevant isomer. Thus our SMC calculations for deoxyribose and the larger backbone fragment 2′-deoxyribose-5′-monophosphate [95] were carried out on the furanoses. As with the nucleotide and nucleoside calculations discussed in Sects. 5.3.1.1 and 5.3.1.2, only static–exchange calculations were carried out for these larger systems. However, by comparing to both the static–exchange and the full results for THF, we can estimate what the actual resonance positions are, provided the cross sections are too dissimilar. In fact, as shown in Fig. 5.7, the qualitative behavior of the integral elastic cross section changes only slightly as we pass from THF to deoxyribose, though the magnitude increases considerably.

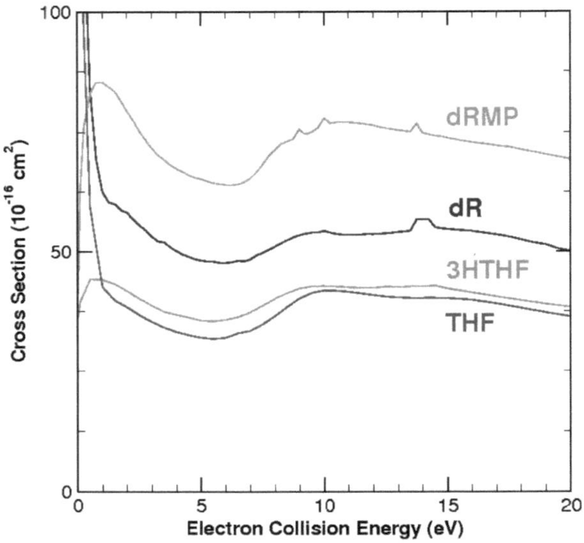

Fig. 5.7 Integral cross sections computed in the static–exchange approximation for low-energy electron scattering by, from bottom to top, tetrahydrofuran [95], 3-hydroxytetrahydrofuran [106], 2′-deoxyribose [95], and 2′-deoxyribose-5′-monophosphate [95]

Adding a phosphate group causes the two broad maxima in the cross section to merge into a single peak, possibly reflecting contributions from resonances localized on the phosphate group (see Sect. 5.3.2.2 immediately below). As in THF and 3HTHF, there is no indication of long-lived, low-energy resonances or of any resonances at all in the very-low-energy region (the narrow spikes visible in Fig. 5.7 are numerical artifacts).

An absence of shape resonances below ~5 eV that might promote dissociative attachment is, as mentioned earlier, consistent with the observation of only weak dissociative attachment in THF [31, 42], but it is rather surprising for 3HTHF and deoxyribose because of the strong low-energy dissociative-attachment yields in those molecules [31, 40]. Unfortunately, additional scattering data against which we can test our results for these molecules are limited. For 3HTHF, Możejko and Sanche have calculated elastic cross sections [92], but the independent-atom model they use is only applicable at higher impact energies, while the combined experimental–theoretical study of Milosavljević and coworkers [108] is likewise limited to higher energies. For the sugars, Gianturco and coworkers have performed calculations for low-energy electron collisions with both deoxyribose and ribose. They see no narrow low-energy shape resonances for either the pyranose or the furanose structure of ribose [102, 109, 110], though in one paper they do report a broad resonance at 4.1 eV for the furanose form [110]. For deoxyribose the situation is less clear. One paper [111] reports that "a complete analysis of all the shape resonances" below 20 eV "found shape resonances only above 7 eV" in either the pyranose or furanose; a nearly simultaneous paper [110] reports a fairly sharp (1.00 eV width) resonance at 3.13 eV in the furanose; while a third paper [109] reports both results: "the lowest resonance of the furanosic conformer" is "located at 6.41 eV in ribofuranose case and at 7.13 eV for its deoxygenated analogue," and, further down the same page, "Both compounds show marked resonant characters in their ICS [integral cross sections] between 5 and 20 eV in the ribose case and between 3 and 20 eV in the deoxyribose case." At any rate, it seems fair to say that there is no unambiguous evidence for narrow shape resonances in elastic electron scattering by furanose-type compounds, including deoxyribose itself.

How, then, might one account for the observation of strong low-energy dissociative attachment to 3HTHF [31], even if one discounts similar observations for ribose [43, 102] and deoxyribose [40] on the grounds that they do not reflect the furanose isomer? One possibility is that the slow-electron attachment dynamics in these systems are intrinsically non-Born–Oppenheimer [102]. Another possibility is that vibrational Feshbach resonances built on dipole-bound states, such as are thought to promote dissociative attachment to the pyrimidine nucleobasess [16], are involved [105]. Yet a third possibility is suggested by a recent study of electron interactions with a series of alcohols—from methanol through various THF and cyclopentane diols—by Allan and coworkers [112]. Each alcohol exhibits a low-energy (2 to 3 eV) dissociative attachment process involving loss of H. Studies of deuterated isotopomers of the smaller alcohols indicate that specifically the hydroxyl hydrogen is lost, while the strength of the isotope effect suggests that the resonance involved is short-lived. Together with electron-impact spectra for ethanol showing a sharp onset at threshold for excitation of the O–H stretching vibration, these data suggest that the low-energy dissociative attachment process involves an O–H σ^* resonance. Also supporting this picture is a still more recent analysis of low-energy dissociative electron attachment to formic acid [113, 114], which shows that the observed spectrum can be explained by a numerical model that has the electron trapped directly by an O–H σ^* resonance, rather than by the C–O π^* resonance that is perhaps the more obvious candidate. In this model, the σ^* resonance is indeed broad and high-lying at the vertical geometry, but it drops steeply and narrows rapidly along the dissociation coordinate. As its proponents point out, the model appears to be consistent with the spectrum for electron-impact vibrational excitation of formic acid [115], which shows resonant excitation of the O–H stretch with a profile that does not match that of the π^* resonance. This picture explains how low-energy dissociative attachment can occur in 3HTHF and deoxyribose despite the absence of shape resonances below about 7 eV in the SMC and other elastic cross sections. Even if this picture is correct, of course, it is not certain that $C_{3,5}$–O and/or P–O phosphoesteric σ^* orbitals in the DNA backbone will behave in the same way as OH σ^* orbitals, but such a direct attachment mechanism appears to deserve serious consideration and further study.

5.3.2.2 The Phosphate Group

Structural diagrams conventionally depict the phosphorus atom of the phosphate group, PO_4, as being singly bonded to three oxygens and doubly bonded to the fourth, thereby suggesting the existence not only of a P–O π orbital making up part of the double bond but also of an empty P–O π^* antibonding valence orbital conjugate to it. This π^* orbital has been mentioned as a potential site of electron attachment to the backbone [49, 116] and has been invoked to explain observations of low-energy dissociative attachment to the dibutyl and triethyl esters of phosphoric acid [46]. On the other hand, as discussed by Burrow and coworkers [117], computational studies question the conventional picture of the P–O bond and thus the existence of such a π^* orbital; and indeed, their electron-transmission spectra for several phosphate-containing compounds do not show evidence of it.

To explore the possible role of the phosphate group in attaching low-energy electrons, we studied the simplest model compound, phosphoric acid (H_3PO_4), and its three methyl esters, mono-, di-, and trimethylphosphate [118]. Although two of the larger DNA fragments discussed above (Sects. 5.3.1.2 and 5.3.2.1) included a phosphate group, letting methyl stand in for deoxyribose avoids obscuring details specific to the phosphate and, more importantly, allows us to carry out more accurate calculations that include polarization effects for H_3PO_4 and monomethylphosphate, against which static–exchange results for the di- and trimethyl esters can be calibrated.

Our integral elastic cross sections for all four molecules are qualitatively similar. While broad shape resonances are visible at higher energy, the calculations including polarization indicate that the lowest resonance lies at about 7 eV, much too high to be a π^* shape resonance. Other scattering calculations on phosphoric acid reach similar conclusions: that of Tonzani and Greene [98] places resonances at 7.7 and 12.5 eV, while the recent study of Bryjko and coworkers [119] indicates a broad shape resonance at about 7 eV as well as nearby Feshbach resonances, but no lower-energy resonances. Thus none of the cross-section calculations gives evidence of a low-energy π^* resonance in H_3PO_4. For trimethylphosphate, Simons and coworkers [120] reached essentially the same conclusion using bound-state methods. They found that, although a clear P–O π^* orbital exists in a trivalent phosphorus compound, CH_3–O–PO, there is no identifiable P–O π^* orbital in the pentavalent trimethyl ester. Likewise, absolute measurements of the dissociative attachment cross section of trimethylphosphate indicate it to be very small [31].

Still, the situation at low energy is not completely clear. Although the scattering calculations [98, 118, 119] and our own estimates based on minimal-basis-set orbital energies [118] concur that there should be no shape resonances below about 7 eV in phosphoric acid or its alkyl esters, the electron transmission spectrum of trimethylphosphate does show two resonances, albeit weak ones, at 2.1 and 4.6 eV, and a procedure of empirically scaling virtual orbital energies also suggests shape resonances in the 2–5 eV range [117]. Moreover, while dissociative attachment to dibutyl phosphate yields a number of peaks in the 0–3 eV range, in various ion channels, dissociative attachment to triethylphosphate appears to be much weaker [46], suggesting that, as was discussed for sugars in Sect. 5.3.2.1, the hydroxyl group may play a significant role in promoting very-low-energy dissociative attachment. Electron-stimulated desportion measurements [45] on the monosodium salt of phosphoric acid, $NaPO_2(OH)_2$, found peaks in the ion yield only between 7 and 9 eV; however, only light ions (H^-, O^-, OH^-) were detected in that work, while the anions seen at low energy in the gas-phase measurements on dibutylphosphate were heavier species such as PO_3^- and $PO_4(C_4H_9)_2^-$. Likewise, measurements on short strands of physisorbed DNA showed desorption of OH^- that, it was argued, resulted from dissociative attachment to the phosphate [7], but only at higher energies (peaks at 5.5 and 6.7 eV). Much more work will be needed to elucidate the importance of, and the mechanisms involved in, possible DNA damage pathways associated with the phosphate group.

5.4 Summary and Discussion

To summarize our main results, SMC electron scattering calculations on the nucleobases support the original assignments of the features seen in the electron transmission spectrum as π^* shape resonances [16, 65]. This has consequences for our understanding of the dissociative attachment mechanisms: the prominent feature at 1.0 eV in attachment to the pyrimidines cannot be a π^* resonance, supporting its interpretation as a vibrational Feshbach resonance [16]. On the other hand, the broader feature seen at about 1.7 eV does appear to be due to a π^* resonance, in particular π_2^*, and may be more important in the condensed phase, and thus in DNA itself [87], than the peak at 1.0 eV. The third π^* resonance in both the purines and pyrimidines is likely not a pure elastic-channel shape resonance but instead contains an admixture of core-excited character, raising the possibility that resonance-enhanced decay into one or more low-lying triplet states forms an alternative pathway to low-energy electron-driven dissociation [85]; this possibility deserves further study. Scattering calculations on deoxyribose and various model compounds indicate that there are only broad shape resonances at energies of about 7 eV or higher, and thus little likelihood of direct attachment to the furanose ring of the DNA backbone. This result is supported by experiments showing only weak dissociative attachment to tetrahydrofuran [31, 42]. However, experimental results for hydroxyl-containing molecules such as 3-hydroxytetrahydrofuran [31] provide a caution: scattering calculations appear to show no shape resonances below 5 eV there either, yet low-energy dissociative attachment does occur. Experiments with tetraacetylfuranose [44] confirm that other substituent groups besides OH can also promote low-energy dissociative attachment. These results suggest two conclusions. First, bond-breaking between the furanose ring and attached groups is driven by attachment to those substituents, rather than to the ring itself; thus in DNA and RNA, the likely sites of attachment are the nucleobases and/or phosphates. In particular, the model of indirect strand breaking by electrons that originally attach to nucleobase π^* orbitals [50, 62] continues to have appeal; the importance of the alternative mechanism, attachment to the phosphate, is, as just discussed in Sect. 5.3.2.2, not yet clear. Second, future scattering calculations will have to account for nuclear dynamics, and in particular for the possibility that some σ^* resonances that are broad and high-lying at the equilibrium geometry undergo rapid changes in position and width as the nuclei move and thereby become important to low-energy dissociation [113].

Besides considering nuclear dynamics, future electron-scattering calculations will also need to consider model systems that more closely approximate conditions in the condensed phase. Base pairing and base stacking may well affect the energies and characters of the nucleobase π^* resonances, and solvation effects are likely to be important for both the bases and the backbone. Moreover, the ordered arrangement of the subunits within DNA raises the possibility of collective or diffractive multiple-scattering effects. Suggestive work on this possibility has already been done [121–123], although with somewhat approximate representations of the individual scattering events. Given the capabilities of present computers and codes, high-level calculations on nucleosides, base pairs, and perhaps nucleotides appear well within reach. A "stretch goal" for the intermediate term would be a similarly high-level calculation on paired oligonucleotides, preferably with some representation of solvation effects; such a calculation would require both careful design to obtain maximum accuracy at minimal computational cost and, most likely, exploitation of advanced hardware such as graphical-processing units. In the meantime, there is much that can be learned from further electron-scattering studies of the isolated base, sugar, and phosphate moieties, both through coupling the electron collision dynamics to the nuclear dynamics, as described in the preceding paragraph, and in exploring electron-impact excitation of the bases, for which very little work has yet been done [71, 75].

Acknowledgements This work was supported by the Chemical Sciences, Geosciences, and Biosciences Division, Office of Basic Energy Sciences, Office of Science, US Department of Energy. Use of the Jet Propulsion Laboratory's Supercomputing and Visualization Facility is gratefully acknowledged. We thank our experimental and theoretical collaborators, in particular Stephen Buckman (Australian National University), Michael Brunger (Flinders University), and Sergio Sanchez (Federal University of Paraná) for their valuable contributions to the work reported here.

References

1. B. C. Garrett et al., Chem. Rev. **105**, 355 (2005)
2. L. Sanche, Eur. Phys. J. D **35**, 367 (2005)
3. B. Boudaïffa, P. Cloutier, D. Hunting, M. A. Huels, L. Sanche, Science **287**, 1658 (2000)
4. B. Boudaïffa, P. Cloutier, D. Hunting, M. A. Huels, L. Sanche, Radiat. Res. **157**, 227 (2002)
5. M. A. Huels, B. Boudaïffa, P. Cloutier, D. Hunting, L. Sanche, J. Am. Chem. Soc. **125**, 4467 (2003)
6. Y. Zheng, P. Cloutier, D. J. Hunting, L. Sanche, J. R. Wagner, J. Am. Chem. Soc. **127**, 16592 (2005)
7. X. Pan, L. Sanche, Phys. Rev. Lett. **94**, 198104 (2005)
8. M.-A. Hervé du Penhoat, M. A. Huels, P. Cloutier, J.-P. Jay-Gerin, L. Sanche, J. Chem. Phys. **114**, 5755 (2001)
9. M. A. Huels, I. Hahndorf, E. Illenberger, L. Sanche, J. Chem. Phys. **108**, 1309 (1998)
10. H. Abdoul-Carime, M. A. Huels, E. Illenberger, L. Sanche, J. Am. Chem. Soc. **123**, 5354 (2001)
11. H. Abdoul-Carime, P. Cloutier, L. Sanche, Radiat. Res. **155**, 625 (2001)

12. R. Abouaf, J. Pommier, H. Dunet, Int. J. Mass Spectrom. **226**, 397 (2003)
13. S. Denifl, S. Ptasinska, M. Cingel, S. Matejcik, P. Scheier, T. D. Märk, Chem. Phys. Lett. **377**, 74 (2003)
14. S. Gohlke, H. Abdoul-Carime, E. Illenberger, Chem. Phys. Lett. **380**, 595 (2003)
15. G. Hanel, B. Gstir, S. Denifl, P. Scheier, M. Probst, B. Farizon, M. Farizon, E. Illenberger, T. D. Märk, Phys. Rev. Lett. **90**, 188104 (2003)
16. A. M. Scheer, K. Aflatooni, G. A. Gallup, P. D. Burrow, Phys. Rev. Lett. **92**, 068102 (2004)
17. H. Abdoul-Carime, S. Gohlke, E. Illenberger, Phys. Rev. Lett. **92**, 168103 (2004)
18. S. Denifl, S. Ptasińska, M. Probst, J. Hrušak, P. Scheier, T. D. Märk, J. Phys. Chem. A **108**, 6562 (2004)
19. S. Denifl, S. Matejcik, S. Ptasinksa, B. Gstir, M. Probst, P. Scheier, E. Illenberger, T. D. Mark, J. Chem. Phys. **120**, 704 (2004)
20. S. Denifl, S. Ptasińska, G. Hanel, B. Gstir, M. Probst, P. Scheier, T. D. Märk, J. Chem. Phys. **120**, 6557 (2004)
21. S. Feil, K. Gluch, S. Matt-Leubner, P. Scheier, J. Limtrakul, M. Probst, H. Deutsch, K. Becker, A. Stamatovic, T. D. Märk, J. Phys. B **37**, 3013 (2004)
22. H. Abdoul-Carime, J. Langer, M. A. Huels, E. Illenberger, Eur. Phys. J. D **35**, 399 (2005)
23. R. Abouaf, H. Dunet, Eur. Phys. J. D **35**, 405 (2005)
24. K. Aflatooni, A. M. Scheer, P. D. Burrow, Chem. Phys. Lett. **408**, 426 (2005)
25. A. M. Scheer, C. Silvernail, J. A. Belot, K. Aflatooni, G. A. Gallup, P. D. Burrow, Chem. Phys. Lett. **411**, 46 (2005)
26. S. Ptasińska, S. Denifl, V. Grill, T. D. Märk, P. Scheier, S. Gohlke, M. A. Huels, E. Illenberger, Angew. Chem. Int. Ed. **44**, 1647 (2005)
27. S. Ptasińska, S. Denifl, B. Mróz, M. Probst, V. Grill, E. Illenberger, P. Scheier, T. D. Märk, J. Chem. Phys. **123**, 124302 (2005)
28. S. Ptasinska, S. Denifl, P. Scheier, E. Illenberger, T. D. Märk, Angew. Chem. Int. Ed. **44**, 6941 (2005)
29. S. Ptasińska, S. Denifl, V. Grill, T. D. Märk, E. Illenberger, P. Scheier, Phys. Rev. Lett. **95**, 093201 (2005)
30. P. D. Burrow, G. A. Gallup, A. M. Scheer, S. Denifl, S. Ptasinska, T. Märk, P. Scheier, J. Chem. Phys. **124**, 124310 (2006)
31. K. Aflatooni, A. M. Scheer, P. D. Burrow, J. Chem. Phys. **125**, 054301 (2006)
32. D. Huber, M. Beikircher, S. Denifl, F. Zappa, S. Matejcik, A. Bacher, V. Grill, T. D. Märk, P. Scheier, J. Chem. Phys. **125**, 084304 (2006)
33. I. I. Shafranyosh, M. I. Sukhoviya, M. I. Shafranyosh, J. Phys. B **39**, 4155 (2006)
34. S. Denifl, F. Zappa, I. Mähr, J. Lecointre, M. Probst, T. D. Märk, P. Scheier, Phys. Rev. Lett. **97**, 043201 (2006)
35. S. Denifl, F. Zappa, A. Mauracher, F. F. da Silva, A. Bacher, O. Echt, T. D. Märk, D. K. Bohme, P. Scheier, ChemPhysChem **10**, 1387 (2008)
36. S. Denifl, P. Sulzer, F. Zappa, S. Moser, B. Kräutler, O. Echt, D. K. Bohme, T. D. Märk, P. Scheier, Int. J. Mass Spectrom. **277**, 296 (2008)
37. J. Kopyra, C. Koenig-Lehmann, E. Illenberger, Int. J. Mass Spectrom. **281**, 89 (2009)
38. D. Antic, L. Parenteau, M. Lepage, L. Sanche, J. Phys. Chem. B **103**, 6611 (1999)
39. D. Antic, L. Parenteau, L. Sanche, J. Phys. Chem. B **104**, 4711 (2000)
40. S. Ptasińska, S. Denifl, P. Scheier, T. D. Märk, J. Chem. Phys. **120**, 8505 (2004)
41. S.-P. Breton, M. Michaud, C. Jäggle, P. Swiderek, L. Sanche, J. Chem. Phys. **121**, 11240 (2004)
42. P. Sulzer, S. Ptasinska, F. Zappa, B. Mielewska, A. R. Milosavljevic, P. Scheier, T. D. Märk, I. Bald, S. Gohlke, M. A. Huels, E. Illenberger, J. Chem. Phys. **125**, 044304 (2006)
43. I. Bald, J. Kopyra, E. Illenberger, Angew. Chem. Int. Ed. **45**, 4851 (2006)
44. I. Bald, J. Kopyra, I. Dabkowska, E. Antonsson, E. Illenberger, J. Chem. Phys. **126**, 074308 (2007)
45. X. Pan, L. Sanche, Chem. Phys. Lett. **421**, 404 (2006)
46. C. König, J. Kopyra, I. Bald, E. Illenberger, Phys. Rev. Lett. **97**, 018105 (2006)

47. H. Abdoul-Carime, S. Gohlke, E. Fischbach, J. Scheike, E. Illenberger, Chem. Phys. Lett. **387**, 267 (2004)
48. S. Ptasińska, S. Denifl, S. Gohlke, P. Scheier, E. Illenberger, T. D. Märk, Angew. Chem. Int. Ed. **45**, 1893 (2006)
49. X. Li, M. D. Sevilla, L. Sanche, J. Am. Chem. Soc. **125**, 13668 (2003)
50. J. Berdys, I. Anusiewicz, P. Skurski, J. Simons, J. Phys. Chem. A **108**, 2999 (2004)
51. J. Simons, Acc. Chem. Res. **39**, 772 (2006)
52. X. Li, M. D. Sevilla, Adv. Quantum Chem. **52**, 59 (2007)
53. R. H. D. Lyngdoh, H. F. Schaefer III, Acc. Chem. Res. **42**, 563 (2009)
54. K. Takatsuka, V. McKoy, Phys. Rev. A **24**, 2473 (1981)
55. K. Takatsuka, V. McKoy, Phys. Rev. A **30**, 1734 (1984)
56. J. Schwinger, Phys. Rev. **72**, 742 (1947)
57. N. Ostlund, Chem. Phys. Lett. **34**, 419 (1975)
58. D. K. Watson, V. McKoy, Phys. Rev. A **20**, 1474 (1979)
59. P. G. Hipes, C. Winstead, V. McKoy, in *Proceedings of the Fifth Distributed Memory Computing Conference,* ed. by D. W. Walker, Q. F. Stout (IEEE Computer Society, Los Alamitos, 1990), p. 498
60. C. Winstead, V. McKoy, Adv. At., Mol., Opt. Phys. **36**, 183 (1996)
61. C. Winstead, V. McKoy, Comput. Phys. Commun. **128**, 386 (2000)
62. R. Barrios, P. Skurski, J. Simons, J. Phys. Chem. B **106**, 7991 (2002)
63. J. Berdys, P. Skurski, J. Simons, J. Phys. Chem. B **108**, 5800 (2004)
64. I. Anusiewicz, J. Berdys, M. Sobczyk, P. Skurski, J. Simons, J. Phys. Chem. A **108**, 11381 (2004)
65. K. Aflatooni, G. A. Gallup, P. D. Burrow, J. Phys. Chem. A **102**, 6205 (1998)
66. F. A. Gianturco, R. R. Lucchese, J. Chem. Phys. **120**, 7446 (2004)
67. S. Tonzani, C. H. Greene, J. Chem. Phys. **124**, 054312 (2006)
68. S. Tonzani, C. H. Greene, J. Chem. Phys. **122**, 014111 (2005)
69. P. D. Burrow, J. Chem. Phys. **122**, 087105 (2005)
70. G. A. Gallup I. I. Fabrikant, Phys. Rev. A **83**, 012706 (2011)
71. C. Winstead, V. McKoy, J. Chem. Phys. **125**, 174304 (2006)
72. F. A. Gianturco, F. Sebastianelli, R. R. Lucchese, I. Baccarelli, N. Sanna, J. Chem. Phys. **128**, 174302 (2008)
73. C. Winstead, V. McKoy, J. Chem. Phys. **129**, 077101 (2008)
74. F. A. Gianturco, F. Sebastianelli, R. R. Lucchese, I. Baccarelli, N. Sanna, J. Chem. Phys. **131**, 249901 (2009)
75. A. Dora, J. Tennyson, L. Bryjko, T. van Mourik, J. Chem. Phys. **130**, 164307 (2009)
76. C. Winstead, V. McKoy, S. d'A. Sanchez, J. Chem. Phys. **127**, 085105 (2007)
77. B. H. Lengsfield III, T. N. Rescigno, C. W. McCurdy, Phys. Rev. A **44**, 4296 (1991)
78. T. N. Rescigno, D. A. Byrum, W. A. Isaacs, C. W. McCurdy, Phys. Rev. A **60**, 2186 (1999)
79. C. Winstead, V. McKoy, Phys. Rev. A **57**, 3589 (1998)
80. R. Abouaf, J. Pommier, H. Dunet, P. Quan, P.-C. Nam, M. T. Nguyen, J. Chem. Phys. **121**, 11668 (2004)
81. R. Abouaf, J. Pommier, H. Dunet, Chem. Phys. Lett. **381**, 486 (2003)
82. C. M. Marian, F. Schneider, M. Kleinschmidt, J. Tatchen, Eur. Phys. J. D **20**, 357 (2002)
83. I. Nenner, G. J. Schulz, J. Chem. Phys. **62**, 1747 (1975)
84. M. Allan, Helvet. Chim. Acta **65**, 2008 (1982)
85. C. Winstead, V. McKoy, Phys. Rev. Lett. **98**, 113201 (2007)
86. C. Winstead, V. McKoy, Phys. Rev. A **76**, 012712 (2007)
87. F. Martin, P. D. Burrow, Z. Cai, P. Cloutier, D. Hunting, L. Sanche, Phys. Rev. Lett. **93**, 068101 (2004)
88. C. Winstead, V. McKoy, J. Chem. Phys. **125**, 244302 (2006)
89. W. Chin, M. Mons, I. Dimicoli, F. Piuzzi, B. Tardivel, M. Elhaine, Eur. Phys. J. D **20**, 347 (2002), and references therein

90. Chr. Plützer, E. Nir, M. S. de Vries, K. Kleinermanns, Phys. Chem. Chem. Phys. **3**, 5466 (2001)
91. A. R. Milosavljević, F. Blanco, D. Šević, G. García, B. P. Marinković, Eur. Phys. J. D **40**, 107 (2006)
92. P. Możejko, L. Sanche, Radiat. Phys. Chem. **73**, 77 (2005)
93. P. Możejko, A. Domaracka, E. Ptasińska-Denga, Cz. Szmytkowski, Chem. Phys. Lett. **429**, 378 (2006)
94. V. M. Rayón, J. A. Sordo, J. Chem. Phys. **122**, 204303 (2005)
95. C. Winstead, V. McKoy, J. Chem. Phys. **125**, 074302 (2006)
96. C. N. Trevisan, A. E. Orel, T. N. Rescigno, J. Phys. B **39**, L255 (2006)
97. D. Bouchiha, J. D. Gorfinkiel, L. G. Caron, L. Sanche, J. Phys. B **39**, 975 (2006)
98. S. Tonzani, C. H. Greene, J. Chem. Phys. **125**, 094504 (2006)
99. C. J. Colyer, V. Vizcaino, J. P. Sullivan, M. J. Brunger, S. J. Buckman, New J. Phys. **9**, 41 (2007)
100. M. Allan, J. Phys. B **40**, 3531 (2007)
101. M. Dampc, A. Milosavljević, I. Linert, B. P. Marinković, M. Zubek, Phys. Rev. A **75**, 042710 (2007)
102. I. Baccarelli, F. A. Gianturco, A. Grandi, N. Sanna, R. R. Lucchese, I. Bald, J. Kopyra, E. Illenberger, J. Am. Chem. Soc. **129**, 6269 (2007)
103. L. P. Guler, Y.-Q. Yu, H. I. Kenttämaa, J. Phys. Chem. A **106**, 6754 (2002)
104. M. Jetzki, R. Signorell, J. Chem. Phys. **117**, 8063 (2002)
105. T. Sommerfeld, J. Chem. Phys. **126**, 124301 (2007)
106. V. Vizcaino, J. Roberts, J. P. Sullivan, M. J. Brunger, S. J. Buckman, C. Winstead, V. McKoy, New J. Phys. **10**, 053002 (2008)
107. A. R. Milosavljević, A. Giuliani, D. Šević, M.-J. Hubin-Franskin, B. P. Marinković, Eur. Phys. J. D **35**, 411 (2005)
108. A. R. Milosavljević, F. Blanco, J. B. Maljković, D. Šević, G. García, B. P. Marinković, New J. Phys. **10**, 103005 (2008)
109. I. Baccarelli, N. Sanna, F. A. Gianturco, F. Sebastianelli, J. Phys. Conf. Ser. **115**, 012009 (2008)
110. I. Baccarelli, F. Sebastianelli, F. A. Gianturco, N. Sanna, Eur. Phys. J. D **51**, 131 (2009)
111. I. Baccarelli, F. A. Gianturco, A. Grandi, N. Sanna, Int. J. Quantum Chem. **108**, 1878 (2008)
112. B. C. Ibănescu, O. May, A. Monney, M. Allan, Phys. Chem. Chem. Phys. **9**, 3163 (2007)
113. G. A. Gallup, P. D. Burrow, I. I. Fabrikant, Phys. Rev. A **79**, 042701 (2009)
114. G. A. Gallup, P. D. Burrow, I. I. Fabrikant, Phys. Rev. A **80**, 046702 (2009)
115. M. Allan, J. Phys. B **39**, 2939 (2006)
116. J. Berdys, I. Anusiewicz, P. Skurski, J. Simons, J. Am. Chem. Soc. **126**, 6441 (2004)
117. P. D. Burrow, G. A. Gallup, A. Modelli, J. Phys. Chem. A **112**, 4106 (2008)
118. C. Winstead, V. McKoy, Int. J. Mass Spectrom. **277**, 279 (2008)
119. L. Bryjko, T. van Mourik, A. Dora, J. Tennyson, J. Phys. B **43**, 235203 (2010)
120. B. Gamoke, D. Neff, J. Simons, J. Phys. Chem. A **113**, 5677 (2009)
121. T. M. Orlando, D. Oh, Y. Chen, A. B. Aleksandrov, J. Chem. Phys. **128**, 195102 (2008)
122. L. Caron, L. Sanche, S. Tonzani, C. H. Greene, Phys. Rev. A **78**, 042710 (2008)
123. L. Caron, L. Sanche, S. Tonzani, C. H. Greene, Phys. Rev. A **80**, 012705 (2009)

Chapter 6
Resonances in Electron Collisions with Small Biomolecules Using the R-Matrix Method

Lilianna Bryjko, Amar Dora, Tanja van Mourik, and Jonathan Tennyson

Abstract It is now widely accepted that collisions with low-energy electrons are the major cause of radiation damage in living cells and that it is the capture of these electrons into long-lived quasi-bound states, resonances, that is responsible for this damage. We have undertaken a set of systematic calculations using the UK Molecular R-matrix codes to study electron collisions with DNA and RNA bases. Here we summarise the results of our calculations for electron collisions with adenine, guanine, uracil, cytosine and thymine. These studies aim to characterize not only low-lying shape resonances, which have been relatively well-studied, but also to detect longer-lived Feshbach resonances which are associated with simultaneous electronic excitation of the target molecule. The results of these calculations are dependent on the model chosen: only the more sophisticated, and computationally expensive, models give Feshbach resonances.

6.1 Introduction

The processes that follow the creation of thousands of low-energy electrons are now recognised as being extremely important [8]. These electrons are stripped off from molecules in the cell either directly by radiation or else by its first products, highly energetic primary electrons that can cause electron-impact ionisation. As a result, in the past few years, a growing literature has emerged concerning the damage to

L. Bryjko • T. van Mourik
EaStCHEM School of Chemistry, University of St Andrews, North Haugh, St. Andrews, Fife KY16 9ST, UK
e-mail: lilabe@o2.pl; tanja.vanmourik@st-andrews.ac.uk

A. Dora • J. Tennyson (✉)
Department of Physics and Astronomy, University College London, Gower St., London WC1E 6BT, UK
e-mail: amardora@gmail.com; j.tennyson@ucl.ac.uk

G.G. Gómez-Tejedor and M.C. Fuss (eds.), *Radiation Damage in Biomolecular Systems*, Biological and Medical Physics, Biomedical Engineering,
DOI 10.1007/978-94-007-2564-5_6,

nucleic acids by low-energy electrons (LEE) with energies between 0 and 20 eV [1,4–8,12,18,20,30,33–36,38,39,57] produced by ionising radiation.

The mechanism of DNA-LEE interaction is important because the low-energy secondary electrons are the most abundant radiolysis species generated following the impact of high-energy radiation [12, 31, 32] and therefore highly pertinent to issues such as radiation damage and the development of radio therapy. Nucleic acids can be ionised and damage produced through the dissociation of the anion when the electron energy is higher than the ionisation threshold for DNA (between 7.85 and 9.4 eV, as measured for the DNA bases [22]). If the electron energy is lower, damage can still be generated, but through a negative anion-mediated mechanism, which starts with the capture of the electron in a molecular resonance, followed by the transfer of energy and electron density towards a weak bond that subsequently ruptures.

There are many controversial issues that concern the location of the initial capture site [8, 40], the dynamics of the metastable anion generated by electron capture (called transient molecular anion), and the identification of the final bond that ruptures [5–7, 23, 27–29, 41, 57].

There is a wide agreement that the electron capture is mainly due to the DNA and RNA bases, because these molecules have extended aromatic systems. The scattering electron can temporarily be captured by an unoccupied π^* orbital giving rise to a shape resonance [1, 5, 25]. When scattering is connected with electron excitation, Feshbach or core-excited resonances can occur. It has been suggested that electron attachment to the phosphate group also contributes to DNA strands breaks [5, 6, 28, 37].

Simons and co-workers [5–7] performed model calculations which showed that electrons with energies of about 1.0 eV can attach to a base to form a π^* anion, which then can break a σ bond connecting the phosphate to a sugar group. Li et al. [21] performed calculations on a sugar-phosphate-sugar model system using the ONIOM layer method and found that the activation barrier for bond rupture of the anion's phosphate-sugar C–O bond is only 0.5 eV, indicating that very-low-energy electrons can induce DNA strand breaks. Berdys at al. [6, 7] found that near zero energy, electrons may not easily attach directly (i.e. vertically) to the phosphate units, but can produce a metastable $P=O$ π^* anion above 2 eV.

Resonances were observed by Pan et al. [27, 28] in linear and super-coiled DNA. They also observed desorption of H^- as the result of temporary capture of electrons by the bases, with a small contribution from a core-excited resonance on the sugar group, OH^- desorption by the localisation of electrons on the protonated form of the phosphate group, and production of O^- via the temporary localisation of electrons on the π^* double bond of the phosphate group. Pan and Sanche [29] measured dissociative-electron attachment (DEA) to the monosodium salt of phosphoric acid, Na_2PO_4, in the condensed phase, confirming DNA damage can be induced by low-energy electrons. A single broad peak whose maximum fell at 8.8, 8.0 or 7.3 eV, depending on whether the anion detected was H^-, O^-, or OH^-, was observed. König et al. [19] measured DEA spectra for the dibuthyl and triethyl phosphate ester, and observed a variety of anionic fragments. Using dibuthyl phosphate they found that the compound undergoes effective DEA within a low-energy resonant feature at 1 eV and a further resonance peaking at 8 eV. The DEA reactions are associated with the direct cleavage of the C-O and the P-O bonds but also the excision of the PO^-, PO_3^- and $H_2PO_3^-$ units. They propose that the most direct mechanism of single strand breaks occurring in DNA is due to DEA directly to the phosphate group.

Low-energy electron collisions with molecules can result in a variety of different processes [46]. A common feature of all these processes is that they can be considered to go via an intermediary, AB^-. One of the methods used to consider low-energy electron-molecule collisions is the R-matrix method [10, 11, 45], which is built around obtaining accurate wave functions for this intermediary and hence gives a theoretical framework capable of modelling all the above processes.

In this chapter we describe the application of R-matrix calculations to DNA and RNA constituents. A particular advantage of this method is its ability to treat not only shape resonances, which have been widely studied by a variety of methods, but also Feshbach resonances. Feshbach resonances have been found to be important experimentally [2, 3] but have received much less attention theoretically.

The size of these molecules and, in particular, the complexity of their electronic wavefunctions makes such calculations very challenging. So far the study of each system has necessitated the consideration of a number of models in order to obtain reliable results. We have completed studies on electron collisions with uracil [13], phosphoric acid [9], adenine, guanine, cytosine and thymine which consider these issues in turn for each molecule. Here, instead, we present a set of results for electron collisions with the bases calculated using a single model, which facilitates intercomparison between these species.

6.2 The R-matrix method

The basis of our calculations is the R-matrix method. The topic of electron-molecule collision calculations has been extensively reviewed by one of us [45] and only a flavor of the method will be given here.

The underlying physical model used in R-matrix calculations is the division of space into an inner region, contained within a sphere of radius a and centered on the target center-of-mass, and an outer region. The inner region is designed to be large enough to contain the entire electron density of the molecular target, including those for excited states in calculations which consider electronic excitation. This can be an issue for the biomolecules considered since they are considerably larger than the small molecules traditionally studied using this methodology. In the inner region the scattering electron and target electrons are treated as being indistinguishable and all interactions, including electron-electron correlation and exchange, are explicitly considered. Conversely, in the outer region the scattering electron is only affected by the long-range potential of the target. As all the molecules considered here have permanent dipole moments, the long-range effect of the resulting dipole potentials need to be treated.

The R-matrix, which relates the scattering wavefunctions to its derivative at a given distance, is constructed on the boundary of the inner and outer region. The R-matrix itself is energy-dependent but can be built from the results of energy-independent inner-region calculations. This has the significant advantage that the energy dependence is entirely obtained from rapid, outer-region calculations. This is particularly useful for calculations aimed at characterizing resonances since such calculations usually require the use of a dense grid of energies.

In this work the R-matrix method is used to obtain the eigenphase sums. To obtain resonance positions and widths, these eigenphase sums were automatically fitted to a Breit-Wigner form by the recursive resonance fitting [47]. The calculations also give elastic electron collision cross sections and, for models which include electronically excited states in the close-coupling expansion, electron impact electronic excitation cross sections.

The UK Molecular R-matrix codes [26] are designed to be very flexible and a number of models have been tested in the course of the work considered here. The simplest of these models is the so-called static exchange (SE) approximation in which a full collision treatment is used for a target which is not allowed to relax during the collision. Polarisation effects can be included using the static exchange plus polarisation (SEP) approximation which allows for single excitations of the target wavefunction. The only resonances that can be detected in an SE calculation are shape resonances where the electron is temporarily trapped behind a centrifugal barrier. The SEP model moves these resonances to lower energy and also can, at least in principle, give Feshbach resonances. Feshbach resonances, which are associated with simultaneous electronic excitation of the target and trapping of the electron, are best given by calculations which explicitly include the parent target state(s) for a given resonance, These are best considered by using several states in a close-coupling (CC) expansion. Our best results presented below were all obtained using CC models.

The use of CC methods for molecules with many valence electrons, such as the ones considered here, leads to very large Hamiltonian matrices. That has led to the development of special methods for treating the problems both in terms of Hamiltonian construction [42] and diagonalisation [16, 44]. In particular the use of the so-called partitioned R-matrix [44] means that it is not necessary to explicitly obtain all the eigenvalues and eigenvectors of the scattering Hamiltonian. The CC results presented below are based on the use of all eigenvalues and vectors (typically 4000 to 6000) for "contracted" calculations but only the 5000 lowest solutions for "uncontracted" calculations which lead to very large Hamiltonian matrices, typically of dimension significantly bigger than 100 000.

The biomolecules considered here have also been the subject of R-matrix calculations by Tonzani and Greene [50]. The R-matrix method developed by these workers [48, 49] has significant differences from the one used by us. Their method involves using R-matrices to solve the electron-molecule scattering problem defined entirely by potentials. Even if one ignores the issue that interactions such as exchange cannot be written as a simple, energy-independent potential, the method is only capable of performing calculations within the SE and SEP models.

6.3 Target calculations

The geometries of all DNA bases except uracil were determined using B3LYP/6-31+G* density functional theory and Gaussian 03 [24]. Except for thymine, the molecules were constrained to be in C_s symmetry. The geometry of uracil was optimised with the MP2 method and using a 6-31G* basis set as reported previously [52].

There are various issues in choosing a suitable target wavefunction for use in scattering calculations, which depend on what aspect of the problem is of interest. For the SE and SEP models the use of Hartree-Fock wavefunctions is standard; for CC calculations there is more choice. In particular these calculations usually attempt to represent several low-lying states of the target. This can be difficult given the constraints that all states must be represented by a single orbital set and that it must be possible to use in a balanced and tractable scattering calculation.

Our favoured method is to represent CC target wavefunctions using a complete active space (CAS) configuration interaction (CI) model. In the CAS-CI model, core electrons are frozen in the self-consistent field (SCF) orbitals and the active electrons are distributed amongst all the valence orbitals, subject only to the constraints of overall space-spin symmetry. This approach has significant advantages in terms of performing a balanced treatment between the target and the scattering wavefunction [43].

Even within the CAS-CI model there is considerable flexibility over the precise choice of molecular orbitals used. Here we use CAS-SCF orbitals generated by MOLPRO. Table 6.1 presents results for the bases considered. The calculations were performed using cc-pVDZ Gaussian Type Orbital (GTO) basis sets and an active space of 14 electrons in 10 orbitals for uracil, cytosine and thymine, 12 electrons in 10 orbitals for adenine and 12 electrons in 9 orbitals for guanine. Full details, including the actual configurations used, are given in Table 6.2. Besides the calculated energies for the ground and 15 lowest excited states, the table gives our calculated permanent dipole moment for each molecule. This property is important since the long-range nature of this moment has a strong effect on the scattering.

6.4 Scattering calculations

In the scattering calculations the target basis set was augmented with sets of GTOs with $\ell \leq 4$ (up to g-wave) to represent the continuum wavefunctions [14]. All calculations were performed for an R-matrix radius $a = 13\ a_0$ except those on thymine which used $a = 15\ a_0$.

Table 6.1 The CASSCF $X\ ^1A'$ (or $X\ ^1A$) ground state energies (in a.u.) and dipole moments (in Debye), and the relative energies (in eV) of the lowest 15 excited states of the nucleic acid bases

Adenine	Guanine	Uracil	Cytosine	Thymine
−464.6212766	−539.476656	−412.563491	−392.691128	−451.606027
4.61 ($^3A'$)	3.93 ($^3A'$)	3.87 ($^3A'$)	3.53 ($^3A'$)	3.86 (^3A)
5.47 ($^3A'$)	5.02 ($^3A'$)	4.75 ($^3A''$)	4.63 ($^3A'$)	4.85 (^3A)
5.87 ($^3A'$)	5.53 ($^3A''$)	4.92 ($^1A''$)	4.64 ($^3A''$)	5.03 (^1A)
6.13 ($^1A'$)	5.66 ($^1A''$)	5.49 ($^3A'$)	4.75 ($^1A'$)	5.54 (^3A)
6.39 ($^3A'$)	5.75 ($^3A'$)	6.29 ($^3A''$)	4.75 ($^1A''$)	6.36 (^3A)
7.09 ($^1A'$)	6.07 ($^1A'$)	6.36 ($^3A'$)	5.35 ($^3A''$)	6.38 (^3A)
7.34 ($^3A''$)	6.58 ($^1A'$)	6.49 ($^1A''$)	5.39 ($^3A'$)	6.56 (^1A)
7.41 ($^1A''$)	6.85 ($^3A''$)	6.59 ($^1A'$)	5.55 ($^3A''$)	6.62 (^1A)
7.82 ($^1A'$)	7.11 ($^1A''$)	7.00 ($^1A'$)	5.57 ($^1A''$)	7.20 (^1A)
7.85 ($^3A''$)	7.14 ($^3A'$)	7.70 ($^3A'$)	5.64 ($^1A''$)	7.78 (^3A)
8.09 ($^1A''$)	7.84 ($^3A''$)	7.78 ($^3A''$)	6.48 ($^1A'$)	7.92 (^3A)
9.38 ($^3A''$)	7.98 ($^1A''$)	7.89 ($^1A''$)	6.92 ($^3A'$)	7.98 (^1A)
9.43 ($^1A''$)	8.12 ($^3A''$)	7.93 ($^3A''$)	7.93 ($^3A'$)	7.98 (^3A)
11.21 ($^3A''$)	8.14 ($^1A''$)	7.96 ($^1A''$)	7.98 ($^3A''$)	8.05 (^1A)
11.23 ($^1A''$)	8.14 ($^1A'$)	8.56 ($^3A'$)	7.98 ($^1A''$)	8.71 (^1A)
3.06	1.58	4.06	6.33	3.96[a]

[a] Calculated ground state dipole moments.

Table 6.2 Configurations used in the CASSCF to generate the orbitals and in the final CAS-CI model

Molecule	CAS	Configurations
Adenine	(12,10)	(1a'-29a')^{58},(30a',1a"-9a")^{12}
Guanine	(12,9)	(1a'-30a')^{60},(1a"-3a")^{6},(31a',32a',4a"-10a")^{12}
Uracil	(14,10)	(1a' - 22a')^{44}, (23a', 24a', 1a" - 8a")^{14}
Cytosine	(14,10)	(1a' - 22a')^{44}, (23a', 24a', 1a" - 8a")^{14}
Thymine	(14,10)	(1a - 26a)^{52}, (27a- 36a)^{14}

In the course of this work we have performed many scattering calculations and this section will only summarise the results, see Table 6.3. In doing this we have chosen to select calculations performed in a uniform manner so that trends become apparent, rather than taking calculations individually optimized for each system. In particular we note that there are quite a number of ways of performing CC calculations. Of particular significance is the choice as to whether to "contract" or "uncontract" the configurations which involves placing the scattering electron in a target virtual orbital [43].

Although the difference between these two approaches may seem technical they actually differ quite significantly. The contracted method is standard but, without

Table 6.3 Low-lying resonant state positions (and widths) of the nucleic acid bases in eV

Bases	Models	$A''(\pi)$-resonances				$A'(\sigma)$-resonances		
Purine bases								
A	SE	3.23 (0.53)	4.02 (0.33)	5.06 (1.11)	10.62 (0.40)	9.85 (0.59)	10.60 (0.63)	
	SEP	1.30 (0.14)	2.12 (0.09)	3.12 (0.28)	7.07 (0.24)	8.60 (0.002)	8.87 (0.006)	
	CC	3.14 (0.42)	3.96 (0.31)	5.03 (0.62)	7.52 (0.22)	7.74 (1.21)	9.34 (0.38)	
	Ref. [51,55][a]	1.10	1.80	4.10				
	Obs. [1][b]	0.54	1.36	2.17				
G	SE	3.00 (0.33)	4.47 (0.35)	5.66 (0.80)	10.12 (1.26)	8.58 (0.21)	9.83 (0.39)	
	SEP	1.83 (0.16)	3.30 (0.24)	4.25 (0.33)	7.36 (0.27)	8.75 (0.004)	9.47 (0.004)	
	CC	1.68 (0.13)	3.19 (0.21)	4.74 (0.43)	6.96 (0.42)	6.44 (0.97)	7.49 (1.72)	
	Ref. [51,55][a]	1.55	2.40	3.75				
	Obs. [1][b]	0.46	1.37	2.36				
Pyrimidine bases								
U	SE	2.25 (0.26)	4.43 (0.41)	8.62 (2.69)		10.82 (3.13)		
	SEP	0.31 (0.015)	2.21 (0.16)	5.21 (0.72)		7.71 (0.01)	8.78 (0.01)	8.89 (0.01)
	CC	1.00 (0.05)	2.94 (0.29)	7.51 (2.38)				
	CC[c]	0.134 (0.0034)	1.94 (0.168)	4.95 (0.38)		6.17 (0.15)	7.62 (0.11)	8.12 (0.14)
	Ref. [52][d]	0.32 (0.018)	1.91 (0.16)	5.08 (0.40)		1.45	8.5	
	Obs. [1][b]	0.22	1.58	3.83				
C	SE	2.65 (0.34)	4.69 (0.67)	9.85 (2.08)		9.90 (0.61)		
	SEP	0.71 (0.05)	2.66 (0.33)	6.29 (0.72)		7.86 (0.003)		
	CC	1.20 (0.06)	3.02 (0.42)	6.08 (1.39)		6.38 (0.59)	6.51 (0.33)	
	CC[c]	0.36 (0.016)	2.05 (0.30)	6.04 (1.65)		6.64 (0.74)	6.87 (0.22)	
	Ref. [54,55][e]	0.50	2.40	6.30				
	Obs. [1][b]	0.32	1.53	4.50				

(continued)

Table 6.3 (continued)

Bases	Models	$A''(\pi)$-resonances			$A'(\sigma)$-resonances
T^f	SE	2.46 (0.36)	4.60 (0.60)	9.12 (0.75)	
	SEP	1.38 (0.19)	3.41 (0.16)	7.81 (0.18)	
	CC	1.50 (0.15)	3.54 (0.19)	4.60 (1.86)	
	CC^c	1.21 (0.13)	3.25 (0.16)	5.42 (1.56)	
	Ref. [54, 55][g]	0.30	1.90	5.70	
	Obs. [1][b]	0.29	1.71	4.05	

[a]Ref. [51, 55] used the SMC method with the SEP model. Only π resonance positions (and no widths) were reported for energies below 5 eV.
[b]Observed (ETS) resonance positions only.
[c]Uncontracted CC model, see Ref. [13].
[d]SMC calculation with the SEP model and forced C_s symmetry.
[e]SMC method with the SEP model. Only π resonance positions (and no widths) were reported.
[f]No symmetry employed in our thymine calculations because of the presence of the methyl group.
[g]Ref. [54] forced C_s symmetry in their thymine calculations. Used SMC method with the SEP model.

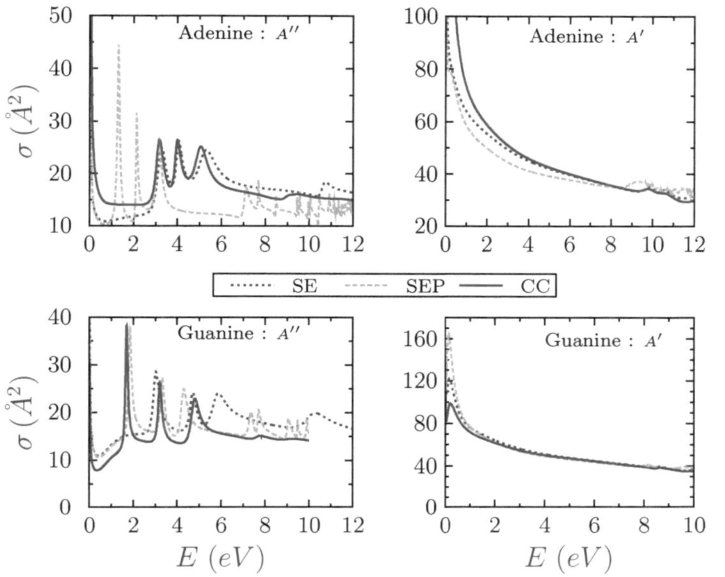

Fig. 6.1 Calculated elastic cross sections for the purine bases adenine and guanine. See text for details of the models used

performing exceptionally large CC expansions, tends to overestimate resonance positions. The uncontracted approach leads to systematically lower resonance positions. However it is not clear that this approach is completely balanced [43] and therefore it is possible to obtain resonances which are too low in energy or even become bound states. Furthermore, fully uncontracted calculations can become very large, requiring either unacceptably long times for the calculations or further compromises to be made on the size of the model wavefunctions chosen. For a more thorough discussion of the differences between these models, including full technical details, the reader is referred to our study on uracil [13]. The CC calculations reported here retained the lowest 32 states for uracil (U), 16 states for guanine (G), adenine (A) and thymine (T) and 20 states for cytosine (C).

Our previous study of resonances in phosphoric acid [9] revealed that the resonances showed little sensitivity to isomerisation. Our calculations suggest that this insensitively even extends between molecules with similar structures. Thus our calculations show a total of four π-type shape resonances for the purine bases, while the pyrimidines all have three.

The effect of these resonances can clearly be seen in the elastic cross sections. Figures 6.1 and 6.2 summarize our calculated elastic cross sections for the purine and pyrimidine DNA bases respectively. It should be noted that for strongly dipolar systems, such as the ones considered here, higher partial waves have a profound effect on the total elastic cross section, particularly at low energy. It is possible to correct for this using the Born approximation [56], which we have indeed done elsewhere [13]. However this correction is so big that it obscures the resonance

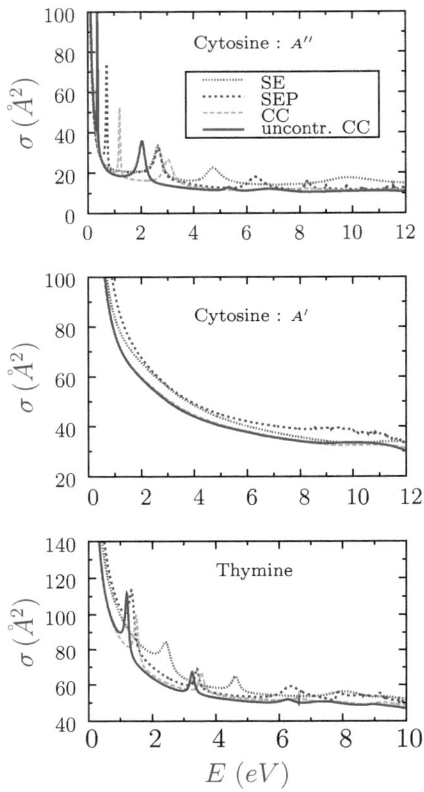

Fig. 6.2 Calculated elastic cross sections for the pyramidine bases cytosine and thymine

structures. For this reason the cross sections we present here are uncorrected. They are also, where possible, separated by symmetry to again emphasize the presence of the resonances.

6.5 Conclusion

This chapter summarizes our approach to identifying resonances in electron collisions with nucleic acid bases. The calculations presented here show that any of the nucleic acid base sites in DNA/RNA is capable of capturing low-energy electrons into π molecular orbitals to form a resonant π anionic state. This observation is in line with the suggestion that such π captures are followed by energy transfer to the σ molecular orbitals which in turn results in strand breaks [8].

A similar R-matrix approach to the one taken here has been used to consider resonances and electron collision cross sections in other biological important molecules such as phosphoric acid [9]. In this context it is worth noting that the R-matrix approach has been systematically applied to the study of low-energy electrons with water, perhaps the most important biomolecule, obtaining results of exceptional accuracy [15]. However the only resonances supported by water are at relatively high electron impact energies [17] and are therefore unlikely to play a significant role in DNA strand breaks and other processes resulting in radiation damage.

Acknowledgements This project was funded by the UK Engineering and Physical Sciences Research Council. LB and TvM thank EaStCHEM for computational support via the EaStCHEM Research Computing Facility.

References

1. Aflatooni, K., Gallup, G.A., Burrow, P.D.: J. Phys. Chem. A **102**, 6205 (1998)
2. Antic, D., Parenteau, L., Lepage, M., Sanche, L.: J. Phys. Chem. **103**, 6611–6619 (1999)
3. Antic, D., Parenteau, L., Sanche, L.: J. Phys. Chem. B **104**, 4711–4716 (2000)
4. Bao, X., Wang, J., Gu, J., Leszczyński, J.: Proc. Natl. Aca. Sci. **103**, 5658 (2006)
5. Barrios, R., Skurski, P., Simons, J.: J. Phys. Chem. B. **106**, 7991 (2002)
6. Berdys, J., Anuszkiewicz, I., Skurski, P., Simons, J.: J. Am. Chem. Soc. **126**, 6441 (2004)
7. Berdys, J., Anuszkiewicz, I., Skurski, P., Simons, J.: J. Phys. Chem. A **108**, 2999 (2004)
8. Boudaiffa, B., Clutier, P., Hunting, A., Hules, M.A., Sanche, L.: Science **287**, 1658 (2000)
9. Bryjko, L., van Mourik, T., Dora, A., Tennyson, J.: J. Phys. B: At. Mol. Opt. Phys. **43**, 235203 (2010)
10. Burke, P.G., Berrington, K.A.: Atomic and Molecular Processes - An R-matrix Approach. Bristol and IOPP (1993)
11. Burke, P.G., Tennyson, J.: Mol. Phys. **103**, 2537–2548 (2005)
12. Cobut, V., Fongillo, Y., Patau, J.P., Goulet, T., Fraser, M.J., Jay-Gerin, J.P.: Radiat. Phys. Chem. **51**, 229243 (1998)
13. Dora, A., Bryjko, L., van Mourik, T., Tennyson, J.: J. Chem. Phys. **130**, 164307 (2009)
14. Faure, A., Gorfinkiel, J.D., Morgan, L.A., Tennyson, J.: Computer Phys. Comms. **144**, 224–241 (2002)
15. Faure, A., Gorfinkiel, J.D., Tennyson, J.: J. Phys. B: At. Mol. Opt. Phys. **37**, 801–807 (2004)
16. Galiatsatos, P.G.: (2010). Private communication
17. Gorfinkiel, J.D., Morgan, L.A., Tennyson, J.: J. Phys. B: At. Mol. Opt. Phys. **35**, 543–555 (2002)
18. Gu, J., Wang, J., Rak, J., Leszczyński, J.: Ang. Chem. **119**, 3549 (2007)
19. König, C., Kopyra, J., Bald, I., Illenberger, E.: Phys. Rev. Letts. **97**, 018105 (2006)
20. Kumar, A., Sevilla, M.D.: J. Phys. Chem. B **111**, 5464 (2007)
21. Li, X., Sevilla, M.D., Sanche, L.: J. Am. Chem. Soc **108**, 19013 (2004)
22. Lias, S.G., Levin, R.D., Kafafi, S.A.: In NIST Chemistry WebBook, NIST standard reference database number 69. National Institute of Stand ards and Technology, Gaithersburg, MD, httpand//webbook.nist.gov (2005)
23. Loos, P.F., Dumont, E., Laurent, A.D., Assfeld, X.: Chem. Phys. Lett. **475**, 120 (2009)
24. M. J. Frisch et al: (2004). Gaussian 03, Revision E.01, Wallingfort, CT
25. Martin, F., Burrow, P.D., Cai, Z., Clutier, P., Hunting, D., Sanche, L.: Phys. Rev. Lett. **93**, 068101 (2004)
26. Morgan, L.A., Tennyson, J., Gillan, C.J.: Computer Phys. Comms. **114**, 120–128 (1998)
27. Pan, X., Cloutier, P., Hunting, D., Sanche, L.: Phys. Rev. Letts. **90**, 208102 (2003)
28. Pan, X., Sanche, L.: Phys. Rev. Letts. **94**, 198104 (2005)
29. Pan, X., Sanche, L.: Chem. Phys. Lett. **421**, 404 (2006)

30. Panajotovic, R., Michaud, M., Sanche, L.: Phys. Chem. Chem. Phys. **9**, 138 (2007)
31. Pimblott, S.M., LaVerne, J.A.: Radiat. Res. **141**, 208 (1995)
32. Pimblott, S.M., LaVerne, J.A.: Radiat. Phys. Chem. **76**, 1244 (2007)
33. Ptasińska, S., Denifl, S., Gohlke, S., Scheier, P., Illenberger, E., Märk, T.D.: Angev. Chem., Int. Ed. **45**, 1893 (2006)
34. Ptasińska, S., Denifl, S., Scheier, P., Märk, T.D.: J. Chem. Phys. **120**, 8505 (2004)
35. Ptasińska, S., Sanche, L.: Phys. Chem. Chem. Phys. **9**, 1730 (2007)
36. Ptasińska, S., Sanche, L.: Phys. Rev. E **75**, 31915 (2007)
37. Rak, J., Kobylecka, M., Storoniak, P.: J. Phys. Chem. B **115**, 1911–1917 (2011)
38. Sanche, L., Schulz, G.: J. Phys. Rev. A **5**, 1672 (1972)
39. Scheer, A.M., Aflatooni, K., Gallup, G.A., Burrow, P.D.: Phys. Rev. Lett. **92**, 068102 (2004)
40. Shukla, M.K., Leszczyński, J. (eds.): Radiation Induced Molecular Phenomena in Nucleic Acids. Springer, Jackson (2007)
41. Takayanagi, T., Asakura, T., Motegi, H.: J. Phys. Chem. A **113**, 4795 (2009)
42. Tennyson, J.: J. Phys. B: At. Mol. Opt. Phys. **29**, 1817–1828 (1996)
43. Tennyson, J.: J. Phys. B: At. Mol. Opt. Phys. **29**, 6185–6201 (1996)
44. Tennyson, J.: J. Phys. B: At. Mol. Opt. Phys. **37**, 1061–1071 (2004)
45. Tennyson, J.: Phys. Rep. **491**, 29–76 (2010)
46. Tennyson, J., Gorfinkiel, J.D., Rozum, I., Trevisan, C.S., Vinci, N.: Radiat. Phys. Chem. **68**, 65–72 (2003)
47. Tennyson, J., Noble, C.J.: Computer Phys. Comms. **33**, 421–424 (1984)
48. Tonzani, S.: Computer Phys. Comm. **176**, 146–156 (2007)
49. Tonzani, S., Greene, C.H.: J. Chem. Phys. **122**, 014111 (2005)
50. Tonzani, S., Greene, C.H.: J. Chem. Phys. **124**, 054312 (2006)
51. Winstead, C., McKoy, V.: J. Chem. Phys. **125**, 244302 (2006)
52. Winstead, C., McKoy, V.: J. Chem. Phys. **125**, 174304 (2006)
53. Winstead, C., McKoy, V.: J. Chem. Phys. **125**, 074302 (2006)
54. Winstead, C., McKoy, V.: J. Chem. Phys. **127**, 085105 (2007)
55. Winstead, C., McKoy, V.: Rad. Phys. Chem. **77**, 1258 (2008)
56. Zhang, R., Faure, A., Tennyson, J.: Physica Scripta **80**, 015301 (2009)
57. Zheng, Y., Cloutier, P., Hunting, D.J., Wagner, J.R., Sanche, L.: J. Chem. Phys. **124**, 064710 (2006)

Chapter 7
A Multiple-Scattering Approach to Electron Collisions with Small Molecular Clusters

Jimena D. Gorfinkiel and Stefano Caprasecca

Abstract We present a method based on multiple-scattering to determine elastic cross sections for electron collisions with molecular clusters. The method is based on the calculation of accurate collisional information for the molecules constituting the cluster that is then combined to obtain a cross section for interaction with the whole system. The method provides a computationally cost-effective way of treating low energy electron scattering from (homogeneous and heterogeneous) molecular clusters and aggregates. Results for $(H_2O)_n$ ($n = 2{,}5$) and $(HCOOH)_2$ are presented; the cross sections agree well with more accurate *ab initio* data.

7.1 Introduction

Clusters are aggregates of atoms or molecules with properties that are often between that of a bulk and a gas but that, on occasions, are different from both. An extensive literature is available on structural and electronic properties of clusters. In molecular clusters, the interactions between sub-units (also called monomers) is limited to the weak intermolecular forces: Van de Waals and dipole – dipole interaction, hydrogen bonding, etc. [1].

Molecular clusters are models for solvation and are also important in atmospheric processes. In the Earth's atmosphere, cosmic rays generate ions which then stabilise by reacting with other atmospheric molecules and through charge-driven clustering with polar molecules H_2O, NH_3 [2, 3]. Such clusters grow to form ultrafine aerosols

J.D. Gorfinkiel (✉)
Department of Physics and Astronomy, The Open University, Walton Hall, MK7 6AA Milton Keynes, UK
e-mail: J.Gorfinkiel@open.ac.uk

S. Caprasecca
Dipartimento di Chimica e Chimica Industriale, Università di Pisa, 56126 Pisa, Italy
e-mail: stefano.caprasecca@for.unipi.it

G.G. Gómez-Tejedor and M.C. Fuss (eds.), *Radiation Damage in Biomolecular Systems*, Biological and Medical Physics, Biomedical Engineering,
DOI 10.1007/978-94-007-2564-5_7,

which are responsible for starting the nucleation into water droplets and can lead to the formation of clouds, fog, the deposition of nitrogen, acidification of precipitations, etc.. In addition, molecular clusters have been observed in interstellar media.

In general, clusters are present in a variety of dense gaseous media, where they form spontaneously; in this environment, many studies have analysed the interaction between electrons and clusters, particularly focusing on the effect clusters have in electron transport and on the attachment, ionisation and fragmentation processes. Clusters of biophysically relevant molecules solvated by water molecules are being investigated as models for *in vivo* proteins, RNA, DNA, etc. (see, for example, [4]).

In spite of the pervasiveness of molecular clusters, and the importance of low energy electron scattering in a variety of environments as initiators of chemical processes [5], theoretical work on electron collisions with molecular clusters is scarce. Studies have been restricted mostly to dimers [6–8]. An exception to this is the work of Fabrikant and collaborators, who have concentrated on electron attachment and vibrational Feshbach resonances [9].

It is therefore highly desirable to have available a theoretical approach for electron collisions with molecular clusters. Evidently, standard electron scattering techniques can be used. However, as the size of the cluster increases, the computational requirements of these calculations, particularly *ab initio* ones, increases significantly: currently, the biggest *ab initio* calculations tackle collisions with targets formed by $\simeq$20-25 atoms. A 10 monomer cluster of triatomic or larger molecules will be bigger and require more resources.

A method has been developed [8, 10], based on multiple-scattering ideas, to treat elastic low energy electron collisions with molecular clusters within the fixed-nuclei approximation. This approach was initially proposed by Caron and Sanche ([11] and references therein). The method is based on the assumption that the potential of a complex target can be separated into non-overlapping regions each of which contains a single scatterer. In the current approach, and in contrast to earlier work on electron scattering [12], each molecule in the cluster (i.e. each monomer) is taken as a scatterer.

In this chapter, we will review the theory underlying the technique, summarise some practical issues regarding its application and present some results.

7.2 Multiple-scattering approach

The purpose of the method is to provide information on the electron-molecular cluster collision process by combining scattering information for each of the molecular sub-units. In order to do this, we need to derive an expression linking cluster scattering data with monomer scattering data.

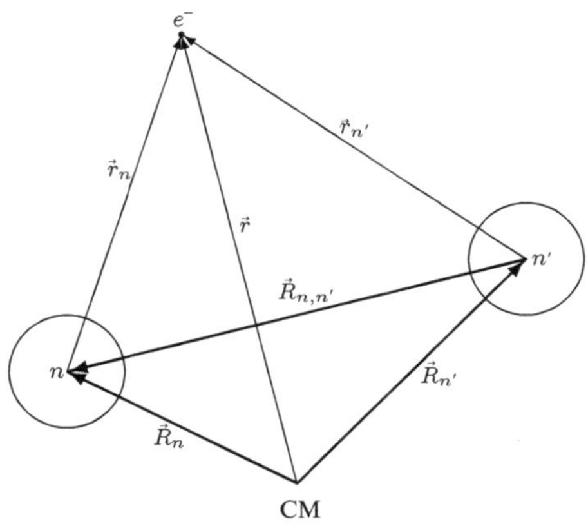

Fig. 7.1 Vector notation used in this section. The centre of mass of the cluster is indicated by CM, the circles labelled n and n' represent two generic sub-units whose centre of mass positions are identified by vectors $\vec{R}_n$ and $\vec{R}_{n'}$ respectively. The vectors $\vec{r}$, $\vec{r}_n$ and $\vec{r}_{n'}$ indicate the position of the scattering electron with respect to the different centres of mass

7.2.1 Theory

The asymptotic form of the total wavefunction describing the electron scattered from the whole cluster may be written as:

$$\psi(\vec{r}) = \sum_{L,L'} Y_{L'}(\hat{r}) \left[j_{l'}(kr)\delta_{L,L'} + \frac{1}{2} h_{l'}^{+}(kr) T^{\mathrm{G}}_{L',L} \right] \widehat{f}_L, \tag{7.1}$$

where the first term in the square brackets represents the incident plane wave of momentum $k = \sqrt{E}$, with E the kinetic energy of the scattering electron. The second term represents the scattered wave having the same energy (since the scattering process is elastic). The index L indicates collectively (l, m), $Y_L(\hat{r})$ are spherical harmonics and j_l and h_l^{+} are, respectively, the spherical Bessel and Hankel functions of the first kind. Here, the coordinates are referred to the centre of mass of the cluster. The matrix $\boldsymbol{T}^{\mathrm{G}}$ contains the information of the scattering process with the whole target (the cluster); collisional data, for example cross sections, can be easily obtained from it.

The incident plane wave from the above equation:

$$\psi_{\mathrm{in,PW}}(\vec{r}) = \sum_{L'} Y_{L'}(\hat{r}) j_{l'}(kr) \widehat{f}_{L'}, \tag{7.2}$$

needs to be re-expanded around the centre of mass of each molecular sub-unit. Following the vector notation indicated in Fig. 7.1, we can write:

$$\psi_{\text{in,PW}}(\vec{r}_n) = \sum_{L'}\sum_{L_1,L_2} \mathrm{i}^{l_1+l_2-l'}(-1)^{m'}F^{l_1,l_2,l'}_{m_1,m_2,-m'}Y_{L_1}(\hat{r}_n)j_{l_1}(kr_n)Y_{L_2}(\hat{R}_n)j_{l_2}(kR_n)\widehat{f}_{L'}$$
$$= \sum_{L',L_1} Y_{L_1}(\hat{r}_n)j_{l_1}(kr_n)M^n_{L_1,L'}\widehat{f}_{L'} = \sum_{L_1} Y_{L_1}(\hat{r}_n)j_{l_1}(kr_n)\widehat{g}^n_{L_1}, \tag{7.3}$$

where we have taken into account that $\vec{r}_n = \vec{r} - \vec{R}_n$ and defined:

$$M^n_{L_1,L'} \equiv \sum_{L_2} \mathrm{i}^{l_1+l_2-l'}(-1)^{m'}F^{l_1,l_2,l'}_{m_1,m_2,-m'}Y_{L_2}(\hat{R}_n)j_{l_2}(kr_n), \tag{7.4}$$

and

$$\widehat{g}^n_{L_1} \equiv \sum_{L'} M^n_{L_1,L'}\widehat{f}_{L'}. \tag{7.5}$$

Here we have used the re-expansion formula [12, 13] for spherical harmonics as defined by Messiah [14]; the expression for $F^{l_1,l_2,l_3}_{m_1,m_2,m_3}$ can be found in [8].

Equation (7.3) represents the wave incident on the sub-unit n due to the incoming plane wave only, expressed as a function of $\vec{r}_n$, the scattering electron coordinate with respect to the centre of mass of this sub-unit. The total incoming wave, $\psi_{\text{in},n}$, should include the contribution of both the plane wave and the waves scattered from the neighbouring sub-units. If we define the amplitude functions $\{\boldsymbol{g}^n\}$, containing all these contributions, we can write:

$$\psi_{\text{in},n}(\vec{r}_n) = \sum_{L_1} Y_{L_1}(\hat{r}_n)j_{l_1}(kr_n)g^n_{L_1}. \tag{7.6}$$

The wavefunction scattered from sub-unit n, $\psi_{\text{sc},n}$, can be expressed, taking into account the incoming wavefunction (7.6), in terms of the sub-unit's T-matrix $\boldsymbol{T}^n$:

$$\psi_{\text{sc},n}(\vec{r}_n) = \frac{1}{2}\sum_{L,L'} Y_{L'}(\hat{r}_n)h^+_{l'}(kr_n)T^n_{L',L}g^n_L. \tag{7.7}$$

$\psi_{\text{sc},n}(\vec{r}_n)$ can be re-expanded around the centre of mass of sub-unit n', considering the relation $\vec{r}_{n'} = \vec{r}_n - \vec{R}_{n',n}$:

$$\psi_{\text{sc},n}(\vec{r}_{n'}) = \sum_{L,L',L_2} Y_{L_2}(\hat{r}_{n'})j_{l_2}(kr_{n'})X^{n',n}_{L_2,L'}T^n_{L',L}g^n_L, \tag{7.8}$$

where we have defined:

$$X^{n',n}_{L_2,L'} \equiv \frac{1}{2}\sum_{L_1} \mathrm{i}^{l_1+l_2-l'}(-1)^{m'}F^{l_1,l_2,l'}_{m_1,m_2,-m'}Y_{L_1}(\hat{R}_{n',n})h^+_{l_1}(kR_{n',n}),$$
$$X^{n,n}_{L_1,L_2} = 0, \tag{7.9}$$

with $R_{n,n'}$ the distance between the centres of mass of sub-units n and n'. Equation (7.8) expresses the wave scattered off sub-unit n and incoming onto sub-unit n'.

We now need to find a relation linking g^n_L (the total incoming wave amplitude) to $\widehat{g}^n_{L_1}$ (the plane wave amplitude), so we will re-write the total incoming wavefunction from equation (7.6) explicitly as:

$$\psi_{\text{in},n}(\vec{r}_n) = \psi_{\text{in,PW}}(\vec{r}_n) + \sum_{n' \neq n} \psi_{\text{sc},n'}(\vec{r}_n). \tag{7.10}$$

Substituting equations (7.3) for $\psi_{\text{in,PW}}(\vec{r}_n)$ and (7.8) for $\psi_{\text{sc},n'}(\vec{r}_n)$:

$$\psi_{\text{in},n}(\vec{r}_n) = \sum_{L_1} Y_{L_1}(\hat{r}_n) j_{l_1}(kr_n) \left[\widehat{g}^n_{L_1} + \sum_{n' \neq n} \sum_{L,L'} X^{n,n'}_{L_1,L'} T^{n'}_{L',L} g^{n'}_L \right], \tag{7.11}$$

and comparing this expression with equation (7.6) we obtain the following relation:

$$g^n_{L_1} = \widehat{g}^n_{L_1} + \sum_{n' \neq n} \sum_{L,L'} X^{n,n'}_{L_1,L'} T^{n'}_{L',L} g^{n'}_L. \tag{7.12}$$

This equation can be rewritten in matrix form:

$$\mathbb{G} = \widehat{\mathbb{G}} + \mathbb{X}\mathbb{T}\mathbb{G}, \tag{7.13}$$

if one defines the following vectors and matrices:

$$\widehat{\mathbb{G}} \equiv \begin{bmatrix} \widehat{\boldsymbol{g}}^1 \\ \widehat{\boldsymbol{g}}^2 \\ \vdots \\ \widehat{\boldsymbol{g}}^N \end{bmatrix}; \quad \mathbb{G} \equiv \begin{bmatrix} \boldsymbol{g}^1 \\ \boldsymbol{g}^2 \\ \vdots \\ \boldsymbol{g}^N \end{bmatrix}; \quad \mathbb{T} \equiv \begin{bmatrix} \boldsymbol{T}^1 & \boldsymbol{0} & \cdots & \boldsymbol{0} \\ \boldsymbol{0} & \boldsymbol{T}^2 & & \boldsymbol{0} \\ \vdots & & \ddots & \\ \boldsymbol{0} & \boldsymbol{0} & & \boldsymbol{T}^N \end{bmatrix}; \tag{7.14}$$

$$\mathbb{X} \equiv \begin{bmatrix} \boldsymbol{0} & \boldsymbol{X}^{1,2} & \cdots & \boldsymbol{X}^{1,N} \\ \boldsymbol{X}^{2,1} & \boldsymbol{0} & \cdots & \boldsymbol{X}^{2,N} \\ \vdots & & \ddots & \\ \boldsymbol{X}^{N,1} & \boldsymbol{X}^{N,2} & \cdots & \boldsymbol{0} \end{bmatrix}, \tag{7.15}$$

where N is the number of sub-units in the cluster. Notice that $\boldsymbol{T}^n$ and $\boldsymbol{X}^{n,n'}$ in these expressions are matrices with dimensions[1] $\left((l_{\max}+1)^2, (l_{\max}+1)^2\right)$, while $\widehat{\boldsymbol{g}}^n$ and $\boldsymbol{g}^n$ are vectors with dimension $(l_{\max}+1)^2$, since the index $L = (l, m)$ runs from

[1] $\boldsymbol{0}$ has the same dimensions.

$(0,0)$ to $(l_{\max}, l_{\max})$. The dimension of vectors $\widehat{\mathbb{G}}$ and $\mathbb{G}$ is then $N \times (l_{\max}+1)^2$; matrices $\mathbb{T}$ and $\mathbb{X}$ have dimension $\big(N \times (l_{\max}+1)^2, N \times (l_{\max}+1)^2\big)$.

We now have an expression for the g_L^n introduced in equation (7.6): they have been expressed, in matrix form, in terms of the plane wave amplitudes $\widehat{\mathbb{G}}$ and a multiple-scattering term $(\mathbb{1} - \mathbb{X}\mathbb{T})^{-1}$. We can now write the total scattered function in the following way:

$$\begin{aligned}\psi_{\mathrm{sc,TOT}} &= \sum_n \psi_{\mathrm{sc},n} = \frac{1}{2}\sum_n \sum_{L,L'} Y_{L'}(\hat{r}_n) h_{l'}^{+}(kr_n) T_{L',L}^n g_L^n \\ &= \frac{1}{2}\sum_n \sum_{L,L',L_1} Y_{L_1}(\hat{r}) h_{l_1}^{+}(kr) N_{L_1,L'}^n T_{L',L}^n g_L^n,\end{aligned} \tag{7.16}$$

where a re-expansion around $\vec{r}$ has been carried out and we have defined:

$$N_{L_1,L'}^n \equiv \sum_{L_2} \mathrm{i}^{l_1+l_2-l'} (-1)^{m'} F_{m_1,m_2,-m'}^{l_1,l_2,l'} Y_{L_2}(-\hat{R}_n) j_{l_2}(kR_n). \tag{7.17}$$

Comparing equations (7.16) and (7.1) and using the vector form for:

$$\mathbb{N} \equiv \begin{bmatrix} \boldsymbol{N}^1 \\ \boldsymbol{N}^2 \\ \vdots \\ \boldsymbol{N}^N \end{bmatrix}; \quad \mathbb{M} \equiv \big[\boldsymbol{M}^1\, \boldsymbol{M}^2 \cdots \boldsymbol{M}^N\big], \tag{7.18}$$

one can write:

$$\mathbb{N}\mathbb{T}\mathbb{G} = \mathbb{N}\mathbb{T}(\mathbb{1} - \mathbb{X}\mathbb{T})^{-1}\widehat{\mathbb{G}} = \mathbb{N}\mathbb{T}(\mathbb{1} - \mathbb{X}\mathbb{T})^{-1}\mathbb{M}\widehat{\boldsymbol{f}} = \boldsymbol{T}^{\mathrm{G}}\widehat{\boldsymbol{f}}, \tag{7.19}$$

where we have used expressions (7.5) and (7.13). This provides us with the desired expression for the T-matrix for the whole cluster, $\boldsymbol{T}^{\mathrm{G}}$, in terms of the T-matrices for each of the sub-units:

$$\boldsymbol{T}^{\mathrm{G}} = \mathbb{N}\mathbb{T}(\mathbb{1} - \mathbb{X}\mathbb{T})^{-1}\mathbb{M}. \tag{7.20}$$

$\boldsymbol{T}^{\mathrm{G}}$ has, as expected, the same dimensions as $\boldsymbol{T}^n$.

In summary, we have expressed the T-matrix for the target cluster as a product of: (i) the monomer T-matrices $\boldsymbol{T}^n$; (ii) the *re-expansion terms* $\mathbb{M}$ (from $\vec{r}$ to $\vec{r}_n$) and $\mathbb{N}$ (from $\vec{r}_n$ to $\vec{r}$) that are determined taking into account the position of the monomers in the cluster with respect to its centre of mass; (iii) a *multiple-scattering* term $(\mathbb{1} - \mathbb{X}\mathbb{T})^{-1}$ that describes the multiple-scattering contribution. The matrix $\mathbb{X}$ includes the "interference" involving the waves scattered by the various sub-units. If $\mathbb{X}$ were null, there would be no interference effect at all and the final T-matrix would just be a superposition of sub-units' T-matrices. In this case, the cross section for electron scattering from the whole cluster would be approximately equal to the sum of the cross sections for each monomer: $\sigma^{\mathrm{G}} \approx \sum_n \sigma^n$.

7.2.2 *Monomer input data*

We have derived an expression for the T-matrix that describes the collisions between an electron and a molecular cluster in terms of $\boldsymbol{T}^n$, the T-matrices describing the collision with the individual molecules. What remains to be discussed is how to obtain these $\boldsymbol{T}^n$: is a standard *ab initio* calculation appropriate, or is there a need to modify it? Use of the MS-based technique to describe the electron band structure of cubic ice [15], electron scattering from water dimers [8, 10] and from DNA sections [16] has shown that two different monomer T-matrices should be used in expression (7.20). The equation can therefore be rewritten as:

$$\boldsymbol{T}^{\mathrm{G}} = \mathbb{N}\mathbb{T}\,(\mathbb{1} - \mathbb{X}\mathbb{T}_{\mathrm{c}})^{-1}\,\mathbb{M}. \tag{7.21}$$

To determine the monomer T-matrices that are the building blocks of both $\mathbb{T}$ and $\mathbb{T}_{\mathrm{c}}$ we use the R-matrix method , described in some detail elsewhere in the book (see also [17]). $\mathbb{T}$ is constructed from sub-unit T-matrices calculated in the standard way. However, the T-matrices going into $\mathbb{T}_{\mathrm{c}}$ need to be calculated in a slightly different way. We will now discuss how both monomer T-matrices are calculated.

7.2.2.1 Rotation of the T-matrices

When the cluster to be studied is homogeneous, there is no need to run N separate R-matrix calculations to obtain the sub-unit T-matrices needed to build matrix $\mathbb{T}$ of equation (7.14). Provided the internal coordinates of the monomers are very similar (and this tends to be the case), it is possible to calculate only two T-matrices (one going into $\mathbb{T}$ and the other into $\mathbb{T}_{\mathrm{c}}$) with the monomer in some convenient orientation and then apply a rotation to it, in order to account for the various orientations of each monomer in the cluster. The required rotation angles can be found by choosing three independent geometrical parameters (e.g. position of atoms, direction of bonds, atom – atom vectors) expressed with respect to the coordinate systems employed in the respective calculations and then comparing the orientations of the isolated monomer and the monomer in the cluster. The rotation is actually applied to the spherical harmonics in whose representation the T-matrix has been originally built. The rotation matrices needed to perform the transformation are constructed according to the formula given by Messiah [14].

7.2.2.2 Partial wave expansion

The partial wave expansion in the R-matrix calculations is carried out up to $l = 4$ or 5. However, earlier work has shown the need for a cut-off on the number of partial waves included in the T-matrices forming $\mathbb{T}_c$: use of the full angular momentum content of the T-matrix produces an incorrect MS cross section. Such behaviour may be explained in terms of the angular momentum energy barrier $E(l,r) = \frac{l(l+1)}{r^2}$: an electron scattering from one sub-unit is able to reach another one only if its energy is large enough to overcome the barrier, i.e., only if $E_e > E(l, R_{n,n'})$, where E_e is the scattering electron kinetic energy. Therefore, the angular momenta to be included in $\mathbb{T}_c$ are those for which the following relation (where the energy is in Rydberg) holds:

$$\frac{l(l+1)}{R^2_{n,n'}} \leq E_e. \tag{7.22}$$

The partial wave cut-off can be achieved by truncating the monomer T-matrix, including in $\mathbb{T}_c$ only those elements corresponding to channels associated with partial waves $l \leq l_c$ such that equation (7.22) holds[2]. This procedure is approximate and leads to the use of T-matrices that do not retain the correct symmetry properties. However, it has the advantage that the R-matrix calculation is not modified and needs to be run only once. A more correct, but more time consuming, alternative is to run one R-matrix calculation for each l_c. Tests for the water dimer have shown that the MS results obtained by using the truncated T-matrices are virtually identical to those in which "correct symmetry" T-matrices have been used. However, the results for the formic acid dimer seem to indicate that this is not always the case (see Section 7.3.2).

Equation (7.22) implies that the partial wave cut-off depends on the kinetic energy of the scattering electron. This leads to discontinuities in the MS cross section at the "cut-off energies" $E_c^{l_c} = \frac{l_c(l_c+1)}{R^2_{n,n'}}$. We use a two-point interpolation of the cross sections to eliminate these: at any energy E lying between two cut-off energies, $E_c^{l_c} < E < E_c^{l_c+1}$, the cross section is a weighted (according to the differences $|E - E_c^l|$ and $|E - E_c^{l+1}|$) combination of $\sigma^l(E)$ and $\sigma^{l+1}(E)$. Here σ^{l_c} indicates the MS cross section, obtained from $\boldsymbol{T}^G$, where $\mathbb{T}_c$ includes partial waves up to l_c only.

In practice [10, 16], an *ad hoc* energy correction parameter γ is needed to eliminate non-physical structures that may arise in the MS cross section. Equation (7.22) is rewritten as:

$$\frac{l(l+1)}{R^2_{n,n'}} \leq \gamma E_e. \tag{7.23}$$

[2] Those for $l > l_c$ are set to zero.

We have found (see below) that a value of $\gamma = 0.75$ is appropriate for all the water clusters studied so far. For the formic acid dimer we found that $\gamma > 1.0$ is required, as is the case for the work on electron scattering from a DNA fragment [16].

7.2.2.3 Dipole moment truncation

Electron collisions with polar molecules are strongly affected by the long range electron-dipole moment interaction, particularly at low energies. However, our multiple-scattering implementation assumes implicitly that the monomer data will only describe the interaction between the electron and an individual molecule in a relatively small spherical space around the centre of mass of the target. This would imply that a cut-off should be introduced to limit the range of action of the monomer's dipole interaction: one should include it up to a certain distance a_c from the sub-unit's centre of mass and neglect it for larger distances. The R-matrix method, with its separation into an inner and outer region and the ability to describe the projectile-target interaction potential to different levels of accuracy, is particularly appropriate when this cut-off needs to be applied.

The choice of a_c must forcibly be done on a target by target basis, as it is dependent on the dipole moment of the individual sub-units. Caron *et al* [15] found that the electronic band structure of water ice was better reproduced when a small range of a_c values was used. However, our scattering calculations for the targets presented in Section 7.3 show that the effect is much less significant than expected. This has also been observed for scattering from DNA [16], and previously in [18], where it was shown that no great differences are encountered for fairly wide variations of the sub-unit sphere radii.

7.3 Some results

The MS technique described in this chapter has been applied so far to small water clusters, ($(H_2O)_n$, $n = 2, 5$) and to the formic acid dimer, $(HCOOH)_2$. We will summarise in this section the most relevant results obtained for these systems. In the case of dimers and trimers, we have been able to compare the MS data with results obtained with more accurate methods. This has allowed us to ascertain the validity of our technique.

7.3.1 Small water clusters

Water is fundamental to the vast majority of life forms found on Earth and for the biological processes taking place in the human cell. Codes of practice for dosimetry recommend using ionisation chambers calibrated in terms of absorbed dose to water to characterise the radiation beams for radiotherapy [19].

Table 7.1 Symmetries (point group) and dipole moments of the water molecule and the dimer geometries considered in this work: (a) present R-matrix calculations

Label	Point group	Dipole moment / D		
		(a)	[23]	[24]
H_2O	C_{2v}	1.86	1.85	-
EQ	C_s	2.73	2.56	2.70
Z_1	C_i	0.00	0.00	0.00
Z_2	C_{2h}	0.00	0.00	0.00
S	C_2	1.77	1.65	1.72
L	C_{2v}	4.36	4.12	4.15

In addition, a wide range of scattering data for water is used in energy deposition simulation software. Therefore, the interaction of low energy electrons with water in all its forms is of fundamental interest to the understanding of radiation interaction with biological material. However, with very few exceptions that concentrated on the formation of anions [20–22], no work on electron collisions with water clusters is available.

The characteristics of the R-matrix calculations performed to generate the input for the MS equations and the geometry parameters of the clusters investigated can be found elsewhere [10]. It should be noted however, that the water molecules in the clusters studied are virtually undistorted from their equilibrium geometry by clusterization: bond-lengths change by less than 1% and the $\widehat{\mathrm{HOH}}$ angle by 3% at most.

7.3.1.1 Water dimer

We have studied the collision with $(H_2O)_2$ for five different geometries of the dimer: the ground state equilibrium one and four corresponding to relative minima of the potential energy surface. The main difference between these geometries lies in the different total dipole moment of the cluster. Table 7.1 summarises these values, as well as the point group to which each dimer geometry belongs.

Figure 7.2 shows the MS and R-matrix elastic cross sections for the dimers studied, in the energy interval $1 - 10$ eV. As can be seen, the MS method is generally able to reproduce very well the more accurate R-matrix results in this energy range. Given that the different dimer geometries studied are characterised by very similar parameters (including the inter-monomer distance, that varies between 5.34 a_0 and 5.70 a_0) the difference in the cross sections is virtually only due to the relative orientation of the two water molecules. However, the way in which this orientation is taken into account by the MS method does not completely reproduce the effect of the electron interaction with the true dipole moment of the dimer. For this reason, agreement between the MS and R-matrix cross sections for $E < 1$ eV is much poorer than at higher energy. This disagreement is most notable for those dimer geometries, Z_1 and Z_2, for which the dipole moment is zero. Figure 7.3 shows the comparison

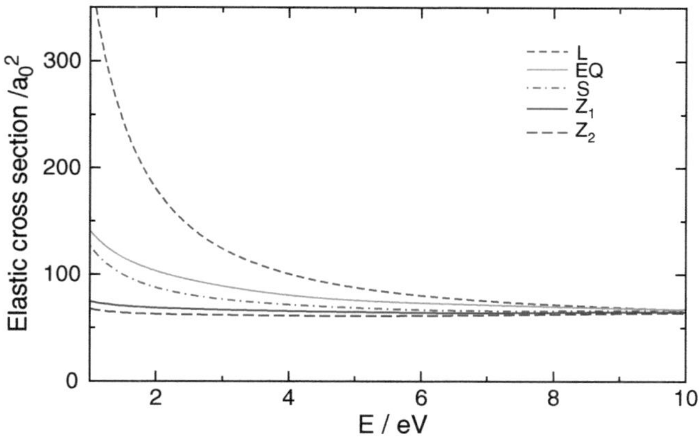

Fig. 7.2 Elastic cross sections for the five dimer geometries specified in the text, calculated with (upper panel) the R-matrix and (lower panel) the multiple-scattering methods

between cross sections down to 0 eV for the Z_2 geometry. We observe that while the behaviour of the R-matrix results is that expected of a cross section for electron collisions with a non-polar molecule, the MS cross section displays the behaviour to be expected from scattering with a polar target. We conclude that the MS treatment is not able to properly account for the total dipole moment of the cluster when it is very different from that of the sub-units that constitute it. It should be noted that the agreement below 1 eV for the dimer geometry 'L', with a dipole moment more than twice that of the isolated molecule, is much better than that for Z_2.

The need for the *ad hoc* parameter γ in equation (7.23) mentioned in Section 7.2.2.2 is illustrated in Fig. 7.3. The MS cross section shows a spurious peak for $\gamma = 1.00$ that is eliminated by setting $\gamma = 0.75$. Figure 7.3 shows the MS cross section for both γ together with the σ^l defined in Section 7.2.2.2: we can see how the spurious peak in the interpolated cross section for $\gamma = 1.00$ is due to the inclusion of σ^4 in the 6-8 eV range. When the thresholds for inclusion of partial waves, $E_c^{l_c}$,

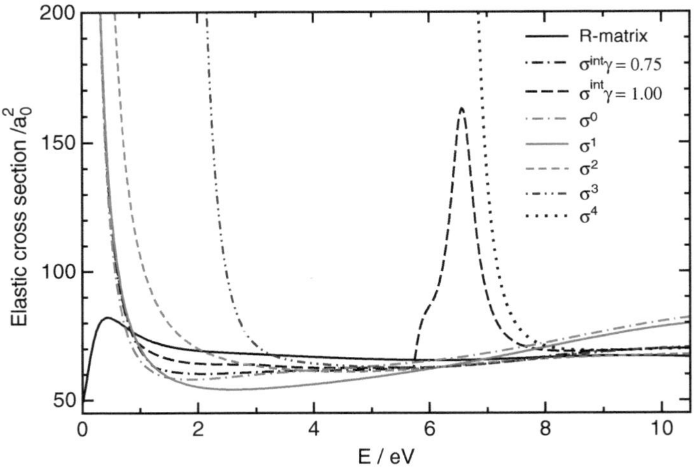

Fig. 7.3 Elastic cross sections for the Z_2 dimer geometry calculated with the R-matrix and the multiple-scattering method. The interpolated (final) MS cross section for both $\gamma = 1.00$ and 0.75 is shown. The cross sections labelled σ^l, $l = 0,4$ correspond to those described in section 7.2.2.2

are shifted to higher energies by the use of $\gamma < 1.0$, σ^4 no longer contributes in the energy range where it is unphysically large.

7.3.1.2 $(H_2O)_n$, $n = 3, 4, 5$

The MS technique allows us to calculate cross sections for bigger clusters easily: whereas an *ab initio* calculation will require significantly more computational resources, the MS calculation does not[3]. All that is required is the cluster geometry (also needed for *ab initio* work) and the sub-unit T-matrices already calculated for the dimers.

We have therefore performed calculations for the trimer, tetramer and pentamer of water in their equilibrium geometry [25, 26]. In our calculations, the trimer has a dipole moment of 0.25 D, the tetramer has no dipole moment while the pentamer has a dipole moment of 0.9 D. Once again, the differences between the geometry of the isolated water molecule and that of the cluster sub-units are small.

Results for the MS calculations are shown in Fig. 7.4: an increase in the size of the cross section as the number of molecules in the cluster, n, increases can be observed for most of the energy range displayed. This increase, however, is not proportional to n. It is still possible to perform R-matrix calculations for the water

[3]A quick inspection of the MS equations shows that as n increases, it is only the the size of the matrices to be multiplied that increases approximately as n^2. In an *ab initio* calculation, the number of integrals to be calculated, for example, grows much faster.

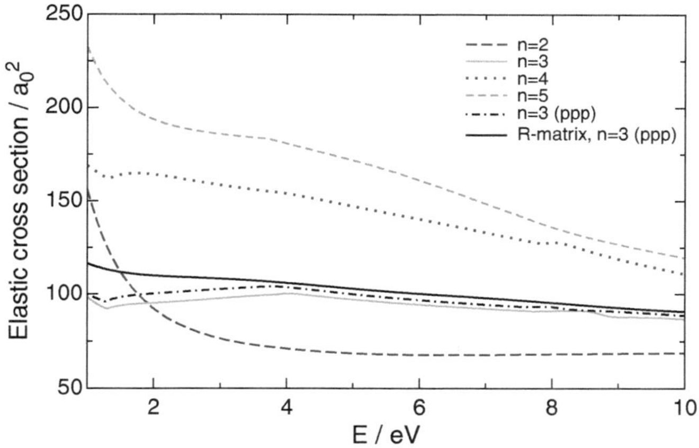

Fig. 7.4 Elastic cross sections for $(H_2O)_n$ for the number of monomers indicated in the figure; all are calculated with $\gamma = 0.75$ and, unless indicated, correspond to the equilibrium geometry of the cluster. Also shown are the R-matrix and multiple-scattering cross section for a non-equilibrium geometry of the trimer labelled ppp. See text for details

trimer to compare with the MS results. For this, we chose a geometry ('ppp') with no dipole moment in which the three water monomers lie on a plane perpendicular to a C_3 axis [27]. The comparison between the R-matrix and MS results for this cluster is also shown in Fig. 7.4: the agreement is reasonable above 3 eV.

7.3.2 Formic acid dimer

Formic acid, HCOOH, is a dipolar molecule that can be found in many different environments, among them the interstellar medium. The dimers of carboxylic acids are characterised by two hydrogen bonds forming an eight-membered ring [28, 29]; their behaviour, under certain conditions, resembles that of aggregates of larger molecules, among which are biologically relevant ones. This suggests that the formic acid dimer can serve as a simple model system to study the effect of hydrogen bridges, for example towards electron attack. The resonances present at low energies when an electron scatters from HCOOH and $(HCOOH)_2$ have been observed to lead to fragmentation; these systems are therefore ideal to analyse electron-induced dissociation patterns in the biological environment. The theoretical and experimental work on electron collisions with aggregates of HCOOH is significant: calculations [6] and experiments [30] for the dimer, as well as experimental work on films [31] and mixed-size clusters [32] have been carried out.

Formic acid is a planar molecule with a dipole moment of around 1.41 D [33]. Its *trans* configuration is the most stable and also forms the most stable dimers; these are non-polar in their equilibrium geometry. We chose $(HCOOH)_2$ as a model to

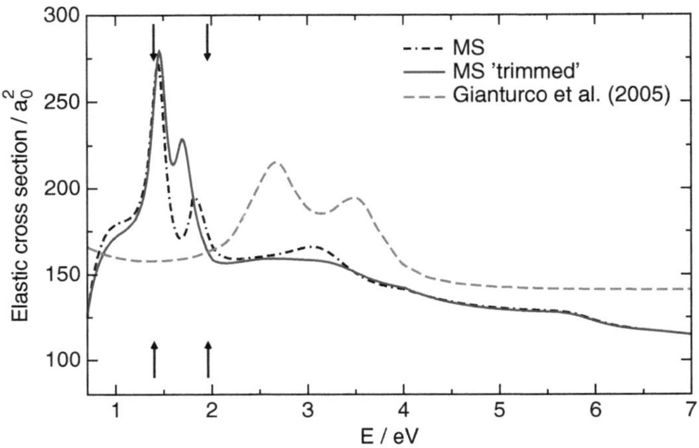

Fig. 7.5 Elastic cross section for $(HCOOH)_2$ collision, calculated with the Multiple Scattering method and $\gamma = 1.25$; the monomer data was generated using a static-exchange plus polarization model. Comparison with prior calculations [6] is shown; the arrows indicate where the two shape resonances are experimentally found [30]

study how well the multiple-scattering technique is able to reproduce the presence of resonances in a collisional process.

The characteristics of the R-matrix calculations performed in order to obtain the monomer input for the MS calculations can be found elsewhere [34]. The model, in this case, was chosen so as to best represent a shape resonance present in the 1.4-2.0 eV range (see [35] and references therein) in electron-HCOOH scattering. The formic acid dimer is rich in resonances: Gianturco and collaborators [6] identify seven in the energy range up to 15 eV. The two lowest have also been observed by Allan [30] at 1.4 and 1.96 eV: they can be interpreted as originating from the split of the monomer resonance mentioned above. It is these two resonances that we expect the MS technique to be able to represent.

Figure 7.5 shows the MS cross section for $(HCOOH)_2$: for this system, the use of 'trimmed' monomer T-matrices (see Sect. 7.2.2.2) produces a cross section that is visibly different to the one generated from 'correct' T-matrices. Both of them display a peak centred around 1.45 eV, but the second peak appears at different energies: using the 'trimmed' T-matrices produces a peak centred around 1.7 eV whereas the latter calculation produces a smaller peak centred around 1.83 eV, in better agreement with experimental results. Calculations with other models also show the presence of two peaks in the low energy region. Notice that for this system and model, $\gamma = 1.25$ was required. We conclude that the MS technique is able to reproduce the presence of two resonances.

7.4 Summary

The method presented here, based on multiple-scattering and the use of accurate R-matrix scattering data for isolated molecules, provides a good description of the elastic cross sections for electron collisions with small molecular cluster: the agreement with more accurate results for the examples discussed above is good. The method is easily scalable to bigger clusters.

Acknowledgements The work on the multiple-scattering treatment of electron collisions with small molecular cluster was initiated in collaboration with L Caron, D Bouchiha and L Sanche. We are indebted to them for their contributions to the development and testing of this technique and for their encouragement. This work was supported by the EPSRC.

References

1. R.L. Johnston, *Atomic and Molecular Clusters* (CRC Press, London, 2002)
2. R.G. Harrison, K.S. Carslaw, Rev. Geophys. **41**, 2 (2003)
3. K.L. Aplin, R.A. McPheat, J. Atmos. Sol-Terr. Phy. **67**, 775 (2005)
4. M.T. Sykes, M. Levitt, Proc. Natl. Acad. Sci. U. S. A. **104**, 12336 (2007)
5. H. Hotop, M. Ruf, M. Allan, I. Fabrikant, Adv. At. Mol. Phys. **49**, 85 (2003)
6. F.A. Gianturco, R.R. Lucchese, J. Langer, I. Martin, M. Stano, G. Karwasz, E. Illenberger, Eur. Phys. J. D **35**, 417 (2005)
7. T.C. Freitas, M.A.P. Lima, S. Canuto, M.H.F. Bettega, Phys. Rev. A **80**, 062710 (2009)
8. D. Bouchiha, L.G. Caron, J.D. Gorfinkiel, L. Sanche, J. Phys. B **41**, 045204 (2008)
9. I. Fabrikant, J. Phys. B-At. Mol. Opt. Phys. **38**, 1745 (2005)
10. S. Caprasecca, J.D. Gorfinkiel, D. Bouchiha, L.G. Caron, J. Phys. B **42**, 095205 (2009)
11. L. Caron, L. Sanche, Phys. Rev. A **73**, 062707 (2006)
12. D. Dill, J.L. Dehmer, J. Chem. Phys. **61**, 692 (1974)
13. M. Danos, L.C. Maximon, J. Math. Phys. **6**, 766 (1965)
14. A. Messiah, *Quantum Mechanics* (Wiley, New York, 1962)
15. L. Caron, D. Bouchiha, J.D. Gorfinkiel, L. Sanche, Phys. Rev. A **76**, 032716 (2007)
16. L. Caron, L. Sanche, S. Tonzani, C.H. Greene, Phys. Rev. A **78**, 042710 (2008)
17. J. Tennyson, Phys. Rep. **491**, 29 (2010)
18. D.A. Case, Ann. Rev. Phys. Chem. **33**, 151 (1982)
19. P. Andreo, D.T. Burns, K. Hohlfeld, M.S. Huq, T. Kanai, F. Laitano, V. Smyth, S. Vynckier, Technical Report Series 398, IAEA International Atomic Energy Agency (2000)
20. M. Knapp, O. Echt, D. Kreisle, E. Recknagel, J. Chem. Phys. **85**, 636 (1986)
21. M. Knapp, O. Echt, D. Kreisle, E. Recknagel, J. Phys. Chem. **91**, 2601 (1987)
22. J. Weber, E. Leber, M.W. Ruf, H. Hotop, Eur. Phys. J. D **7**, 587 (1999)
23. X. Huang, B.J. Braams, J.M. Bowman, J. Phys. Chem. **110**, 445 (2006)
24. G.S. Tschumper, M.L. Leininger, B.C. Hoffman, E.F. Valeev, H.F. Schaefer, M. Quack, J. Chem. Phys. **116**, 690 (2002)
25. M. Schütz, W. Klopper, H. Lüthi, S. Leutwyler, J. Chem. Phys. **103**, 6114 (1995)
26. S.S. Xantheas, T.H. Dunning, Jr, J. Chem. Phys. **99**, 8774 (1993)
27. W. Klopper, M. Schütz, H.P. Lüthi, S. Leutwyler, J. Chem. Phys. **103**, 1085 (1995)
28. F. Madeja, M.M. Havenith, J. Chem. Phys. **117**, 7162 (2002)
29. J. Chocholousova, J. Vacek, P. Hobza, Phys. Chem. Chem. Phys. **4**, 2119 (2002)
30. M. Allan, Phys. Rev. Lett. **98**, 123201 (2007)

31. T. Sedlacko, R. Balog, A. Lafosse, M. Stano, S. Matejcik, R. Azria, E. Illenberger, Phys. Chem. Chem. Phys. **7**, 1277 (2005)
32. I. Martin, T. Skalicky, J. Langer, H. Abdoul-Carime, G. Karwasz, E. Illenberger, M. Stano, S. Matejcik, Phys. Chem. Chem. Phys. **7**, 2212 (2005)
33. NIST Computational Chemistry Comparison and Benchmark DataBase (2010). http://cccbdb.nist.gov/. *National Institute of Standards and Technology — USA*
34. S. Caprasecca, J.D. Gorfinkiel, (2011). To be submitted
35. M. Allan, J. Phys. B **39**, 2939 (2006)

Chapter 8
Positronium Formation and Scattering from Biologically Relevant Molecules

G. Laricchia, D.A. Cooke, and S.J. Brawley

Abstract Recent progress in our experimental studies of positronium formation and scattering from simple atomic and molecular systems are reviewed. The former are used to highlight key features of ionizing collision by positrons before considering recent phenomena observed in the case of molecular targets, including positron impact excitation-ionization and the electron-like scattering of positronium. The guiding theme of this review arises from the role that repeated cycles of formation and dissociation of positronium are expected to play in the accurate description of positron interaction with matter.

8.1 Introduction

The study of the interaction of positrons and positronium (Ps, the bound-state of an electron and a positron) with atoms and molecules is variously motivated, for example, by the need to understand basic matter-antimatter interactions, to assist the development of accurate scattering theories, to aid the analysis of astrophysical events and to support tests of QED bound-state problems (e.g. [1]). Recently, the outstanding success of positron-emission-tomography (PET) in imaging human pathologies and physiological functions through the visualization of metabolic pathways, has resulted in its proliferation across major hospitals around the world [2]. In turn, this has highlighted the lack of accurate β^+ dosimetry, most protocols presently being based on macroscopic doses which are computed by dividing the total energy deposited in the body by its total mass (sometimes allowing for the non-uniform dose patterns arising from different permanence times in various organs) [3]. Thus, as discussed elsewhere in this book (Chaps. 13,14,16), event-by-event

G. Laricchia (✉) • D.A. Cooke • S.J. Brawley
University College London, UCL Department of Physics and Astronomy,
Gower Street, London WC1E 6BT
e-mail: g.laricchia@ucl.ac.uk; david.cooke@ucl.ac.uk; s.brawley@ucl.ac.uk

G.G. Gómez-Tejedor and M.C. Fuss (eds.), *Radiation Damage in Biomolecular Systems*, Biological and Medical Physics, Biomedical Engineering,
DOI 10.1007/978-94-007-2564-5_8, © Springer Science+Business Media B.V. 2012

simulations are being developed to describe the microscopic distribution of energy deposits along the positron tracks. These are needed both to limit damage to healthy tissue (and indeed to target tumours by the same means) as well as to extract their range, an important parameter for imaging-resolution to define with best accuracy e.g. tumour volumes. The modelling requires accurate information of how positrons and Ps interact at the molecular level: total cross-sections, (integral and differential) partial cross-sections for elastic and inelastic processes. Notably, the total ionization cross-section by positron impact generally exceeds that by electrons at low and intermediate energies, primarily due to the formation of Ps, e.g. [4]. The latter is readily formed in encounters of positrons with matter, and its formation (especially at high energies) and its subsequent interactions with biomolecules (including its fragmentation) plays a crucial role in determining the track-end. This typically extends several mm in biological materials for β^+ emitted from PET isotopes and thus seriously influences spatial resolution. In general, positron annihilation is not a significant effect except at very low energies [5] and so investigations relying on γ-ray detection, as in PET, are dominated by Ps events. Whilst detailed knowledge of positron and Ps interactions with biomolecules remains a significant challenge experimentally and theoretically, the last few years have seen a significant endeavour to begin to remedy this deficiency by addressing molecules of biological relevance, foremost amongst these H_2O , commonly used to simulate the biological medium [6–9]. Below we review some of our contributions to this progress, starting off with simple atoms in order to highlight key features of ionizing collision by positron impact and concluding with recent unexpected findings concerning excitation-ionization of molecules by positrons and the electron-like scattering of Ps.

8.2 Positron interactions

8.2.1 Basics

Although annihilation is its ultimate fate, a positron in matter may undergo any number of scattering events. Two processes which may occur at any energy are direct annihilation (which results in the emission of γ-rays) and elastic scattering (which leaves the internal energy of the target unchanged). If a positron possesses sufficient energy, other inelastic scattering processes become feasible. Listed in approximately ascending order of energy threshold, they are: Ps formation, excitation, direct ionization, ionization–excitation, transfer ionization (Ps formation simultaneous to direct ionization), and multiple direct ionization.

Positronium has a ground-state binding energy of 6.8 eV and can exist in two spin-states, arising from the possible orientations of the spins of its constituent particles. These may be parallel, leading to the triplet state (ortho-Ps), or antiparallel, leading to the singlet state (para-Ps). The triplet ground state ($1\ ^3S_1$) has a lifetime of 142 ns whereas the corresponding singlet state ($1\ ^1S_0$) has a lifetime of just

125 ps. In order to conserve charge-parity, the decay of ground-state Ps proceeds by an odd number of γ-rays for ortho-Ps or an even number for para-Ps . Thus, a unique signal of ground-state ortho-Ps is the coincident detection of three γ-rays whose energies must total the sum of twice the rest mass energy of the electron (1.022 MeV) and any initial kinetic energy. As with 'direct' annihilation, which occurs principally by the emission of two γ-rays, decay via a higher number of photons is reduced by several orders of magnitude. At the coarsest level of description, Ps is structurally similar to atomic hydrogen, with the Bohr energy levels halved as a result of the lower mass of the positron in comparison to the proton. Beyond this similarity, the fine structure is quite different (e.g. [10]) due to the ratio of magnetic moment of the positron to that of the proton ($\sim$658), which elevates the hyperfine structure observed in Ps to the order of the fine structure observed in H [11].

In addition to the formation of Ps, the positron may ionize a target directly by releasing one (or more) electron(s). Overall, the total ionization cross-section (Q_i^t) is defined by: $Q_i^t = Q_{\text{Ps}} + Q_i^+ + Q_{\text{ti}} + Q_{\text{ann}} + Q_i^{n+}$, where the elements of the sum are the cross-sections for Ps formation, direct ionization, transfer ionization, annihilation and multiple ionization, respectively.[1] For atomic targets, this can be approximated to $Q_i^t \approx Q_{\text{Ps}} + Q_i^+$, the cross-sections for the other processes often being comparatively negligible [4]. At their maxima, the probabilities for Ps formation and direct ionization account for roughly half of the overall scattering probability.

8.2.2 Ionization of atoms

The atomic targets examined in this section are used to introduce key features of positron-impact ionization before considering molecules. (A recent review of positron–impact ionization of the inert atoms can be found in [4]).

Figure 8.1a shows the partitioning of Q_i^t for He into the contributions from the two dominant ionization processes, Ps formation and direct ionization, and serves to highlight some common features for atoms. While at high energies electron data may be used to approximate corresponding positron results,[2] at energies below 1 keV major differences arise due to the diverse nature of the interactions and reactions of the different projectiles. These include exchange for electrons (and Ps—see Section 8.3), the equal and opposite static interaction for electrons and positrons, as well as the possibility of electron capture and annihilation for positrons (and Ps). In general, the total ionization cross-section by positron impact exceeds that by electrons at low and intermediate energies, primarily due to Ps formation. For low-Z atoms, Q_i^+ also exceeds that for electrons due to polarization effects [20], a situation which is reversed at the lowest impact energies due to a combination

---

[1] Ionization–excitation events are a subset of the above-identified ionization events.

[2] As per the charge-sign independence of the cross-sections within the First Born Approximation.

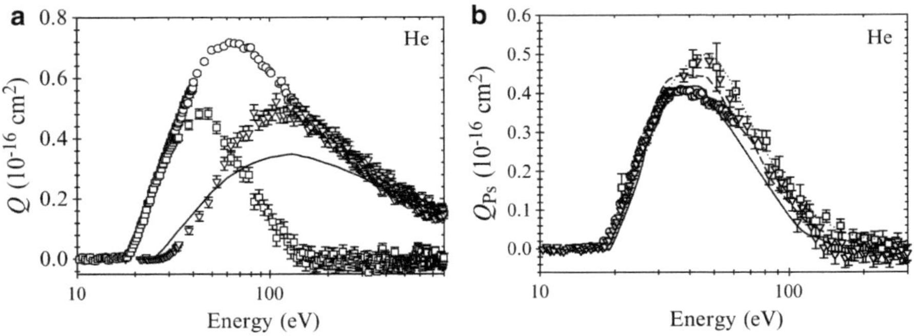

Fig. 8.1 (**a**) partitioning of the total ionization cross-section for He into direct ionization and Ps formation components. ○—Q_i^t [12], ∇—Q_i^+ [13], □—Q_{Ps} [12], solid line—electron Q_i^t [14]. (**b**) a comparison of recent experimental and theoretical results for He Q_{Ps}. Experimental: ○—[15], ∇—[12], □—[16]. Theoretical: solid line—[17], dotted line—[18], dashed line—[19]

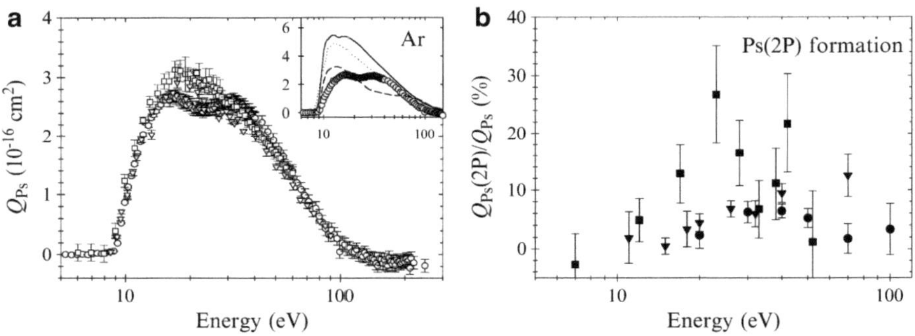

Fig. 8.2 (**a**) experimental and theoretical values of Q_{Ps} for Ar □—[22], ∇—[23], ○—[24]. Also shown in the inset are calculations: dotted line—[26], solid line—[27], dashed line—[28] compared with [24]. (**b**) Proportions of Ps formed in the $2P$ state for He (•), Ar (▼) and Xe (■) [25]

of trajectory effects and the sharing of flux between direct ionization and Ps formation for positrons [21]. Also of note is that Ps formation in atoms appears to be not very significant above 100–150 eV [4]. Displayed on Fig. 8.1b are the most recent experimental and theoretical results for Q_{Ps} from He, illustrating the level of consensus achieved for this simple target. The situation for more complex targets is less satisfactory, especially between experiment and theory but also, despite improved convergence, among recent experiments (e.g. [22–24]) where discrepancies remain especially around the peak, as exemplified in Fig. 2(b) for Ar. The cross-sections for excited-state Ps formation from the noble gases have been recently measured using coincidences between the Lyman-α photon and the residual ion [25]. The proportions of Ps formed in the $2P$ state are shown in Fig. 8.2(b), suggesting increasing excited-state Ps formation with increasing Z, the maximum value for $(2P)$ formation rising from 0.06 ± 0.01 for He to 0.26 ± 0.09 for Xe.

8.2.3 Ionization of molecules

The situation for molecules is less well established, as illustrated in Fig. 8.3 for simple molecules such as CO_2, N_2, H_2O and O_2. Even so, interesting differences emerge in comparison with atoms (e.g. significant formation of Ps at several hundred eV and ionization–excitation) which may provide useful insights into the interaction of positrons with larger molecular systems. As with atoms, the contribution this process makes to Q_i^t is high. Convergence among different experiments has not yet been achieved and theoretical results are absent with the exception of H_2O for which one high-energy calculation exists [29]. Both determinations of Q_{Ps} for O_2 display a distinctive early peak followed by a second peak at higher energies. The local minimum between these peaks is thought to arise from the coupling between Ps formation and excitation of O_2 to the Schumann–Runge band, as present in Q_i^t shown in the inset [30].

It has been recently found [37] that significant ionization–excitation occurs during positron-impact ionization of CO_2 and N_2 (see Fig. 8.4). For both these

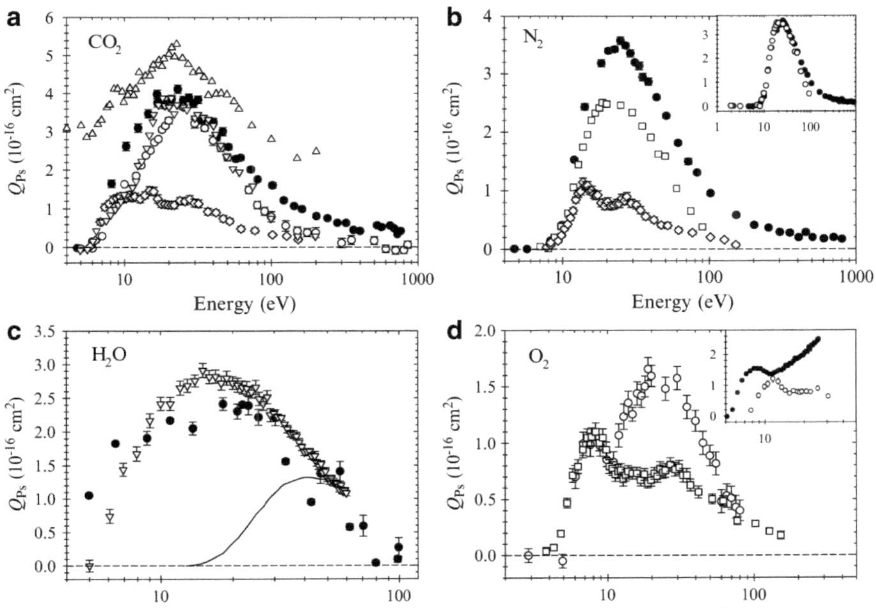

Fig. 8.3 **(a)** measurements of Q_{Ps} for CO_2: [31] (•), [7] (○), the upper (△) and lower (▽) limits of [32] and the early work of [33] (◇). **(b)** three experimental determinations of Q_{Ps} for N_2: •—[31], □—[34], ◇—[33]. Inset: [31] compared those of [34] renormalized to demonstrate energy dependence similarity. **(c)** Q_{Ps} for H_2O: [7] (•), [9] (▽), calculation of [29] (solid line). **(d)** measurements of Q_{Ps} for O_2: [34] (○) and [33] (□). Inset shows local minimum in Q_i^t (○—[35]) for this target coinciding with peak in the excitation cross-section [36]

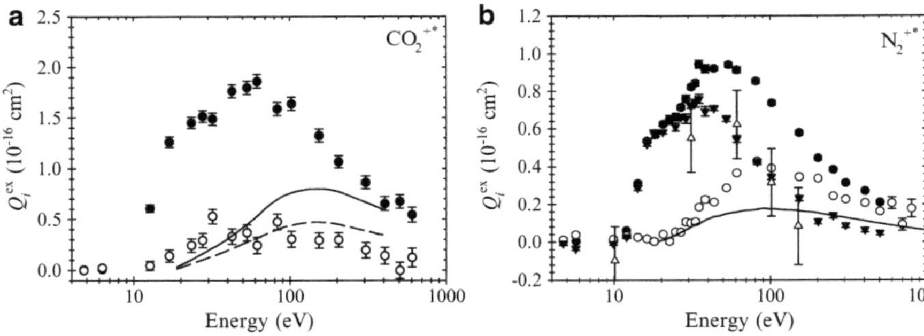

Fig. 8.4 (**a**) measurement of Q_i^{ex} for the $A\ ^2\Pi_u \rightarrow X\ ^2\Pi_g$ (•) and $B\ ^2\Sigma_u^+ \rightarrow X\ ^2\Pi_g$ (○) transitions in CO_2^+ [37], compared with electron measurements of [38] (solid and dotted curves, respectively). (**b**) measurements of Q_i^{ex} for the $B\ ^2\Sigma_u^+ \rightarrow X\ ^2\Sigma_g^+$ transition in N_2^+ [37], showing the total (•), contributions from direct ionization (○) and Ps formation (measured using both the difference between Q_i^{ex} and $Q_i^{ex/+}$ (▼) and photon–γ-ray coincidences (△))

molecules, the cross-section for ionization–excitation (Q_i^{ex}) has been found to be much greater than for electron impact. This enhancement derives largely (in fact, exclusively, in the case of CO_2) from Ps formation, and also extends to several hundred eV. It has been suggested that this occurs because of an accidental resonance between a neutral excited state of the molecule and Ps formation, leaving the resultant ion in an excited state [35, 37]. This phenomenon might be a common occurrence in molecules, the near-degeneracy made probable (in contrast with atoms) by the variation of the molecular interaction energy with internuclear distance and the fine energy-structure associated with vibrational and rotational excitations.

8.3 Positronium interactions

Investigations of Ps scattering generally employ either annihilation techniques (e.g. ACAR, Angular Correlation of Annihilation Radiation or DB, Doppler broadening), or a beam of variable-energy. Whereas the former probe the low energies (i.e. typically less than 3 eV), Ps beams operate at intermediate energies (i.e. between ~7 and 400 eV). Total cross-sections (Q_T^{Ps}) have been measured using the beam method for the inert atoms and simple molecules including H_2O and SF_6 (see [39] and references therein) whilst momentum transfer cross-sections have been measured using ACAR or DB, e.g. [40, 41]. Forward scattering effects have also been investigated [42]. Integral and differential cross-sections for Ps fragmentation have been determined together with first measurements of target ionisation [43–45].

Examples of available measurements for Q_T^{Ps} are shown with corresponding theories in Fig. 8.5. In helium, a fair agreement is achieved between theory and experiment. Recent measurements for He at the lowest available energy, 10 eV,

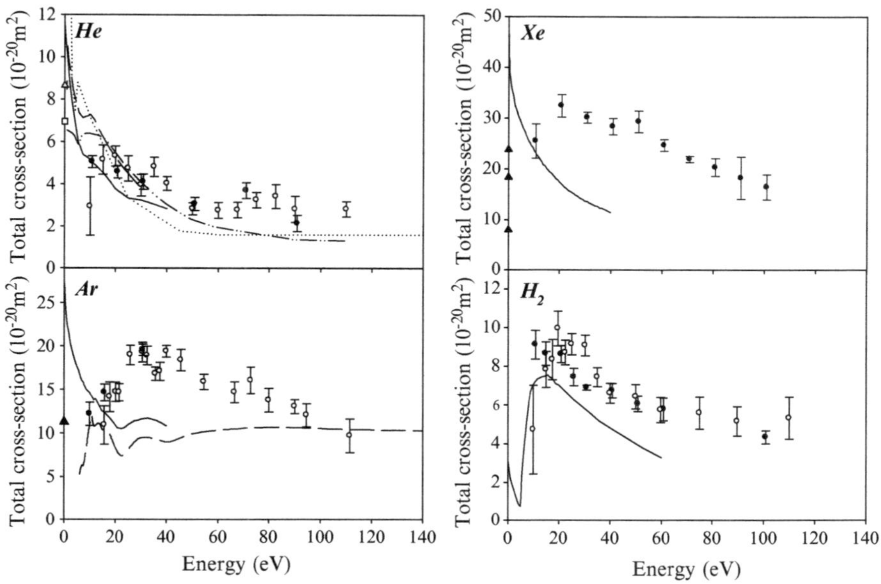

Fig. 8.5 A comparison of Q_T^{Ps} with available theories. All graphs: •, measurements of [39]. He: ○, measurements of [48]; Δ, Q_{el}^{Ps} stochastic variational theory; □, Q_{el}^{Ps} diffusion Monte Carlo approach; ———, target-elastic 22-state close-coupling calculation; — — —, 3-state close-coupling calculation; — · · —, 22-state close coupling target-inelastic calculation; ·········, 3-state close-coupling calculation. Ar: ○, measurements of [49]; ▲, Q_{el}^{Ps} fixed-core stochastic variational method; ——— coupled-pseudostate theory; — — —, 3-state close-coupling calculation. Xe: ▲, Q_{el}^{Ps} fixed-core stochastic variational method; ———, static-exchange approximation. H_2: ○, measurements of [48]; ———, 3-state coupled channel calculation with an exchange model potential [50]

favour a 22-pseudostate close-coupled calculation [46]. With increasing energy, the experimental Q_T^{Ps} approaches another theory [47] which supplements the results of [46] with target-inelastic contributions. Greater discrepancies develop with target complexity as the explicit description of exchange and correlation effects becomes increasingly challenging.

Recently, a comparison of Q_T^{Ps} with corresponding results for equivelocity electrons Q_T^- has revealed a strong similarity between them [39,52] as illustrated in Figs. 8.6 and 8.7, the CO_2 study demonstrating that the similarity may extend to the formation of shape resonances. These findings may be of import for PET dosimetry, since low energy (<15 eV) electron projectiles have been shown to cause single- and double-strand breaks in DNA molecules via the formation of transient anions [53].

The most biologically relevant molecule investigated to-date with Ps projectiles is H_2O. The results of [8] are shown in Fig. 8.7 along with Q_T^- (solid symbols) and the positron total cross-section Q_T^+ (hollow symbols). Whether the similarity between Q_T^{Ps} and Q_T^- extends to this target is unclear as considerable scatter remains among experiments due to the strong permanent dipole of H_2O which introduces

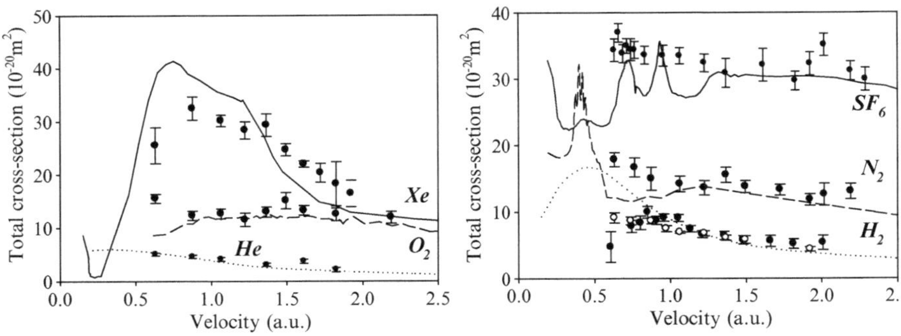

Fig. 8.6 Examples of similarities between equivelocity Q_T^{Ps} and Q_T^- measurements. •, Q_T^{Ps} [39]; ○, [48]; lines, Q_T^- [51]

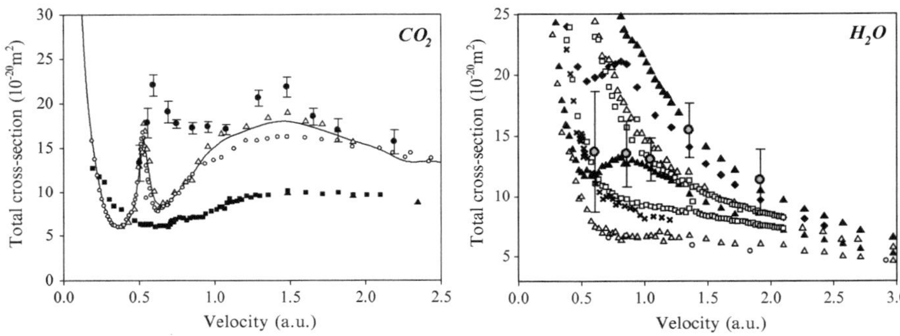

Fig. 8.7 Comparison of Ps, electron and positron total cross-section measurements, Q_T^{Ps}, Q_T^- and Q_T^+, respectively for collisions with CO_2 (left) and H_2O (right). CO_2: •, Q_T^{Ps} [52]; Δ, Q_T^- [54]; ———, average of previous Q_T^- [55]; ○, Q_T^- [56]; ■, Q_T^+ [54]; ▲, Q_T^+ [57]. H_2O: grey symbols, Q_T^{Ps} [8]; white symbols, Q_T^+; black symbols, Q_T^-. Measurements corrected for forward scattering effects are shown as the upper data set where symbols are used more than once. Triangles, [58,59]; squares, [9]; crosses, [6]; diamonds, [60]; ○, [8]

severe forward-scattering effects for electrons and positrons (see Fig. 8.7). Such errors are expected to be smaller for the Ps measurements due to a better experimental angular discrimination ($\sim 1^0$).

Above 6.8 eV impact energy, Ps may break-up into its constituent particles. This process was first investigated theoretically by Ludlow and Walters [61] in which the doubly differential cross-sections of the ejected electrons and positrons for high energy Ps collisions with Xe were calculated. Whilst these conditions are not yet amenable to experimental investigations, much lower energy (15–33 eV) studies have been carried out with He and Xe. Experiments and theories have not yet addressed molecules. The integrated fragmentation cross-section (Q_f^+), encompassing reactions (a), (d) and (f) of Table 8.1, has been determined by

Table 8.1 The lowest energy reactions for Ps–He scattering in which a free positron or electron is produced in the final state. TI and TE denote target-ionisation and target-excitation, respectively

	Reaction	Threshold (eV)	Description
(a)	$Ps + He \longrightarrow e^+ + e^- + He$	6.8	Ps fragmentation
(b)	$Ps + He \longrightarrow Ps + e^- + He^+$	24.2	Ps negative ion formation
(c)	$Ps + He \longrightarrow Ps + e^- + He^+$	24.5	TI
(d)	$Ps + He \longrightarrow e^+ + e^- + He^*$	27.0	Ps fragmentation and TE
(e)	$Ps + He \longrightarrow Ps^* + e^- + He^+$	29.7	Ps excitation and TI
(f)	$Ps + He \longrightarrow e^+ + 2e^- + He^+$	31.4	Ps fragmentation and TI

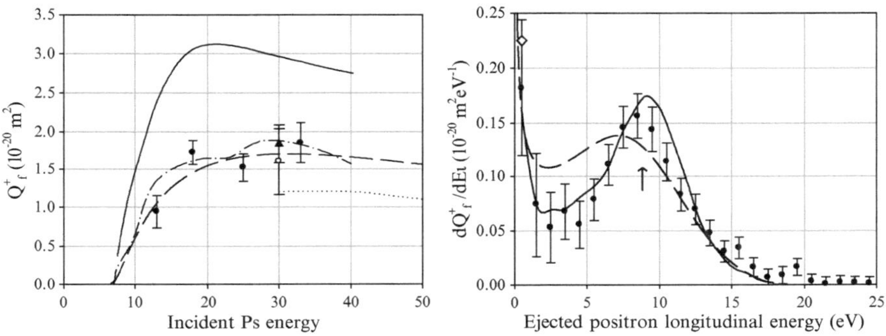

Fig. 8.8 Integrated (left), Q_f^+ and Q_f^-, and differential (right), dQ_f^+/dE_ℓ, fragmentation cross-sections for He. Left: •, Q_f^+ [43]; ▲, Q_f^+ [44]; ∘, Q_f^- [44]; ———, CTMC Q_f^+ [62]; —·—·—, Q_f^+ coupled-state calculation [46]; — — —, Q_f^+ impulse approximation [63]; ·········, Q_f^+ Coulomb–Born approximation [64]. Right: •, dQ_f^+/dE_ℓ [43]; ———, CTMC dQ_f^+/dE_ℓmultiplied by 0.5 [62]; — — —, impulse approximation [63]; ◇, average $dQ_f^+/dE_\ell <$ 1 eV [63]. Half residual energy position: ↑

detecting the ejected positron, a unique signature of Ps fragmentation [43]. The probability of fragmentation , Q_f^+/Q_T^{Ps}, has been found to range in He and Xe from ~20% at 13 eV to ~30-40% at 30 eV, Q_f^+ for Xe being around four times those of He, the latter shown in Fig. 8.8 [44, 45]. Measurements carried out by detecting ejected electrons suggest that the cross-section for ionization of Xe, corresponding to reactions (a) to (f), might be appreciable [45].

Differential cross-sections with respect to the longitudinal energy of the ejected positrons (dQ_f^+/dE_ℓ) have also been measured via a time-of-flight technique (see Fig. 8.8) [43]. The peak observed near half of the residual energy, E_r, where, $E_r = E_{Ps} - 6.8$eV, indicates electron-loss-to the continuum, a process first observed in ion–atom collisions in which the ionised projectile and the ejected electron remain strongly correlated with a small relative velocity. From the peak position and shape, it was deduced that the positron is emitted in the laboratory frame within a ~20° cone around the beam axis, a result supported by classical trajectory Monte Carlo (CTMC) [62] and impulse approximation [63] calculations. The former agrees in shape and the latter in shape and magnitude with the experimental data.

8.4 Outlook

It would be interesting to extend our methods to study positron and Ps induced ionization of molecules relevant to life (e.g. H_2O, CH_4, C_2H_4, $(CH_2)_4O$, CH_2O_2) at incident energies from a few eV up to around 1 keV. Both integral and differential cross-sections could be determined, and the yield of final state ion(s) could be studied in coincidence with one or more of the residual light particles. Our recent studies of positron and Ps interactions with molecules have revealed unforeseen aspects relating to energy deposition in matter whose repercussions might be of import in biomedical applications of positrons.

Acknowledgements We wish to thank the Engineering and Physical Research Council and The European Union for supporting our positron and positronium research.

References

1. R. Campeanu, J. Darewych, A. Stauffer, *XV International Workshop on Low Energy Positron and Positronium Physics* (IOP Publishing, London, 2010)
2. S. Cherry, et al, *Physics in Nuclear Medicine* (Saunders, Philadelphia, 2003)
3. C. Hayes, et al, J Nucl Med **43**, 210 (2002)
4. G. Laricchia, et al, Adv. At. Mol. Opt. Phys. **56**, 1 (2008)
5. P. Van Reeth, G. Laricchia, J.W. Humberston, Physica Scripta **71**, C9 (2005)
6. A. Zecca, et al, J Phys B **39**, 1597 (2006)
7. D.J. Murtagh, et al, Nucl. Instr. Meth. B **247**, 92 (2006)
8. J. Beale, S. Armitage, G. Laricchia, J Phys B **39**, 1337 (2006)
9. C. Makochekanwa, et al, New J. Phys. **11**, 103036 (2009)
10. A. Rich, Rev. Mod. Phys. **53**, 127 (1981)
11. S.D. Benedetti, H.C. Corben, Ann. Rev. Nucl. Sci. **4**, 191 (1954)
12. D.J. Murtagh, et al, J. Phys. B **38**, 3857 (2005)
13. J. Moxom, P. Ashley, G. Laricchia, Can. J. Phys. **74**, 367 (1996)
14. R. Rejoub, B.G. Lindsay, R.F. Stebbings, Phys. Rev. A **65**, 042713 (2002)
15. P. Caradonna, et al, Phys. Rev. A **80**, 032710 (2009)
16. N. Overton, R.J. Mills, P.G. Coleman, J. Phys. B **26**, 3951 (1993)
17. R. Utamuratov, et al, J. Phys. B **43**, 125203 (2010)
18. Y. Cheng, Y. Zhou, Phys. Rev. A **76**, 012704 (2007)
19. C.P. Campbell, et al, Nucl. Instr. Meth. B **143**, 41 (1998)
20. H. Knudsen, J.P. Reading, Phys. Rep. **212**, 107 (1992)
21. K. Paludan, et al, J. Phys. B **30**, L581 (1997)
22. A.C.L. Jones, et al, J. Phys. Conf. Ser. **194**, 072012 (2009)
23. J.P. Marler, J.P. Sullivan, C.M. Surko, Phys. Rev. A **71**, 022701 (2005)

24. G. Laricchia, et al, J. Phys. B **35**, 2525 (2002)
25. D.J. Murtagh, et al, Phys. Rev. Lett. **102**, 133202 (2009)
26. L.J.M. Dunlop, G.F. Gribakin, Nucl. Instr. Meth. B **247**, 61 (2006)
27. S. Gilmore, et al, Nucl. Instr. and Meth. B **221**, 129 (2004)
28. M. McAlinden, H. Walters, Hyperfine Interactions **73**, 65 (1992)
29. P.A. Hervieux, et al, J. Phys. B **39**, 409 (2006)
30. G. Laricchia, J. Moxom, M. Charlton, Phys. Rev. Lett. **70**, 3229 (1993)
31. D.A. Cooke, et al, J. Phys. Conf. Ser. **199**, 012006 (2010)
32. C.K. Kwan, et al, Nucl. Instr. Meth. B **143**, 61 (1998)
33. T.C. Griffith, *Positron Scattering in Gases* (Plenum, New York, 1984), p. 53
34. J.P. Marler, C.M. Surko, Phys. Rev. A **72**, 062713 (2005)
35. G. Laricchia, J. Moxom, Phys. Lett. A **174**, 255 (1993)
36. Y. Katayama, O. Sueoka, S. Mori, J. Phys. B **20**, 1645 (1987)
37. D.A. Cooke, et al, Phys. Rev. Lett. **104**, 073201 (2010)
38. S. Tsurubuchi, T. Iwai, J. Phys. Soc. Jap **37**, 1077 (1974)
39. S. Brawley, et al, Science **330**, 789 (2010)
40. M. Skalsey, et al, Phys. Rev. A **67**, 022504 (2003)
41. F. Saito, Y. Nagashima, T. Hyodo, J Phys B **36**, 4191 (2003)
42. A. Garner, A. Özen, G. Laricchia, J Phys B **33**, 1149 (2000)
43. S. Armitage, et al, Phys Rev Lett **89**, 173402 (2002)
44. S. Armitage, et al, Nucl. Instr. Meth. B **247**, 98 (2006)
45. S.J. Brawley, et al, Nucl. Instr. Meth. B **266**, 497 (2008)
46. J. Blackwood, et al, Phy. Rev. A **60**, 4454 (1999)
47. A. Basu, P. Sinha, A. Ghosh, Physical Review A **63**, 052503 (2001)
48. A. Garner, G. Laricchia, A. Özen, J Phys B **29**, 5961 (1996)
49. A. Garner, A. Özen, G. Laricchia, Nucl. Instr. Meth. B **143**, 155 (1998)
50. P. Biswas, S. Adhikari, J. Phy. B. **33**, 1575 (2000)
51. W. Kauppila, T. Stein, *Adv. At. Mol. Optic. Phys. 26* (Academic Press, San Diego, 1990)
52. S. Brawley, et al, Phys Rev Lett **105**, 263401 (2010)
53. B. Boudaïffa, et al, Science **287**, 1658 (2000)
54. K.R. Hoffman, et al, Phys. Rev. A **25**, 1393 (1982)
55. Y. Itikawa, J. Phys. Chem. Ref. Data **31**, 749 (2002)
56. C.e.a. Szmytkowski, J Phys B **20**, 5817 (1987)
57. C.K. Kwan, et al, Physical Review A **27**, 1328 (1983)
58. O. Sueoka, et al, J Phys B **19**, L373 (1986)
59. M. Kimura, et al, Adv. Chem. Phys. **111**, 537 (2000)
60. C. Szmytkowski, Chemical Physics Letters **136**, 363 (1987)
61. J. Ludlow, H. Walters, *Many-Particle Spectroscopy of Atoms, Molecules, Clusters, and Surfaces* (Kluwer/Plenum, 2001)
62. L. Sarkadi, Phys. Rev. A **68**, 032706 (2003)
63. C. Starrett, M. McAlinden, H. Walters, Phys. Rev. A **72**, 012508 (2005)
64. H. Ray, J Phys B **35**, 3365 (2002)

Chapter 9
Total Cross Sections for Positron Scattering from Bio-Molecules

Luca Chiari, Michael J. Brunger, and Antonio Zecca

Abstract We review recent (2005–2010) results from our total cross section measurements for positron scattering from bio-molecules. In particular we systematically analyse those data in an attempt to understand the role that the long-range dipole interactions (i.e. dipole moment and dipole polarisability) have on the interaction dynamics, when positrons respectively scatter from water, formic acid, tetrahydrofuran, 3-hydroxy-tetrahydrofuran, pyrimidine and di-hydropyran.

9.1 Introduction

Particular interest has grown in recent years, within the scientific community, to investigate the effect that low energy charged particles may cause when entering the human body [1], specifically during medical therapies or diagnostic tests. While most medical devices initially start with very high-energy photons (e.g. X-rays), electrons or positrons (e.g. in positron emission tomography), this high-energy radiation quickly thermalises in the body through processes such as

L. Chiari (✉)
Department of Physics, University of Trento, 38123 Povo (Trento), Italy and ARC Centre for Antimatter-Matter Studies, School of Chemical and Physical Sciences, Flinders University, GPO Box 2100, Adelaide, SA 5001, Australia
e-mail: chiari@science.unitn.it; luca.chiari@flinders.edu.au

M.J. Brunger
ARC Centre for Antimatter-Matter Studies, School of Chemical and Physical Sciences, Flinders University, GPO Box 2100, Adelaide, SA 5001, Australia
e-mail: Michael.Brunger@flinders.edu.au

A. Zecca
Department of Physics, University of Trento, 38123 Povo (Trento), Italy
e-mail: zecca@science.unitn.it

G.G. Gómez-Tejedor and M.C. Fuss (eds.), *Radiation Damage in Biomolecular Systems*, Biological and Medical Physics, Biomedical Engineering,
DOI 10.1007/978-94-007-2564-5_9, © Springer Science+Business Media B.V. 2012

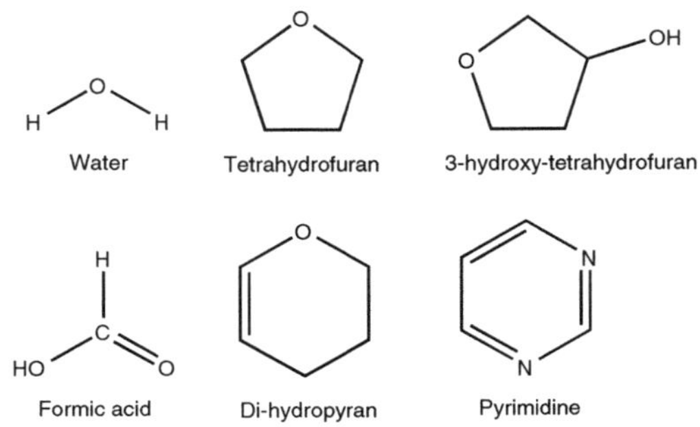

Fig. 9.1 Schematic diagrams representing the structures of the bio-molecules pertaining to this study

direct ionisation which in turn leads to the liberation of significant numbers of lower energy secondary electrons. Those secondary electrons may subsequently attach to the various components of DNA, causing important cell and tissue damage [2,3]. This is the reason why it is interesting to study the interaction between positrons and those molecules which can be considered the "building blocks" of DNA.

Here we review the experimental total cross section (TCS) results on positron scattering from biologically relevant molecules, performed at the University of Trento in the last six years [4–9]. Water has an undoubtedly important role in chemical, bio-physical and environmental processes and is at the base of the smallest constituent of life, namely the biological cell. The structure of the pyrimidine molecule (see Fig. 9.1) is very similar to those of the nucleobases cytosine, uracil and thymine, the component units of RNA and DNA. Indeed those nucleobases can all be thought of as being pyrimidine derivatives. Tetrahydrofuran (THF), 3-hydroxy-tetrahydrofuran (3-h-THF) and di-hydropyran all belong to the same class of cyclic ethers, and can be thought of as "sub-units" or moieties to the nucleotides (again see Fig. 9.1). Therefore these aforementioned five molecules may be considered as sort of prototypes for living matter. The remaining molecule, formic acid, is the simplest organic acid and it is a key component in the formation process of larger bio-molecules such as glycine and acetic acid. Indeed formic acid is at the base of more complex bio-molecules including some of the amino acids. In addition, in this paper we also seek to investigate if there are any trends in the energy dependence of the TCSs for these species, and if so can those trends be related to some of the most important physico-chemical properties (see Table 9.1) of the bio-molecules in question.

In the next section we give some relevant experimental details, by briefly describing the experimental apparatus and our measurement techniques and procedures. We then, in our results and discussion section, report on our data for the total cross section as a function of energy for each bio-molecule, before systematically comparing those results between the various species. Finally, some conclusions are drawn.

Table 9.1 Summary of some of the important physico-chemical properties for the present bio-molecules. Note that the molecular diameter of di-hydropyran was estimated by using Gaussian software [49] with a B3LYP/6-31 G model chemistry. It is likely to be underestimated

Bio-molecule	Molecular diameter (Å)	Dipole polarisability (a.u.)	Dipole Moment (D)	First Ionisation Potential (eV)	Positronium formation Energy (eV)
Water	2.72 [23]	9.78 [24]	1.85 [24]	12.6 [25]	5.8
Formic acid	3.8 [30]	22.5 [30]	1.41 [30]	11.4 [31]	4.6
THF	4.63 [26]	1st and 2nd conformer 47.08 [6]	1st and 2nd conformer 1.63 [27]	9.74 [28]	2.94
3-h-THF	4.63 [26]	1st conformer 50.68 [6] 2nd conformer 50.98 [6]	1st conformer 1.74 [6] 2nd conformer 2.88 [6]	9.8 [29]	3.0
Di-hydropyran	2.8	65.0 [46]	1.38–1.48 [32, 33]	8.6 [34]	1.8
Pyrimidine	5.5 [35]	$\alpha_{xx} \sim 69$ [36, 37] $\alpha_{yy} \sim 71$ [36, 37] $\alpha_{zz} \sim 38$ [36, 37] $\overline{\alpha} \sim 59$	2.28–2.39 [38–40]	9.33–9.73 [41–45]	2.53–2.93

9.2 Experimental apparatus and measurement techniques

The present measurements have been performed with a positron spectrometer whose details have already been provided elsewhere in the literature [4]. Therefore we only briefly remind the reader of its main functions here. The slow positron beam is produced by a ~2 mCi radioactive source (^{22}Na) in conjunction with a 1 μm thick tungsten moderator [10] and a set of electrostatic optics. Our linear transmission experiments are based on the Beer-Lambert law:

$$I_1 = I_0 \exp\left(\frac{-(P_1 - P_0)L\sigma}{kT}\right), \tag{9.1}$$

relating the total cross section σ at each energy to the positron beam count rate I_1 at pressure P_1 (the target molecules' pressure inside the scattering cell), the count rate I_0 at pressure P_0 (the background pressure), the scattering cell length L, Boltzmann's constant k and the temperature T of the target molecules in the scattering cell.

The geometrical length of the scattering cell employed here is $L = 22.1 \pm 0.1$ mm. End effects, due to having entrance and exit apertures of finite diameter at the scattering cell, were also considered. However, such effects are known to minimise if both apertures are of the same small size [11], as in the present case. In our typical application of equation (1), the value of L is always corrected to account for the path increase caused by the gyration of the positrons due to the presence of a focussing axial magnetic field in the scattering region. The magnitude of this magnetic field has usually been in the range $B \sim 3$–12 G, leading to a correction (increase) in L of 1.6–6%.

The angular acceptance at our detector is 4°, corresponding to a solid angle of 4×10^{-3} sr. A recent study by the Australian National University (ANU) group [12], showed that earlier measurements from the Detroit group [13, 14], on argon and xenon, could be reconciled with their own provided a 20° angular resolution correction was applied to the Detroit data at each energy. We note here that our own measurements for these same atomic targets [15] agree with the ANU TCS results to typically $\sim \pm 3\%$. This suggests that our angular resolution effect correction is very close to that for the ANU apparatus in the overlap energy range, a quite remarkable result given that the two apparatus have very different designs [16]. We believe that the Trento apparatus will have a similar performance with other targets, except perhaps for cases where a strongly forward (in angle) peaked elastic differential cross section is in play. In any event this effect can be corrected for [17] provided the elastic differential cross section (DCS) values for the target of interest are available at each relevant energy. Unfortunately these DCSs are not typically available in the literature, so that the TCSs we report represent a lower bound to the true values.

The pressure readings (for P_1 and P_0) have been performed with a MKS Baratron capacitance manometer (Model 628B or 627BX), operated at 100°C or

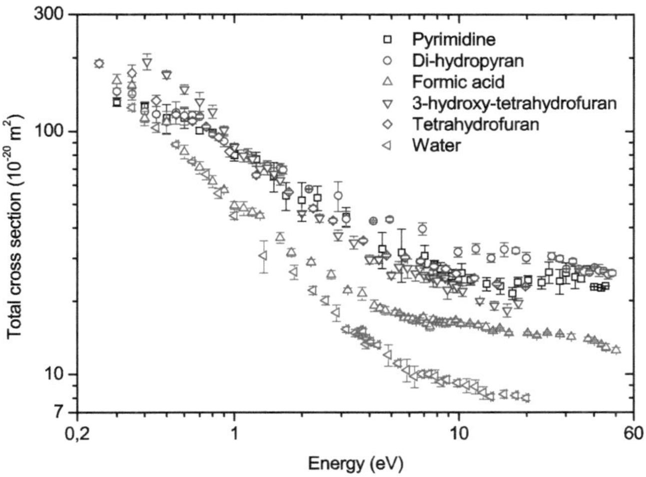

Fig. 9.2 Total cross sections ($\times 10^{-20}$ m^2) as a function of positron impact energy (eV) for the six molecules of this study. See legend in figure for further details

45°C respectively. Since the scattering cell was maintained at room temperature ($T = 24 \pm 2$°C), a thermal transpiration correction was applied to the data. This correction was made according to the model of Takaishi and Sensui [18] and was typically less than 10% of the TCS values over the entire energy range.

The positron beam stability was excellent, usually the positron current only varying within a few percent over times of the order of months. The energy of the beam was calibrated periodically during that period, with a retarding potential analysis (RPA) of the beam following [19]. Measurements throughout the last 6 years have shown a remarkable stability in terms of the variation in the positron energy calibration (<0.05 eV) result, allowing us to estimate that the error in the energy scale calibration is $\sim\pm 0.1$ eV. The RPA also allows us to determine the energy width of the positron beam as ~0.25 eV (FWHM) [10], with an uncertainty of less than 0.05 eV. The typical energy range of the measurements presented here spans from ~0.3 to 20 eV or ~0.3 to 50 eV. The TCS values we give below ~0.5 eV stem from the convolution of the "real" TCS with our present energy beam width. As such caution must be exercised, the true TCS below 0.5 eV is likely to be somewhat larger in magnitude than shown here. We also note that the energy scale of the TCS for water, THF and di-hydropyran, as given in Fig. 9.2, differs very slightly from those given in their respective original papers [4, 5, 8]. This was due to a re-assessment of the original calibrations as a part of this study. In any event, these small energy shifts (~0.15 eV) are compatible with the uncertainty in our energy calibration procedure.

In all of the work reported here [4–9] we have cited the total errors on our TCS data as being typically in the 5–12% range, depending on the actual positron energy being considered. Measurements from the ANU [20] group, for formic acid and water, and the San Diego group [21], for molecular hydrogen, are over the common energy range of interest, in good accord with the present [5,7,22] to typically ±5%. We believe this comparison provides some veracity for the total errors we cite.

9.3 Results and Discussion

In Fig. 9.2 we present our measured total cross sections as a function of energy for positron scattering from the six bio-molecules we have studied [4–9]. The results in this figure should be studied in conjunction with Table 9.1, where some of the important physico-chemical properties [6, 8, 23–46] of those species are summarised. Please note that the errors plotted in Fig. 9.2 are statistical only, and all are at the one standard deviation level. For each molecule the trend in the TCS as a function of energy is quite clear, namely the TCS decreases in magnitude as the incident positron energy increases until the respective positronium formation thresholds (see Table 9.1) are reached. With the opening of the positronium formation channel there is a dramatic change in the slope of the TCS, so that its magnitude at higher energies (at least to the upper limit in energy of our investigations) essentially sits on a plateau. The low energy behaviour of each species, where the elastic, rotational and vibrational channels are open, has been discussed by us previously and is thought to be due to the long-range dipole interaction between the respective species and the incident positrons. Above the energy for positronium formation (E_{Ps}), the electronic-state and direct ionisation channels open progressively and also contribute to the plateau in the TCS that we observe in each species (see Fig. 9.2).

If we now compare the behaviour between the respective species, then several general trends also emerge. Below the relevant E_{Ps} and down to about 0.6 eV, the magnitude of each species' TCS seems strongly correlated to the value of its dipole polarisability (α). For instance both the conformers of THF [47] and 3-h-THF, as well as pyrimidine, have similar values for their dipole polarisabilities, and their corresponding TCSs in this energy range are almost identical. Below 0.6 eV the TCS of 3-h-THF is rather larger than those of the other species (THF and pyrimidine), but this can be understood in terms of its next-lowest-energy conformer (which is present in our 3-h-THF sample) having a much bigger permanent dipole moment (μ) than the others. Similarly, these three species, in this same energy range, have significantly larger TCSs than those for water and formic acid, whose dipole polarisabilities are both much lower in value (see Table 9.1). While it might appear that water (H_2O) and formic acid (HCOOH) do not entirely fit the trend we are espousing, as for $E < 1$ eV their TCSs are almost identical yet $\alpha_{HCOOH} \gg \alpha_{H_2O}$, this can be understood in terms of a sort of "compensation effect" due to $\mu_{H_2O} > \mu_{HCOOH}$. Hence both these long-range interactions are important when looking at the comparative behaviour of these systems. The behaviour of di-hydropyran (see Fig. 9.2) can also be understood in this same light, for energies between 0.6 eV and its E_{Ps} (see Table 9.1). While it has the largest dipole polarisability of all the species in this study, its TCS is almost identical to that for THF, 3-h-THF and pyrimidine. We believe this simply reflects that di-hydropyran has a permanent dipole moment which is amongst the smallest of the bio-molecules we consider, so that a "compensation effect" between the two long-range interactions is prevalent again here. Below $\sim$0.6 eV the situation is complicated by the energy resolution convolution effect on our measured TCSs. As a consequence we do not consider that energy regime further in this paper.

In their study on positron scattering from H_2O and HCOOH, Makochekanwa *et al.* [20] were able to directly measure the positronium formation integral cross sections for both species. They found similar qualitative behaviour for the shapes of both TCSs as a function of energy, in particular no structure was noted. Of great interest from this study was that the ratio of the maximum values of the TCSs for these species seemed to scale with the ratio of their dipole polarisabilities. We believe that this observation, at least to some extent, helps explain the behaviour in Fig. 9.2 for positron energies (E_0) greater than E_{Ps} in each molecule. In particular, the conformers of THF and 3-h-THF, as well as pyrimidine, all have (see Table 9.1) similar values for α and this is reflected in Fig. 9.2 by them all having similar TCS values for $E_0 > E_{Ps}$. Di-hydropyran has a larger dipole polarisability than those three aforementioned molecules, and so in Fig. 9.2 we observe that it has the largest TCSs (remember the y-axis is a log scale) of all the species we consider for $E_0 > E_{Ps}$. Finally, for water and formic acid, the trend in their TCSs is also consistent with the idea that the dipole polarisability is playing a major role in the scattering dynamics for these molecules in this energy regime. The above conclusions are entirely consistent with the findings in Danielson *et al.* [48]. In that study the dependence of positron-molecule binding energies on the molecular properties of the species considered were investigated, with similar results to those we have outlined here being found.

A "curious" feature of the present measurements, which has puzzled us (in the absence of theoretical guidance) for some time, is that the energy dependence of our positron TCSs at energies below a couple of eV, mainly seems to fall like $1/E_0$ or $1/\sqrt{E_0}$. In the $1/E_0$ case this behaviour is seen with the targets pyrimidine, formic acid, 3-h-THF and water. On the other hand, di-hydropyran shows a $1/\sqrt{E_0}$ dependence while the behaviour of THF lies somewhere in between. We have also seen these $1/E_0$ or $1/\sqrt{E_0}$ trends in our measurements on atoms [15] and other molecules (e.g. [22]), so that we do not think it can simply be a coincidence.

9.4 Conclusions

We have systematically examined our positron–bio-molecule TCS scattering data, in an attempt to uncover any underlying trends in that data. We believe that a strong semi-quantitative link between the value of the target dipole polarisability and the strength of the total cross section has been established, with the permanent molecular dipole moment also playing an important, if perhaps more secondary, role. The exception to this last statement was at very low positron impact energies, where there was clear evidence for the important role being played by the dipole moment on the scattering dynamics.

In hindsight, perhaps this effect of the dipole polarisability on the value of the TCSs should not be considered too surprising. This follows as the target dipole polarisabiltiy is in some sense a measure of the spatial extent of the target molecular charge cloud, so it makes intuitive sense that the larger is this electron charge distribution the more likely an incident positron will interact with it in some fashion.

Acknowledgements This work was supported under a Memorandum of Understanding between the University of Trento and the Flinders University node of the ARC Centre for Antimatter-Matter Studies. MJB thanks Flinders University for his Outside Studies Programme (OSP) support, while LC thanks the Department of Physics, University of Trento, for financial support.

References

1. B. Boudaiffa, P. Cloutier, D. Hunting, M.A. Huels, L. Sanche, Science **287**, 1658 (2000)
2. A.-C. Hassan, P.-C. Dugal, L. Sanche, Radiat. Res. **153**, 23 (2000)
3. P.L. Levesque, M. Michaud, L. Sanche, J. Chem. Phys. **122**, 094701 (2005)
4. A. Zecca, C. Perazzolli, M.J. Brunger, J. Phys. B **38**, 2079 (2005)
5. A. Zecca, D. Sanyal, M. Chakrabarti, M.J. Brunger, J. Phys. B **39**, 1597 (2006)
6. A. Zecca, L. Chiari, A. Sarkar, M.J. Brunger, J. Phys. B **41**, 085201 (2008)
7. A. Zecca, L. Chiari, A. Sarkar, M.A.P. Lima, M.H.F. Bettega, K.L. Nixon, M.J. Brunger, Phys. Rev. A **78**, 042707 (2008)
8. A. Zecca, L. Chiari, K.L. Nixon, M.J. Brunger, S. Chattopadhyay, D. Sanyal, M. Chakrabarti, J. Phys. Chem. A **113**, 14251 (2009)
9. A. Zecca, L. Chiari, G. Garcia, F. Blanco, E. Trainotti, M.J. Brunger, J. Phys. B **43**, 215204 (2010)
10. A. Zecca, L. Chiari, A. Sarkar, S. Chattopadhyay, M.J. Brunger, Nucl. Instrum. Methods B **268**, 533 (2010)
11. H.J. Blaauw, F.J. de Heer, R.W. Wagenaar, D.H. Barends, J. Phys. B **10**, L299 (1977)
12. J.P. Sullivan, C. Makochekanwa, A. Jones, P. Caradonna, D.S. Slaughter, J. Machacek, R.P. McEachran, D.W. Mueller, S.J. Buckman, J. Phys. B **44**, 035201 (2011)
13. W.E. Kauppila, T.S. Stein, G. Jesion, Phys. Rev. Lett. **36**, 580 (1976)
14. M.S. Dababneh, W.E. Kauppila, J.P. Downing, F. Laperriere, V. Pol, J.H. Smart, T.S. Stein, Phys. Rev. A **22**, 1872 (1980)
15. A. Zecca, L. Chiari, E. Trainotti, D.V. Fursa, I. Bray, M.J. Brunger, Eur. Phys. J. D, in press (2011) doi:10.1140/epjd/e2011-20333-7
16. C. Surko, G. Gribakin, S.J. Buckman, J. Phys. B **38**, R57 (2005)
17. A. Hamada, O. Sueoka, J. Phys. B **27**, 5055 (1994)
18. T. Takaishi, Y. Sensui, Trans. Faraday Soc. **59**, 2503 (1963)
19. A. Zecca, M.J. Brunger, in *Nanoscale Interactions and Their Applications: Essays in Honour of Ian McCarthy*, ed. by F. Wang, M.J. Brunger (Research Signpost, Trivandrum, India, 2007), p. 21
20. C. Makochekanwa, A. Bankovic, W. Tattersall, A. Jones, P. Caradonna, D.S. Slaughter, K. Nixon, M.J. Brunger, Z. Petrovic, J.P. Sullivan, S.J. Buckman, New J. Phys. **11**, 103036 (2009)
21. J.P. Sullivan, S.J. Gilbert, S.J. Buckman, C.M. Surko, J. Phys. B **34**, L467 (2001)

22. A. Zecca, L. Chiari, A. Sarkar, K.L. Nixon, M.J. Brunger, Phys. Rev. A **80**, 032702 (2009)
23. J.C. Mathai, S. Mori, B.L. Smith, G.M. Preston, N. Mohandas, M. Collins, P.C.M. van Zijl, M.L. Zeidel, P. Agre, J. Bio. Chem. **271**, 1309 (1996)
24. C. Millot, B.J. Costa Cabral, Chem. Phys. **460**, 466 (2008)
25. C.I.M. Beenakker, F.J. de Heer, H.B. Krop, G.R. Möhlmann, Chem. Phys. **6**, 445 (1974)
26. M. Dampc, A.R. Milosavljevic, I. Linert, B.P. Marinkovic, M. Zubek, Phys. Rev. A **75**, 042710 (2007)
27. D. Bouchiha, J.D. Gorfinkiel, L.G. Caron, L. Sanche, J. Phys. B **40**, 1259 (2007)
28. H. Inoue, T. Sakurai, T. Hoshi, J. Okubo, J. Photochem. Photobiol. A: Chem. **72**, 41 (1993)
29. A. Giardini, D. Catone, S. Stranges, M. Satta, M. Taccani, S. Piccirillo, S. Turchini, N. Zema, G. Contini, T. Prospari, P. Decleva, D. DiTommaso, G. Fronzoni, M. Stener, Chem. Phys. Chem. **6**, 1164 (2005)
30. V. Vizcaino, M. Jelisavcic, J.P. Sullivan, S.J. Buckman, New. J. Phys. **8**, 85 (2006)
31. D.R. Lide, *CRC Handbook of Chemistry and Physics*, 79th edn. (CRC, Boca Raton, Fl, USA, 1998)
32. S. Soundararajan, S.V. Anantakrishnan, Proc. Ind. Acad. Sci. **A37**, 278 (1953)
33. L.G. Scanlon Jr, W. Krawiec, A.T. Yeates, D.S. Dudis, Electrochim. Acta **40**, 2431 (1995)
34. A.A. Planckaert, J. Doucet, C. Sanderfy, J. Chem. Phys. **60**, 4846 (1974)
35. K.G. Spears, M. El-Manguch, Chem. Phys. **24**, 65 (1977)
36. C. Hattig, O. Christiansen, S. Coriani, P. Jorgensen, J. Chem. Phys. **109**, 9237 (1998)
37. B. Jansik, D. Jonsson, P. Salek, H. Agren, J. Chem. Phys. **121**, 7595 (2004)
38. G.L. Blackman, R.D. Brown, F.R. Burden, J. Mol. Spectrosc. **35**, 444 (1970)
39. P. Chen, R.A. Holroyd, J. Phys. Chem. **100**, 4491 (1996)
40. Z. Kisiel, L. Pszczólkowski, J.C. López, J.L. Alonso, A. Maris, W. Caminati, J. Mol. Spectrosc. **195**, 332 (1999)
41. E.D. Bergmann, Ann. N. Y. Acad. Sci. **158**, 140 (1969)
42. R. Buff, J. Dannacher, Int. J. Mass Spectrom. Ion Process. **62**, 1 (1984)
43. M.J.S. Dewar, S.D. Worley, J. Chem. Phys. **51**, 263 (1969)
44. M.N. Piancastelli, P.R. Keller, J.W. Taylor, J. Am. Chem. Soc. **105**, 4235 (1983)
45. A.W. Potts, D.M.P. Holland, A.B. Trofimov, J. Schirmer, L. Karlsson, K. Siegbahn, J. Phys. B **36**, 3129 (2003)
46. http://www.chemspider.com/Chemical-Structure.7789.htm
47. A. Giuliani, P. Limão-Vieira, D. Duflot, A.R. Milosavljević, B.P. Marinković, S.V. Hoffmann, N. Mason, J. Dewiche, M..J. Hubin-Franskin, Eur. Phys. J. D **51**, 97 (2009)
48. J.R. Danielson, J.A. Young, C.M. Surko, J. Phys. B **42**, 235203 (2009)
49. M.J. Frisch *et al.*, *Gaussian 03, revision C.02* (Gaussian Inc., Wallingford, CT, USA, 2004)

Chapter 10
Soft X-ray Interaction with Organic Molecules of Biological Interest

P. Bolognesi, P. O'Keeffe, and L. Avaldi

Abstract Molecular fragmentation following absorption of X-ray radiation is one of the effects of radiation damage. To investigate this process in a class of organic molecules of biological interest the inner shell excitation, ionisation, electron decay and fragmentation of the pyrimidine molecule have been fully characterised by electron and ion spectroscopy techniques and theoretical calculations. These techniques have allowed us to observe site-selective molecular fragmentation of the N and C (1 s) core excited states and to explain that it is governed by the final singly charged ion state reached by resonant Auger electron decay.

10.1 Inner shell mechanisms for radiation damage

It has long been known that soft and hard X-ray exposure causes dramatic effects on living organisms, producing both reparable and irreparable damage, which lead to alterations, malfunctioning and even mutations and cellular death. There is nowadays substantial evidence that macroscopic damage may be initiated at the microscopic scale of the DNA chain contained in living cells [1]. These DNA alterations are due to significant energy deposition either in the DNA constituents or in its neighbouring molecules, causing direct and indirect effects respectively.

On the other hand, the same pathogenic effects of X-ray radiation are proficiently used in radiotherapy for cancer treatment. In this respect, the harmful action of ionising radiation on tumour cells can be further enhanced via the use of properly designed radiosensitisers [2], aiming at a more selective and amplified damage of the tumour rather than healthy cells. Pyrimidine and halogenated pyrimidines play an important role in this context. The former is the building block of cytosine, thymine and uracil DNA/RNA bases, while the latter group of molecules, which mimic the

P. Bolognesi (✉) • P. O'Keeffe • L. Avaldi
CNR-IMIP, Area della Ricerca di Roma 1, Montelibretti Scalo, Italy
e-mail: paola.bolognesi@imip.cnr.it; patrick.okeefffe@imip.cnr.it; lorenzo.avaldi@imip.cnr.it

G.G. Gómez-Tejedor and M.C. Fuss (eds.), *Radiation Damage in Biomolecular Systems*, Biological and Medical Physics, Biomedical Engineering,
DOI 10.1007/978-94-007-2564-5_10,

structure of pyrimidine, apart from the inclusion of a halogen atom, represents an important class of prototype radiosensitisers, such as for example the bromo- or iodo-deoxyuridine (UdR) or the 5-fluorouracil (5-FU).

Until recently, little was known about how these drugs produce the radiosensitising effect [3]. Among the physical mechanisms proposed to explain the radiosensitising effect of halogenated pyrimidines [4], inner shell excitation/ionisation are considered as the primary source of the electron and/or the responsible for the enhanced and selective molecular fragmentation induced in the radiosensitiser rather than on the unsubstituted molecule. This has led to an extensive investigation of the electronic structure and chemical properties of pyrimidine and some of its halogenated substitutes via inner shell photoabsorption [5], photoionisation [6, 7], electron decay [8, 9] and photofragmentation [10] both from an experimental and theoretical point of view.

On the experimental side, soft X-ray tunable radiation coupled to electron/ion spectroscopy on gas phase prototype organic molecules of biological interest provides a unique tool to i) investigate fundamental chemical physical mechanisms of radiation damage active at the single-molecule level and ii) disentangle such mechanisms from the indirect effects, due to the surrounding environment hosting the DNA. The highly differential and selective results provided by electron/ion spectroscopic techniques necessarily require theoretical support for the interpretation of the complex spectroscopy and dynamics taking place on polyatomic molecules. Density Functional Theory (DFT) [11] is an extremely successful approach for the description of the electronic properties of many electron systems. It has allowed the calculation of the energy of valence [7] and core ionised cations as well as the accurate evaluation of the chemical shifts of DNA/RNA bases [12, 13] and it has been recently used to predict time-resolved fragmentation of dications of isolated uracil and uracil embedded in water environment [14].

The use of complementary experimental and theoretical methods is essential in providing a comprehensive description of the target molecules, which is in turn the key knowledge required for the understanding of the dynamical processes. The present work establishes a clear characterisation and understanding of the behaviour of the pyrimidine molecule exposed to soft X-ray radiation. This has the dual purpose of providing the benchmark data for the study of the radiosensitising effect in the halogenated pyrimidines and determining the methodology of the work. In section 2 an overview of the spectroscopic characterisation of the pyrimidine molecule with brief glimpses at the halogenated cases is summarised. The mechanisms of photofragmentation following the creation of inner shell vacancies are presented and explained in section 3. In section 4 the conclusions and future perspectives are discussed.

10.2 The pyrimidine molecule

The six-member ring of pyrimidine with two nitrogen and four carbon atoms as well its Cl and Br compounds are shown in Fig. 10.1.

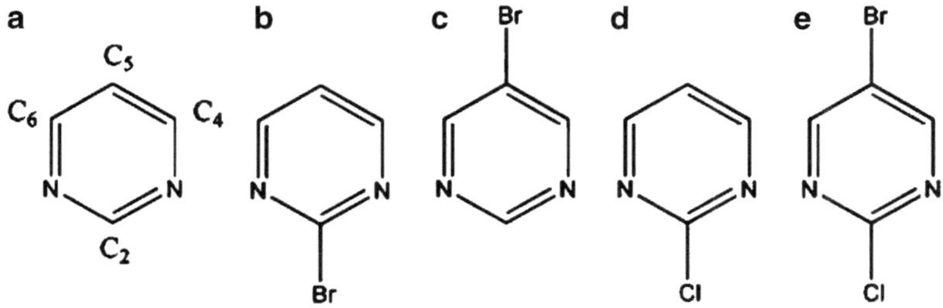

Fig. 10.1 Schematic representation of the pyrimidine molecule **(a)** and some halogenated pyrimidines **(b–d)**. These molecules are selected with the purpose of investigating the effect of the halogenation on the pyrimidine ring as function of the halogen atom (chlorine/bromine) or atomic site of halogenation (2/5)

Fig. 10.2 The C (1 s) XPS spectrum of pyrimidine

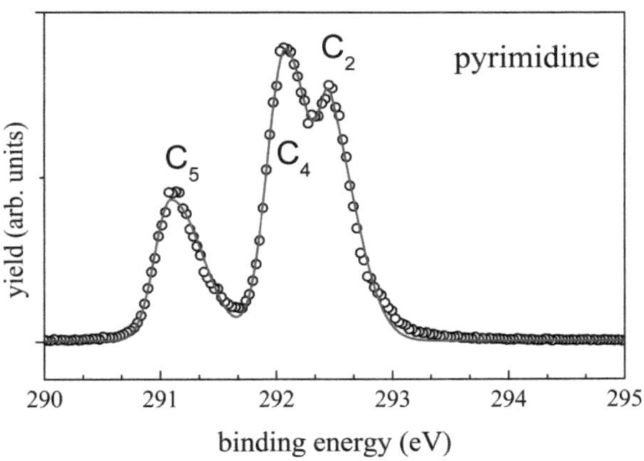

Photoionisation/excitation and subsequent electron decay processes involving the absorption of soft X-rays, hυ, by the target molecule, M, can be outlined in the following scheme:

$$h\nu + M \overset{ionisation}{\longrightarrow} M^{+}(1s^{-1}) + e^{-}_{ph} \overset{Auger\text{-}decay}{\longrightarrow} M^{2+}(valence^{-2}) + e^{-}_{ph} + e^{-}_{AE} \tag{10.1}$$

$$h\nu + M \overset{excitation}{\longrightarrow} M^{*}(1s^{-1}\pi/\sigma) \overset{Resonant\text{-}Auger\text{-}decay}{\longrightarrow} M^{+}(valence^{-1}) + e^{-}_{RAE} \tag{10.2}$$

where e^{-}_{ph}, e^{-}_{AE} and e^{-}_{RAE} represent the photoelectron, the Auger and Resonant Auger electrons, respectively.

The inner shell ionisation of pyrimidine has been recently studied by X-ray photoemission spectroscopy (XPS) [6], showing significant chemical shifts of about 1 and 1.4 eV among the three non-equivalent carbon atoms, see Fig. 10.2.

Table 10.1 Experimental and theoretical binding energies of the C and N(1 s) of pyrimidine [6]

	Binding energy (eV) Experiment	Binding energy (eV) Theory
C_5	291.09 ± 0.04	291.09
$C_{4/6}$	292.08 ± 0.02	292.08
C_2	292.48 ± 0.03	292.48
N	405.23 ± 0.04	405.23

This can be qualitatively explained in terms of the different chemical environments experienced by the three C atoms involved in different chemical bonds. So for example, the C_2 atom, positioned in site 2 of the ring, is the most affected by the higher electronegativity of the two neighbouring N atoms. This results in the highest binding energy as compared to the $C_{4/6}$ and C_5 atoms, which are further and further away from the N atoms (see Table 10.1).

A more quantitative analysis and deeper understanding of the XPS spectra of pyrimidine and its halocompounds have been achieved via DFT calculations [6] which matched very well all the measured experimental binding energies. In the halogenated pyrimidine cases, where different halogen atoms/sites of halogenation are considered, the binding energy of the carbon atom involved in the halogenation systematically shows the largest increase in binding energy, due to the higher electronegativity of the newly bound halogen atom. The DFT calculations have also allowed an explanation of these chemical shifts as being the result of the interplay between the inductive and resonant effects in the charge redistribution screening the core hole [6]. Similar effects due to the different chemical environment of the non-equivalent C_2, $C_{4/6}$ and C_5 atoms along the pyrimidine ring have been also observed and explained [5] in core excitation via near-edge x-ray fine structure spectroscopy (NEXAFS), see Fig. 10.3a. In this case in addition to the pure assignment of all spectral features in the spectrum, the DFT study has also provided insight into the electronic and geometrical structure of the molecules, accessing information such as bond lengths for example.

The core-hole states are highly unstable and rapidly decay, most likely via non-radiative processes that end up in doubly/singly charged ion states in the case of the Auger/Resonant Auger decay processes, respectively (Eqs. 10.1 and 10.2). The electron emission produced by these non-radiative decays is a significant source of electrons which may lead to radiation damage [15]. The energy distribution of these electrons has been characterised by measuring the Auger (AE) and resonant Auger electron (RAE) spectra respectively [8,9]. The RAE decays, Fig. 10.3b [9] populate one-hole and two-hole-one-particle final states in the so called 'participator' and 'spectator' transitions, respectively. These states correspond to increasingly excited states of the singly charged ion. The RAE spectra resemble the valence photoelectron spectrum (PES), Fig. 10.3c and Table 10.2 [7]. However, even though the final electronic configuration is the same, the intensity distribution is different depending on whether the electronic state has been reached by direct photoionisation or by decay of an intermediate, resonantly excited state. This is clearly seen in

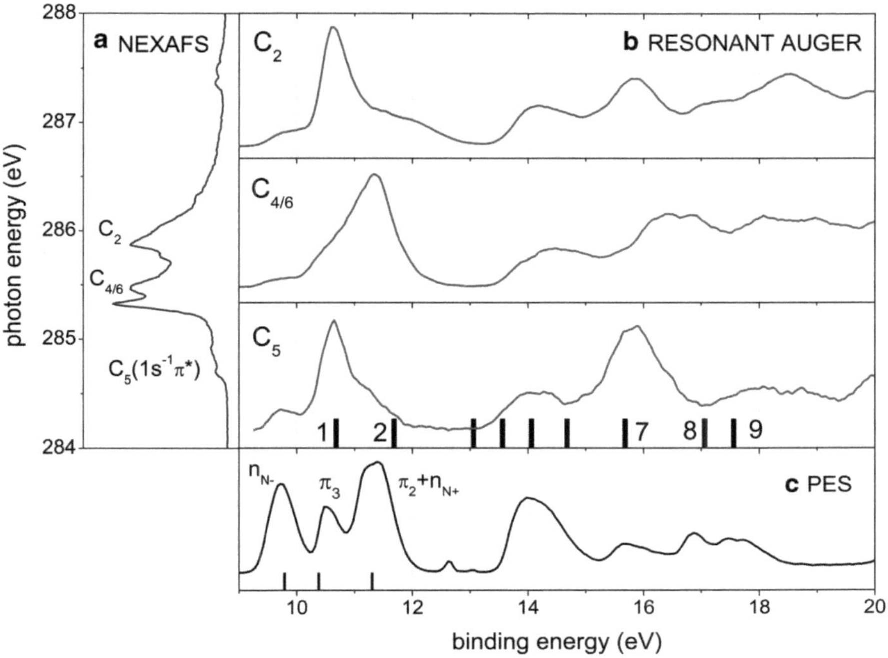

Fig. 10.3 For the pyrimidine molecule **(a)** C(1 s) NEXAFS spectrum in the region of the π^* transitions for the three non-equivalent C sites. Finer structure attributed to vibrational features is clearly visible [5]. **(b)** The C RAE spectra reported in binding energy scale of the final singly charged ion state. The vertical bars labelled from 1 to 9 represent the selected RAE transitions discussed in section 3. **(c)** Photoelectron spectrum measured at 30 eV photon energy. The assignment of the four lowest ionic states is also reported in the figure (see also Table 10.2 and Fig. 10.4) [7]

Table 10.2 Experimental and theoretical [7, 9] binding energies, assignment and corresponding Mulliken population of the four lowest valence shell states of pyrimidine

MO	Mulliken population	Experiment (eV)	OVGF//cc-pVTZ
n_{N-}	$N_{1,3}$ (66.8%); $C_{4,6}$(16.8%)	9.73 ± 0.02	9.79
π_3	C_5(42.9%); C_2(22.2%); $C_{4,6}$(19.4%); $N_{1,3}$ (15.4%)	10.53 ± 0.02	10.38
π_2	$N_{1,3}$ (65.8%); $C_{4,6}$(33.8%)	11.20 ± 0.20	11.30
n_{N+}	$N_{1,3}$ (57.6%); $C_{4,6}$(13.8%); C_5(13.4%);	11.4 ± 0.30	11.31

Fig. 10.3, where the three RAE spectra corresponding to the decay of the C_2, $C_{4/6}$ and $C_5(1\,s^{-1}\pi^*)$ excited states of pyrimidine are reported on the binding energy scale of the final state and compared to the PES spectrum obtained in the direct photoionisation. The three RAE spectra are different among each other and with respect to the PES spectrum.

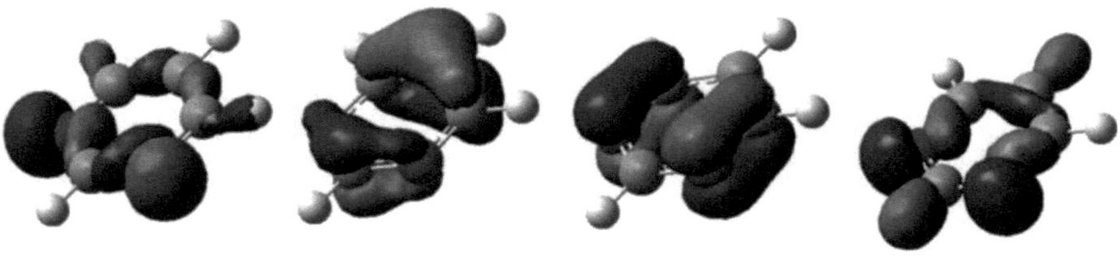

Fig. 10.4 The Hartree–Fock molecular orbitals of pyrimidine for the four lowest molecular orbitals [7]

The intensities of the RAE spectra depend on the overlap of the wave functions of the intermediate and final states, i.e. the RAE decay favours the population of final ionic states with similar localisation of the charge distribution as in the intermediate state. This can be deduced on the basis of the electronic assignment and charge distribution of the intermediate and final states reported in Table 10.2 and Fig. 10.4, respectively. So for example the decay to the n_{N-} state is very weak in all cases of Fig. 10.3b, while the decay to the π_2 state is strongly favoured from $C_{4/6}(1\,s^{-1}\pi^*)$.

10.3 The photofragmentation of pyrimidine

The pioneering work of Sanche and co-workers has shown that low-energy electrons of even less than 3 eV can cause single- and double-strand breaks of plasmid DNA [16] via dissociative electron attachment (DEA) processes. Other efficient mechanisms for molecular fragmentation involve the direct collision with heavy ions [17, and references therein] and the absorption of high energy radiation [18, and references therein]. The latter process, with formation of inner shell vacancies, whose subsequent decay inevitably lead to the molecular fragmentation with one or more charged fragments in the final state, is the subject of the present discussion.

The photon energy dependency of the molecular fragmentation of pyrimidine has been investigated by measuring the time-of-flight (TOF) mass spectrum from the appearance potential regions [19] up to the N(1 s) ionization threshold, see Fig. 10.5. The mass spectrum of pyrimidine can be roughly divided into the six regions labelled A–F in Fig. 10.5. As a rule of thumb, these regions can be interpreted as charged fragments with one to six ring atoms, going from A to F respectively. Atomic fragments $H^+/C^+/N^+$ form group A while the parent ion $C_4N_2H_4{}^+$ belongs to group F. Within each group several fragmentation channels, each one with multiple hydrogen losses are present and make the detailed analysis far more complicated than the rough assignment just presented. As a general feature, we can observe that *i*) fragments with five-members of the original pyrimidine ring (group E) are not observed at any of the investigated photon energies. This could be explained by the unstable species corresponding to these masses, not living long

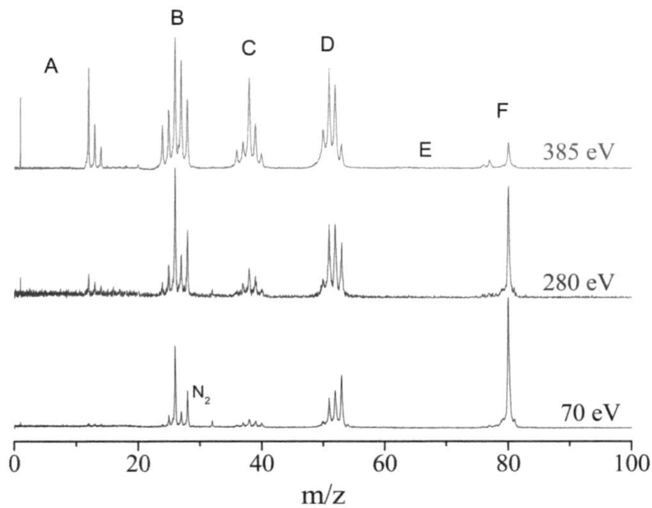

Fig. 10.5 Pyrimidine time-of flight mass spectra measured at several photon energies using tunable synchrotron radiation at the Gas Phase photoemission beamline of the Elettra storage ring, Trieste, and a pulsed TOF spectrometer using a 5 kHz pulse generator. The spectra are reported on an arbitrary intensity scale. Contamination of N_2 has been signalled

enough to reach the detector within the typical few μs time of flight; *ii*) there is a significant reduction in parent ion formation (F) with increasing intensity of groups B–D and higher probability for multiple H losses as the photon energy increases; *iii*) groups A and C, not observed up to 20 eV [19] and only barely visible at 70 eV photon energy show the most significant increase near the opening of C/N(1 s) inner-hole vacancies. Then the first, quite predictable observation is that the fragmentation pattern of pyrimidine, as already observed in thymine [18], depends on photon energy, with the molecule being more fragmented, i.e. more damaged, as the photon energy increases.

Due to the 'atomic-like' nature of inner shell electrons, which makes them rather localised on specific molecular sites, the question arises whether the molecular fragmentation induced by inner shell processes is site-selective, i.e. affected by the 'localisation' of the core hole [20]. The pyrimidine molecule has two equivalent N atoms and three non-equivalent C sites chemically shifted and well resolved in the XPS spectrum, see Fig. 10.2. However, at photon energies above the ionisation thresholds we cannot localise the inner hole, because all the energy allowed states contribute to the measured ion spectrum. To add site-selectivity to localise the atomic site of the molecule where the energy of the photoabsorption process has been 'deposited', we can use inner shell resonant excitation. Indeed, at variance with the core ionisation (10.1) the core excitation (10.2) being a resonant process allows for an unambiguous control of the core hole localisation and corresponding electronic symmetry in the photon absorption process just by a suitable choice of the photon energy, see Fig. 10.3a.

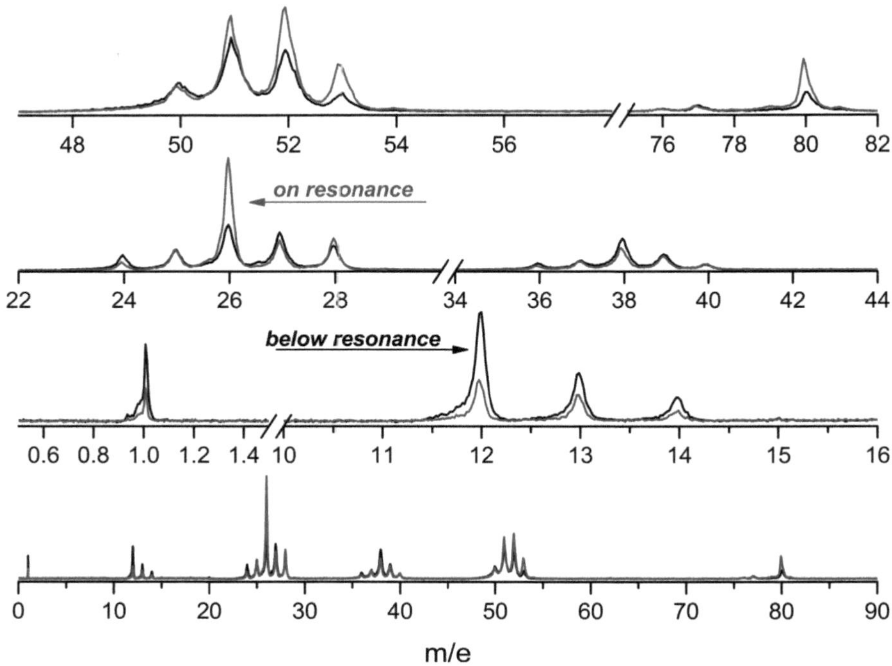

Fig. 10.6 The TOF spectra of pyrimidine measured at the N(1 s^{-1} π*) resonance, 398.8 eV photon energy [5] (red) and in the ionization continuum below the resonance, at 385 eV (black). In the bottom plot there is an overview, while expanded views are reported in the upper plots

Now, how does this affect the molecular fragmentation? It is well established that there is an increased total ion yield production, due to enhanced photoabsorption cross section on resonance. Furthermore, we observed for the pyrimidine case that the fragmentation pattern itself depends on the location of the inner shell hole, so the formation of some specific charged fragments is enhanced on resonance, with respect to others. The experimental procedure that led to this conclusion consisted of measuring TOF spectra at selected photon energies on/off specific NEXAFS resonances, specifically the C_2, $C_{4/6}$, C_5 and N(1 s^{-1}π*) transitions. For each case, after a background subtraction, the on/off-resonance spectra were normalised to each other so that the total area is made to be the same and they can be compared. As an example of such procedure, the TOF spectra measured at the photon energy of the N(1 s^{-1}π*) resonance and just below it, are reported in Fig. 10.6.

This data analysis shows *relative intensities*, stressing the differences in the fragmentation patterns at photon energies on/off inner shell resonant excitations. As already mentioned, on an absolute scale all fragmentation channels gain intensity on resonance. Figure 10.6 shows that some channels gain more than others so that the fragmentation is not just more intense on resonance, but also qualitatively different.

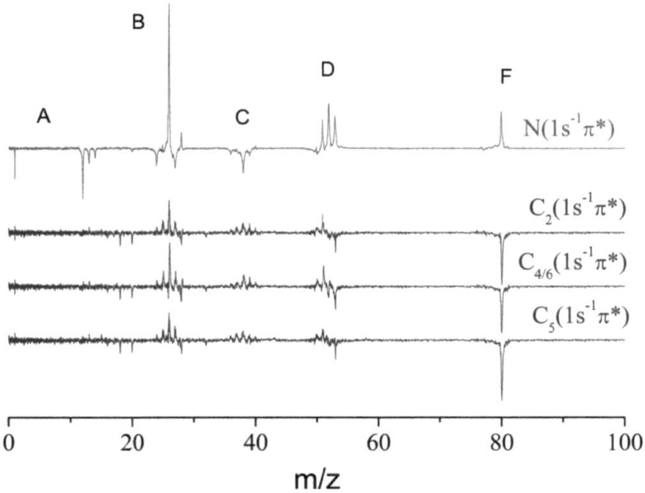

Fig. 10.7 The 'difference' TOF mass spectra for the N and the three non-equivalent C sites of pyrimidine obtained as difference between the on/off-resonance spectra (see text). The off-resonance spectra are measured at 280 and 385 eV for the C and N cases respectively. The on-resonance spectra are measured at the C_2, $C_{4/6}$, C_5 and $N(1\,s^{-1}\pi^*)$ transitions energies

So for example the loss of one or two neutral HCN groups (m/z around 53 and 26 respectively) as well as parent ion formation ($m/z = 80$) is enhanced at the $N(1\,s^{-1}\pi^*)$ resonance, while less atomic species are observed. This behaviour is quite different from the general photon energy dependency discussed in Fig. 10.5. Furthermore, the two spectra of Fig. 10.6 have been measured within only about 14 eV of each other. So this is very unlikely to be an effect due to the photon energy, but has to be related to the resonant excitation itself. A more direct comparison of the on/off resonance spectra reported in Fig. 10.6 can be made by subtracting one spectrum from the other. The results for the N as well as for the three C cases are reported in Fig. 10.7, where features above the zero-line represent fragmentation channels that have been further enhanced on resonance with respect to the fragmentation channels having features below the zero-line.

Large differences between the C and N cases are clearly visible in the region of the parent ion, F, as well as in the region of the atomic species, A. Interestingly, within some groups, like B and D for example, there are peaks showing positive as well as negative features probably corresponding to the different channels contributing to the same group. The observed site-selectivity of Fig. 10.7 can be summarised stating that an energy deposition on the N sites favours parent ion formation, with loss of one/two HCN groups, while an energy deposition on the C sites is more likely to produce a molecular breaking, with a relative reduction of parent ion formation. Negligible differences are observed among the three non-equivalent C sites.

These inner-shell excited neutral states can decay either to the ground and excited states of the singly charged ion or to doubly charged ion states. Depending on the final state and the internal energy available to the ion, fragmentation will occur following different patterns [14, 21]. The results presented so far are concerned only with the charged fragments produced, regardless of the particular decay path of the neutral inner-shell excited state. Measuring the mass spectrum in coincidence with a specific resonant Auger electron which defines the final ionic state gives the full control of both intermediate state, selected by photon excitation, and final ion state, selected by the kinetic energy of the resonant Auger.

At some selected resonant Auger transitions following the C_5 $1s \rightarrow \pi^*$ excitation, marked in Fig. 10.3b by the bars labelled 1 to 9 moving towards increasing binding energies of the residual ion, we measured in coincidence the mass spectrum. The results, reported in Fig. 10.8, show that the 'coincidence spectra' are very different with respect to those measured at the same photon energy but without any selection of the final state. First of all the coincidence mass spectra are less complex. Parent ion formation dominates the spectrum up to binding energy of about 12 eV. Above 12 eV, as we move towards more excited states of the singly charged ion molecular fragmentation occurs, giving rise to the smaller fragment of group D and then B. This is consistent with the measurement of the ion appearance energy of the different fragmentation channels of pyrimidine [19], measured at 12.27 ± 0.05 eV, 13.75 ± 0.1 and 14.2 ± 0.2 eV for m/z 53, 52 and 26 respectively. The corresponding fragmentation is reported in Fig. 10.8.

No appreciable changes are observed changing the site of excitation. So these more selective measurements suggest that the fragmentation does not depend on the intermediate, but only on the final singly charged ion state. In other words, as long as we fix the final state reached by the electron decay of the intermediate, inner shell excited state, it does not matter anymore whether the energy was initially deposited in the C or N atoms, as the molecular fragmentation follows the same pattern.

Indeed, as already discussed in the previous section and Fig. 10.3b, the electron decay strongly depends on the intermediate inner shell excited state. So we can infer that the site-selectivity observed in the inner shell resonant excitation is due to the selective population of the different final ion states in their electron decay.

Preliminary investigations of the halogenated pyrimidine [22] cases have shown promising results, revealing how the halogen atom can play a significant role in the molecular fragmentation as well as in the site-selectivity of the molecular bond breaking.

10.4 Summary and conclusions

A broad investigation of the photofragmentation of the pyrimidine molecule at several photon energies from the VUV to the soft X-ray energy range via tunable synchrotron radiation has allowed the identification of the site-selective character of the fragmentation of this important molecule, which is the building block of some

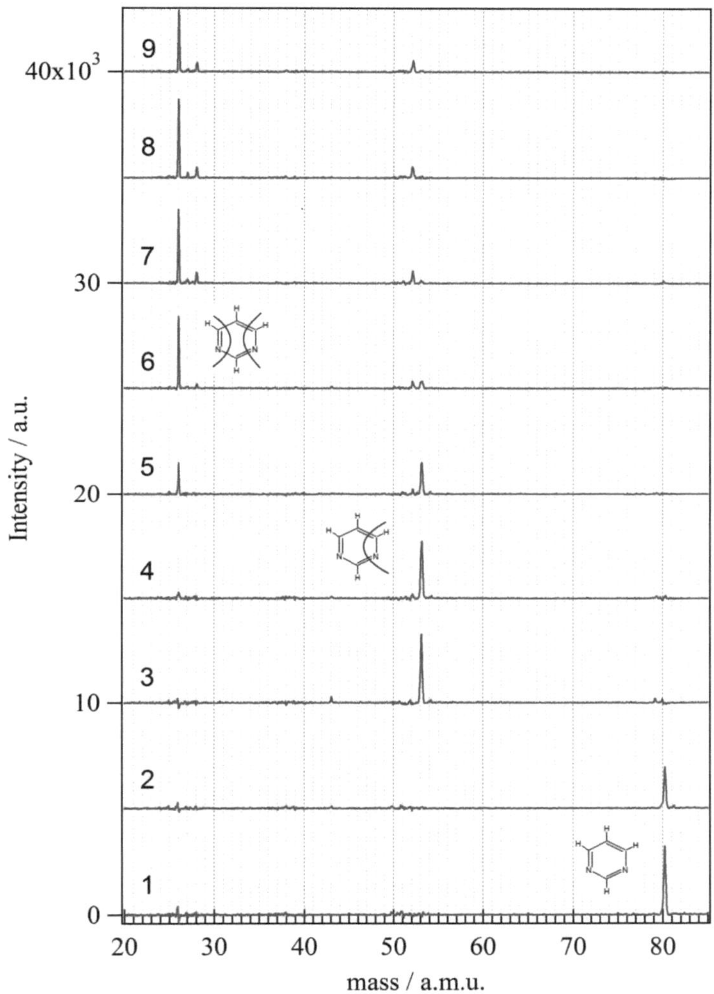

Fig. 10.8 The TOF mass spectra of pyrimidine measured in coincidence with the resonant Auger electrons of selected kinetic energies, labelled in Fig. 10.3b emitted following the excitation of the C_5 1 s → π^* transition

DNA/RNA bases playing a relevant role in the alphabet of life. The experiments have shown that the fragmentation pattern significantly depends on the atomic site of the energy deposition. More selective coincidence experiments combined with a broad experimental and theoretical characterisation of the pyrimidine molecule have allowed us to fully explain this observation as being governed by the final singly charged ion state reached by the fast resonant Auger electron decay rather than the intermediate state itself.

The present work provides a comprehensive characterisation of the pyrimidine molecule and suggests an effective methodology for the investigation and interpretation of the dynamics of its fragmentation. These results provide benchmark data for the interpretation of the fragmentation dynamics of the halogen substituted pyrimidine molecules and are expected to shine some light on possible fundamental chemical physical mechanisms of the radiosensitising effects.

References

1. M.A. Hervé du Penhoat, et al., Int. J. Radiat. Biol. **86**, 205–219 (2010)
2. P. Wardman, Clin. Oncol. **19**, 397–417 (2007)
3. C.J. McGinn, et al., J. Natl. Cancer Inst. **88**, 1193–1203 (1996)
4. R. Watanabe, H. Nikjoo, Int. J. Radiat. Biol. **78**, 953–966 (2002)
5. P. Bolognesi, et al., J. Chem. Phys. **133**, 034302 (2010); G. Vall-llosera, et al., J. Chem. Phys. **128**, 044316 (2008)
6. P. Bolognesi, et al., J. Phys. Chem. **113**, 13593–13600 (2009)
7. A.W. Potts, et al., J. Phys B: At. Mol. Opt. Phys. **36**, 3129-3143 (2003); P. O'Keeffe, et al., Mol. Phys. **107**, 2025–2037 (2009)
8. L. Storchi, et al., J. Chem. Phys. **129**, 154309 (2008)
9. P. Bolognesi, et al, to be submitted
10. P. Bolognesi, et al., J. Phys. Conf. Ser. **212**, 012002 (2010)
11. G. Tu, et al., J. Chem. Phys. **127**, 174110 (2007)
12. F. Wang, J. Phys. Conf. Ser. **141**, 012019 (2008)
13. Y. Takahata, et al., Int. J. Quantum Chem. **106**, 2581–2586 (2006)
14. P. López-Tarifa, Fragmentation dynamics of biomolecules in gas phase and water environment. PhD Thesis, Universidad Autonoma de Madrid (2011)
15. K.G. Hofer, Acta Oncol. **39**, 651-657 (2000); J.A. O'Donoghue, T.E. Wheldon, Phys. Med. Biol. **41**, 1973-1992 (1996); A.I. Kassis, et al., Rad. Res. **109**, 78–89 (1987)
16. B. Boudaïffa, et al., Science **287**, 1658 (2000)
17. T. Schlathölter, et al., Chem. Phys. Chem. **7**, 2339–2345 (2006)
18. G.G.B. deSouza, et al., J. Phys. Conf. Ser. **88**, 012005 (2007)
19. M. Schwell, et al., Chem. Phys. **353**, 145–162 (2008)
20. K. Kooser, et al., J. Phys. B: At. Mol. Opt. phys. **43**, 0235103 (2010); E. Kukk, et al., Int. J. Mass. Spec. **279**, 69–75 (2009); D. Céolin, et al., J. Chem. Phys. **128**, 024306 (2008)
21. O. Plekan, et al., Phys. Scr. **78**, 058105 (2008)
22. P. Bolognesi, et al., private communications

Chapter 11
Ion-Induced Radiation Damage in Biomolecular Systems

Thomas Schlathölter

Abstract The interaction of keV ions with building blocks of DNA and proteins is of fundamental interest to proton and heavy ion therapy. During the last decade, ion-induced ionization and fragmentation was studied for isolated biomolecules, biomolecular clusters, nanosolvated isolated biomolecules and solid thin biomolecular films. This article gives a brief overview over the research on biomolecular mechanisms underlying ion-induced radiation damage with a focus on the different target systems.

11.1 Introduction

Triggered by the ideas of Wilson [1] who first recognized the great potential of the unique dose distribution of fast protons, first clinical trials with proton therapy were already carried out in the 1950s [2]. Over the last decades, proton therapy of malignant tumors has developed into an established form of radiotherapy. With a small delay, the even more promising heavy-ion therapy emerged which allows for a better damage localization within the body and can give access to otherwise untreatable tumors. Heavy ion therapy went through a phase of clinical trials first at Lawrence Berkeley Laboratory starting in the late 1970s with He and Ne irradiations of skull base tumors [3]. In the 1990s, clinical trials were performed with C ions e.g. for non-small cell lung tumors at the Research Center Hospital for Charged Particle Therapy in Chiba, Japan [4] and for skull base tumors at the Gesellschaft für Schwerionenforschung (GSI) in Darmstadt, Germany [5]. Heavy ion therapy is nowadays also considered an established form of cancer treatment. Presently, we

T. Schlathölter (✉)
KVI Atomic and Molecular Physics, University of Groningen,
9747AA Groningen, The Netherlands
e-mail: tschlat@kvi.nl

G.G. Gómez-Tejedor and M.C. Fuss (eds.), *Radiation Damage in Biomolecular Systems*, Biological and Medical Physics, Biomedical Engineering,
DOI 10.1007/978-94-007-2564-5_11, © Springer Science+Business Media B.V. 2012

witness the emergence of enormous social and corporate interest in both proton and heavy ion therapy, resulting in the planning and construction of many public and first commercial particle treatment centers.

The main advantage of ion therapy as compared to conventional radiation therapy is its finite and energy-dependent penetration depth in biological tissue, leading to the phenomenon of the so-called Bragg-peak. At the Bragg-peak, fast ions are slowed down to MeV energies and below. Ion velocities become comparable to velocities of molecular valence electrons and the energy deposition into the medium maximizes. Furthermore, the biological effectiveness (RBE) is strongly enhanced.

Undesired effects of proton and heavy ion irradiation due to naturally occuring solar particle events and galactic cosmic rays are one of the constraints on manned space exploration. Even for heavy shielding, a trip to Mars would lead to an additional risk of cancer of about 1% per year which may extend up to 5% per year [6]. The August 1972 solar proton storm (in between two Apollo missions), which lasted about 15 hours only would have exposed astronauts to a severe dose of 1 Gy of radiation (heaviest shielding). With light shielding, a lethal dose of more than 4 Gy would have been accumulated by the astronauts [7].

Another example in which energetic ions act as primary quanta of radiation are α-emitters incorporated into the human body: For instance inhaled Radon and its decay products emit α-particles of several MeV directly into the lung tissue where they pose a large damage risk.

But the biological effects fast ions and ionizing radiation in general have on living cells are not a mere result of the direct impact of high energy quanta of radiation. The primary interaction will lead to molecular excitation, ionization and fragmentation. In the process, the primary particle loses energy and secondary particles such as low energy electrons, radicals but also (multiply charged) ions are formed within the track. The interaction of these secondary particles with biologically relevant molecules is responsible for a large fraction of biological radiation damage to a cell [8].

To date, molecular mechanisms underlying the exceptional cell killing efficiency of ions as compared to e.g. photons are largely unknown. The investigation of this issue in macroscopic systems (*in vivo* and *in vitro*) is generally hindered by the enormous complexity of the systems. Also, in living systems relevant timescales range from the fs-scale of primary ionization and excitation processes over μs timescales of diffusion limited radical chemistry up to second and even year timescales of biological processes and biological endpoints.

For a deep understanding of biological radiation damage on the level of individual molecules it is important to quantify excitation, ionization and fragmentation cross sections as well as kinetic energies of the various primary and secondary species. Low energy electrons (0 - 20 eV) are the most abundant secondary particles [9] and the first studies on molecular mechanisms underlying biological radiation damage have focused on low energy electron interactions with DNA. Low energy electrons were found to cause single and double strand breaks of plasmid DNA [10, 11] but very remarkably even at almost zero kinetic energy electrons can induce single strand breaks, with pronounced maxima between 0 and 3 eV [12].

A series of gas phase experiments did prove that dissociative electron attachment to isolated nucleobases exhibits resonances at very similar energies [13–17]. This indicates that dissociative electron attachment is likely to be the process underlying low energy electron induced formation of DNA single strand breaks.

Triggered by the very first electron attachment studies, pioneering experiments of keV ion interaction with nucleobases were performed: Coupier *et al.* compared dissociative ionization of uracil induced by keV protons and by 200 eV electrons [18] and de Vries *et al.* investigated collisions of keV multiply charged carbon ions with uracil [19]. Since then the field of ion-biomolecule research has matured rapidly and in the following I will briefly review key experiments and discuss their relevance in the context of biological radiation damage.

11.2 Biomolecular targets

Many studies on low energy ion-induced biomolecular radiation damage employ relatively simple target molecules that can be brought into the gas phase by mere evaporation. The great advantage of the evaporation technique is the easy production of dense gas phase targets that allow application of most established mass-spectrometric or spectroscopic techniques. The main disadvantage is the limitation to thermally stable molecules that stay intact during the evaporation process. Also it has to be kept in mind that evaporated molecules have elevated internal energies corresponding to the evaporation temperature, typically in the range between 100 and 200° C. Many DNA and RNA building blocks such as the nucleobases uracil, thymine, cytosine, adenine and to some extent guanine and the sugar deoxyribose are thermally stable. (Fig. 11.1 sketches how these building blocks relate to the structure of double stranded DNA.) Thermal stability is also observed for the proteinaceous amino acids glycine, valine, and alanine. The relevance of proteins and thus also of amino acids in the context of biological radiation damage is due to the fact that in chromatin, DNA is tighly wound around disk-like protein spools, the so-called histones, which represent about half of the chromatin mass.

The response of this repertoire of biomolecules upon low energy ion impact has been extensively studied over the recent years. The strength of these gas phase experiments lies in the possibility of studying ionization and fragmentation dynamics in finite systems without the perturbing effects of a surrounding medium. The relevance of gas phase studies for biomolecular radiation damage, however, relies on the assumption of similar fundamental ionization and dissociation dynamics for isolated molecules and for molecules embedded in their natural biological environment. Obviously, this is not *a priori* the case. In aqueous solution the vertical ionization energies for DNA building blocks are altered. Furthermore, vibrational excitation might be transferred to the environment before fragmentation can occur. Protonation of intact molecules or their fragments, as observed for instance in ion stimulated desorption experiments on thin films of thymine [20] is obviously ruled out in the gas phase.

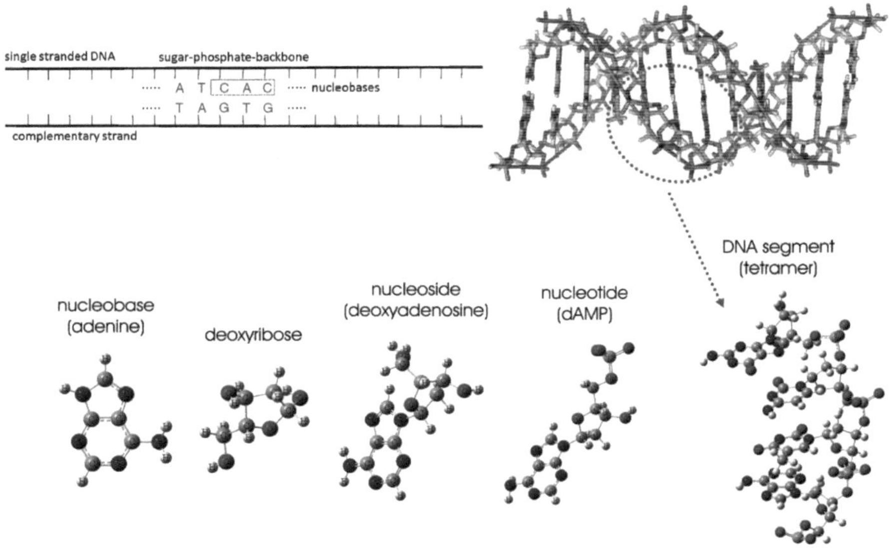

Fig. 11.1 Top left: schematic 2D structure of double stranded DNA; top right: DNA double helix; bottom row: zoom into an oligonucleotide (right), the nucleotide deoxyadenosinemonophosphate, the nucleoside deoxyadenosine, the sugar deoxyribose and the nucleobase adenine

A straightforward step to add a chemical environment to a gas phase molecule is the use of clusters. In many fields clusters are used to study energetics and dynamics of intermediate states of matter as cluster systems bridge the gap between the gas phase and condensed phase [21]. In this sense, biomolecular radiation damage studies on clusters can be seen as a step towards molecules embedded e.g. in aqueous solution. Experiments have until now focused on clusters of nucleobases, mixed clusters of nucleobases, clusters of amino acids and mixed amino acid-water clusters. The limitation of cluster studies however, is their limitation to relatively small biomolecules which can be brought into the gas phase by evaporation. Last but not least, ion interactions with thin solid films of biomolecules can be studied. This approach has the advantage of allowing a relatively wide range of target systems. Small biomolecules such as nucleotides can be deposited *in vacuo* on a substrate as thin films and ion-induced damage of these films can be studied by mass spectrometric techniques [22] and to some extent on an event by event basis. On the other hand, very large systems, such as plasmid DNA, can be deposited for instance by freeze drying under a dry nitrogen atmosphere before being transferred into ultra high vacuum for the irradiation. The accumulated damage to the sample is then typically investigated under athmospheric conditions. In any case, the solid target allows investigation of processes with small cross sections at the cost of the presence of an infinite surrounding medium. It is for instance difficult to quantify possible effects due to radical action in the target.

Larger, usually much more fragile biomolecules such as DNA oligonucleotides or peptides and proteins require different approaches. The two approaches applied in biomolecular radiation damage studies have been desorption techniques (e.g. matrix assisted laser desorption ionization (MALDI, [23]) or laser induced accoustic desorption (LIAD, [24])) and spraying techniques (e.g. electrospray ionization [25]). Electrospray ionization has been employed for ion-induced radiation damage studies by a number of groups and until now, mainly peptides, nucleotides and oligonucleotides have been investigated. The electrospray technique is an ideal approach to go one step further and surround the biomolecule under study by solvation shells of water molecules.

11.2.1 Small isolated biomolecules

The very first studies focused on the fragmentation of the nucleobase uracil upon keV proton [18] and C^{q+} [19] impact. In both proof of principle experiments, time-of-flight mass spectrometry was employed to identify relative fragment yields. In the proton experiments by the Lyon group the charge state of the outgoing projectile was determined, too, to quantify the relative contributions of direct ionization and electron capture. In all cases, no single-atom abstraction from the molecule was observed but rather complete shattering of the ring. Non-dissociative ionization turns out to be a weak channel, only. For the case of C^{2+}, where due to the absence of ionic states resonant with the highest occupied uracil molecular orbitals, resonant electron capture is inhibited, almost zero non-dissociative ionization was observed. First theoretical studies on the C^{q+}-uracil system [26] based on *ab initio* molecular calculation of the potential energies and corresponding coupling matrix elements followed by semiclassical dynamics calculations indeed yielded very low electron capture cross sections for the C^{2+} projectile (6×10^{-19} cm^2 for v = 0.1 a.u.). For C^{4+} on the other hand, much larger cross sections ($\approx 4 \times 10^{-16}$ cm^2 for v = 0.1 a.u.) were calculated. The electron capture probabilities strongly depend on the alignment of the molecules with respect to the ion trajectories [27]. Such anisotropies could not be observed experimentally, yet.

In subsequent studies on C^{q+} [30,31] and Xe^{q+} [28,32] induced fragmentation of uracil and thymine (a nucleobase where a methyl group replaces the H atom attached to the 5-C in uracil), kinectic energies of fragmentation products were determined in coincidence. For moderate degrees of ionization of the molecule, i.e. $q = 1 - 6$, fragmentation proceeds by prompt emission of protons followed by the release of heavier fragments in a subsequent fragmentation cascade. Fragment kinetic energies can exceed 20 eV (see Fig. 11.2a). Already for $q \approx 8$ prompt complete Coulomb fragmentation occurs. Protons with up to 80 eV are formed and fragment kinetic energies strongly depend on the molecular geometry and composition. The production of energetic fragment ions is of particular relevance in the context of biological radiation damage, since such fragments may have the potential to induce subsequent DNA damage. In low energy ion irradiation studies on nucleobase thin films, Huels and coworkers could indeed show that atomic and molecular ions with

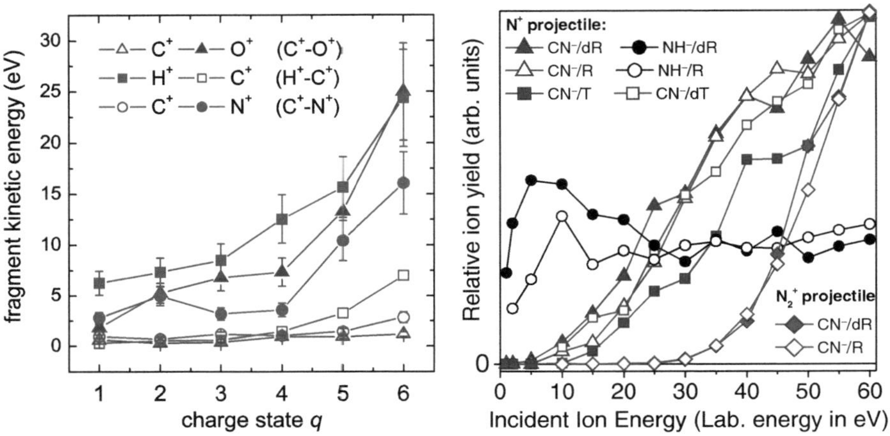

Fig. 11.2 Left: Kinetic energies for atomic fragment ions from 4 keV/amu (C^+: 2 keV/amu) C^{q+} induced fragmentation of thymine [28]. Right: Desorption energy thresholds of NH^- and CN^- anions during N^+ and $N2^{2+}$ ion irradiation of deoxyribose dR, ribose R, thymine T and thymidine dT films on Pt substrate [29]

only a few eV of kinetic energy can efficiently damage nucleobases collisionally (see Fig. 11.2b, [20, 22]). Nucleobase damage due to reactive scattering even seems to occur at almost zero kinetic energy of the impinging ion (see Fig. 11.2b, for ribose and deoxyribose, NH^- can only be formed in a reactive hydrogen abstraction process involving the projectile N atom, [29]). KeV ion-induced biomolecular fragmentation in a biological environment is thus likely to trigger a whole avalanche of damage, which is interesting because biological effects of ionizing radiation are closely related to the complexity of the induced DNA strand breaks. The relative biological effectiveness (RBE) for double strand break production is typically higher for densely ionizing radiation (e.g. heavy ions) than for sparsely ionizing (low linear energy transfer (LET)) radiation, e.g. photons. A possible explanation for this observation could be that densely ionizing radiation leads to formation of clustered double strand breaks [33]. Psonka-Antonczyk *et al.* found evidence for formation of such clustered lesions in atomic force microscopy studies of heavy ion irradiated plasmid DNA [34].

A motivation for ion-biomolecule collision studies is the need for absolute ionization and fragmentation cross sections which can serve as input for track structure calculations. By determination of the layer thickness deposited on a cooled aluminum plate opposite the sublimation oven, Tabet *et al.* [35] recently were able to determine densities for various nucleobase targets and report absolute cross sections for direct ionization and for electron capture following 80 keV proton collisions. Typical cross sections were $\approx 6 \times 10^{-15}$ cm^{-2} for capture from and $\approx 17.5 \times 10^{-15}$ cm^{-2} for direct ionization of the nucleobases uracil, thymine and adenine. For cytosine, surprisingly 3 times smaller cross sections were obtained. First quantum mechanical predictions of these ionization cross sections (performed within the first-order Born approximation with the nucleobases described at the restricted HartreeFock level with geometry optimization) are systematically underestimating the experimental data, yet [36]. Tabet *et al.* also determined fragmentation branching ratios for keV proton induced direct ionization and electron capture of the same nucleobases [37, 38].

Doubly differential cross sections for electron emission following 25-100 keV proton collisions with uracil have been performed by Moretto-Capelle and Le Padellec [39]. Here, the target density was derived from the measured number of protons elastically scattered under 90° which can be directly related to the sum of the atomic scattering cross sections of the uracil constituents using e.g. screened Coulomb interaction potentials. The cross sections have relatively constant values of $36 \times 10^{-19}\,cm^2\,eV^{-1}\,sr^{-1}$ with weak dependence on the proton energy. For electron energies exceeding 20-40 eV, an exponential decrease of the doubly differential cross sections was observed, in agreement with the Bethe-Born approximation. It is estimated that low energy electrons are typically responsible for about half of the biological damage induced by ionizing radiation [8]. Accurate cross sections for electron emission are thus of crucial importance for realistic track structure calculations. To investigate the fragmentation dynamics of nucleobases upon keV ion impact in more detail, various groups have focused on adenine ($C_5H_5N_5$). From photofragmentation studies, it is known that adenine cation fragmentation can proceed via successive loss of neutral HCN units [41]. Using a tailored electric field in their time-of-flight system, Martin *et al.* [42] could show that successive emission of 3 HCN units from adenine$^+$ competes with a process where after a μs delay, an intermediate (adenine-HCN)$^+$ complex loses an H_2CNCN fragment. Alvarado *et al.* [43] compared collisions of 14 keV He^+ and neutral He with adenine using a reflectron time-of-flight spectrometer. For neutral projectiles, for which electron capture is ruled out, e.g. direct HCN loss from the adenine cation was found to compete with a channel where prompt CN loss is followed by μs delayed H loss. For the alanine^{2+} dication formed in 100 keV proton collisions, Moretto-Capelle *et al.* [44] observed e.g. delayed two body fragmentation where a $HCNH^+$ molecular ion is formed with a 120 ns delay.

Very recently, Brédy *et al.* have applied their "collision-induced dissociation under energy control" (CIDEC) approach to 3 keV Cl^+-adenine collisions [45]. The method is suitable e.g. for double electron capture into an initially singly charged ion. If the first and second ionization potential of the target are known and the projectile ion is in a well defined state before *and* after the interaction, then determination of projectile energy loss *after* the collision process allows to determine the amount of deposited energy that is linked to a specific fragmentation channel. The authors could show that the mean excitation energy of the (intact) adenine dication was 5.8 eV. A mean excitation of 6.7 eV leads to loss of a $HCNH^+$ ion under release of about 1 eV of kinetic energy. Similarly, many other single and multi-step fragmentation channels could be fully characterized.

Besides nucleobases, research has also focused on deoxyribose, the sugar molecule located in the DNA backbone (see Fig. 11.1). The CIDEC technique [40] proved that for different proton energies (3 keV and 7 keV) , double electron capture

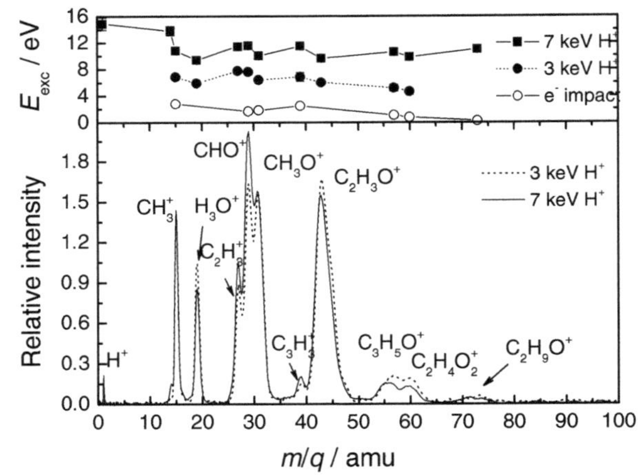

Fig. 11.3 Lower panel: fragment ion mass spectra from collisions of 3 and 7 keV protons with 2-deoxy-D-ribose. Upper panel: most probable excitation energy E_{exc} associated to major fragments for two proton impact energies (3 keV and 7 keV) and excitation energy obtained by electron impact ionization [40]

leads to similar fragmentation patterns even though average internal energies are very different (see Fig. 11.3. This finding supports a statistical fragmentation model for deoxyribose fragmentation for this range of internal energies. Complementary high resolution time-of-flight spectrometry studies [46] already yielded fragment distributions by and large following power lawsa fingerprint of statistical fragmentation. Furthermore, virtually complete disintegration of the molecule was observed, independent of ion charge state, atomic number and kinetic energy. The latter finding suggests that deoxyribose is much more sensitive to ion impact than nucleobases.

Very recently, ion collisions with amino acids have been investigated [47, 48]. The α and β isomers of alanine where found to exhibit a fundamentally different fragmentation pattern upon multiply charged ion impact. The fragment kinetic energies were found to be similar to those observed in ion-nucleobase collisions and thus sufficient to induce damage to neighboring molecules in a biological environment. An interesting finding was the formation of H_3^+ ions from various amino acids, which requires substantial rearrangement of the molecule before fragmentation.

11.2.2 Biomolecular clusters

First studies on the response of larger biomolecular complexes upon ion impact focused on nucleobase clusters. Such targets are easily produced if sublimation of the DNA building blocks takes place in the cooled He atmosphere of a cluster aggregation source. Figure 11.4 compares the low mass spectrum due to fragmentation of isolated thymine molecules with the fragment spectrum of neutral thymine clusters [49]. For the clusters the peak around $m/q = 109$ (OH loss) is one of the dominating ones whereas it is completely absent for isolated molecules. Using a coincidence technique, it could be shown that already for the smallest nucleobase clusters, i.e. dimers and trimers, the OH loss channel opens up. Interestingly, this channel is

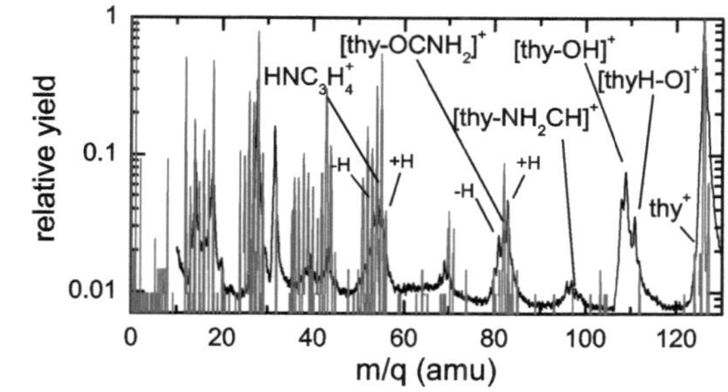

Fig. 11.4 Typical mass spectrum for collisions of 60 keV C5+ with isolated thymine molecules (grey) and of 50 keV O^{5+} with neutral thymine clusters (black) [49]

observed in ion-induced desorption studies on thymine thin films, too [20], i.e. it is likely that OH loss is a fingerprint of a chemical environment. Similar additional fragmentation channels were also observed for clusters of other nucleobases. It is concluded that these channels are probably due to intermolecular hydrogen bonding between O and H atoms of neighboring thymine molecules that weaken the intramolecular bonds. Similar experiments have recently been performed for clusters of amino-acids [50]. Also here, the chemical environment drastically alters fragmentation pathways due to rapid energy and charge redistribution inside the cluster. In contrast to nucleobase clusters, where ion impact leads to formation of nucleobase fragment ions or nucleobase cluster ions, for amino acid clusters also polymerization is observed. In particular, loss of single or multiple units of mass 18 are observed, which could be explained by stabilization of peptide bonds within the cluster [50]. Note that this finding is of great interest in the context of astrochemistry and astrobiology, since it hints at a possible pathway to formation of complex biomolecules in astrophysical environments. In the context of biomolecular radiation damage, however, the important outcome of the performed experiments is: Ion-induced biomolecular fragmentation is strongly influenced by the presence of a chemical environment. It is thus neccessary to study this influence in more details.

11.2.3 Biomolecules on surfaces

A complementary approach to gas phase studies is the investigation of ion-interactions with thin biomolecular films. Such studies are appealing for a number of reasons: First of all, in chromatin DNA is typically found in a relatively closely packed structure, which in certain aspects resembles the solid phase. Furthermore, interactions of biomolecules with a chemical environment are inherently included. Last but not least, targets are dense and relatively easy to produce. On the other hand fragment charge states and kinetic energies can change during the desorption process, rendering the observed data to some extent indirect. Over the last years, surface studies have contributed greatly to the knowledge on very low energy ion-biomolecule collisions. Huels and coworkers could show that Ar^+ ions with kinetic energies exceeding 10 eV can induced fragmentation of the nucleotide thymidine [22]. This result is important since it clearly shows that the product ions from typical fragmentation events induced by keV ions have sufficient kinetic energies to induce damage in neighboring molecules. Reactive scattering, i.e. abstraction of a molecule constituent by a projectile ion, was found to occur even for ion kinetic energies below 10 eV. Examples are NH^- formation in collisions of N^+ with deoxyribose or CN^- production in N^+ collisions with thymidine [29]. For dexoyribose it could furthermore be shown that for reactive N^+ scattering from deoxyribose, the C5-site is most vulnerable [24,51]. In DNA, C5 abstraction would lead to a single strand break.

A different approach is the irradiation of complex biomolecular targets *in vacuuo* followed by offline analysis using established biochemical tools. In a first study of this type, Lacombe and coworkers have investigated the response of lyophilized plasmid DNA upon irradiation with keV Ar^+ ions [52]. Agarose gel electrophoresis analysis revealed a sizeable fraction of single and double strand breaks. In a similar experiment Hunniford *et al.* [53] compared the action of C^+ and C^{2+} ions and found enhanced damage for the doubly charged projectile ions. The latter result suggest that potential energy plays a crucial role in biomolecular radiation damage.

11.2.4 Complex isolated biomolecules

Assemblies of DNA building blocks and amino acids with structures also found in the nuclei of living cells, i.e. oligonucleotides and peptides are even more realistic targets. In pioneering studies, Nielsen, Hvelplund and coworkers have investigated ionization and fragmentation of electrosprayed peptide cations and oligonucleotide anions upon collisions with noble gas atoms [54]. For instance total destruction cross sections were obtained for multiply deprotonated $(d(A)_7\text{-}nH)^{n-}$ septamers colliding at 50 keV with He, Ne and Ar atoms [55]. Cross section values around 1.5×10^{-14} cm^2 were observed which increase with the oligonucleotide charge state and with the noble gas atomic number. Cross sections for electron loss were found to be at least an order of magnitude smaller, in agreement with a model in which electrons are removed mainly from the independent O^- ions located in the phosphate groups.

Very recently, Bari *et al.* introduced an experiment in which electrosprayed biomolecular ions are collected in a radiofrequency (RF) trap and serve as a target for collisions with singly and multiply charged keV ions [56]. First experiments dealt with keV H^+ and He^{q+} collisions with a series of protonated peptides [56,57]. Very extensive fragmentation was observed, in line with the expectations from a simple electronic stopping model. The fragment mass spectrum was found to be dominated by amino acid side chains (see Fig. 11.5). The sidechain can be lost as a neutral (e.g. for substance P) or as a cation (e.g. for leucine enkephalin). Cleavage of the side chain linkage seems to be a fast process competing with backbone cleavage, which typically occurs after internal vibrational redistribution of the excitation energy.

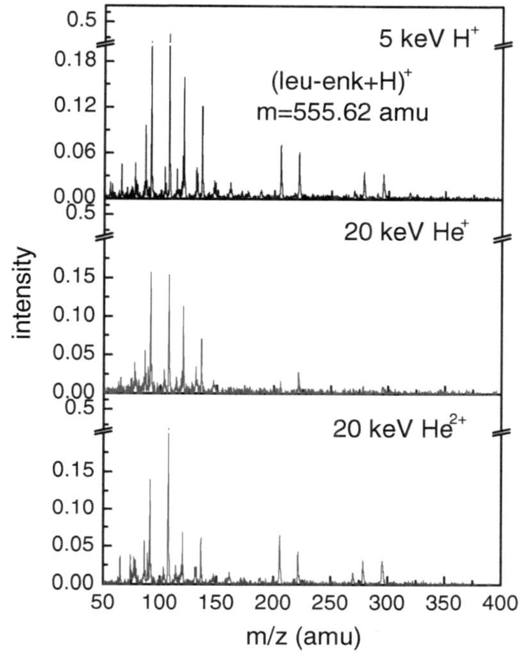

Fig. 11.5 Fragmentation spectra of trapped protonated leucine enkephalin cations after collision with 5 keV/nucleon H^+, He^+ and He^{2+}. Sidechain fragments dominate the spectrum for $m/z < 150$. Fragments resulting from backbone scission are found at higher masses. [49]

In agreement with earlier studies on isolated DNA building blocks, fragmentation was found to be most severe for He^+ projectile, for which resonant electron capture is only possible at very small impact parameters. At present, first experiments on ion-induced fragmentation of oligionucleotides are in progress.

11.2.5 Nanosolvated biomolecules

To mimick the chemical environment within the nucleus of a living cell, it is straightforward to surround oligonucleotides or even DNA-protein complexes with solvation shells of water molecules and/or to attach them to proteins. Until now, ion interaction studies with such systems have not been performed. A pioneering experiment has been performed by Liu *et al.* [25] who investigated collisions of 50 keV adenosine 5-monophosphate (AMP) anions with neutral Ne and Na atoms. It was observed, that collision-induced fragmentation can be quenched by nanosolvation with water molecules (see Fig. 11.6). Addition of at least 13 water molecules to the nucleotide fully protects the DNA building block, even though the water molecules are expected to stick predominantly to the phosphate group. This implies that water even after a direct hit, the nucleotide can release its energy by evaporation of water molecules *before* fragmentation sets in. Interestingly,

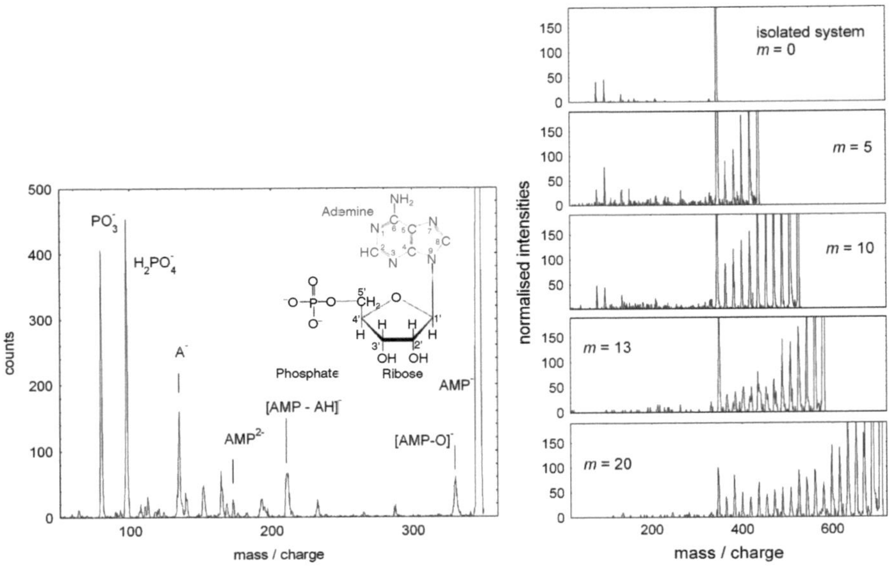

Fig. 11.6 Left: Fragmentation mass spectrum obtained in 50 keV collisions of AMP anions with neutral sodium atoms. Right: Fragmentation spectra of $AMP^{-}(H2O)_m$ ions for different m numbers of attached water molecules [25]

in case of dissociation-induced by electron capture from Na atoms, the opposite behaviour is observed: electron capture always leads to loss of at least a hydrogen atom from the AMP anion and the cross section for this fragmentation process *increases* with increasing number of attached water molecules [58]. For the case of protonated peptide cations, the influence of the type of nanosolvent (water, ammonia, crown ether) upon electron capture induced dissociation was studied. Crown ether was for instance found to strongly increase peptide backbone scission due to a shift of the singly occupied molecular orbital towards the backbone [59]. The message from these first experiments is that solvation shell strongly affect the ion-induced fragmentation dynamics, often in a non-obvious way. It is clear that more experimental and theoretical data is needed, to get a good understanding of the dynamics of biomolecule-solvent interactions in ion-induced radiation damage.

11.3 Conclusion

Over the last decade, the field of ion-induced biomolecular radiation damage has booked impressive progress. On the one hand, the study of isolated DNA building blocks and amino acids has led to a deep understanding of the relevant processes. Experimental studies on more realistical complex biomolecular systems such as nanosolvated oligonucleotides are within reach. Challenges for the coming years will be the production of reproducible targets for such studies, the application of state of the art detection techniques and last but not least the development of theoretical tools to handle dynamics in strongly excited complex systems.

References

1. R. R. Wilson, Radiology **47**, 487 (1946)
2. C. A. Tobias, J. H. Lawrence, J. L. Born, R. K. McCombs, J. E. Roberts, H. O. Anger, B. V. A. Low-Beer, C. B. Huggins, Cancer Res. **18**, 121 (1958)
3. J. R. Castro, D. E. Linstadt, J.-P. Bahary, P. L. Petti, I. Daftari, J.M. Collier, P. H. Gutin, G.ant Gauger, T. L. Phillips, Int. J. Radiat. Oncol. Biol. Phys. **4**, 647 (1994)
4. T. Miyamotoa, N. Yamamotoa, H. Nishimuraa, M. Kotoa, H. Tsujiia, J. Mizoea, T. Kamadaa, H. Katoa, S. Yamadaa, S. Moritaa, K. Yoshikawaa, S. Kandatsua, T. Fujisawa, Radiother. Oncol. **66**, 127 (2003)
5. D. Schulz-Ertner, A. Nikoghosyan, C. Thilmann, T. Haberer, O. Jäkel, C. Karger, G. Kraft, M. Wannenmacher, J. Debus, Int. J. Rad. Oncol. Biol. Phys. **2**, 631 (2004)
6. F.A. Cucinotta, W. Schimmerling, J. W. Wilson, L. E. Peterson, G. D. Badhwar, P. B. Sagantia, John F. Dicelloe, Radiat. Res. **156**, 682 (2001)
7. J.L. Parsons, L. W. Townsend, Radiat Res. **153**, 729 (2000)
8. C. von Sonntag, *The chemical basis for radiation biology*, Taylor and Francis, London (1987)
9. V. Cobut, Y. Frongillo, J. P. Patau, T. Goulet, M. J. Fraser, J. P. Jay-Gerin, Radiat. Phys. Chem. **51**, 229 (1998)
10. B. Boudaïffa, P. Cloutier, D. Hunting, M. A. Huels, L. Sanche, Science **287**, 1658 (2000)
11. X. Pan, C. Cloutier, D. Hunting, L. Sanche, Phys. Rev. Lett. **90**, 208102 (2003)
12. F. Martin, P. D. Burrow, Z. Cai, P. Cloutier, D. Hunting, L. Sanche, Phys. Rev. Lett. **93**, 068101 (2004)
13. M. A. Huels, I. Hahndorf, E. Illenberger, L. Sanche, J. Chem. Phys. **108**, 1309 (1998)
14. H. Abdoul-Carime, M. A. Huels, F. Brüning, E. Illenberger, L. Sanche, J. Chem. Phys. **113**, 2517 (2000)
15. G. Hanel, B. Gstir, S. Denifl, P. Scheier, M. Probst, B. Farizon, M. Farizon, E. Illenberger, T. D. Märk, Phys. Rev. Lett. **90**, 188104 (2003)
16. S. Denifl, S. Ptasinska, M. Cingel, S. Matejcik, P. Scheier, T. Märk, Chem. Phys. Lett. **377**, 74 (2003)
17. H. Abdoul-Carime, S. Gohlke, E. Illenberger, Phys. Rev. Lett. **92**, 168103 (2004)
18. B. Coupier, B. Farizon, M. Farizon, M.J. Gaillardand F. Gobet, N.V. de Castro Faria, G. Jalbert, S. Quaskit, M. Carre, B. Gstir, G. Hanel, S. Denifland L. Feketova, P. Scheier, T. Märk, Eur. Phys. J. D **20**, 459 (2002)
19. J. de Vries, R. Hoekstra, R. Morgenstern, T. Schlathölter, J. Phys. B.: At. Mol. Opt. Phys. **35**, 4373 (2002)
20. Z. Deng, M. Imhoff, M. A. Huels, J. Chem. Phys. **123**, 144509 (2005)
21. A. W. Castleman K. H Bowen, J. Phys. Chem. **100**, 12911 (1996)
22. Z. Deng, I. Bald, E. Illenberger, M. A. Huels, Phys. Rev. Lett. **95**, 153201 (2005)
23. T. L. Merrigan, C. A. Hunniford, D. J. Timson, M. Catney, R. W. McCullough, Biochem. Soc. Trans. **37**, 905 (2009)
24. I. Bald, I. Dabkowska, E. Illenberger, Angew. Chem. Int. Ed. **47**, 8518 (2008)
25. B. Liu, S. Brondsted Nielsen, P. Hvelplund, H. Zettergren, H. Cederquist, B. Manil, B. A. Huber, Phys. Rev. Lett. **97**, 133401 (2006)
26. M. C. Bacchus-Montabonel, M. Labuda, Y. S. Tergiman, J. E. Sienkiewicz, Phys. Rev. A **72**, 052706 (2005)
27. M. C. Bacchus-Montabonel, Y. S. Tergiman, Phys. Rev. A **74**, 054702 (2006)

28. T. Schlathölter, R. Hoekstra, S. Zamith, Y. Ni, H. G. Muller, M. J. J. Vrakking, Phys. Rev. Lett. **94**, 233001 (2005)
29. Z. Deng, I. Bald, E. Illenberger, M. A. Huels, Phys. Rev. Lett. **96**, 243203 (2006)
30. J. de Vries, R. Hoekstra, R. Morgenstern, T. Schlathölter, Eur. Phys. J. D **24**, 161 (2003)
31. T. Schlathölter, R. Hoekstra, R. Morgenstern, Int. J. Mass Spectr. **233**, 173 (2004)
32. J. de Vries, R. Hoekstra, R. Morgenstern, T. Schlathölter, Phys. Rev. Lett. **91**, 053401 (2003)
33. J. Kiefer, New J. Phys. **10**, 075004 (2008)
34. K. Psonka-Antonczyka, Th. Elsässer, E. Gudowska-Nowak, G. Taucher-Scholz, Radiat. Res. **172**, 288 (2009)
35. S. Tabet, S. Eden, S. Feil, H. Abdoul-Carime, B. Farizon, M. Farizon, S. Quaskit, T. D. Märk, Phys. Rev. A **82**, 022703 (2010)
36. C. Champion, H. Lekadir, M. E. Galassi, O. Fojón, R. D. Rivarola, J. Hanssen, Phys. Med. Biol. **55**, 6053 (2010)
37. S. Tabet, S. Eden, S. Feil, H. Abdoul-Carime, B. Farizon, M. Farizon, S. Quaskit, T. D. Märk, Int. J. Mass Spectr. **292**, 53 (2010)
38. S. Tabet, S. Eden, S. Feil, H. Abdoul-Carime, B. Farizon, M. Farizon, S. Quaskit, T. D. Märk, Phys. Rev. A **81**, 012711 (2010)
39. P. Moretto-Capelle, A. Le Padellec, Phys. Rev. A **74** (2006)
40. F. Alvarado, J. Bernard, B. Li, R. Bredy, L. Chen, R. Hoekstra, S. Martin, T. Schlathölter, ChemPhysChem **9**, 1254–1258 (2008)
41. H.-W. Jochims, M. Schwell, H. Baumgärtel, S. Leach, Chem. Phys. **314**, 263 (2005)
42. S. Martin, R. Bredy., A. R. Allouche., J. Bernard., A. Salmoun., B. Li, L. Chen, Phys. Rev. A **77**, 062513 (2008)
43. F. Alvarado, S. Bari, R. Hoekstra, T. Schlathölter, J. Chem. Phys. **127**, 034301 (2007)
44. P. Moretto-Capelle, A. Le Padellec, G. Briére, S. Massou, F. Franceries, J. Chem. Phys. **127**, 234311 (2007)
45. Richard Bredy, Jerome Bernard, Li Chen, Guillaume Montagne, Bin Li, Serge Martin, J. Chem. Phys. **130**, 114305 (2009)
46. F. Alvarado, S. Bari, R. Hoekstra, T. Schlathölter, Phys. Chem. Chem. Phys **8**, 1922 (2006)
47. S. Bari, P. Sobocinski, J. Postma, F. Alvarado, R. Hoekstra, V. Bernigaud, B. Manil, J. Rangama, B. Huber, T. Schlathölter, J. Chem. Phys. **128**, 074306 (2008)
48. S. Bari, F. Alvarado, J. Postma, P. Sobocinski, R. Hoekstra, T. Schlathlter, Eur. Phys. J. D **51**, 81 (2009)
49. T. Schlathölter, F. Alvarado, S. Bari, A. Lecointre, R. Hoekstra, V. Bernigaud, B. Manil, J. Rangama, B. Huber, ChemPhysChem **7**, 2339 (2006)
50. S. Maclot, M. Capron, R. Maisonny, A. Lawicki, A. Méry, J. Rangama, J.-Y. Chesnel, S. Bari, R. Hoekstra, T. Schlathlter, B. Manil, L. Adoui, P. Rousseau, B. A. Huber, ChemPhysChem **(in press)** (2011)
51. I. Bald, Z. Deng, E. Illenberger, M. Huels, Phys. Chem. Chem. Phys **8**, 1215 (2006)
52. S. Lacombe, C. Le Sech, V. Esaulov, Phys. Med. Biol **49**, N65 (2004)
53. C. A. Hunniford, D. J. Timson, R. J. H. Davies, R. W. McCullough, Phys. Med. Biol. **52**, 3729 (2007)
54. S. B. Nielsen, J. U. Andersen, P. Hvelplund, B. Liu, S. Tomita, J. Phys. B **37**, R25 (2004)
55. B. Liu, P. Hvelplund, S. B. Nielsen, S. Tomita, Int. J. Mass Spectr. **230**, 19 (2003)
56. S. Bari, R. Hoekstra, T. Schlathölter, Phys. Chem. Chem. Phys. **12**, 3360 (2010)
57. S. Bari, R. Hoekstra, T. Schlathölter, Int J. Mass. Spectr. **299**, 64 (2011)
58. B. Liu, N. Haag, H. Johansson, H. T. Schmidt, H. Cederquist, S. Brondsted Nielsen, H. Zettergren, P. Hvelplund, B. Manil, B. A. Huber, J. Chem. Phys. **128**, 075102 (2008)
59. Henning Zettergren, Lamri Adoui, Virgile Bernigaud, Henrik Cederquist, Nicole Haag, Anne I. S. Holm, Bernd A. Huber, Preben Hvelplund, Henrik Johansson, Umesh Kadhane, Mikkel Koefoed Larsen, Bo Liu, Bruno Manil, Steen Brondsted Nielsen, Subhasis Panja, Jimmy Rangama, Peter Reinhed, Henning T. Schmidt, Kristian Stochkel, ChemPhysChem **10**, 1619 (2009)

Chapter 12
Theory and Calculation of Stopping Cross Sections of Nucleobases for Swift Ions

Stephan P.A. Sauer, Jens Oddershede, and John R. Sabin

Abstract The effects of energy transfer from swift ion radiation to molecules are best described by the stopping cross section of the target for the projectile ion. In turn, the mean excitation energy of the target is the determining factor in the stopping cross section. Using polarization propagator methodology, the mean excitation energies of the five DNA nucleobases have been calculated, and subsequently used to determine the stopping cross sections of the bases.

12.1 Introduction and History

As biological material is primarily composed of water, most damage resulting from the interaction of fast ion radiation with biological material is the result of attack by $O\cdot$ and $\cdot OH$ free radicals formed in the ion-water interaction [1, 2]. However there are other biomolecules, for example proteins and the various components of RNA and DNA that are also present in the mix. Were one to offer a complete theoretical description of the interaction of swift ion radiation with a bio system, one would have to follow the ion molecule interactions of all components, including fragmentation products and secondary (δ) electrons to the terminus of the process, an obviously impossible task at the moment.

S.P.A. Sauer
Department of Chemistry, University of Copenhagen, Copenhagen, Denmark
e-mail: sauer@kiku.dk

J. Oddershede
Department of Physics and Chemistry, University of Southern Denmark, Odense Denmark
e-mail: jod@sdu.dk

J.R. Sabin (✉)
Quantum Theory Project, Department of Physics, University of Florida, Gainesville, Florida, USA sabin@qtp.ufl.edu

G.G. Gómez-Tejedor and M.C. Fuss (eds.), *Radiation Damage in Biomolecular Systems*, Biological and Medical Physics, Biomedical Engineering,
DOI 10.1007/978-94-007-2564-5_12,

However, one can approach the problem piecemeal. In this contribution, we consider theoretically the question as to what the result of direct interaction of fast ion radiation impinging on nucleobases might be, and whether it might be possible to selectively interact with one nucleobase in a DNA or RNA strand.

Of the crucial descriptors necessary to the understanding of the interaction of fast ions and biomolecules, the energy transfer characteristics between the two are of utmost importance: In particular, the ability of the target molecule to absorb energy from the projectile.

We consider a swift ion passing through a target composed of molecules. The energy deposited in the target per unit projectile path length, or the stopping power, of the target is given by [3]:

$$-\frac{dE}{dx} = nS(v) \tag{12.1}$$

where n is the number density of scattering centers, and $S(v)$ is the stopping cross section as a function of the projectile (ion) velocity [4–6]. In the Bethe-Bloch version of the theory [7, 8], the stopping cross section is given as a function of the ion charge (Z_1) and number of target electrons (Z_2), multiplying the stopping number $L(v)$,

$$S(v) = \frac{4\pi e^4 Z_1^2 Z_2}{mv^2} L(v) \tag{12.2}$$

which is a Born series expansion in powers of the ion charge:

$$L(v) = \sum_{i=0} Z_1^i L_i(v) \tag{12.3}$$

It is in the stopping number that the physics of the problem is located.

The first term in the Born series for the non-relativistic case, the Bethe term ($\propto Z_1^2$), was derived by Bethe using a quantum mechanical perturbation treatment [7], and he determined the stopping number to be the log of the first energy weighted moment of the dipole oscillator strength distribution (DOSD) of the molecule, I_0, which is referred to as the mean excitation energy.

$$\ln I_0 = \frac{\sum\limits_k^{\text{bound}} f_{0k} \ln E_{0k} + \int\limits^{\text{continuum}} \ln E_{0k} \frac{df_{0k}}{dE_{0k}} dE_{0k}}{\sum\limits_k^{\text{bound}} f_{0k} + \int\limits^{\text{continuum}} \frac{df_{0k}}{dE_{0k}} dE_{0k}} \tag{12.4}$$

Here $\{E_{0k}\}$ is the complete set of electronic excitation energies for the target, $E_{0k} = E_k - E_0$, and $\{f_{0k}\}$ is the complete set of the corresponding dipole oscillator strengths, which in the dipole length formulation and Hartree atomic units, is defined as:

$$f_{0k} = \tfrac{2}{3}\left(E_k - E_0\right)\left|\langle\Psi_0|\sum_i \hat{\vec{r}}_i\,|\Psi_k\rangle\right|^2 \tag{12.5}$$

If the square of the norm is "interpreted" as the dot product of the indicated matrix element and its adjoint, then (12.5) represents the oscillator strength of the directionally averaged dipole oscillator strength.

During Bethe's derivation, the assumption was made that the projectile velocity was much greater than that of the target electrons. As the assumption can be violated, especially on collision with inner shell electrons, a shell correction, $-C(v)/Z_2$, was added as compensation by Fano [9]. The Bethe term including shell corrections is shown in (12.6), and the definition of the mean excitation energy was given in (12.4).

$$L_0(v) = \ln\frac{2mv^2}{I_0} - \frac{C(v)}{Z_2} \tag{12.6}$$

Although shell corrections are normally calculated from either electron gas, hydrogenic, or Hartree-Fock models [10], all curves of $C(v)/Z_2$ *vs.* v have similar shapes, and can be fit to a function of the form [11]:

$$\frac{C(v)}{Z_2} = C_1 v e^{-C_2 v} \tag{12.7}$$

Summing the shell corrections for the N atoms comprising a molecule in the spirit of the Bragg Rule [12–14] results in an expression for the shell corrections for a polyatomic molecule as:

$$\frac{C(v)}{Z_2} = \frac{1}{N}\sum_i n_i \frac{C_i(v)}{Z_2} = \frac{1}{N}\sum_i n_i c_1^i e^{-c_2^i v} v \tag{12.8}$$

where n_i is the number of atoms of type i.

The second term in the Born series, the Barkas term $\left(\propto Z_1^3\right)$, accounts for polarization of the target electrons by approach of the projectile ion. Numerical work [15] has determined atomic Barkas corrections, but for this application, it is most useful to utilize the analytic form introduced by Lindhard [16];

$$L_1(v) = Z_1 \frac{3\pi I_0}{2v^3} \ln\frac{2mv^2}{I_0} \tag{12.9}$$

Finally, the Bloch term $\left(\propto Z_1^4\right)$ can be written [8]:

$$L_2(v) = -Z_1^2 \frac{1.202}{v^2} \tag{12.10}$$

Putting these terms into (12.2) with $i = 0 \rightarrow 2$ in (12.3) leads to an analytic expression for the stopping power of the nucleobases for swift ions:

$$S(v) = \frac{4\pi e^2 Z_1^2 Z_2}{mv^2} \times \left[\ln \frac{2mv^2}{I_0} - \frac{1}{N} \sum_{j=atoms} n_j c_1^j e^{-c_2^j v} v + Z_1 \frac{3\pi I_0 e^2}{2\hbar v^3 m} \ln \frac{2mv^2}{I_0} - Z_1^2 \frac{1.202}{v^2} \right] \quad (12.11)$$

The only input needed is the mean excitation energy of the base and the coefficients for the atomic shell corrections introduced in (12.6).

12.2 Calculation of Mean Excitation Energies

The most important quantity in (12.11) is the molecular mean excitation energy for the target, in this case the nucleobases. In order to calculate the mean excitation energy of a molecule, one needs the full set of vertical electronic excitation energies and associated electronic transition dipole moments. These can conveniently be extracted from the linear response function or polarization propagator [17] as can be seen from its spectral representation in the basis of the eigenstates $\{|\Psi_0\rangle, |\Psi_k\rangle\}$ of the molecular Hamiltonian $\hat{H}$:

$$\langle\langle \hat{P}; \hat{Q} \rangle\rangle_E = \sum_{k \neq 0} \left[\frac{\langle \Psi_0 | \hat{P} | \Psi_k \rangle \langle \Psi_k | \hat{Q} | \Psi_0 \rangle}{E - E_k + E_0} + \frac{\langle \Psi_0 | \hat{Q} | \Psi_k \rangle \langle \Psi_k | \hat{P} | \Psi_0 \rangle}{E + E_k - E_0} \right] \quad (12.12)$$

Here the sum is over all excited states $\{|\Psi_k\rangle\}$ of the system. The poles and residues of the propagator give the excitation energies, $E_{0k} = E_k - E_0$, and transition matrix elements, $\langle \Psi_0 | \hat{P} | \Psi_k \rangle$, of the operators $\hat{P}$ and $\hat{Q}$. If we choose $\hat{P} = \hat{Q} = \sum_i \hat{\vec{r}}_i$, then the residues and poles may be used to compute the dipole oscillator strengths of the system in the length formulation, while choices of $\hat{P} = \hat{Q} = \sum_i \hat{\vec{p}}_i$ or $\hat{P} = \sum_i \hat{\vec{r}}_i$ and $\hat{Q} = \sum_i \hat{\vec{p}}_i$ lead to dipole oscillator strengths in the velocity and mixed formulations, respectively. All the quantities reported here have been calculated in the dipole length formulation.

The polarization propagator, i.e. the left hand side of (12.12), is calculated without using the right hand side of the equation. In practice, this is done utilizing that the polarization propagator is the linear response to an external perturbation. This method generates a set of equations from which one may retrieve poles and residues, that is the quantities of interests in this calculation. If the linear response is calculated through first order in the electron repulsion, then one obtains the Random Phase Approximation (RPA) to the polarization propagator, and this is the level of approximation we use in this communication. Another way of viewing the method is to say that the exact ground state of the system $|\Psi_0\rangle$ is replaced by some approximate wave function $|\Phi\rangle$ which is a linear combination of antisymmetrized products of molecular orbitals, so-called Slater determinants, while the excited states $|\Psi_k\rangle$ are also approximated by linear combinations of Slater determinants obtained by replacing one or more of the molecular orbitals occupied in the ground state Slater determinants $|\Psi_0\rangle$ by virtual orbitals[1].

This approach yields a finite number of excitations, the number of which is determined by the size of the basis set, but no continuum states. As a result, the integrations over the continuum states in (12.4) are done numerically using the excitation energies with energies larger than the first ionization energy of the system, called pseudo-states, as integration points. We have found that this discretization of the continuum works well when – and only when - sums over the entire excitation spectrum are taken [18]. It should be noted that energy weighted oscillator strength sums can also be obtained directly from matrices used in the polarization propagator [19, 20], but the direct sum over states was used in this application as it was more convenient. The mean excitation energies in (12.4) are then obtained by explicit summation of the oscillator strengths to all bound states and to the discrete continuum pseudo-states. Experience indicates that the molecules in question about 12% of the mean excitation energy is due to excitation to bound states, while the remaining 88% comes from transitions into the pseudo-states [21].

Experience also shows [22–24] that some amount of electron correlation is needed in order to calculate reliable spectral moments of the DOSD. One needs to calculate the propagator at least at the level of the time-dependent Hartree-Fock, also called the random phase approximation (RPA) [25,26]. The RPA adds correlation in both ground and excited states in a balanced way [27]. On the other hand, in a recent QM/MM calculation on glycine surrounded by 511 water molecules we found that the effect of solvation by water is rather small [28].

One must also choose a atomic orbital basis for the calculations. In a complete basis set and in the exact case as well as for certain approximate methods such as the RPA, the dipole oscillator strength sum fulfills the Thomas-Reiche-Kuhn (TRK) sum rule [29–31]:

$$S_0 = \sum_k f_{0k} = Z_2 \tag{12.13}$$

That is, the oscillator strengths from the ground state to all excited stated, including those discrete states representing the continuum, will sum to the number of target electrons (Z_2). Also in the RPA, used here, the dipole oscillator strengths calculated in dipole velocity, dipole length, or mixed representation and all sum rules would be identical would be fulfilled exactly *i.e.* be equal to the number of electrons, if the

[1]For details of the scheme, see [17]

computational basis were complete [32, 33]. Comparison of the oscillator strengths calculated in the different formulations thus gives a measure of the completeness of the computational basis.

12.3 Calculation of Shell Corrections

In order to obtain C_1 and C_2 for calculation of the shell corrections in (12.8), (12.7) was fitted to calculated shell correction data for H, C, N, and O [34]. That data [11] is given in Table 12.1.

12.4 Calculational Details

In order to obtain ground state geometries, we have optimized the geometries of the five nucleobases [14] at the density functional theory (DFT) level with the B3LYP functional [35] and the 6-31+G(d,p) [36] one-electron basis set using the Gaussian program, starting from geometries which had already been optimized at the MP2/6-31G* level [37]. After the geometry optimization the molecules were oriented in such a way that the heterocycles are placed in the xy-plane as in Fig.12.1 and Fig. 12.2.

Using the minimum energy molecular geometries of the nucleobases, the vertical singlet excitation energies and associated electronic transition dipole moments were calculated with the TURBOMOLE program [38, 39] using linear response or polarization propagator methods [17] at the RPA level [14]. As the cc-CVTZ+(3df,p)& s + p + d-recontracted basis developed for glycine [24] produced excellent results,

Table 12.1 Atomic Shell Correction Parameters for use in (12.8)

Atom	C_1	C_2
H	1.50	1.00
C	0.14	0.20
N	0.15	0.21
O	0.26	0.33

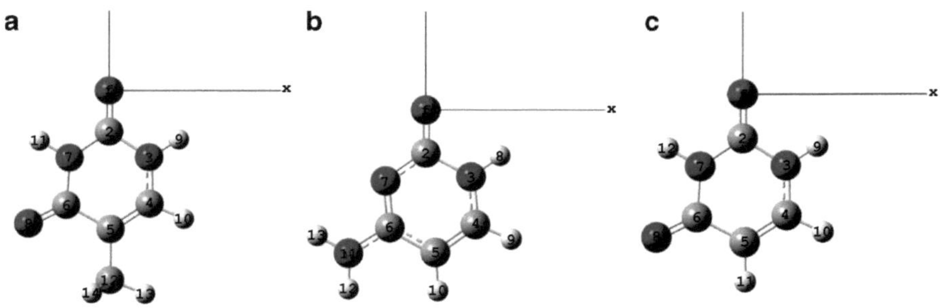

Fig. 12.1 Optimized geometries of the pyrimidine nucleobases thymine (**a**), cytosine (**b**), and uracil (**c**). The molecules lie in the xy-plane with the y-axis point upwards and the x-axis to the right

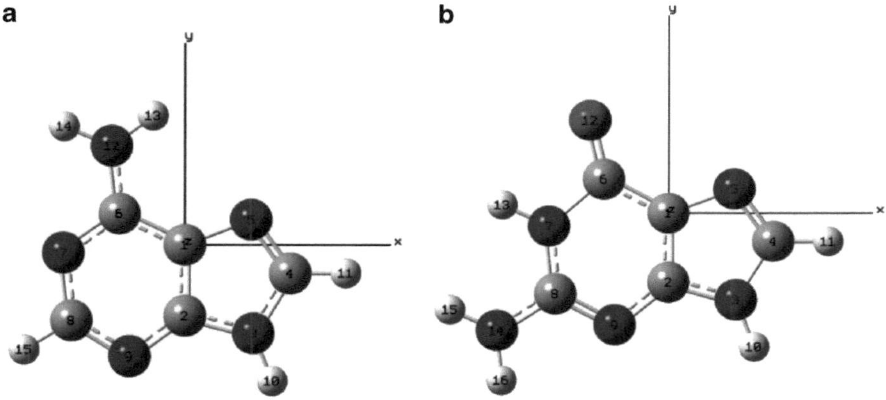

Fig. 12.2 Optimized geometries of the purine nucleobases adenine (**a**) and guanine (**b**). The molecules lie in the xy-plane as indicated

Table 12.2 The number of electrons, the Thomas-Reiche-Kuhn sum rule in the length representation, S_0^L, for the nucleobases [14]

	Z_2	S_0^L
Cytosine	58	58.00
Uracil	58	58.03
Thymine	66	66.01
Adenine	70	70.01
Guanine	78	78.05

we chose to use the same basis also for the present calculations[2]. As noted above, in the complete basis set limit of the RPA the Thomas-Reiche-Kuhn (TRK) sum rule should be equal to the number of electrons Z_2 and the mean excitation energies calculated from oscillator strengths in the length, mixed or velocity representation should have the same values. The fulfillment of (12.6) and the agreement between the mean excitation energies in different representations thus provide figures of merit for the basis set used in calculation. In Table 12.2 we have thus listed the calculated TRK sums in the length representation for each of the nucleobases. Very good agreement is obtained.

In addition, we observe excellent agreement between the length and velocity representation of the mean excitation energy as well as for the Thomas-Reiche-Kuhn sum rule and can therefore conclude that our tailor-made basis set is sufficiently close to the basis set limit.

12.5 Calculated Stopping Cross Sections

Using the scheme outlined above, the mean excitation energies of the nucleobases were calculated [14], and they are presented in Table 12.3.

[2]The orbital exponents and contraction coefficients for the basis are given in Tables 2–5 of Ref. [24]

Table 12.3 The mean excitation energy in the length representation, I_0^L, for the nucleobases [14]

	I_0^L (eV)
Cytosine	69.60
Uracil	73.13
Thymine	70.00
Adenine	69.06
Guanine	71.58

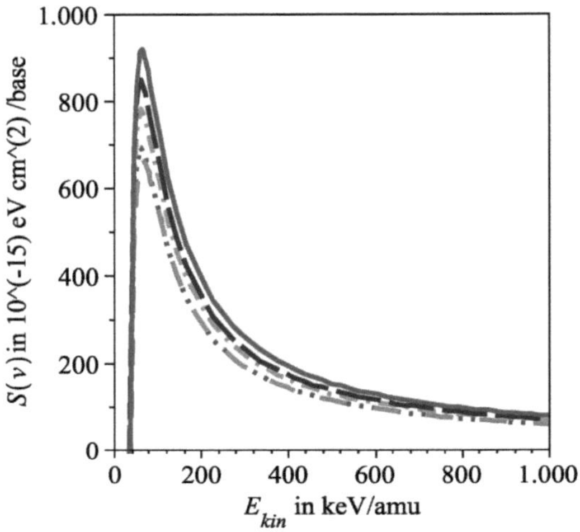

Fig. 12.3 Stopping cross sections of the nucleobases for protons as a function of projectile velocity [green-dash-uracil; red-dot-cytosine; yellow-dash-dot-thymine; blue-long-dash-adenine; lila

Table 12.4 Calculated mean excitation energies of the amino acids [13]

	I_0(eV)		I_0(eV)		I_0(eV)
alanine	67.5	glutamine	68.8	phenylalanine	60.7
arginine	65.3	glycine	71.2	serine	71.3
asparagine	71.0	isoleucine	61.9	threonine	68.5
aspartic acid	73.9	leucine	61.9	tyrosine	63.2
cysteine	84.0	lysine	62.7	valine	63.3
glutamic acid	71.3	methionine	76.3	H_2O	72.2

The mean excitation energies and shell correction coefficients from Table 12.1 were then used in the (12.11) to determine the stopping cross sections for the five nucleobases. These cross sections for a singly charged projectile ion are plotted in Fig. 12.3.

The principal quantity determining the stopping properties of the nucleobases is the mean excitation energy. As the nucleobases all have similar mean excitation energies, it would not be expected that they would differ much in stopping properties. This expectation is borne out in Fig. 12.3. In fact, the mean excitation energies of the nucleobases are very close to calculated mean excitation energies [13, 23, 40] of water (72.2 eV) and those of the amino acids (mean of 68.4 eV for 17 amino acids, see Table 12.4). Thus, the energy absorbing characteristics of water, amino acids, and nucleobases are all very similar.

12.6 Conclusions

The question of whether it would be possible to irradiate a biological material with an ion beam, and tune the beam such that only specific components of the target biomaterial would be affected is interesting. However, the calculations reported above indicate that water, amino acids and nucleobases, and by extension most biomaterials, will absorb energy from swift ions in roughly the same manner. Thus it would seem impossible to selectively deposit energy in biological materials by direct hits by fast ions.

It is important to note that direct hits of swift ions on biomolecules is only one of several ways that fast ion radiation can lead to damage of biomaterials. For example, most of the damage done to cellular DNA results from attack of radicals produced from water which fragments on collision with fast ions [1, 2]. Thus, even though molecules such as water and formaldehyde have similar mean excitation energies $\left(I_0^{H_2O} = 72.2\,eV;\ I_0^{CH_2O} = 71.4\,eV\right)$ [13], they fragment in quite different ways [41, 42], leading to different damage patterns in the biomaterial.

Acknowledgments This work was supported by the Danish Center for Scientific Computing (DCSC), the Carlsberg Foundation and the Danish Natural Science Research Council/The Danish Councils for Independent Research (Grant No. 272-08-0486).

References

1. C. von Sonntag, The Chemical Basis for Radiation Biology, Taylor and Francis, London (1987)
2. C. von Sonntag, *Free-radical DNA damage and its repair - a chemical perspective*, Springer Verlag, Heidelberg (2005)
3. For a general review of the theory see: M. Inokuti, Rev. Mod. Phys. **43**, 297 (1971); J.F. Ziegler, J. Appl. Phys./Rev. Appl. Phys. **85**, 1249 (1999)
4. J.F. Ziegler, J.P. Biersack, U. Littmark, *The Stopping and Range of Ions in Solids. Vol. 1 of the Stopping and Ranges of Ions in Matter*, Pergamon Press, Oxford (1985)
5. J.F. Janni, At. Data Nuc. Data Tables **27**, 150 (1982)
6. E. Bonderup, *Penetration of Charged Particles through Matter*, 2nd edition, University of Århus, Århus (1981)
7. H. Bethe, Ann. Phys. (Leipzig) **5**, 325 (1930)
8. F. Bloch, Ann. Phys. (Leipzig) **16**, 285 (1933)
9. U. Fano, Ann. Rev. Nucl. Sci. **13**, 67 (1963)
10. Cf. e.g. M.C. Walske, Phys. Rev. **88**, 1283 (1952); ibid. **101**, 940 (1956); E. Bonderup, K. Dan. Vidensk. Selsk. Mat. Fys. Medd. **35**, No. 17 (1967); G.S. Khandelwal, Nuc. Phys. A 116, 97 (1968), J. Oddershede, J.R. Sabin, At. Data Nuc. Data Tables **31**, 275 (1984)
11. J.R. Sabin, J. Oddershede, Int. J. Quantum Chem **109**, 2933 (2009)
12. J.R. Sabin, J. Oddershede, Nucl. Inst. Methods B **27**, 280 (1987); J. Oddershede, J.R. Sabin, Nucl. Inst. Methods B **42**, 7 (1989)
13. S. Bruun-Ghalbia, S.P.A. Sauer, J. Oddershede, J.R. Sabin, J. Phys. Chem. B **114**, 633 (2010).
14. S.P.A. Sauer, J. Oddershede, J.R. Sabin, J. Phys. Chem. C **114**, 20335 (2010)
15. J.C. Ashley, R.H. Ritchie, W. Brandt, Phys. Rev. B **5**, 2393 (1972); ibid. Phys. Rev. A 8, 2404 (1973)

16. J. Lindhard, Nucl. Inst. Methods **132**, 1 (1976)
17. For a review of the theory and implementation of the polarization propagator method, see J. Oddershede, P. Jørgensen, D.L.Yaeger, Compt. Phys. Rep. **2**, 3 (1984); J. Oddershede, Adv. Chem. Phys. **69**, 201(1987); S.P.A. Sauer, *Molecular Electromagnetism: A Computational Chemistry Approach*, Oxford University Press, Oxford, 2011, Chapters 2 & 10
18. J. Geertsen, J. Oddershede, J.R. Sabin, Phys. Rev. A **34**, 1104 (1986)
19. P. Jørgensen, J. Oddershede, N.H.F. Beebe, J. Chem. Phys. **68**, 2527 (1978)
20. S.P.A. Sauer, G.H.F. Diercksen, J. Oddershede, Int. J. Quantum Chem. **39**, 667 (1991)
21. S. Bruun-Ghalbia, S.P.A. Sauer, J. Oddershede, J.R. Sabin, Eur. Phys. J. D **60**, 71 (2010)
22. J. Oddershede, J.R. Sabin, Phys. Rev. A **39**, 5565 (1989)
23. S.P.A. Sauer, J.R. Sabin, J. Oddershede, Nucl. Inst. Methods B **100**, 458 (1995)
24. S.P.A. Sauer, J. Oddershede, J.R. Sabin, J. Phys. Chem A **110**, 8811 (2006)
25. A.D. McLachlan, M.A. Ball, Rev. Mod. Phys. **36**, 844 (1964)
26. J.D. Rowe, Rev. Mod. Phys. **40**, 153 (1968)
27. Aa. E. Hansen, T.D. Bouman, Mol. Phys. **37**, 1713 (1979)
28. K. Aidas, J. Kongsted, J.R. Sabin, J. Oddershede, K.V. Mikkelsen, S.P.A. Sauer, J. Phys. Chem. Lett. **1**, 242 (2010)
29. W. Thomas, Naturwiss. **13**, 627 (1925)
30. W.Z. Kuhn, Phys. **33**, 408 (1925)
31. F. Reiche, F., W. Thomas, Z. Phys. **34**, 510 (1925)
32. R.A. Harris, J. Chem. Phys. **50**, 3947 (1969)
33. P. Jørgensen, J. Oddershede, J. Chem Phys. **78**, 1898 (1983).
34. J. Oddershede, J.R. Sabin, At. Data Nuc. Data Tables **31**, 275 (1984)
35. C. Lee, W. Yang, R.G. Parr, Phys. Rev. B **37**, 785 (1988); A.D. Becke, J. Chem. Phys. **98**, 5648 (1993)
36. W.J. Hehre, R. Ditchfield, J.A. Pople, J. Chem. Phys. **56**, 2257(1972); T. Clark, J. Chandrasekhar, P. Schleyer, J. Comp. Chem. **4**, 294 (1983)
37. M. Schreiber, M.R. Silva-Junior, S.P.A. Sauer, W. Thiel, J. Chem. Phys. 128, 134110 (2008)
38. R. Ahlrichs, M. Bär, M. Häser, H. Horn, C. Kölmel, Chem. Phys. Lett. **162**, 165 (1989); M. Häser, R. Ahlrichs, J. Comput. Chem., **10**, 104 (1989); O. Treutler, R. Ahlrichs, J. Chem. Phys. **102**, 346 (1995); R. Bauernschmitt, R. Ahlrichs, Chem. Phys. Lett. 256, 454 (1996); S. Grimme, F. Furche, R. Ahlrichs, Chem. Phys. Lett. **361**, 321 (2002); F. Furche, D. Rappoport, Computational Photochemistry. In Computational, Theoretical Chemistry; M. Olivucci, Ed.; Elsevier: Amsterdam, 2005; Vol. 16; Chapter 3; M.v.Arnim, R. Ahlrichs, J. Chem. Phys. **111**, 9183 (1999)
39. TURBOMOLE V6.0 2009, a development of University of Karlsruhe and Forschungszentrum Karlsruhe GmbH, 1989-2007, TURBOMOLE GmbH, since 2007; available from http://www.turbomole.com
40. S.P.A. Sauer, J.R. Sabin, J. Oddershede, Phys. Rev. A **47**, 1123 (1993)
41. J.R. Sabin, R. Cabrera-Trujillo, Y. Öhrn, E. Deumens, J. Phys.: Conf. Series **101**, 012009 (2008)
42. J.R. Sabin, R. Cabrera-Trujillo, N. Stolterfoht, E. Deumens, Y. Öhrn, Nucl. Inst. Methods B **267**, 196 (2009)

Part II
Modelling Radiation Damage

Chapter 13
Monte Carlo Methods to Model Radiation Interactions and Induced Damage

Antonio Muñoz, Martina C. Fuss, M.A. Cortés-Giraldo, Sébastien Incerti, Vladimir Ivanchenko, Anton Ivanchenko, J.M. Quesada, Francesc Salvat, Christophe Champion, and Gustavo García Gómez-Tejedor

Abstract This review is devoted to the analysis of some Monte Carlo (MC) simulation programmes which have been developed to describe radiation interaction with biologically relevant materials. Current versions of the MC codes Geant4 (GEometry ANd Tracking 4), PENELOPE (PENetration and Energy Loss of Positrons

---

A. Muñoz
Centro de Investigaciones Energéticas, Medioambientales y Tecnológicas (CIEMAT), 28040 Madrid, Spain
e-mail: roldan@ciemat.es

M.C. Fuss • G.G. Gómez-Tejedor (✉)
Instituto de Física Fundamental, Consejo Superior de Investigaciones Científicas, 28006 Madrid, Spain
e-mail: mc.fuss@iff.csic.es; g.garcia@iff.csic.es

G.G. Gómez-Tejedor
Departamento de Física de los Materiales, UNED, 28040 Madrid, Spain

M.A. Cortés-Giraldo • J.M. Quesada
Departamento de Física Atómica, Molecular y Nuclear, Universidad de Sevilla, E-41080 Sevilla, Spain
e-mail: miancortes@us.es; quesada@us.es

S. Incerti
Université Bordeaux 1, CNRS/IN2P3, Centre d'Etudes Nucléaires de Bordeaux Gradignan, CENBG, BP120, 33175 Gradignan, France
e-mail: incerti@cenbg.in2p3.fr

V. Ivanchenko
Laboratoire de Physique Moléculaire et des Collisions, Institut de Physique, Université Paul Verlaine-Metz, 57078 Metz Cedex 3, France
e-mail: Vladimir.Ivantchenko@cern.ch

A. Ivanchenko
Université Bordeaux 1, CNRS/IN2P3, Centre d'Etudes Nucléaires de Bordeaux Gradignan, CENBG, BP120, 33175 Gradignan, France

Geant4 Associates International Ltd, Manchester, United Kingdom
e-mail: Anton.Ivantchenko@cern.ch

G.G. Gómez-Tejedor and M.C. Fuss (eds.), *Radiation Damage in Biomolecular Systems*, Biological and Medical Physics, Biomedical Engineering,
DOI 10.1007/978-94-007-2564-5_13,

and Electrons), EPOTRAN (Electron and POsitron TRANsport), and LEPTS (Low-Energy Particle Track Simulation) are described. Mean features of each model, as the type of radiation to consider, the energy range covered by primary and secondary particles, the type of interactions included in the simulation and the considered target geometries are discussed. Special emphasis lies on recent developments that, together with (still emerging) new databases that include adequate data for biologically relevant materials, bring us continuously closer to a realistic, physically meaningful description of radiation damage in biological tissues.

13.1 Introduction

Decades ago, the Monte Carlo (MC) method started to be used for the simulation of radiation transport [1–4]. It consists in the numerical determination of multiple particle trajectories, so-called particle "histories", across an absorber material. This is achieved by following each incident particle through the subsequent collisions it undergoes and applying specific rules each time one of the expected interaction processes occurs. This approach is particularly useful in complex conditions where deterministic calculations would be unfeasible. By sampling a sufficiently large number of tracks and averaging over the ensemble obtained, MC simulation can – at least in principle – predict radiation-matter interactions exactly.

While the precision (given by the statistical uncertainty) of a MC result depends on computational parameters, the accuracy achieved in practice depends on the existing knowledge of the elementary collisional processes involved and on their implementation in the interaction model. Since the first computer programmes became available to a wider public [5–8], succesive codes or versions have steadily evolved in order to include new insights regarding the underlying physics (e.g. through database updates or a more differentiated treatment of the given collision types). Also through improvements in practical aspects such as the definition of composition and geometry of the simulated absorbers, graphical output, etc., using MC codes in a steadily broader field of applications and by a wider community of users was facilitated. At the same time, there have been computational advances aimed mainly at reducing calculation time (thus increasing a code's "efficiency"), the most notable developments probably being the introduction of the

F. Salvat
Facultat de Física (ECM), Universitat de Barcelona, Societat Catalana de Física (IEC), 08028, Barcelona, Spain
e-mail: cesc@ecm.ub.es

C. Champion
Université Bordeaux 1, CNRS/IN2P3, Centre d'Etudes Nucléaires de Bordeaux Gradignan, CENBG, BP120, 33175 Gradignan, France

Laboratoire de Physique Moléculaire et des Collisions, Institut de Physique, Université Paul Verlaine-Metz, 57078 Metz Cedex 3, France
e-mail: champion@univ-metz.fr

"condensed history" technique [9] (an approximation for multiple scattering which is the base of many modern MC codes) and the application of variance-reduction techniques [10,11] which reduce statistical uncertainty. Concerning radiation effects in biomolecular systems, the adaptation of the interaction model for use at lower energies [12, 13, later 14, 15] and modifications in electron physics in order to distinguish different inelastic processes [16,17] have constituted important advances in MC modelling.

Today, a variety of MC codes are available for simulating radiation-matter interactions. Amongst the most important ones, there are Geant4 [18] which origins are at the European Organization for Nuclear Research (CERN) and is being developed and maintained by an international collaboration, EGS4 [19], developed at Stanford Linear Accelerator Centre and now maintained as the EGSnrc [20] version by the National Research Council of Canada, ITS [21], MCNP [22] from Los Alamos National Laboratory, PENELOPE [15], PARTRAC [23], EPOTRAN [24], and LEPTS (Low-Energy Particle Track Simulation, [16, 25]). Principal differences can be found in the interaction models utilized, the origin and nature of the input databases, and the kind of output data and representation. Their applications currently range from high-energy physics (e.g. detector responses and shielding requirements) over electron microscopy to medical physics.

There, rapid developments – especially in the last decade – have led from initially general programmes used only for comparative studies to user-friendly, specialized codes that are clinically applicable to a growing range of situations. Additionally to treatment verification or planning systems, MC calculation is extending to uses including diagnostic imaging and linac beam simulation (Chap. 19). However, many efforts in the biomedical context still focus on modelling radiation transport and damage induced in patient tissue.

A complete (from the physics point of view) simulation tool for modelling radiation damage in biomaterials should be able to model all possible combinations of:

- different tissues such as muscle, bone, tumour, lung, etc.
- different radiation types (photons, electrons, heavy charged particles)
- different energies/spectra, including that of secondary particles
- incident radiation geometries

Ultimately, it would be interesting to include the effects of reactive molecular species produced and biological response factors. This would provide a detailed view of the molecular damage induced and the related bio-functional impact at different scales.

In the following sections, we describe the current versions of the MC codes Geant4 (GEometry ANd Tracking 4), PENELOPE (PENetration and Energy LOss of Positrons and Electrons), EPOTRAN (Electron and POsitron TRANsport), and LEPTS (Low-Energy Particle Track Simulation). Special emphasis lies on recent developments that, together with (still emerging) new databases that include adequate data for biologically relevant materials, bring us continuously closer to a realistic, physically meaningful description of radiation damage in biological tissues.

13.2 Extension of the Geant4 Monte Carlo simulation toolkit for the modeling of radiation biological damages

13.2.1 The Geant4 Monte Carlo simulation toolkit

The Geant4 Monte Carlo simulation toolkit offers a general-purpose platform for the simulation of particle-matter interactions [18, 26, 27]. It includes a significant set of components for geometry description, particle definition, navigation and tracking, electromagnetic fields, physics models for electromagnetic, hadronic, optical, photo-nuclear and electro-nuclear interactions, event scoring, visualization and management of software components. Improvements and extensions of Geant4 capabilities continue, while its physics models are refined and results are accumulated. Improvement of Geant4 simulation performance is provided thanks to validation efforts and physics comparisons against data in collaboration with different experiments and user groups. The toolkit is developed by an international collaboration of about a hundred physicists and software engineers and Geant4 users participate actively in the verification and in the validation of the software. The development of Geant4 follows the open-source strategy: the source software is entirely accessible and freely downloadable via the Geant4 web site (http://www.geant4.org).

The toolkit is capable of modeling a large variety of physical processes for electromagnetic, optical and hadronic interactions. Electromagnetic interactions of photons and charged particles with matter are available in two alternative sub-packages, the "standard" electromagnetic sub-package which covers interactions from 1 keV up to 10 PeV, and the "low energy" electromagnetic sub-package which is applicable down to ~100 eV. These sub-packages include processes describing ionization, bremsstrahlung, electron-positron pair creation, Compton scattering, single and multiple Coulomb scattering, Rayleigh scattering, photoelectric effect, and other processes [27].

Geant4 hadronic physics sub-packages include a large variety of physics models, which are complementary or sometimes alternative to each other [27] including models of high energy inelastic interactions based on string theories and models for moderate energy cascades, pre-equilibrium, low energy precise neutron transport, nuclear deexcitation, elastic and quasi-elastic scattering. At run time, depending on projectile particle type and energy, corresponding electromagnetic and hadronic models are selected.

Originally the development of Geant4 physics models focused on the simulation of High Energy Physics experiments, such as the experiments at the Large Hadron Collider at the European Organization for Nuclear Research (CERN, Geneva, Switzerland). Thanks to the use of object-oriented technology (C++), the toolkit was designed to be universal, taking into account requirements from other application domains involving particle-matter interactions. With time, new physics models and other software relevant to various medical and space applications were added. These domains also include biomedical physics and space physics, from micrometer biological cells [28] up to planetary scales [29].

Geant4 is currently being extended for the modeling of biological damage from ionizing radiation in the framework of the "Geant4-DNA" project which was initiated in 2001 by the European Space Agency [30].

13.2.2 The Geant4-DNA project

The Geant4-DNA project proposes to extend Geant4 for radiobiological damage simulation at the DNA scale in the context of long duration exposition to cosmic radiation, for example during long stays aboard the International Space Station or for future manned exploration missions of the Solar System. Applications in medical physics, such as track structure simulations for high-LET proton and hadrontherapy, are also foreseen. This project is an activity of the Geant4 collaboration and its developments are part of the electromagnetic package of Geant4. Since the latest release of the Geant4 toolkit (release 9.4, December 2010) Geant4-DNA proposes to Geant4 users a set of physical models for microdosimetry describing the interactions of electrons, protons, neutral hydrogen atoms, helium particles of different charged states and a few heavier ions (C, N, O, Fe - preliminary) in liquid water, the main component of biological tissues. Physical interactions include elastic scattering, electronic excitation, ionization and charge exchange. These models take into account a fine description of the water molecule, including its five electronic excitation levels, as well as the four valence and K shells for ionization. Vibrational excitation and dissociative attachment for electrons have also been included for the simulation of thermalization of sub-excitation electrons [31]. The lowest reachable energies are a few eVs for electrons, $\sim$100 eVs for protons and neutral hydrogen, $\sim$1 keV for Helium and a few MeV for heavier ions. These models have been recently compared to experimental cross sections in the vapor phase as well as to international recommendations [32]. For illustration, total cross sections for electrons in the [4 eV – 100 keV] range are shown in Fig. 13.1. The corresponding models are described elsewhere [31, 32].

Geant4-DNA models do not use any condensed history approximation and simulate explicitly step-by-step all interactions with water, a necessary requirement for track structure simulations at the scale of the DNA molecule. Consequently, their usage requires significant CPU resources and is not recommended at high energies. However, since the 9.4 release of the Geant4 toolkit, these models can be combined with other electromagnetic physics models available in Geant4. It is for example possible to apply them in conjunction with "standard" electromagnetic physics models which use condensed history techniques, are very CPU efficient especially at higher energies ($>$10 MeV), and are adapted to larger scales. Using this combination, Geant4-DNA models can be applied to selected geometrical regions of the simulated setup within a specified energy range. A dedicated so-called "advanced example" named "microdosimetry" is available in the Geant4 toolkit in order to explain to users how to perform this multi-scale combination of models.

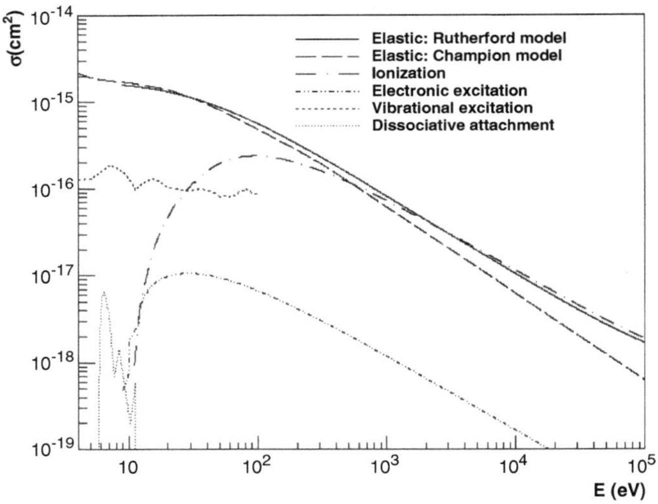

Fig. 13.1 Total cross section models for electrons in the [4 eV – 100 keV] range in liquid water available in the Geant4-DNA extension of the Geant4 toolkit (release 9.4, December 2010). The corresponding physical processes are: elastic scattering (two alternative models are available in Geant4-DNA, the screened Rutherford model – solid line – and the Champion's model – long-dashed curve –), ionization (dashed-dotted curve), electronic excitation (dashed-dotted-dotted curve), vibrational excitation (short-dashed curve) and dissociative attachment (dotted curve)

13.2.3 Modeling of biological damages

Geant4 combines simulations of direct effects of high energy radiation and low-energy secondary effects. High energy hadronic elastic and inelastic models sample interactions of primary particles with nucleons in the nucleus. After generating high energy secondaries, the nucleus is left in an excited state, which should be de-excited to a thermalized state by a pre-equilibrium model. The final de-excitation is provided by de-excitation models. These low-energy final state models are responsible for sampling the multiplicity of neutrons, protons, light ions and isotopes, which affects the overall picture of hadron transport and defines the major part of direct hadron/ion effects inside biological objects.

The modeling of direct biological damages due to excitation and ionization on biological targets such as DNA requires physics models down to very low energies as described above. First estimations of direct DNA single-strand breaks and double-strand breaks yields obtained with the Geant4-DNA physics models have been recently published for MeV protons in liquid water [33] and show a general agreement with reference Monte Carlo and experimental data.

The modeling of ionizing radiation non-direct effects requires specific physico-chemical and chemical processes for the production, diffusion and mutual interactions of molecular species produced from water radiolysis. These molecular species are responsible for non-direct damages to the DNA molecule which become dominant for low-LET particles such as electrons. Although Geant4 cannot handle mutual interactions between particles, Geant4-DNA is currently being extended for the simulation of such physico-chemistry processes.

These developments should become publicly available soon in upcoming releases of the general-purpose Geant4 simulation toolkit. We expect they will offer users an open-source alternative to already existing advanced codes [34] usually designed for specific applications and not easily accessible.

13.3 PENELOPE

PENELOPE is a general-purpose Monte Carlo code system for the simulation of coupled electron-photon transport in arbitrary materials, which has been developed at the University of Barcelona over the last 15 years [14, 35, 36]. The name, an acronym for "PENetration and Energy LOss of Positrons and Electrons", was inherited from earlier works of the authors on the transport of low-energy electrons in solids, where the conventional detailed simulation (i.e., interaction by interaction) is applicable. This background naturally lead to the adoption of mixed simulation schemes (class II schemes in the terminology of Berger [9]) for electrons and positrons, which is the most characteristic feature of PENELOPE.

PENELOPE allows the simulation of electron-photon showers in material systems consisting of homogeneous bodies with arbitrary chemical compositions, for an energy range from 50 eV to 1 GeV (although the interaction database extends down to 50 eV, results for energies less than about 1 keV should be regarded as semi-quantitative). The interaction models implemented in the code are based on the most reliable information currently available, limited only by the required generality of the code. These models combine results from first-principles calculations, semi-empirical models and evaluated databases.

The core of the code system is a Fortran subroutine package that generates electron-photon showers in homogeneous materials. These subroutines are invoked from a main steering program, to be provided by the user, which controls the evolution of the tracks and keeps score of the relevant quantities. The code system also includes a flexible subroutine package for automatic tracking of particles within quadric geometries (i.e. systems consisting of homogeneous bodies limited by quadric surfaces) and a geometry viewer and debugger. A generic main program, called PENMAIN, allows the simulation of a variety of radiation sources in arbitrary quadric geometries; the user can define impact detectors and energy-deposition detectors to extract information from the simulation. The operation of PENMAIN is completely controlled from an input file. The latest public version of PENELOPE, released in 2008, is available from the OECD Nuclear Energy Agency Data Bank (http://www.nea.fr).

PENELOPE has been applied to a wide variety of problems in dosimetry, microdosimetry, radiotherapy, radiation protection, nuclear spectroscopy, electron microscopy, electron probe microanalysis, etc. A comprehensive comparison of simulation results with experimental data available from the literature [37] for electrons with initial energies ranging from a few keV up to 1 GeV demonstrated the reliability of the adopted interaction models and tracking algorithm.

13.3.1 Interaction models

Interactions with the material change the energy and direction of movement of the transported particle, and may also produce secondary particles. PENELOPE combines numerical and analytical total and differential cross sections (DCS) for the different interaction mechanisms. These cross sections are necessarily approximate. For example, the cross sections for photoelectric absorption pertain to free atoms and, therefore, possible extended x-ray absorption fine structure effects are disregarded. Similarly, the x-ray energies and the transition probabilities of excited atoms with inner-shell vacancies are those of free atoms and, consequently, the effect of aggregation on these quantities is neglected. Nevertheless, the structure of the code is flexible enough to allow the use of alternative, more elaborate physical models when needed. Details on the physics models can be found in the code manual [36] and in the review article by Salvat and Fernández-Varea [38].

The interaction mechanisms considered in PENELOPE, and the corresponding DCSs, are the following:

- Elastic scattering of electrons and positrons: numerical DCSs obtained from Dirac partial-wave analysis for the electrostatic potential derived from Dirac-Fock atomic electron densities, with the exchange potential of Furness and McCarthy for electrons. These cross sections were calculated using the program ELSEPA [39, 40].
- Inelastic collisions of electrons and positrons: Born DCS obtained from the Sternheimer-Liljequist generalised oscillator strength model [41, 42], with the density-effect correction. The excitation spectrum is modelled by a discrete set of delta oscillators, whose resonance energies are scaled so as to reproduce the mean excitation energies recommended in the ICRU Report 37 [43]. Thus, collision stopping powers calculated from this model agree closely with the tabulations in [43]. Optionally, the DCS can be renormalised to reproduce the collision stopping power read from an input file.
- Electron impact ionisation: numerical total cross sections for ionisation of K, L and M electron shells of neutral atoms, calculated by means of the distorted-wave (first) Born approximation with the Dirac-Hartree-Fock-Slater self-consistent potential [44].
- Bremsstrahlung emission by electrons and positrons: the energy of the emitted photons is sampled from numerical energy-loss spectra derived from the scaled cross-section tables of Seltzer and Berger [45, 46], optionally renormalised to reproduce the radiative stopping power read from the input file. The intrinsic angular distribution of emitted photons is described by an analytical expression –

an admixture of two "boosted" dipole distributions – [47] with parameters determined by fitting the benchmark partial-wave shape functions of Kissel, Quarles and Pratt [48].

- Positron annihilation: Heitler DCS for two-photon annihilation in flight.
- Coherent (Rayleigh) scattering of photons: Born DCS with atomic form factors and angle-independent effective anomalous scattering factors taken from the LLNL Evaluated Photon Data Library [49].
- Incoherent (Compton) scattering of photons: DCS calculated using the relativistic impulse approximation with analytical one-electron Compton profiles [50].
- Photoelectric absorption of photons: total atomic cross sections and partial cross sections for the K-shell and L- and M- subshells from the LLNL Evaluated Photon Data Library [49]. The initial direction of photoelectrons is sampled from Sauter's [51] K-shell hydrogenic DCS.
- Electron-positron pair production: total cross sections obtained from the XCOM program of Berger and Hubbell [52]. The initial kinetic energies of the produced particles are sampled from the Bethe-Heitler DCS, with exponential screening and Coulomb correction, empirically modified to improve its reliability for energies near the pair-production threshold.

Most of these interaction models pertain to free atoms. Usually, they are extended to compounds and mixtures by assuming the additivity approximation, that is, molecular cross sections are obtained by adding the cross sections of the atoms in a molecule. An exception occurs for inelastic collisions of electrons and positrons, where molecular binding effects can be accounted for appropriately by using the mean excitation energy of the compound material.

The current version of the code allows the simulation of polarized photon beams, with the state of polarization described by the Stokes parameters. However, secondary photons (i.e., characteristic x-rays and Auger electrons, as well as bremsstrahlung photons emitted by electrons or positrons) are assumed to be unpolarized.

13.3.2 Simulation algorithm and geometry routines

Particle histories are simulated from the initial energy down to the absorption energies selected by the user, at which particles are considered to be effectively absorbed in the medium. Secondary electrons and photons emitted with initial energy larger than the corresponding absorption energy are simulated after completion of each primary track. Secondary particles are produced in direct interactions (hard inelastic collisions, hard bremsstrahlung emission, positron annihilation, Compton scattering, photoelectric absorption and pair production) and as radiation (characteristic x rays and Auger electrons) following inner-shell ionisation. PENELOPE simulates the emission of characteristic x-rays and Auger electrons that result from vacancies produced in K-shells and L- and M-subshells by photoelectric absorption and Compton scattering of photons and by electron or positron impact. The relaxation of these vacancies is followed until all vacancies have migrated to N and outer shells. The adopted transition probabilities were extracted from the LLNL Evaluated Atomic Data Library [53].

The simulation of photon tracks follows the usual detailed procedure, i.e., all the interaction events in a photon history are simulated in chronological order. Detailed simulation of high-energy electrons and positrons is impractical, because these particles undergo a large number of interactions in the course of their slowing down. Electron and positron trajectories are generated by means of a mixed (class II) algorithm [54] that allows the generation of electron tracks with a relatively small number of computational steps. Hard interactions, with scattering angle or energy loss larger than certain cutoff values, are simulated in a detailed way, i.e., by random sampling from the corresponding restricted DCS. The path length to the next hard interaction is sampled according to the energy-dependent mean free path for hard interactions. The combined effect of all soft interactions that occur along the trajectory segment between two consecutive hard interactions is simulated as a single "artificial" soft event (a random hinge) where the particle loses energy and changes its direction of motion. The energy loss and angular deflection at the hinge are generated according to a multiple-scattering approach that yields energy-loss distributions and angular distributions with the correct mean and variance.

The manual contains a detailed description of the sampling algorithms adopted to simulate the different interactions. Continuous distributions are sampled by means of the adaptive algorithm RITA (Rational Inverse Interpolation with Aliasing); Walker's [55] aliasing method is adopted to sample discrete distributions with large numbers of possible outcomes. These sampling methods are both robust and fast.

The subroutine package PENGEOM tracks particles in material systems consisting of homogeneous bodies limited by quadric surfaces. These subroutines have been tailored to minimize the numerical work required to locate the particle (i.e., to find the body where it is moving) and to determine intersections of the particle trajectory with limiting surfaces. PENGEOM can describe very complicated systems with up to 5,000 bodies and 10,000 limiting surfaces. Material bodies can be grouped in modules, which in turn are organized in a genealogical tree structure. When the tree of modules is properly defined, the speed of the geometry operations is largely independent of the complexity of the whole material system.

13.4 EPOTRAN: a full-differential Monte Carlo code for electron and positron transport in liquid and gaseous water

EPOTRAN (an acronym for Electron and POsitron TRANsport in water) is a home-made full-differential Monte Carlo (MC) simulation developed by Champion [24] for modelling electron and positron histories in liquid and gaseous water for impact energies ranging from 10 eV to 100 keV. All the induced collisional processes are studied in detail via theoretical differential and total cross sections calculated

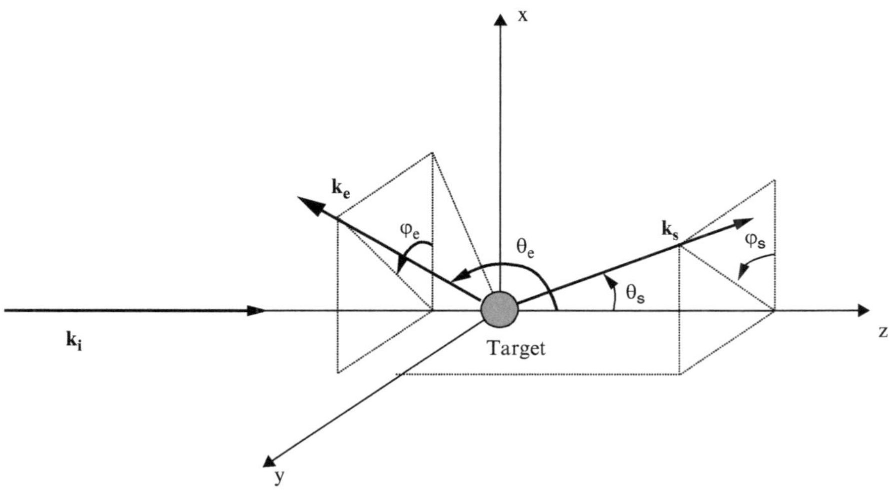

Fig. 13.2 Reference frame of the ionizing collision of a water target. $\boldsymbol{k}_i$, $\boldsymbol{k}_s$, and $\boldsymbol{k}_e$ represent the wave vectors of the incident, scattered and ejected electrons, respectively. The corresponding polar and azimuthal angles are denoted θ_s, φ_s and θ_e, φ_e, respectively

within the quantum mechanical framework by using partial wave methods. Water molecules are treated as point-like targets and therefore any potential energy associated with ionization, excitation or Positronium formation event is assumed to be locally deposited.

Under these conditions, EPOTRAN represents an event-by-event charged particle transport simulation which consists in a series of random samplings determining successively *i)* the distance travelled by the charged particle between two collisions, *ii)* the type of interaction that occurred and finally *iii)* the complete kinematics of the resultant particles (the primary - scattered - projectile as well as the potentially created secondary electron often called δ-ray).

Thus, if the selected interaction is an elastic scattering, the corresponding singly differential cross sections are sampled in order to determine the scattering direction, while the electron incident energy E_{inc} remains quasi unchanged, the energy transfer induced during elastic process being very small (of the order of meV). In the case of ionization, the kinetic energy of the ejected electron E_e is first determined by random sampling among the singly differential cross sections $d\sigma/dE_e$, while the ejection and scattering directions are respectively determined from the triply and doubly differential cross sections hereafter denoted $d^3\sigma/d\Omega_s d\Omega_e dE_e$ and $d^2\sigma/d\Omega_s dE_e$, where Ω_s refers to the scattering direction, Ω_e to the ejection direction and E_e to the energy transfer (see Fig. 13.2). The incident particle energy is finally reduced by $E_e + IP_j$, where the latter term corresponds to the ionization potential of the jth molecular subshell (with j ranging from 1 to 5 for the five water molecule subshells referred as $1b_1$, $3a_1$, $1b_2$, $2a_1$ and $1a_1$, respectively). As mentioned above, let us note that IP_j is considered as locally deposited, except if the selected interaction is inner-shell ($1a_1$) ionization. In this case, Auger electrons with a kinetic energy $E_{Auger} = 467.6\,\text{eV}$ are produced and isotropically emitted, the remaining energy $(IP_5 - E_{Auger})$ being considered as locally deposited. If an excitation is selected, the relative magnitudes of all the partial excitation cross sections are randomly sampled for selecting an excitation channel n whose corresponding energy E_n is considered as locally deposited. The incident particle energy is then reduced by E_n whereas no angular deflection is assumed as experimentally observed. Finally, for positron energy $E_{inc} \geq IP_1 - I_{Ps}$ (where I_{Ps} corresponds to the Positronium binding energy), Positronium (Ps) formation may be considered. To do that, we first determine the target molecular subshell concerned by the capture process as well as the final Positronium state ($Ps(1s)$ or $Ps(2s)$) according to the relative magnitude of all the partial capture cross sections. The quantity $(IP_j - I_{Ps})$ is then assumed as locally deposited whereas the kinetic Positron energy E_{Ps} is simply determined from kinematical considerations $(E_{inc} + I_{Ps} = E_{Ps} + IP_j)$. Finally, for computing velocity reasons we here assume that Positronium formation induces no angular deflection and then suppose that Positronium is ejected in a direction collinear to that of the incident positron. However, in a more sophisticated version of the code, Positronium ejection direction could easily be selected from the pre-calculated singly differential cross sections.

All these steps are repeated for all primary and secondary particles until their kinetic energy falls below a predetermined cut-off value, here fixed at $E_{th} = 7.4\,\text{eV}$, which corresponds to the electronic excitation threshold. Sub-threshold (sub-excitation) electrons and positrons are then assumed to deposit their energy where they are created. In fact, these low-energy species essentially induce vibrational and/or rotational excitations as well as elastic collisions whose total cross section becomes very large (about $20 \cdot 10^{-16}\,\text{cm}^2$), leading to a mean free path less than 1 nm. Therefore, assuming that these 'killed' particles stay where they have been created, introduces uncertainties smaller or of the order of 1 nm in the final energetic cartography.

Finally, note that multiple electronic processes are not taken into account in the current version of our code but should be implemented in the near future [56]. Similarly, we have neglected the contribution of Bremsstrahlung in the electronic and positronic slowing-down due to its minor influence in the energy range here considered ($E_{inc} < 1\,\text{MeV}$).

To describe the following-up of electrons and positrons in water, it is essential to address a large set of differential and total cross sections corresponding to the different types of interactions induced by the charged particles in the medium. In the following, we describe all these cross sections, whose calculation has been performed within the quantum-mechanic framework by using the partial wave expansion method.

13.4.1 *The elastic scattering description*

The perturbation potential induced by charged particles in water can be approximated by a spherically symmetric potential $V(r)$ composed of three distinct terms:

a static contribution $V_{st}(r)$ and two fine correction terms corresponding to the correlation-polarization and the exchange interactions, denoted $V_{cp}(r)$ and $V_{ex}(r)$, respectively. The total interaction potential can then be written as

$$\begin{cases} V(r) = V_{st}(r) + V_{cp}(r) + V_{ex}(r) & \text{for electrons,} \\ V(r) = -V_{st}(r) + V_{cp}(r) & \text{for positrons.} \end{cases}$$

In water vapour, the static potential was numerically calculated within the spherical average approximation from the target molecular wave functions. These latter have been taken from Moccia [57] who described the water molecule by means of single-centre wave functions, each of them being expressed in terms of Slater-type-orbital functions, all centered at a common origin (the oxygen atom). For liquid water, the situation being less trivial essentially due to the scarcity of available molecular wave functions (except those provided by theoretical calculations performed in the Dynamic Molecular framework), we have privileged an empirical approach, which consists in extrapolating the static potential from the experimental liquid water electron density recently reported by Neuefeind et al. [58] (see [59] for more details).

To treat the correlation and polarization effects, we have followed the recommendations of [60] who successfully introduced a correlation–polarization potential into the treatment of electron scattering by noble gas and mercury. Whereas the polarization contribution was treated by means of a polarization potential of the Buckingham type [61] for both electrons and positrons, correlation effects were introduced via different potentials, namely, that described by Padial and Norcross [62] for electrons and that reported by Jain [63] for positrons. Finally, the exchange process (only used for electrons) was treated via the phenomenological potential given by Riley and Truhlar [64]. Singly differential cross sections $d\sigma/d\Omega_s$ were then evaluated within the partial wave framework for both electrons and positrons in gaseous and liquid water whereas total cross sections were simply obtained by means of numerical integration (see Fig. 13.3).

13.4.2 The ionization treatment

In the 1st Born approximation, triply differential cross sections for both electrons and positrons are defined as

$$\begin{aligned} \sigma^{(3)}(\Omega_s, \Omega_e, E_e) &\equiv \frac{d^3\sigma}{d\Omega_s d\Omega_e dE_e} = \sum_{j=1}^{N_{MO}=5} \frac{d^3\sigma_j}{d\Omega_s d\Omega_e dE_e} \\ &= \sum_{j=1}^{N_{MO}=5} (2\pi)^4 \frac{k_1 k_s}{k_i} \left|[T_{ab}]_j\right|^2, \end{aligned} \tag{13.1}$$

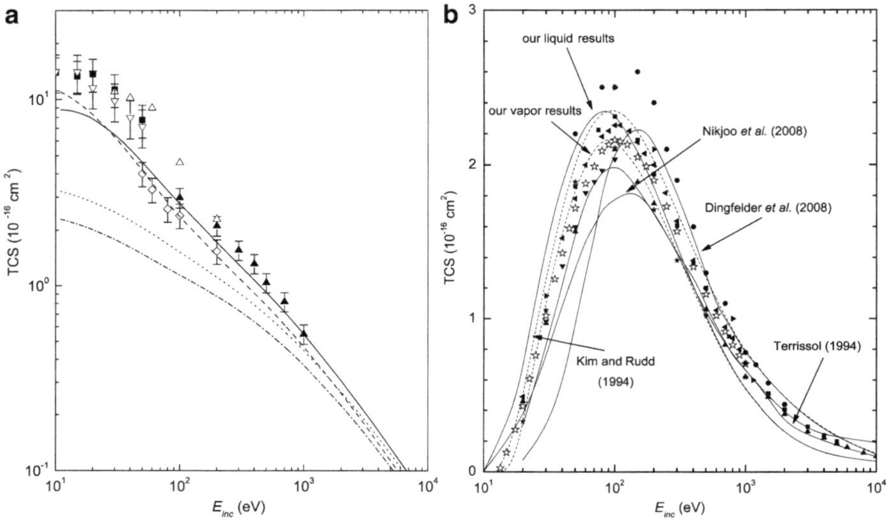

Fig. 13.3 Panel **(a)** Total cross sections (TCS) for elastic scattering of electrons in gaseous water (solid line) and liquid water (dashed line) compared to their equivalents for positrons (dotted and dash–dotted line, respectively). Experimental measurements taken from different sources are also reported for comparison (see [59] for more details). Panel **(b)** Total ionization cross sections versus the incident energy for both phases. Experimental measurements are also reported for comparison (see [56] for more details)

where the transition amplitude between the initial state labeled a and the final state labeled b, denoted $[T_{ab}]_j$, is expressed by

$$[T_{ab}]_j = \left\langle \Psi_b^j(\boldsymbol{r_0},\boldsymbol{r_1}) \middle| V(\boldsymbol{r_0},\boldsymbol{r_1}) \middle| \Psi_a^j(\boldsymbol{r_0},\boldsymbol{r_1}) \right\rangle, \tag{13.2}$$

where $V(\boldsymbol{r_0},\boldsymbol{r_1})$ represents the interaction potential between the incident electron/positron and the target, which can be reduced as

$$V(\boldsymbol{r_0},\boldsymbol{r_1}) = \frac{1}{|\boldsymbol{r_0}-\boldsymbol{r_1}|} - \frac{1}{r_0}. \tag{13.3}$$

Then, by using the well-known partial-wave expansion of the plane wave as well as that of the Coulomb wave, (13.1) may be written in a convenient analytical form from which doubly differential cross sections were obtained by analytical integration over the ejected electron direction whereas singly differential and then total cross sections were finally determined via numerical integrations (for more details, we refer the reader to our previous works [65, 66]).

Furthermore, contrary to the previous case where electronic density was simply needed for describing the elastic process, ionization treatment requires the knowledge of accurate target wave functions, what remains a difficult task essentially due to the multi-centered nature of the water target. In this context, we have recently proposed a unified methodology to express the water molecular wave functions in both phases by means of a single-centre partial-wave description (see [56, 67] for more details). In brief, the wave functions have been carried out by using the Gaussian 03 program and computed at the Hartree-Fock level of theory by using the augmented, correlation-consistent, polarized-valence quadruple-zeta basis set (aug-cc-pvQZ) of Kendall *et al.* [68]. Geometry optimization has been done by including electronic correlation energy at the second-order Møller-Plesset perturbation theory (MP2, [69]). For the computations in the liquid phase we have used the polarizable continuum model (PCM) developed by Tomasi et al. [70] based on the representation of the liquid by a polarizable dielectric continuum having the static dielectric constant of water ($\varepsilon = 78.39$). Thus, a cavity was created in this continuum and a water molecule was placed in it. The molecule was then described quantum mechanically with a Hamiltonian including the electrostatic interactions with the surrounding dielectric medium. Once polarized by the molecular charges, the continuum creates a reaction potential inside the cavity, which in turn polarizes the molecule. The wave function was also obtained by an iterative computation using the so-called self-consistent reaction field approach.

The obtained wave functions were then here used as input data in our theoretical treatment developed for describing the water ionization induced by electron and positron impact in the energy range 10 eV–100 keV (see Fig. 13.3).

13.4.3 The excitation processes

Excitation includes all the processes that modify the internal state of the impacted target molecule (without secondary electron creation), each of them giving a non negligible contribution to the final energetic cartography. They include in particular: *i)* electronic transitions towards Rydberg states or degenerate states ($\tilde{A}_1B_1$, $\tilde{B}_1A_1$, diffuse band), *ii)* dissociative attachment leading to the formation of negative ions, *iii)* dissociative excitation, leading to excited radicals (H^*, O^* et OH^*), and in a minor part *iv)* vibrational and rotational excitations.

In order to account all the processes listed *i), ii)* and *iv)*, we have used the semi-empirical approach of Olivero et al. [71], whereas the dissociative excitation processes were treated via the approach proposed by Green and Dutta [72]. Moreover, following some experimental observations [73], we assume that excitation induces no angular deflection.

Finally, note that we have here assumed that electron- and positron-induced excitation could be treated in the same way considering recent experimental data on Neon which only reported slight differences in terms of total excitation cross sections between the two projectile types (see [74] for more details).

13.4.4 The Positronium formation

Due to its positive charge, the positron has the possibility to capture one of the target electrons leading in the final channel of the reaction to the formation of a bound system - consisting of an electron and a positron - called Positronium (Ps). A continuum distorted-wave final-state approximation was then developed for describing this process [75] in which the final state of the collision was distorted by two Coulomb wave functions associated with the interaction of both the positron and the active electron (the captured one) with the residual ionic target. Thus, singly differential and total cross sections were successively calculated within the well-known frozen-core approximation.

13.5 Low-Energy Particle Track Simulation

Following the discovery of radiation damage in biomolecular systems by low energy, sub-ionising electrons (see Chap. 1), the concept of nanodosimetry (a procedure to quantify radiation damage in nanovolumes) has developed. It aims at a detailed description of the interaction processes occurring in a nano-size target and their implications in terms of radiation damage (number of dissociative events, type of radicals generated, etc…). A molecular-level model suitable to obtain this kind of information requires a reliable and complete set of cross section data, not only as far as integral values are concerned, but also including differential data for all scattering angles. Since elastic processes are relevant to shape the particle track (and thus determine to which extent a particular nano-volume is irradiated), the desired level of detail cannot be achieved by using only high-energy approximations.

This section describes a new low-energy particle track simulation (LEPTS) code, especially designed to provide interaction details at the nano-scale. The simulation procedure is based on an event-by-event Monte Carlo code which uses previously obtained experimental and theoretical electron and positron scattering cross sections and energy loss distribution functions as input parameters.

13.5.1 Interaction processes and input data

The main aspects to evaluate the reliability of a Monte Carlo model to achieve the above mentioned objectives, are the interaction processes considered and the probability distribution functions used to describe these processes. Our model is focused on electrons and positrons and, as far as their energies are concerned, we distinguish two different regions: above and below 10 keV. It has been shown [76] that for different molecules combining the atomic species H, C, N and O, the Born-Bethe approximation applies only for energies above 10 keV. This is due to the overestimation of the elastic cross section, which is even observed at an energy of 5 keV. As such fast electrons can only distinguish the constituent atoms, a similar behaviour can be expected for most biomolecular systems at those energies. We will therefore use, for energies above 10 keV, the Born-Bethe theory [77] with an independent atom model representation [78] to describe the elastic and inelastic processes, as well as the energy deposition, in terms of the corresponding Bethe surfaces.

Below 10 keV, an appropriate description of the various processes and energy loss in terms of the scattering cross sections is required:

- The total scattering cross section provides the mean free path for the simulations.
- Differential and integral elastic cross sections are crucial to define the paths of the particles all along the energy degradation process down to their final thermalisation.
- Differential and integral inelastic cross sections for ionisation (total and partial), electronic excitation, rotational excitation, vibrational excitation, neutral dissociation and electron attachment. In the case of positrons, positronium formation is also considered and is critically important. Differential data is obtained as described in [17].

To illustrate the different approaches used to get a complete input data set, in the following sections we will only account for the case of molecular water.

13.5.1.1 Electron scattering experimental data

Total cross sections and energy loss spectra in the forward direction are derived in a transmission beam experiment [25]. Conventional electron energy loss experimental arrangements at Flinders and Liège Universities have been used to obtain high resolution (50–100 meV) electron energy loss spectra as a function of energy and scattering angle. By integrating the differential inelastic data, electron impact excitation cross sections for a given excitation energy are derived [79]. By combining electron-ion current measurements with time of flight spectrometry of the induced fragments [80], total and partial ionisation cross sections are measured from threshold up to 10 keV.

13.5.1.2 Positron scattering experimental data

Positron data are mainly provided by the Centre for Antimatter-Matter Studies positron beam line facility at the Australian National University. Simultaneous differential and integral cross section data are measured in a transmission beam experiment carried out under intense axial magnetic beam conditions. The design and operation principles of this experiment have been described in detail previously [81]. Additional positronium formation cross section data are also derived from the analysis of the transmission curves.

13.5.1.3 Electron scattering calculations

Differential and integral elastic, as well as integral inelastic, cross sections are calculated with an optical potential method, which is based on an independent atom representation followed by a screening-corrected addition rule procedure to account for molecular targets (IAM-SCAR). The first subjects of these calculations are the constituting atoms. We represent each atomic target by an optical potential, whose real part accounts for the elastic scattering of the incident electrons while the imaginary part represents the inelastic processes which are considered as 'absorption' from the incident beam. To construct this complex potential for each atom the real part of the potential is represented by the sum of three terms: (i) a static term derived from a Hartree-Fock calculation of the atomic charge distribution, (ii) an exchange term to account for the indistinguishability of the incident and target electrons and (iii) a polarisation term for the long-range interactions which depend upon the target dipole polarisability. The imaginary part then treats inelastic scattering as electron collisions. Later improvements [82] finally led to a model which provides a good approximation for electron-atom scattering over a broad energy range.

To calculate the cross sections for electron scattering from molecules, we follow the independent atom model (IAM) by applying what is commonly known as the additivity rule (AR). In this paradigm the molecular scattering amplitude is derived from the sum of all the relevant atomic amplitudes, including the phase coefficients. Alternatively, ICSs can also be derived from the relevant atomic ICSs in conjunction with the optical theorem [78]. An inherent contradiction between the ICSs derived from these two approaches was solved by employing a normalisation procedure during the computation of the DCSs [83]. A limitation of the AR is that no molecular structure is considered, so that it is typically only applicable above $\sim$100 eV. To reduce this limitation we introduced the SCAR method [84], which considers the geometry of a relevant molecule (atomic positions and bond lengths) by employing some screening coefficients. With this correction the range of validity might be extended to incident electron energies as low as $\sim$50 eV. Furthermore, for polar molecules such as water, additional dipole-excitation cross sections can be calculated to further extend the energy range of validity (to $\sim$10 eV). In the present implementation, rotational excitation cross sections for a free electric dipole are calculated by assuming that the energy transferred is low enough, in comparison to the incident energy, to validate the first Born approximation. Under these circumstances, we have calculated approximate rotational excitation cross sections for water at 300 K by weighting the population for the J-th rotational quantum number at that temperature and estimating the average excitation energy from the corresponding rotational constants.

13.5.1.4 Positron scattering calculations

A similar procedure has been followed to calculate positron scattering cross sections. In this case the atomic optical potential used for electrons has been replaced by that proposed by Reid and Wadehra [85]. The other major differences for positron scattering compared to electron scattering, are that the positron scattering potential does not include an exchange term and the positronium formation channel is indirectly introduced by fixing the threshold energy for the absorption potential just at the positronium binding energy, i.e. 6.2 eV.

13.5.2 Monte Carlo simulation procedure

The basis of the Monte Carlo code used in our simulations has been published elsewhere [16]. It is an event-by-event simulation procedure, programmed in C++, which is compatible with other general purpose Monte Carlo codes like GEANT4 [18]. Other related tools, such as the Geant4-based Architecture for Medicine-Oriented Simulations (available from http://fismed.ciemat.es/GAMOS/), have been used to define the target geometries. Photon and high energy electron (above 10 keV) tracks are then simulated with that general code whereas low energy electrons (below 10 keV) and positrons are treated by LEPTS.

For an incoming low energy electron or positron, the free path in the medium is first sampled. Once the location of an event is defined, partial cross sections determine whether an elastic or inelastic process is to take place and call the appropriate interaction routine. For elastic collisions, the programme samples the outgoing particle's angle according to the distribution established by the corresponding differential cross sections. In the case of inelastic collisions, different sub-processes (with their relative frequency given by the corresponding partial cross section values) handle the different types of interactions that are accessible depending on the particles' energy. First, the energy lost in the collision is determined as a fixed value (in the case of vibrational excitation) or from the electron energy loss distributions (for all other inelastic channels). Subsequently, the particle's outgoing direction is sampled using the differential cross section expressed as a function of the momentum transfer (rather than the angle). If ionisation has taken place, a secondary electron is automatically generated and enters the simulation process with an energy given by the energy lost by the primary electron less the ionisation energy and moving in the direction obtained when applying linear momentum conservation. Secondary electron or positron tracks are then fully simulated with the same procedure.

13.5.3 Some results

As an example of the results that can be obtained with our LEPTS procedure, 50 single tracks corresponding to 15 keV electrons in liquid water ($1\ g/cm^3$ density) are shown in Fig. 13.4(a). Each plotted dot represents an interaction event, with the type of interaction being given by the spot size. To simplify the plot, tracks

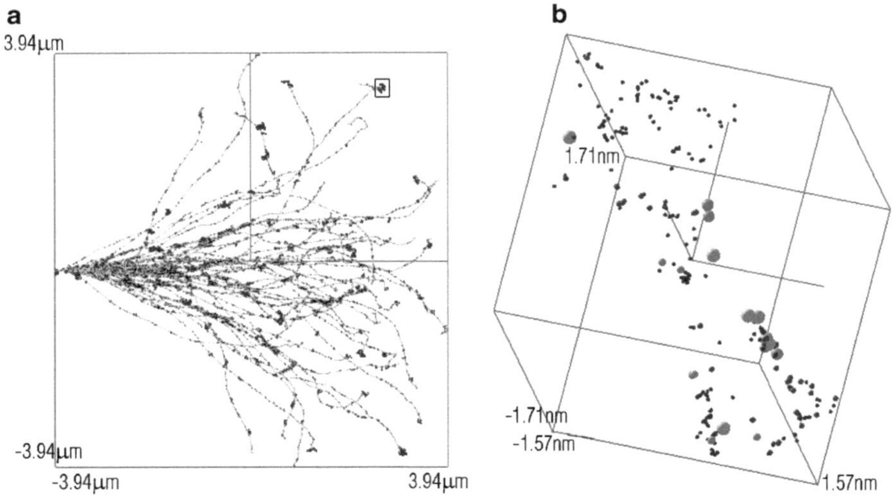

Fig. 13.4 Electron track simulation with the preset LEPTS (Low Energy Particle Track Simulation) code. **(a)** Fully simulated tracks for 50 electrons with energies from 15 keV incident energy to their final thermalisation. **(b)** Nanovolume obtained by "zooming up" the end of one of the tracks as indicated by the small box in (a); the dots' colour indicates the type of interaction event (•, elastic; •, rotational excitation; •, vibrational excitation; •, neutral dissociation; •, ionization)

Table 13.1 Results on the energy deposition and interaction processes derived from the simulation of 50 electron tracks in liquid water for 15 keV incident energy

	Whole irradiated area	End track nanovolume
Volume	4.72×10^{-7} mm^3 (472 μm^3)	5.63×10^{-17} mm^3 (56.3 nm^3)
Total number of interactions:	1490019	273
– Elastic	1083817	214
– Rotational excitation	310899	40
– Vibrational excitation	55692	9
– Electronic excitation	2125	–
– Neutral dissociation	11773	3
– Ionisation	25201	7
– Auger electron generation	197	–
– Electron attachment	309	–
Energy deposition (inelastic)	738.3 keV (1.18×10^{-13}J)	190.4 eV (3.06×10^{-17}J)
Absorbed dose	$2.5 \cdot 10^5$ Gy	$5.4 \cdot 10^{11}$ Gy

have been projected onto the YZ plane. Fig. 13.4(b) shows a magnified region of a three-dimensional nano-volume around the end of one of the tracks. Representative information provided by the LEPTS model, for both target volumes, is summarised in Table 13.1. As can be seen in this table, modelling at the molecular level provides not only information about energy deposition but also a detailed description of the type of interaction taking place in the target volume. By reducing the region of interest to the nano-scale, it is obvious that absorbed dose is not a good quantity to describe radiation effects (see the absorbed dose values in Table 13.1). However the level of detail given by our LEPTS model allows us to develop new tools for nanodosimetry, based on the number of ionisation events or, even more properly, the number of molecular dissociations induced in the nano-volume. Note that for this example we are only showing the total number of ionisations, but this simulation also gives the number and type of ionic fragments produced which together with the information on neutral dissociation and dissociative electron attachment will allow us to characterise radiation effects in terms of structural molecular alterations.

Acknowledgments Work presented in this contribution has been partially supported by the following projects and institutions: Ministerio de Ciencia e Innovación (Project FIS2009-10245), EU Framework Programme (COST Action MP1002) and the Australian Research Council through its Centres of Excellence program.

References

1. R.R. Wilson Phys. Rev. **86**, 261–269 (1952)
2. J.C. Butcher, H. Messel, Phys Rev **112**, 2096–2106 (1958)
3. J.C. Butcher, H. Messel Electron number distribution in electron-photon showers in air and aluminium absorbers. Nucl Phys **20**, 15–128, (1960)
4. A.A. Varfolomeev, I.A. Svetlolobov, Sov Phys JETP **36**, 1263–1270 (1959)
5. C.D. Zerby, H.S. Moran, *A Monte Carlo calculation of the three-dimensional development of high-energy electron-photon cascade showers*. Report ORNL-TM-422, Oak Ridge Natinoal Laboratory (Oak Ridge, Tennessee, 1962)
6. C.D. Zerby, H.S. Moran, J Appl Phys **34**, 2445–2457 (1963)
7. H.H. Nagel, *Die Berechnung von Elektron-Photon-Kaskaden in Blei mit Hilfe der Monte-Carlo Methode*. Dissertation (Rheinische Friedrich-Wilhelms-Universität, 1964)
8. H.H. Nagel, Z Phys **186**, 319–346 (1965)
9. M.J. Berger, Monte Carlo calculation of the penetration and diffusion of fast charged particles. In: B. Alder, S. Fernbach, M. Rotenberg (ed) *Methods in Computational Physics*, vol 1 (Academic, New York, 1963)
10. I. Kawrakow, M. Fippel, Phys Med Biol **45**, 2163 (2000)
11. I. Kawrakow, E. Mainegra-Hing, D.W.O. Rogers, F. Tessier, B.R.B. Walters, *The EGSnrc Code System: Monte Carlo Simulation of Electron and Photon Transport*. Technical report PIRS-701, National Research Council of Canada (Ottawa, 2000)
12. M.J. Berger, S.M. Seltzer, Phys Rev C **2**, 621–631 (1970)
13. M.J. Berger, S.M. Seltzer, *ETRAN Monte Carlo code system for electron and photon transport through extended media*. RSIC Report CCC-107, Oak Ridge National Laboratory (Oak Ridge, 1973)
14. J. Baró, J. Sempau, J.M. Fernández-Varea, F. Salvat, Nucl Instrum Meth Phys Res B **100**, 31–46 (1995)
15. F. Salvat, J.M. Fernández-Varea, J. Sempau, PENELOPE. *A code system for Monte Carlo simulation of electron and photon transport*. OECD-Nuclear Energy Agency (2003)
16. A. Muñoz, J.M. Pérez, G. García, F. Blanco, Nucl Instrum Meth A **536**, 176–188 (2005)
17. M.C. Fuss, A. Muñoz, J.C. Oller, F. Blanco, M.-J. Hubin-Franskin, D. Almeida, P. Limão-Vieira, G. García, Chem Phys Lett **486**, 110–115 (2010)
18. S. Agostinelli et al., Nucl. Instrum. Meth. Phys. Res. A, **506**(3), 250–303 (2003)

19. A.F. Bielajew, H. Hirayama, W.R. Nelson, D.W.O. Rogers, *History, overview and recent improvements of EGS4*. Report SLAC-PUB-6499, Stanford Linear Accelerator Centre (1994)
20. I. Kawrakow, Med Phys **27**, 485–498 (2000)
21. J.A. Halbleib, T.A. Melhorn, *ITS: The integrated TIGER series of coupled electron/photon Monte Carlo transport codes*. Report SAND84–0573, Sandia National Laboratory (Albuquerque, 1984)
22. F.B. Brown, *MCNP — A general Monte Carlo-particle transport code, version 5*. Report LA-UR-03, Los Alamos National Laboratory (Los Alamos, 2003)
23. W. Friedland, P. Jacob, P. Kundrát, Radiat. Res. **173**, 677–688 (2010)
24. C. Champion, C. Le Loirec, B. Stosic, *Int. J. Radiat. Biol.* (2011), doi: 10.3109/09553002.2011.641451
25. Muñoz A, Blanco F, Oller JC, Pérez JM, García G (2007) Adv Quant Chem **52**, 21–57
26. J. Allison et al., IEEE Transactions on Nucl. Sci. **53**(1), 270–278 (2006)
27. J. Apostolakis et al., Radiat. Phys. Chem. **78**(10), 859–873 (2009)
28. S. Incerti, H, Seznec, M. Simon, P. Barberet, C. Habchi, P. Moretto, Radiat. Prot. Dosim. **133**(1), 2–11 (2009)
29. A. Le Postollec, S. Incerti, M. Dobrijévic, L. Desorgher, G. Santin, P. Moretto, O. Vandenabeele-Trambouze, G. Coussot, L. Dartnell, P. Nieminen, Astrobiol. **9**(3), 311–323 (2009)
30. S. Incerti, G. Baldacchino, M. Bernal, R. Capra, C. Champion, Z. Francis, P. Guèye, A. Mantero, B. Mascialino, P. Moretto, P. Nieminen, C. Villagrasa, C. Zacharatou, Int. J. Model. Simul. Sci. Comput. **01**(02), 157 (2010)
31. Z. Francis, S. Incerti, R. Capra, B. Mascialino, G. Montarou, V. Stepan, C. Villagrasa, Appl. Radiat. Isot. **69**(1), 220–226 (2011)
32. S. Incerti, A. Ivanchenko, M. Karamitros, A. Mantero, P. Moretto, H.N. Tran, B. Mascialino, C. Champion, V.N. Ivanchenko, M.A. Bernal, Z. Francis, C. Villagrasa, G. Baldacchino, P. Guèye, R. Capra, P. Nieminen, C. Zacharatou, Med. Phys. **37**(9), 4692 (2010)
33. Z. Francis, C. Villagrasa, I. Clairand, Comput. Meth. Progr. Biomed. **101**(3), 265–270 (2011)
34. H. Nikjoo, S. Uehara, D. Emfietzoglou, F.A. Cucinotta, Radiat. Measurem. **41**(9–10), 1052–1074 (2006)
35. J. Sempau, E. Acosta, J. Baró, J.M. Fernández-Varea, F. Salvat, Nucl Instrum Meth Phys Res B **132**, 377–390 (1997)
36. F. Salvat, J.M. Fernández-Varea, J. Sempau, *PENELOPE-2008: A Code System for Monte Carlo Simulation of Electron and Photon Transport* (Issy-les-Moulineaux, France: OECD/NEA Data Bank 2009). http://www.oecd-nea.org/science/pubs/2009/nea6416-penelope.pdf
37. J. Sempau, J.M. Fernández-Varea, E. Acosta, F. Salvat, Nucl. Instrum. Meth. B **207**, 107–123 (2003)
38. F. Salvat, J.M. Fernández-Varea, Metrologia **46**, S112–S138 (2009)
39. F. Salvat, A. Jablonski, C.J. Powell, Comput. Phys. Commun. **165**, 157–190 (2005)
40. ICRU Report 77, *Elastic Scattering of Electrons and Positrons* (ICRU, Bethesda, MD, 2007)
41. D. Liljequist, J. Phys. D: Appl. Phys. **16**, 1567–1582 (1983)
42. R.M. Sternheimer, Phys. Rev. **88**, 851–859 (1952)
43. ICRU Report 37, *Stopping Powers for Electrons and Positrons*, (ICRU, Bethesda, MD, 1984)
44. D. Bote, F. Salvat, Phys. Rev. A **77**, 042701 (2008)
45. S.M. Seltzer, M.J. Berger, Nucl. Instrum. Meth. B **12**, 95–134 (1985)
46. S.M. Seltzer, M.J. Berger, At. Data Nucl. Data Tables **35**, 345–418 (1986)
47. E. Acosta, X. Llovet, F. Salvat, Appl. Phys. Lett. **80**, 3228–230 (2002)
48. L. Kissel, C.A. Quarles, R.H. Pratt, At. Data Nucl. Data Tables **28**, 381–460 (1983)
49. D.E. Cullen, J.H. Hubbell, L. Kissel, *EPDL97 The evaluated data library, '97 version*, Report UCRL-50400 vol. 6, rev. 5 (Lawrence Livermore National Laboratory, Livermore, 1997)
50. D. Brusa, G. Stutz, J.A. Riveros, J.M. Fernández-Varea, F. Salvat, Nucl. Instrum. Meth. A **379**, 167–175 (1996)
51. F. Sauter, Ann. Phys. **11**, 454–488 (1931)

52. M.J. Berger, J.H. Hubbell, *XCOM: Photon Cross Sections on a Personal Computer*, Report NBSIR 87-3597 (National Bureau of Standards, Gaithersburg, MD, 1987)
53. S.T. Perkins, D.E. Cullen, M.H. Chen, J.H. Hubbell, J. Rathkopf, J. Scofield, *Tables and graphs of atomic subshell and relaxation data derived from the LLNL evaluated atomic data library (EADL), $Z = 1$–100*, Report UCRL-50400 vol. 30 (Lawrence Livermore National Laboratory, Livermore, CA, 1991)
54. J.M. Fernández-Varea, R. Mayol, J. Baró, F. Salvat, Nucl. Instrum. Meth. B **73**, 447–473 (1993)
55. A.J. Walker, ACM Trans. Math. Software **3**, 253–256 (1977)
56. C. Champion, D. Oubaziz, H. Aouchiche, Y.V. Popov, C. Dal Cappello, Phys. Rev. A **81**, 032704 (2010)
57. R. Moccia, J. Chem. Phys., **40**, 2186–2192 (1964)
58. J. Neuefeind, C.J. Benmore, B. Tomberli, P.A. Egelstaff, J. Phys.: Cond. Matter **14**, L429–L433 (2002)
59. H. Aouchiche, C. Champion, D. Oubaziz, Radiat. Phys. Chem. **77**, 107–114 (2008)
60. F. Salvat, Phys. Rev. A, **68**, 012708 (2003)
61. D.R. Bates, H.S.W. Massey, Philos. Trans. R. Soc. London Ser. A **239**, 269–304 (1943)
62. N.T. Padial, D.W. Norcross, Phys. Rev. A **29**, 1742–1748 (1984)
63. A. Jain, Phys. Rev. A **41**, 2437–2444 (1990)
64. M.E. Riley, D.G. Truhlar, J. Chem. Phys. **63**, 2182–2191 (1975)
65. C. Champion, J. Hanssen, P.-A.Hervieux, J. Chem. Phys. **117**, 197–204 (2002)
66. C. Champion, Phys. Med. Biol. **48**, 2147–2168 (2003)
67. C. Champion, Phys. Med. Biol. **55**, 11–32 (2010)
68. R.A. Kendall, T.H.Jr. Dunning, R.J. Harrisson, J. Chem. Phys. **96**, 6796 (1992)
69. C. Møller, M. Plesset, Phys. Rev. A **46**, 618 (1934)
70. J. Tomasi, B. Mennucci, R. Cammi, Chem. Rev. **105**, 2999 (2005)
71. J.J. Olivero R.W. Stagat, A.E.S. Green, J.Geophys. Res. **77**, 4797–4811 (1972)
72. A.E.S. Green, S.K. Dutta, J.Geophys. Res. **72**, 3933 (1967)
73. R.N. Compton, L.G. Christophorou, Phys. Rev. **154**, 110–116 (1967)
74. C. Champion, C. Le Loirec, Phys. Med. Biol. **51**, 1707–1723 (2006)
75. P.-A. Hervieux, O.A. Fojon, C. Champion, R.D. Rivarola, J. Hanssen, J. Phys. B: At. Mol. Opt. Phys. **39**, 409–419 (2006)
76. F. Manero, F. Blanco, G. García, Phys. Rev. A **66**, 032714 (2002)
77. M. Inokuti, Rev. Mod. Phys. **43**, 297 (1971)
78. N.F. Mott, H.S.W. Massey, *The Theory of Atomic Collisions*, 3rd edn. Oxford University Press (1965)
79. A. Muñoz, F. Blanco, G. García, P.A. Thorn, M.J. Brunger, J.P. Sullivan, S.J. Buckman, Int. J. Mass Spectrom. **277**, 175 (2008)
80. M.C. Fuss, A. Muñoz, J.C. Oller, F. Blanco, D. Almeida, P. Limão-Vieira, T.P.D. Do, M.J. Brunger, G. García, Phys. Rev. A **80**, 052709 (2009)
81. J.P. Sullivan, A. Jones, P. Caradonna, C. Makochekanwa, S.J. Buckman, Rev. Scient. Instrum. **79**, 113105 (2008)
82. F. Blanco, G. García, Phys. Lett. A **295**, 178 (2002); Phys. Rev. A **67**, 022701 (2003)
83. J.B. Maljković, A. Milosavljević, F. Blanco, D. Šević, G. García, B.P. Marinković, Phys. Rev. A **79**, 052706 (2009)
84. F. Blanco, G. García, J. Phys. B **42**, 145203 (2009)
85. D.D. Reid, J.M. Wadehra, Phys. Rev. A **50**, 4859 (1994); J.Phys. B **29**, L127 (1996)

Chapter 14
Positron and Electron Interactions and Transport in Biological Media

Modeling Tracks and Radiation Damage

Ronald White, James Sullivan, Ana Bankovic, Sasa Dujko, Robert Robson, Zoran Lj. Petrovic, Gustavo García Gómez-Tejedor, Michael Brunger, and Stephen Buckman

Abstract We present Boltzmann and Monte Carlo models of positron and electron transport in water, in the vapour and liquid states, which are based on measured and calculated cross section data that has been compiled into "complete" cross section

R. White (✉)
Centre for Antimatter-Matter Studies, James Cook University, Townsville, Australia
e-mail: ronald.white@jcu.edu.au

J. Sullivan
Centre for Antimatter-Matter Studies, Australian National University, Canberra, Australia
e-mail: jps107@physics.anu.edu.au

A. Bankovic • Z. Lj. Petrovic
Institute of Physics, Belgrade, Serbia
e-mail: anchuga@ipb.ac.rs; zoran@ipb.ac.rs

S. Dujko
Institute of Physics, Belgrade, Serbia,

Centre for Antimatter-Matter Studies, James Cook University, Townsville, Australia

Centre for Mathematics and Computer Science (CWI), Amsterdam, The Netherlands
e-mail: sasha@ipb.ac.rs

R. Robson
Centre for Antimatter-Matter Studies, James Cook University, Townsville, Australia
e-mail: robert.robson@jcu.edu.au

G.G. Gómez-Tejedor
Instituto de Física Fundamental, Consejo Superior de Investigaciones Científicas, 28006 Madrid, Spain

Departamento de Física de los Materiales, UNED, 28040 Madrid, Spain
e-mail: g.garcia@iff.csic.es

M. Brunger
Centre for Antimatter-Matter Studies, Flinders University, Adelaide, Australia
e-mail: Michael.Brunger@flinders.edu.au

S. Buckman
Centre for Antimatter-Matter Studies, Australian National University, Canberra, Australia
e-mail: stephen.buckman@anu.edu.au

G.G. Gómez-Tejedor and M.C. Fuss (eds.), *Radiation Damage in Biomolecular Systems*, Biological and Medical Physics, Biomedical Engineering,
DOI 10.1007/978-94-007-2564-5_14, © Springer Science+Business Media B.V. 2012

sets for both positrons and electrons. The cross section measurements, transport theory, and implications for the study of charged particle transport in soft matter are discussed.

14.1 Introduction

The transport of charged particles - electrons, positrons, protons – in gaseous and liquid media has been a subject of considerable interest in a number of research and development fields. Charged-particle transport has applications in fields as diverse as gas discharge physics, atmospheric and astrophysical environments, environmental science and most recently, the implications for radiation damage in tissue have been revealed.

We have a program for the measurement and application of both positron and electron interactions with biologically relevant molecules. Our rationale is to establish the best and most accurate cross sections for all relevant collision processes, either through accurate measurement (using both positrons and electrons) or the application of contemporary, state-of-the-art theoretical models. These cross sections are compiled into self-consistent sets that can then be used in our Boltzmann equation and Monte Carlo modeling codes to establish transport parameters such as drift and diffusion coefficients and also to model charge penetration or range, reaction products along particle tracks and energy deposition, amongst other things.

In this paper we consider the interaction of positrons and electrons with water molecules comprising both gases and liquids, and provide a brief background to the measurement of $e^{\pm}$-, H_2O cross sections and the way they are applied in the various approaches to modeling of macroscopic phenomena. Ongoing and future work involving a series of other bio-molecules is also discussed.

14.2 Data Considerations – Measurement and Calculation of Cross Sections

No single experimental apparatus is used for the measurements that are used in our modelling studies described here, but rather a range of techniques that provide absolute scattering cross sections for both positron and electron interactions have been applied. Where measurements are not available, we have used the best available theoretical calculations to obtain the relevant cross sections. In some cases, especially for positrons, neither experimental nor theoretical values are available and we have had to make educated guesses at the value of the cross sections, usually based on electron data.

The positron data for water vapour comprises measurements of the grand total, total elastic and total positronium (Ps) formation cross sections using a high-resolution, trap-based positron beam [1]. Other processes, such as rotational,

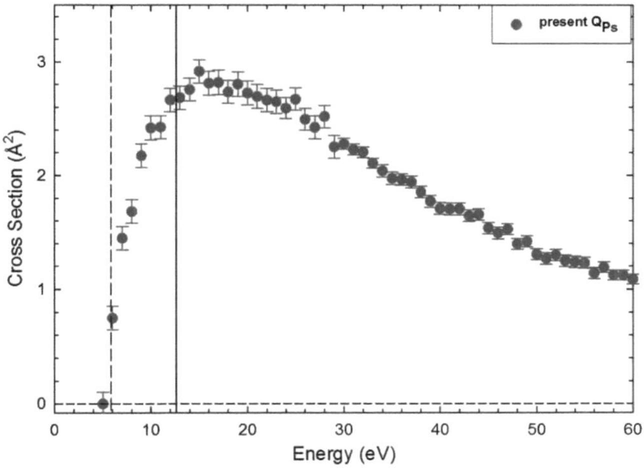

Fig. 14.1 The Positronium formation cross section for water vapour. The vertical lines indicate the Ps threhold (dashed) and the direct ionization threshold (solid)

vibrational and electronic excitation are provided by theory or through comparison with electron data. In the case of electronic excitation one needs to consider only all singlet excitation channels as triplet states are not accessible from the singlet ground state with a positron interaction. The ionization cross section which is used is that of Campeanu and coworkers [2]. Theoretical results were taken as the existing experimental results were not yet published. An example of the measured Ps formation cross section for water is shown in Fig. 14.1.

The electron scattering data set (which is used in both the electron and positron track models) is based on a range of electron interaction data from the literature and includes total, elastic (differential and integral), all excitation, ionization and dissociation channels. It has been discussed previously [3].

14.3 Electron transport in water vapour – a test for accuracy and completeness of cross-section sets

While there has been much recent effort invested in obtaining accurate individual cross-sections using theoretical, beam and other techniques, for modeling of macroscopic systems, one must ensure that the cross-section sets are complete and accurate. One of the tests for completeness and accuracy of cross-section sets is the comparison of experimental swarm transport properties with those calculated using the proposed set of cross-sections. A swarm experiment entails measuring physical properties of an electron swarm (an ensemble of electrons sufficiently dilute so that electron-electron interactions can be neglected), that is generally in a quasi-steady state determined by a balance between power input from an applied electric field E and energy loss rate via collisions between electrons in the swarm and particles of a neutral gas of density n_0. A similar definition applies for positron swarms. Various configurations of swarm experiments exist [4]. It is important to understand that swarm experiments differ significantly from beam-based methodologies: swarm experiments are many scattering experiments while beam experiments are single scattering experiments. By applying a field, the swarm is driven out of thermal equilibrium and the velocity distributions are distinctly non-Maxwellian. The swarm may be in equilibrium with electric field (energy and momentum gained from the field are dissipated in collisions) or be in the so-called non-hydrodynamic (non-equilibrium) regime whereby the distribution is both space and time dependent. It is worth noting that most gas filled traps including the Surko trap start with a mono-energetic distribution of positrons and, after several collisions, develop a broad swarm-type distribution.

Variations in the applied field allow one to selectively assess various energy regions in the cross-sections. Initially swarm experiments were designed to indirectly extract complete sets of cross-sections, and although the number of swarm experiments has declined in recent years, these experiments still continue to provide important information for electron systems. Completeness and accuracy of the cross-section set (which includes the momentum transfer cross-section, rotational cross-sections, vibrational cross-sections etc. for all energetically allowed excitations) is determined by correspondence of the measured transport coefficients with those calculated or simulated. These transport coefficients include e.g. the drift velocity W, transverse and longitudinal diffusion coefficients D_T and D_L respectively and the rate coefficients for a range of applied reduced fields E/n_0 (see e.g. [4] for details). The textbook by Robson [5] gives an overview of modern charged particle transport theory and for a recent review of swarm experiments and swarm transport data the reader is referred to [6].

For modeling radiation damage in biological matter, establishing an accurate and complete set of electron – water cross-sections is paramount. Recently a recommended "best set" of cross-sections in water vapour has been compiled [3]. Of particular importance for radiation damage is the low-energy range up to about 20 eV, corresponding to secondary electrons from primary ionization. In this region, there exists experimental swarm data for electrons in water vapour [7–9]. When combined with the current transport theory or simulation that is of equal (or better accuracy), the process can provide a definitive test on the accuracy and completeness of the current sets of cross-sections for electrons in water vapour. One of the impediments to a true test has been the differences in the transport coefficient definition and measurement techniques that exist between the various groups. Robson et al. [10] have recently reconciled the differences, outlining procedures for true comparisons of recently measured drift velocities of electrons in water vapour [9]. A recent study [11] has made comment on the accuracy of recommended cross-section sets, exploring the impact of variations and sensitivities in the recommended sets of cross-sections including the anisotropic nature of the scattering.

14.4 Positron transport in water vapour

The history and current status of positron (swarm) experiments has been recently reviewed [12, 13]. One motivation for such experiments is clear - a test for the accuracy and completeness of the new generation of positron scattering cross-section sets. Although there are currently no positron swarm experiments of equivalent accuracy to their electron counter part, design considerations were outlined in [12]. As experimental data for complete sets of positron cross-sections become available [13, 14], theoretical and computational techniques developed for electrons have been adapted and applied to positrons. The macroscopic manifestations of the microscopic differences in electron and positron cross-sections and processes are quite striking. Of particular note is the impact of the non-conservative Ps-formation process on the transport properties. In particular, the phenomenon of Ps-induced negative differential conductivity (NDC) (selectively existing only for the bulk transport coefficient- the definition is provided later on the next page), i.e. the decrease in the drift velocity with increasing electric field strength, is now well known in both atomic [15] and molecular gases [16]. Also, of further note is the impact of Ps-formation on the longitudinal diffusion coefficient and excessive skewness of spatial profiles that strongly depart from the expected Gaussians defined by diffusion [14].

Future optimisation of positron-based imaging (Positron Emission Tomography – PET) and therapies is dependent on, amongst other things, an accurate knowledge of positron transport in human tissue or water. Using the set of positron impact cross-sections described earlier, we have performed a study of the transport of positrons in water vapour under the influence of an electric field [16]. In this work, we use and compare two independent techniques – a multi-term solution of Boltzmann's equation (see e.g. the review [17]) and a Monte-Carlo simulation (see e.g. the review [6]). In Fig. 14.2, we present the drift velocity of positrons in water vapour. It is now well known that there are two different types of drift velocities [18] (i) the flux drift velocity, which effectively measures the mean velocity of the positrons within the swarm, and (ii) the bulk drift velocity, which effectively measures the time rate of change of the centre-of-mass of the swarm. It is generally the latter which is measurable in experiment, though both are calculable in theory (although earlier theories provided mostly the flux coefficients). Differences between the two sets manifest themselves when there are energy dependent non-conservative (e.g. annihilation, Ps-formation) processes present and there is a non-symmetric spatial variation in the average energy through the swarm. The strength of the non-conservative Ps-formation processes is such that the differences between the two sets of drift velocities can be as large as two orders of magnitude. The bulk drift velocity is typically less than the flux drift velocity because the loss of positrons due to Ps-formation occurs preferentially at the leading edge of the swarm relative to the tail - a process resulting in a shift in the centre-of-mass of the swarm in a direction opposite to the applied force. Like positron transport in other gases, we again observe the existence of Ps-induced NDC. The combination of the increased

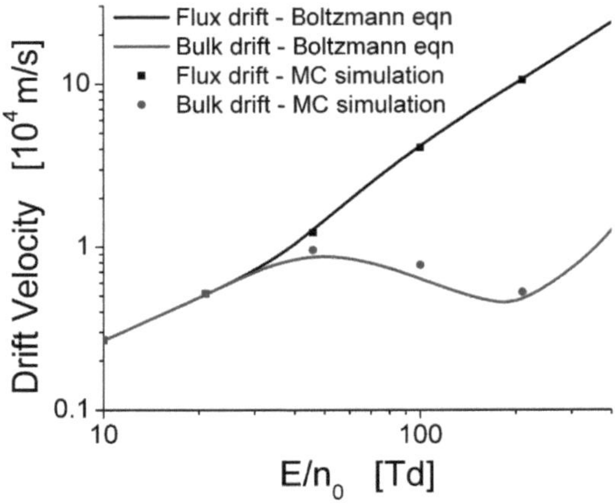

Fig. 14.2 Bulk and flux drift velocities for positrons in water vapour. The solid lines represent the Boltzmann equation solution while the symbols represent the Monte Carlo simulation results. One should be aware that the present results do not include detailed anisotropic scattering represented by differential cross sections

Ps-formation rate (see Fig. 14.6) along with a sufficiently strong spatial variation in the field direction represent conditions sufficient for the presence of this effect. For a detailed discussion on the physical mechanisms for Ps-induced NDC, the reader is referred to [14, 15]. Importantly the MC results and the numerical solutions to the Boltzmann equation are in good agreement over the fields considered, lending support to the observed phenomena, as the two techniques are quite different. This of course holds only if the cross section sets are reliable.

Starting from the basic definitions for the flux drift velocity:

$$w_i = \left\langle \frac{dr_i}{dt} \right\rangle = \langle v_i \rangle,$$

and the bulk drift velocity

$$\mathbf{W} = \frac{d}{dt}\langle \mathbf{r} \rangle,$$

one may obtain the first-order relationship between the two that would involve the rate for the nonconservative process ν_{PF} (positronium formation – PF in this case) [5, 14]:

$$W \approx w - \frac{2\langle \varepsilon \rangle}{3e} \frac{d\,\langle \nu_{PF}(\varepsilon) \rangle}{dE}. \tag{14.1}$$

In Fig. 14.3 we show the calculated flux drift velocity (solid squares), the contribution of the second term in (14.1) based on simulated rates of Ps formation (solid triangles) and the bulk drift velocity as predicted by (14.1) (solid circles). This is only the first order theory assuming symmetric Gaussian profiles and yet it gives an order of magnitude difference from the flux drift velocity. It needs further refinements (higher order theory) as the real bulk property is one order of magnitude lower (open circles) [19].

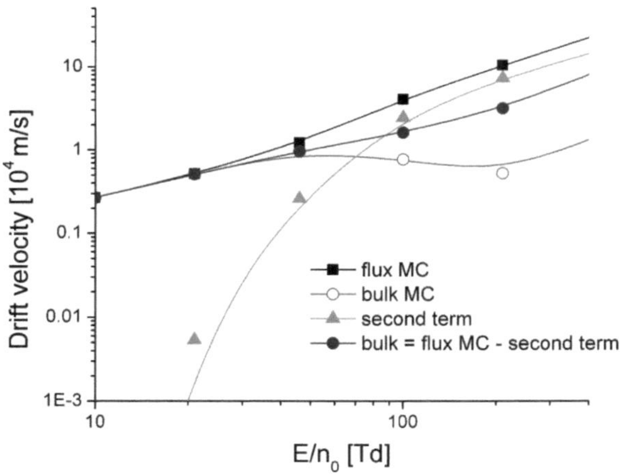

Fig. 14.3 Solid squares are the flux drift velocity calculations. If one takes the PS formation rates (from Fig. 14.6) and uses (14.1) one may obtain the solid circles for the bulk drift velocity. The actual bulk drift velocity is given as open circles

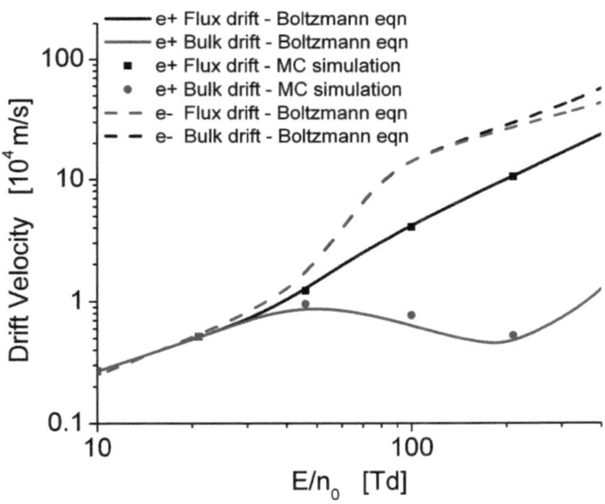

Fig. 14.4 Bulk and flux drift velocities for electrons and positrons in water vapour as a function of the reduced electric field E/n_0

In Fig. 14.4, we present a comparison of drift velocities for electron and positron transport in water vapour. The electron results are taken from [20]. These results highlight the macroscopic differences arising from differences in the microscopic cross-sections (for accessible processes) as well as the differences in the microscopic processes available, i.e. for electrons there are non-conservative processes resulting in both loss (attachment) and gain (ionization) of electrons, while for positrons there is only the loss processes (Ps-formation). Differences between the electron and positron drift velocities can approach two-orders of magnitude. This result illustrates the importance of using accurate positron cross-sections in applications involving positrons – approximations using electron cross-sections to describe positron behaviour can result in considerable error.

14.5 Positron transport in liquid water

One of the most vexing questions within the field of radiation damage at present is how to accurately treat the transport of electrons, positrons and other charged particles in soft-condensed matter. Commonly, one implements a gas-phase assumption, whereby the gas phase results are scaled to the liquid phase through an increase in the density. For soft-condensed systems, this assumption can be quantitatively and qualitatively in error. Amongst other things, specifically the possibility of simultaneous many body coherent scattering from the correlated constituent molecules is neglected. It is consequently important for theories to consider the structure of the material. One of the key elements of our program is to further explore techniques for adapting accurately measured/calculated gas-phase cross-sections for electron and positron interactions for use in the analysis of macroscopic phenomena in soft-condensed systems.

In a recent article [21], a first step towards this goal was made by generalizing the somewhat heuristic Cohen–Lekner two-term kinetic theory, to account for the effects of coherent scattering from correlated molecules in the material. The resulting multi-term solution of Boltzmann's equation is valid for both electrons and positrons in structured matter. The reader is referred to [21, 22] for details on the derivation and solution of the new kinetic equation. It will suffice here to comment that the solution technique adapts much of the mathematical machinery developed previously for treatment of electron and positron swarms in the gas phase. Importantly, the inputs to this model are the measurable single-particle scattering cross-sections and the measureable static structure factor for the medium. The theory has been applied to both real viz., liquid argon [21, 22] and model systems [22]. New phenomena including structure-induced NDC and structure-induced anisotropic diffusion have been predicted.

In Figs. 14.5 and 14.6, we present results for the drift velocity and Ps-formation rate of positron swarms in liquid water at 300K respectively. The results are compared with the corresponding results for the positrons in water vapour. We implement the same set of cross-sections used in the gas-phase case considered above, and utilize the static structure of water detailed in [23] to account for the structural properties of liquid water. The differences between the two different phases correspond to regions where the average de Broglie wavelength is greater than the inter-particle spacing, and coherent scattering effects are thus significant. The manifestation of coherent scattering effects is an effective reduction in the momentum transfer cross-section describing the process. This facilitates enhanced energy transfer from the field into the swarm in the liquid phase. The enhanced drift velocity for positrons in the liquid phase over the vapour phase for a given reduced electric field then follows. As the reduced electric field and hence mean energy of the positrons increase, the average de Broglie wavelength reduces and the impact of coherent scattering effects is reduced and the differences between transport in the two different phases is consequently reduced.

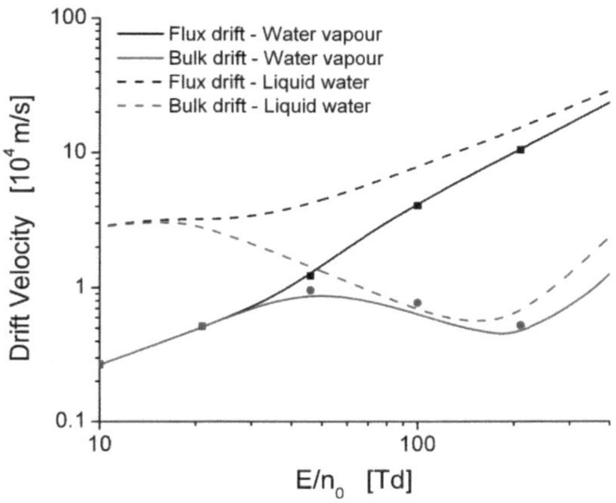

Fig. 14.5 Variation of the bulk and flux drift velocities of positrons in water vapour and liquid water as a function of the reduced electric field E/n_0. Lines denote Boltzmann equation results while symbols denote MC simulation results

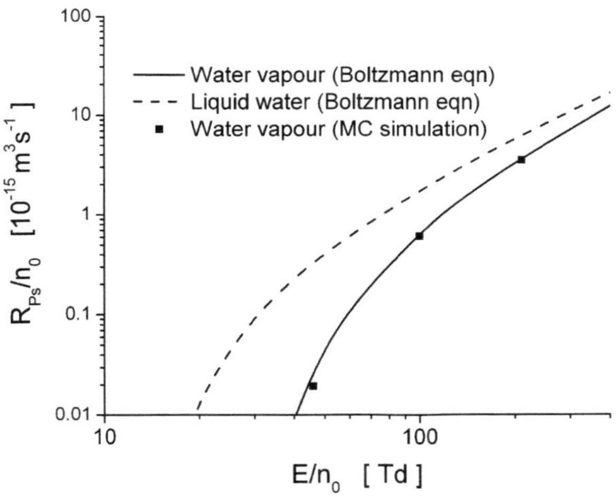

Fig. 14.6 Variation of the reduced positronium formation rate for positrons in water vapour and liquid water as a function of reduced electric field E/n_0

In contrast to other systems investigated previously [21, 22], there is not the pronounced structure-induced NDC region for positrons in liquid water. This is a consequence of the momentum exchange dependence of the structure factor. The differences between the Ps-formation rates in the two different phases shown in Fig. 14.6 are a reflection of the enhanced efficiency of inputting energy from the field due to coherent scattering effects. Note, in this preliminary study, we have assumed that the Ps-formation cross-section is not modified by structural effects, an assumption that may require subsequent justification.

This work represents the first step in overcoming the gas-phase assumption that is commonplace in radiation damage modeling. The structure of the material is included in the theory and the effects of coherent scattering off correlated constituents accounted for through input of measurable gas-phase scattering cross-sections and material structure factors. There are other many-body, multiple scattering and trapping/detrapping effects currently under investigation to improve the accuracy of modeling such soft-condensed systems.

14.6 Positron tracks in water

An example of positron thermalisation in water is shown in Fig. 14.7. It was generated using a new Monte Carlo code (LEPTS: Low Energy Particle Track Simulation) which has been especially designed to describe event-by-event electron and positron interactions [24]. The input parameters used for the simulation are positron and electron interaction cross sections for water, in combination with appropriate angular and energy loss distribution functions that we have previously measured or calculated. Under these conditions, for any selected volume of the target, even for cubic nanometer sizes, the model provides the track structure for positrons and secondary electrons, the total energy deposited in the volume, as well as the type and number of interactions taking place in it. This representation allows not only the study of radiation effects by the traditional method of the absorbed

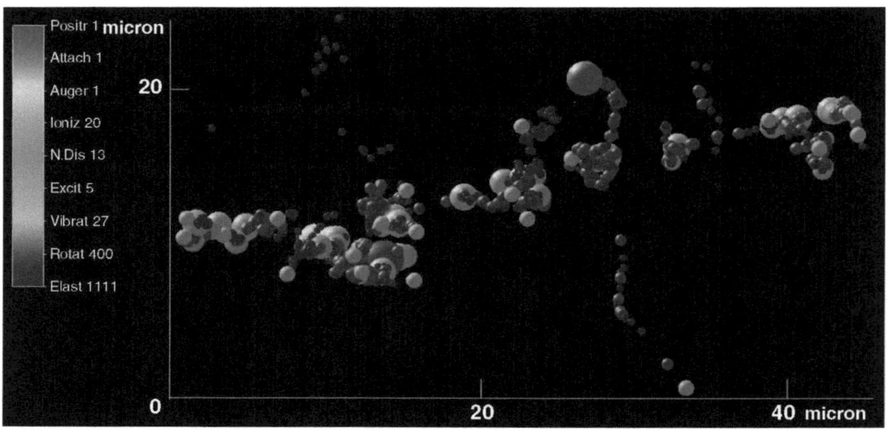

Fig. 14.7 The final few picoseconds in the life of a 10 keV positron in water. The positron quickly thermalises (10's of picoseconds) through collisions, with the colour code broadly indicating the energy loss per collision, and the nature of the processes involved. The picture is a detail of the end of the trajectory with the positron appearing from the left side with a degraded energy of 670 eV which is slowing down to ~1eV by several collisions before being finally annihilated via positronium formation (the red ball). One of these collisions before the positronium formation generates an Auger electron (yellow ball) with an energy close to 500 eV. This secondary electron produces successive electrons via ionization and finally attaches to a water molecule to form a transient negative water ion, which will subsequently dissociate (orange ball)

dose (only valid for macroscopic volumes) but also in terms of induced molecular alterations, which is more realistic to characterise details at the nano-scale.

In this example the last few collisions of a positron, which had an initial energy of 10 keV, are followed as it slows from around 700 eV on entering the frame on the left. This program thus allows one to not only follow the positron through its path to thermalisation and annihilation but also importantly tracks the wake of the copious numbers of secondary electrons that are produced during the thermalisation process.

14.7 Conclusions and future plans

The work described here represents initial, but significant, steps towards a description of positron transport in soft condensed matter, with a view to better informing processes such as Positron Emission Tomography and, ultimately, Positron Dosimetry in the body.

Work is presently underway on developing cross sections sets for both positrons and electrons in a range of other biologically important molecules such as the DNA and RNA bases and the amino acids.

References

1. J.P. Sullivan, A. Jones, P. Caradonna, C. Makochekanwa, S.J. Buckman Rev. Sci. Inst. **79** 113105 (2008)
2. I. Toth, R.I. Campeanu, V. Chis, L Nagy 2010 J. Phys.: Conf. Ser. **199**, 012018
3. Y. Itikawa, N. Mason, J. Phys. Chem. Ref. Data **34**(1), 1 (2005)
4. L. G. H. Huxley, R. W. Crompton, The Drift and Diffusion of Electrons in Gases (Wiley, New York, 1974)
5. R.E. Robson, *Introductory transport theory for charged particles in gases*, (World Scientific Publishing, Singapore, 2006)
6. Z Lj Petrović, S Dujko, D Marić, G Malović, Ž Nikitović, O Šašić, J Jovanović, V Stojanović, M Radmilović-Radenović, J. Phys. D: Appl. Phys. **42**, 194002 (2009)
7. B. Cheung, M. T. Elford, Aust. J. Phys. **43**(6), 755 (1990)
8. M. T. Elford, Aust. J. Phys. **48**, 427 (1995)
9. H. Hasegawa, H. Date, M. Shimozuma, J. Phys. D **40**(8), 2495 (2007)
10. R. E. Robson, R. D. White, K. F. Ness, J. Chem. Phys. **134**, 064319 (2011)
11. K.F. Ness, R.E. Robson, M.J. Brunger and R.D. White, J. Chem. Phys. (submitted)
12. M. Charlton, J. Phys.: Conf. Ser. **162**, 012003 (2009)
13. Z. Lj. Petrović, A. Banković, S. Dujko, S. Marjanović, M. Šuvakov, G. Malović, J. P. Marler, S. J. Buckman, R. D. White, R. E. Robson, J. Phys.: Conf. Ser. **199**, 012016 (2010)
14. M. Šuvakov, Z. Lj. Petrović, J. P. Marler, S. J. Buckman, R. E. Robson, G. Malović, New J. Phys. **10**, 053034 (2008)
15. A. Banković, J. P. Marler, M. Šuvakov, G. Malović, Z. Lj. Petrović, Nuclear Inst. Methods B, **266**, 462–465 (2008)
16. A. Banković, S. Dujko, J.P. Marler, G. Malović, R.D. White, S.J. Buckman, Z. Lj. Petrović, in preparation
17. R.D. White, R.E. Robson, S. Dujko, P. Nicoletopoulos, B. Li, J. Phys. D **42**, 194001 (2009)

18. H. Tagashira, Y. Sakai, S. Sakamoto, J. Phys. D **10**, 1051 (1977)
19. A Bankovic, Z.Lj. Petrovic. RE Robson, JP Marler, S Dujko, G Malovic, Nuc. Instrum. Meth. B **267**, 350 (2009)
20. K. F. Ness, R. E. Robson, Phys. Rev. A **38**, 1446 (1988)
21. R.D. White, R.E. Robson, Phys. Rev. Lett. **102**, 230602 (2009)
22. R.D. White and R.E. Robson, Phys. Rev. E **84**, 031125 (2011)
23. L. Pusztai, Phys. Chem. Chem. Phys. **2**, 2703 (2000)
24. M.C. Fuss, A. Muñoz, J.C. Oller, F. Blanco, P. Limão-Vieira, A. Williart, C. Huerga, M. Téllez, G. García, Eur. Phys. J. D **60**, 203 (2010)

Chapter 15
Energy Loss of Swift Protons in Liquid Water: Role of Optical Data Input and Extension Algorithms

Rafael Garcia-Molina, Isabel Abril, Ioanna Kyriakou, and Dimitris Emfietzoglou

Abstract A short review of the dielectric approach used to describe the energy deposited in liquid water by swift proton beams is presented. Due to the essential role played by the electronic excitation spectrum of the target, characterized by its energy loss function (ELF), we discuss in detail the procedure to obtain a reliable ELF from experimental optical data, which corresponds to zero momentum transfer. We also analyse the influence of the different methods used to extend this optical ELF to non-zero momentum transfers. From these different methods we calculate the stopping power and energy loss straggling of liquid water for proton beams, comparing them with other data available in the literature. In general, a good agreement is found at high projectile incident energy, but differences appear at energies around and lower than the maximum in the stopping power. Finally, the energy delivered to the target as a function of the depth (i.e., the depth-dose distribution) is obtained by means of a simulation code that takes into account the main interactions of the projectile with the target.

R. Garcia-Molina (✉)
Departamento de Física – Centro de Investigación en Óptica y Nanofísica, Universidad de Murcia, E-30100 Murcia, Spain
e-mail: rgm@um.es

I. Abril
Departament de Física Aplicada, Universitat d'Alacant, E-03080 Alacant, Spain
e-mail: ias@ua.es

I. Kyriakou • D. Emfietzoglou
Medical Physics Laboratory, University of Ioannina Medical School, Ioannina 451 10, Greece
e-mail: ikyriak@cc.uoi.gr; demfietz@cc.uoi.gr

G.G. Gómez-Tejedor and M.C. Fuss (eds.), *Radiation Damage in Biomolecular Systems*, Biological and Medical Physics, Biomedical Engineering,
DOI 10.1007/978-94-007-2564-5_15,

15.1 Introduction

After the pioneering proposal by Wilson [1] that malignant cancer cells could be destroyed with energetic proton beams, in less than ten years cancer patients were treated for the first time by proton beams [2]. The first proton therapy facilities soon appeared around the world and nowadays there are more than 25 hadrontherapy centres in the world [3] and the number is expected to double within the next 5–10 years. To a large extend, the success of using ion, instead of photon or electron, beams for cancer treatment lies in the high energy they deposit in a delimited region (where the tumour to be destroyed is located) with a sparing effect to the surroundings (i.e., healthy tissue). The energy delivered by the swift ions as a function of the depth is characterized by the Bragg curve, with a notorious peak almost at the end of the trajectory, whose position and height depend on projectile energy and target nature. Another feature of hadrontherapy is that ions have larger biological effectiveness than other ionizing radiations [4], and lower lateral scattering.

The damage of the cells due to ion irradiation proceeds through a very complex way, involving different stages, mechanisms, as well as time and spatial scales [5]. As most ($\sim$80%) of human body is made of liquid water, a reliable description of the energy loss by the incident projectiles in liquid water is needed as a first step towards an accurate understanding of physical, chemical and biological processes taking place in the irradiated body.

Different frameworks can be used to describe the energy loss of protons in liquid water, all of them relying on the Coulomb interaction between the charged projectile and the electrons of the stopping medium. The Bethe formalism provides a simple analytical expression where the main (non-trivial) parameter is the mean excitation energy I of the target [6, 7]. Other methods are based on a detailed account of the cross sections for the different interaction processes (ionization, excitation, charge-exchanging…) that can take place along the projectile trajectory [8–10]. The energy transfer from the projectile to the target can also be obtained employing the dielectric formalism [11–14] together with a suitable description of the whole electronic excitation spectrum of the target, which can be obtained from optical data and appropriate extension algorithms.

In this work we will use the latter method to analyse in detail how the energy loss of protons in liquid water is affected by the description of the target excitation spectrum in the whole space of momentum ($\hbar k$) and energy ($\hbar\omega$) transfers. The target-dependent main input in the dielectric formalism is the energy loss function, ELF(k, ω), and presently two sets of experimental data for liquid water exist in the optical limit (i.e., at $k = 0$) [15, 16]. Different approaches have been proposed to extend these optical data to non-zero momentum transfers [17–22].

15.2 Dielectric formalism for the electronic energy loss of swift projectiles

A swift projectile travelling through a solid interacts with the target electrons and nuclei, which reduces gradually its energy, and affects its direction of motion as well as its charge state.

For projectile energies in the range of a few keV to several MeV the energy loss due to electronic processes is dominant, whereas the energy loss resulting from nuclear collisions is negligible [23].

In this section we provide the fundamentals of the dielectric formalism to evaluate the relevant magnitudes for the energy loss distribution of swift projectiles in condensed matter. These magnitudes are the stopping power S and the energy loss straggling Ω^2, which are related to the mean value and the variance of the energy loss distribution, respectively. More detailed information on the foundations of the dielectric formalism can be found in [12–14, 24].

When a projectile moves inside a target it can vary its charge state by exchanging (capturing or losing) electrons with the target, reaching an equilibrium charge state after a few femtoseconds. As the energy loss depends on the charge state of the projectile, we write the stopping power, S, and the energy loss straggling, Ω^2, as a weighted sum over the corresponding magnitudes (S_Q and Ω^2_Q, respectively) for the different charge states Q of the projectile:

$$S = \sum_{Q=0}^{Z_1} \phi_Q S_Q, \quad \Omega^2 = \sum_{Q=0}^{Z_1} \phi_Q \Omega^2_Q. \tag{15.1}$$

In the above expressions ϕ_Q represents the probability to find the projectile (with atomic number Z_1) in a given charge state Q. When dynamic equilibrium is attained, ϕ_Q is equivalent to the projectile charge-state fraction, which depends on the target material, the projectile nature and energy. The energy dependence of ϕ_Q for protons ($Z_1 = 1$) in liquid water is obtained by the parameterization to experimental data given by [25].

Based on the first Born approximation, the dielectric formalism provides the following expressions [12] for the stopping power, S_Q, and the energy loss straggling, Ω^2_Q, of a material for a projectile with mass M, kinetic energy E, and charge state Q:

$$S_Q(E) = \frac{M\,e^2}{\pi\,E} \int_0^\infty \frac{\mathrm{d}k}{k} \rho^2_Q(k) \int_0^{k\sqrt{2E/M}} \mathrm{d}\omega\,\omega\,\mathrm{Im}\left[\frac{-1}{\varepsilon(k,\omega)}\right], \tag{15.2}$$

$$\Omega^2_Q(E) = \frac{M\,\hbar e^2}{\pi\,E} \int_0^\infty \frac{\mathrm{d}k}{k} \rho^2_Q(k) \int_0^{k\sqrt{2E/M}} \mathrm{d}\omega\,\omega^2\,\mathrm{Im}\left[\frac{-1}{\varepsilon(k,\omega)}\right]. \tag{15.3}$$

In the above expressions, the projectile nature enters through the Fourier transform of its charge density $\rho_Q(k)$ for a given charge state Q, whereas the target electromagnetic response to the whole spectrum of momentum ($\hbar k$) and energy ($\hbar\omega$) excitations is encoded in its energy loss function, $\mathrm{Im}[-1/\varepsilon(k,\omega)]$, where $\varepsilon(k,\omega)$ is the dielectric function.

The projectile electron-density is described by the statistical Brandt-Kitagawa model [26], where all the bound electrons are characterised by a generic orbital. This model provides analytical expressions for the Fourier transform of the projectile charge density that are easy to implement in the calculations.

Other minor contributions to the stopping power, such as energy loss in the electron capture and loss processes as well as polarization of the projectile cloud could also be included into Eqs. 15.1–15.3, as described in [27].

It is worth mentioning that the dielectric formalism overcomes the most important limitations of the Bethe formula, namely the need for an independent estimate of the shell-corrections and the I-value [28, 29].

The only target-parameter necessary to find the electronic energy loss of a swift projectile is the ELF, $\mathrm{Im}[-1/\varepsilon(\omega,k)]$, which contains the response of the target to the external perturbations. The three-dimensional plot of the ELF over the momentum-energy, $(k-\omega)$, plane is known as the Bethe surface of the material [6]. In what follows, we discuss in detail the procedure to obtain the whole $(k-\omega)$ dependence of the ELF for liquid water starting from optical data (i. e., $k=0$) and a suitable extension scheme for arbitrary $k\neq 0$ values.

15.3 Energy loss function: a review of extended optical-data models

Formally, the ELF of the target depends on the initial and final state many-electron wavefunctions. Thus, apart from the free-electron gas and atomic hydrogen, where close analytic forms exist, the calculation of the ELF from first-principles requires a formidable effort for practical applications [30]. Therefore, most models are based on very simplistic assumptions (which in some cases are far from realistic) or are fed from experimental (optical or energy loss spectroscopy) data at $k=0$ with suitable extension schemes to incorporate the dependence on the momentum transfer k. The latter are the so-called extended optical-data models, which are expected to provide a computationally simple, yet accurate, representation of the ELF over the whole momentum and energy excitation spectrum (i.e. the Bethe surface).

In this work we follow the second procedure and this section is devoted to the discussion of the main methodologies currently used to obtain the Bethe surface of liquid water from available experimental optical data.

15.3.1 Experimental optical data

The first step is the description of the optical energy loss function (OELF hereafter) of the target. Presently there are only two sets of experimental ELF data for liquid water in the optical limit ($k = 0$) that cover a sufficient part of the valence energy-excitation spectrum.

The first set comes from the Oak Ridge group [15], which provides, from reflectance measurements on liquid water surfaces, the real refractive index, n, and the extinction coefficient, κ, of the complex index of refraction $\tilde{n} = n + \mathrm{i}\kappa$ over the excitation-energy range 7.6–25.6 eV. For more than 25 years the Oak Ridge data were the sole source of information on the dielectric response of liquid water in the VUV range. The OELF limit is obtained from the ω-dependent $\tilde{n}$ as follows:

$$\mathrm{Im}\left[\frac{-1}{\varepsilon(k\approx 0,\omega)}\right]=\mathrm{Im}\left[\frac{-1}{(n(\omega)+\mathrm{i}\kappa(\omega))^2}\right]=\frac{2n(\omega)\,\kappa(\omega)}{\left(n^2(\omega)-\kappa^2(\omega)\right)^2+\left(2n(\omega)\,\kappa(\omega)\right)^2}. \tag{15.4}$$

The second set of optical data comes from the Sendai group [16], which used inelastic X-ray scattering spectroscopy (IXSS) to measure the generalized oscillator strength (GOS) of liquid water at nearly vanishing momentum transfer ($k \approx 0$). The IXSS data extend from 6 to 160 eV excitation energies, providing a near complete knowledge of the dielectric response properties of the valence-shells of liquid water. The OELF limit is then obtained as:

$$\mathrm{Im}\left[\frac{-1}{\varepsilon(k\approx 0,\omega)}\right]=\frac{\pi\,\omega_{\mathrm{pl}}^2}{2\,\omega}\frac{\mathrm{d}f(k\approx 0,\omega)}{\mathrm{d}\omega}, \tag{15.5}$$

with $\mathrm{d}f(k \approx 0,\omega)/\mathrm{d}\omega$ being the GOS at $k \approx 0$. The nominal plasmon energy of the material, $\hbar\omega_{\mathrm{pl}}$, is determined from the relationship $\hbar\omega_{\mathrm{pl}} = 4\sqrt{n_{\mathrm{e}}\pi a_0^3}\,\mathrm{Ry}$, where n_{e} is the target electronic density, $a_0 = 0.529\,\text{Å}$ and $\mathrm{Ry} = 13.606\,\mathrm{eV}$ is the Rydberg energy. For liquid water of mass-density $1\,\mathrm{g/cm^3}$, we have $n_{\mathrm{e}} = 3.34 \times 10^{23}\,\mathrm{cm^{-3}}$, so $\hbar\omega_{\mathrm{pl}} = 21.4\,\mathrm{eV}$.

In Fig. 15.1 we present the ELF of liquid water at the optical limit, $k \approx 0$, obtained from the data of both groups, which will be called REF [15] and IXSS [16] data hereafter. Although the shape of the ELF given by both sets (REF, IXSS) of data is similar, there is a sizeable disagreement with respect to the intensity of the main excitation peak at ~21 eV. Specifically, in the IXSS data the peak intensity is reduced by a factor of 1.5 compared to the REF data. A reduction of that magnitude has also been observed in the spectrum of ice water, both in its hexagonal and amorphous forms [31–33]. This has often been used as an argument in favour of the IXSS data, particularly in view of the fact that amorphous ice has characteristics similar to liquid water [34].

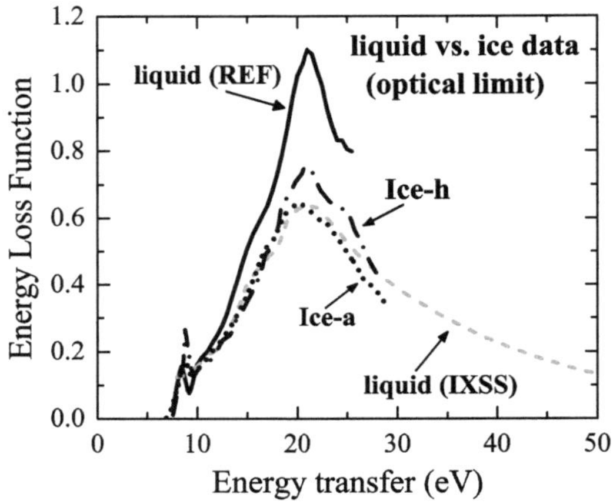

Fig. 15.1 Experimental energy loss function of liquid water at the optical limit obtained from reflectance measurements (REF) [15] and from inelastic X-ray scattering spectroscopy (IXSS) [16]. The ELF of ice water, both in hexagonal (h) and amorphous (a) forms [31–33], is also shown

Previous to introducing the momentum-dependence of the ELF, we will discuss separately the contribution to the ELF due to inner- and outer-shell electron excitations, respectively, of liquid water. The former comes from the K-shell electrons of oxygen, which having a binding energy $E_{\rm K} = 540\,{\rm eV}$ preserve their atomic character and can be described by the generalized oscillator strengths (GOS) in the hydrogenic approach [22, 35]. The contribution of the oxygen K-shell electrons to the OELF of liquid water is given by the following relation [36]

$$\mathrm{Im}\left[\frac{-1}{\varepsilon(k,\omega)}\right]_{\text{K-shell}} = \frac{2\pi^2 N}{\omega}\sum_j \alpha_j \sum_{nl} \frac{\mathrm{d}f_{nl}^{j}(k,\omega)}{\mathrm{d}\omega}, \tag{15.6}$$

where N is the molecular density of the target, $\mathrm{d}f_{nl}^{j}(k,\omega)/\mathrm{d}\omega$ is the hydrogenic GOS corresponding to the (n, l)-subshell of the jth element, and α_j indicates the stoichiometry of the jth-element in the compound target. In the case of liquid water, j refers to oxygen, then $\alpha_{\rm O} = 1/3$, $(n = 1,\ l = 0)$ and $N = 3.34 \times 10^{22}$ molecules/cm^3.

The description of the inner-shell contribution to the OELF is independent of the outer-shell (i.e., valence electron) contributions, where condensed phase effects in the ELF are expected to be important (as is evident in Fig. 15.1) and the dielectric theory most justified. Therefore, the rest of the work is devoted to a detailed discussion of the role played by the input data (resulting from valence excitations) used for the OELF and its extension to non-zero momentum transfer.

All the methods we will present in the following accounting for the valence-electrons contribution to the OELF are based on a fitting to the experimental OELF through a sum of Drude-type ELF

$$\mathrm{Im}\left[\frac{-1}{\varepsilon(k=0,\omega)}\right]_{\text{experimental}} = \sum_i \frac{A_i}{W_i^2}\mathrm{Im}\left[\frac{-1}{\varepsilon_{\rm D}(k=0,\omega;W_i,\gamma_i)}\right], \tag{15.7}$$

where each Drude-type ELF,

$$\mathrm{Im}\left[\frac{-1}{\varepsilon_{\mathrm{D}}(k=0,\omega;W_i,\gamma_i)}\right]=\frac{W_i^2\,\omega\,\gamma_i}{(W_i^2-\omega^2)^2+(\omega\,\gamma_i)^2}, \tag{15.8}$$

is characterized by the position W_i and width γ_i of its peak [37], whose intensity is quantified by the constant A_i.

Besides a satisfactory agreement with the available optical data, the consistency of the fitting procedure must be ensured by the fulfilment of physical constrains, such as several sum rules [38, 39]. The f-sum rule gives the effective number of electrons per molecule that can be excited and guarantees a good behaviour of the ELF at high energy transfers:

$$\frac{m_{\mathrm{e}}}{2\pi^2 e^2 N}\int_0^\infty d\omega\,\omega\left\{\mathrm{Im}\left[\frac{-1}{\varepsilon(k=0,\omega)}\right]_{\mathrm{valence}}+\mathrm{Im}\left[\frac{-1}{\varepsilon(k=0,\omega)}\right]_{\mathrm{K\text{-}shell}}\right\}=Z_2, \tag{15.9}$$

where Z_2 is the number of electrons of the water molecule and m_{e} is the electronmass.

The Kramers-Kronig or perfect screening sum rule is an important test for the accuracy of the ELF at low energy transfer:

$$\frac{2}{\pi}\int_0^\infty \frac{1}{\omega}\mathrm{Im}\left[\frac{-1}{\varepsilon(k=0,\omega)}\right]\mathrm{d}\omega+n^{-2}(\omega=0)=1, \tag{15.10}$$

where $n(\omega=0)$ represents the refractive index at the static limit.

All the procedures to describe the ELF of a material, to be discussed in what follows, satisfy better than 99% both sum rules, Eqs. 15.9 and 15.10. Besides, these methods guarantee the fulfilment of the sum rules for every momentum transfer, provided it is satisfied at $k=0$.

In Fig. 15.2a we show the fitting curves resulting from applying Eq. 15.7 to the experimental OELF derived from both sets of data, REF. [15] and IXSS [16], for the valence-electron excitations of liquid water. The right panel of Fig. 15.2 corresponds to higher energy transfer, where the OELF values have been obtained from the FFAST database of NIST for the water molecule [40], and from the x-ray scattering factors of H and O [41]. The contribution of the inner-shell electrons to the OELF of liquid water, Eq. 15.6, has been added to the valence contribution, Eq. 15.7, for energy transfers greater than the K-shell binding energy of oxygen ($E_{\mathrm{K}}=540\,\mathrm{eV}$).

The fitting curves in Fig. 15.2 exhibit a satisfactory agreement with the experimental OELF, whose main trends are well reproduced. This is one of the main advantages of the approach based on using optical data specific to the material under consideration, which automatically accounts for electronic-structure effects in a realistic manner not always possible within electron gas models.

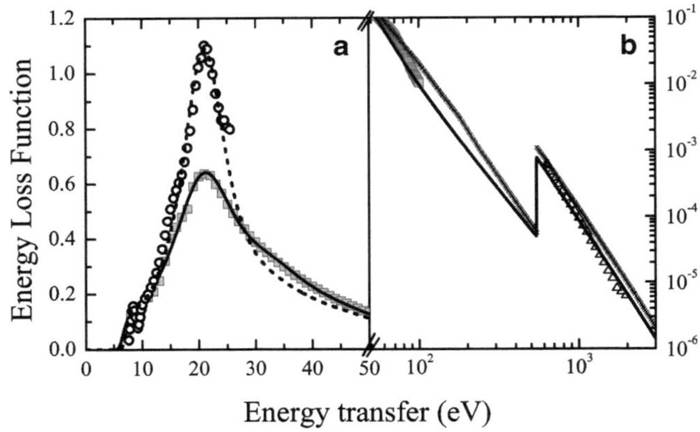

Fig. 15.2 **(a)** Experimental and fitted OELF of liquid water. Continuous and dashed lines represent the extended Drude model fitting to the IXSS data (squares) [16] and REF measurements (circles) [15], respectively. **(b)** At high energy transfers we obtain the OELF from the FFAST database of NIST for the water molecule (triangles) [40], and from the x-ray scattering factors of H and O (crosses) [41]; the solid line represents the extended Drude model fitting to the IXSS data, to which we have added the contribution from the oxygen K-shell electrons through their GOS after the K-shell binding energy ($E_K = 540\,\text{eV}$). See the text for more details

15.3.2 Extension algorithms at k ≠ 0 of the valence excitation spectrum

In order to use Eqs. 15.2 and 15.3 for calculations of the stopping power or the energy-loss straggling, the ELF must be known for arbitrary momentum-transfer (i.e. $k \neq 0$). However, experimental data for the ELF of liquid water at $k \neq 0$ is only available from the Sendai group [42,43] in the range $0.19 \leq k \leq 3.59$ a.u.

Current theories overcome the problem of describing the ELF over the complete Bethe surface by extrapolating the optical data through a suitable extension algorithm. In the simplest case, an extension algorithm is a dispersion relation, i.e. an analytic expression of energy-transfer as a single-valued function of momentum-transfer. In what follows we present a summary of the most used extension algorithms for the calculation of the OELF at finite k.

15.3.2.1 Methods based on the Drude dielectric function

A widely used methodology for the extension algorithm of the ELF to the whole energy-momentum region (k, ω), was proposed by Ritchie and Howie [17]. These authors suggested to incorporate non-zero momentum-transfers through the k-dependence of the parameters $W_i(k)$ and $\gamma_i(k)$:

$$\text{Im}\left[\frac{-1}{\varepsilon(k,\omega)}\right]_{\text{Ritchie-Howie}} = \sum_i \frac{A_i}{W_i^2(k)}\text{Im}\left[\frac{-1}{\varepsilon_{\text{D}}(k=0,\omega;W_i(k),\gamma_i(k))}\right]. \tag{15.11}$$

The dependence of the parameters $W_i(k)$ and $\gamma_i(k)$ on momentum transfer should be obtained from physically motivated dispersion relations, such as the observation that at $k = 0$ (and $W_i \approx \omega_{\text{pl}}$) the Drude-type ELF with $\gamma = 0$ coincides with the Lindhard ELF [12], which is evaluated in the random phase approximation (RPA).

An important property of the extended-Drude ELF, Eq. 15.11, is that it fulfills the sum rules Eqs. 15.9 and 15.10, independent of the dispersion relations, as long as it fulfills them at $k = 0$.

In particular, Ritchie and Howie [17] proposed the following quadratic RPA dispersion relation for the Drude energy coefficient:

$$W_i(k) = W_{i,0} + \alpha_{\text{RPA}}\frac{k^2}{2m_e} \simeq W_{i,0} + \frac{k^2}{2m_{\text{e}}}, \tag{15.12}$$

where $\alpha_{\text{RPA}} = 6\omega_{\text{F}}/(5\omega_{\text{pl}}) = 0.981 \simeq 1$ for liquid water, whose free-electron Fermi and plasmon energies are $\hbar\omega_{\text{F}} = 17.5$ eV and $\hbar\omega_{\text{pl}} = 21.4$ eV, respectively. On the other hand, no dispersion was assumed for the damping coefficient, i.e. $\gamma_i(k) = \gamma_{i,0}$.

The parameters $W_{i,0} = W_i(k = 0)$, $\gamma_{i,0} = \gamma_i(k = 0)$ and A_i are obtained by a fitting at $k = 0$ to the experimental OELF, i.e., through Eqs. 15.7 and 15.8.

Despite its simplicity, the quadratic dispersion relation Eq. 15.12 is adequate for sufficiently fast projectiles by virtue of its correct limiting form at $k \to 0$ and $k \to \infty$ [44, 45]. The latter ensures that, at high k, single-particle effects are accounted for in an approximate manner by a quadratic kinetic term, $k^2/(2m_{\text{e}})$, which represents a free-electron like response and reproduces the characteristic Bethe ridge [46]. Also, for materials where the Fermi and plasmon energies do not differ appreciably, Eq. 15.12 resembles the plasmon dispersion of the free-electron gas at small k. However, for not too fast projectiles (e.g. with energies around the Bragg peak region) improvements upon the quadratic RPA dispersion must be considered [47–50].

Extended-Drude model with damping

The extension of the ELF at finite momentum should consider the plasmon damping as the momentum transfer increases, which can be done to various degrees of sophistication.

A simple extension of the damping coefficients to non-zero k-values is made by taking the linear dispersion relation for the damping constant

$$\gamma_i(k) = \gamma_{i,0} + \text{Ry}\, k, \tag{15.13}$$

proposed in [21], which will be called damped Ritchie model hereafter. In the above expression, $\gamma_{i,0}$ are the damping parameters used to fit the OELF through Eq. 15.7.

Improved extended-Drude model

A further improvement in the extension of the OELF of liquid water to $k \neq 0$ values resulted in the proposal of the new dispersion relations

$$W_i(k) = W_{i,0} + [1 - \exp(-ck^d)]\frac{k^2}{2m_e}, \tag{15.14}$$

and

$$\gamma_i(k) = \gamma_{i,0} + ak + bk^2, \tag{15.15}$$

which were empirically derived by Emfietzoglou *et al.* [20] to better satisfy the experimental trends of the k-dependent IXSS data [42, 51]. The value of the parameters are $a = 10\,\text{eV}$, $b = 6\,\text{eV}$, $c = 1.2$ and $d = 0.4$ (assuming k in a.u. and energies in eV), with practically no dependence upon the method used to parameterize the IXSS optical data [20, 52].

The improved extended-Drude (IED hereafter) model accounts in a phenomenological way for the shifting and broadening of the Bethe ridge as observed experimentally for liquid water [42, 43] and predicted by local-field corrections to the electron-gas theory, because the term $[1 - \exp(-ck^d)]$ provides a reduction of $W_i(k)$ at not too large k, which results in shifting the ELF to lower excitation energies, while the k-dependence of the damping constant leads to the broadening of the ELF.

15.3.2.2 Methods based on the Lindhard dielectric function

The Lindhard [12] dielectric function, $\varepsilon_L(k,\omega)$, can be derived from the quantum perturbation theory [12] or following the random phase approximation (RPA) [24], and provides an analytic expression for the dielectric response function of the homogeneous electron gas. Besides the original work [12], a detailed account of the Lindhard dielectric function appears in [37, 53]. In this scheme the two basic modes of energy absorption by the electrons of the system are single-particle excitations (also called electron-hole pair excitations) and collective or plasmon excitations.

The utility of the Lindhard-type ELF in the context of an extended-optical-data methodology was revealed by the pioneering work of Penn [54], who proposed a scheme where the sum over a finite number of Drude-type ELF was replaced by an integration over Lindhard-type ELF:

$$\mathrm{Im}\left[\frac{-1}{\varepsilon(k,\omega)}\right]_{\mathrm{Penn}} = \int_0^\infty \frac{2}{\pi\,\omega'}\mathrm{Im}\left[\frac{-1}{\varepsilon(k\approx 0,\omega')}\right]_{\mathrm{experimental}} \mathrm{Im}\left[\frac{-1}{\varepsilon_{\mathrm{L}}(k,\omega;\omega')}\right]\mathrm{d}\omega'. \tag{15.16}$$

In this manner, the weight factor A_i of the Drude summation in Eq. 15.7 has been replaced by the spectrum density

$$A(\omega) = \frac{2}{\pi\,\omega}\mathrm{Im}\left[\frac{-1}{\varepsilon(k\approx 0,\omega)}\right]_{\mathrm{experimental}}, \tag{15.17}$$

which can be obtained from the experimental OELF.

Thus, in Penn's model a summation over an infinite number of Lindhard terms is made to obtain the ELF at $k \neq 0$. No parameter fitting is needed in this model, since the k-dependence of the Lindhard dielectric function completely determines $\mathrm{Im}[-1/\varepsilon_{\mathrm{L}}(k,\omega)]_{\mathrm{Penn}}$ from the optical ELF, $\mathrm{Im}[-1/\varepsilon(k=0,\omega)]$.

A simplification of Penn's model [54] was proposed by Ashley [18] based on the single-pole or δ-oscillator approximation to the Lindhard-ELF. Here the energy loss function is connected with the experimental optical energy-loss function ($k = 0$) through

$$\mathrm{Im}\left[\frac{-1}{\varepsilon(k,\omega)}\right]_{\mathrm{Ashley}} = \frac{1}{\omega}\int_0^\infty \mathrm{d}\omega'\omega'\,\mathrm{Im}\left[\frac{-1}{\varepsilon(k\approx 0,\omega')}\right]_{\mathrm{experimental}} \delta\left[\omega - \left(\omega' + \frac{k^2}{2m_{\mathrm{e}}}\right)\right]. \tag{15.18}$$

Since each term is a δ-function, the procedure of extrapolating to the $k \neq 0$ region is greatly simplified. After a straightforward calculation, the following simple expression is obtained

$$\mathrm{Im}\left[\frac{-1}{\varepsilon(k,\omega)}\right]_{\mathrm{Ashley}} = \frac{\omega - k^2/(2m_{\mathrm{e}})}{\omega}\mathrm{Im}\left[\frac{-1}{\varepsilon\left(k\approx 0,\omega - k^2/(2m_{\mathrm{e}})\right)}\right]_{\mathrm{experimental}} \Theta\left(\omega - k^2/(2m_{\mathrm{e}})\right). \tag{15.19}$$

Therefore, the basis of Ashley's model is an expression for the energy-loss function that assumes a simple quadratic dependence on the momentum transfer k.

15.3.2.3 Methods based on the Mermin dielectric function

Lindhard's theory [12] treats plasmons as undamped electronic excitations (i.e. having infinite lifetime or zero linewidth) up to a critical wavevector where they decay to electron-hole pairs. However, such a sharply peaked ELF spectrum does not agree with the body of experimental evidence indicating a strong damping mechanism at all k for most materials [55].

Mermin [56] solved this trouble by providing a phenomenological modification to the Lindhard dielectric function that includes plasmon damping through phonon-assisted electronic transitions. The Mermin dielectric function improvement upon the Lindhard function accounts in a consistent manner for the finite width of the plasmon peak (i.e. the finite plasmon lifetime) and, therefore, provides a more realistic extension to finite momentum transfers.

The Mermin dielectric function ε_{M} is written in terms of the Lindhard dielectric function ε_{L} as follows [56]:

$$\varepsilon_{\mathrm{M}}(k,\omega) = 1 + \frac{(1 + \mathrm{i}\gamma\,\hbar/\omega)\,[\varepsilon_{\mathrm{L}}(k,\omega+\mathrm{i}\gamma) - 1]}{1 + (\mathrm{i}\gamma\,\hbar/\omega)\,[\varepsilon_{\mathrm{L}}(k,\omega+\mathrm{i}\gamma) - 1]/[\varepsilon_{\mathrm{L}}(k,0) - 1]}, \tag{15.20}$$

where the plasmon damping appears through the γ coefficient.

Due to the equivalence between the Mermin-type ELF and the Drude-type ELF at the optical limit ($k = 0$), the complex structure of the ELF for real materials is suitably described in a similar manner to the procedure employed with the Drude-type ELF, Eq. 15.7. Therefore, a linear combination of Mermin-type ELF (MELF), $\mathrm{Im}[-1/\varepsilon_{\mathrm{M}}(k,\omega)]$, is fitted to the experimental OELF resulting from the loosely bound electrons of the target [19, 22]

$$\mathrm{Im}\left[\frac{-1}{\varepsilon(k,\omega)}\right]_{\mathrm{MELF}} = \sum_i \frac{A_i}{W_i^2}\mathrm{Im}\left[\frac{-1}{\varepsilon_{\mathrm{M}}(k,\omega;W_i,\gamma_i)}\right]_{\mathrm{experimental}} \Theta\left(\omega - \omega_{\mathrm{th},i}\right). \tag{15.21}$$

As in Eq. 15.7, the fitting parameters W_i, γ_i and A_i (related, respectively, to the position, width and relative weight of the peaks observed in the experimental optical ELF spectrum) are chosen in such a way that the MELF reproduces the main trends of the experimental OELF and satisfies the sum rules at $k = 0$, Eqs. 15.9 and 15.10; $\omega_{\mathrm{th},i}$ is a threshold energy, which is 7 eV for liquid water.

A great feature of the MELF method is that, unlike the previous models, a prescription for the $k \neq 0$ extension of the peaks in the OELF is not required since the Mermin-type ELF analytically covers the whole k-space.

The MELF methodology has been successfully applied to describe realistically the electronic properties of elemental and compounds targets [19, 57–59], as well as liquid water [60, 61] and dry DNA [52, 62].

15.3.2.4 Comparison between different dielectric descriptions for liquid water

As all the previous models for describing the OELF of a material reproduce the Drude-type ELF at $k = 0$, then the same fitting of the experimental ELF at $k = 0$ is used for all the above models and the parameters A_i, W_i and γ_i are common to all of them. However they employ different approaches to extend the ELF to arbitrary momentum transfer.

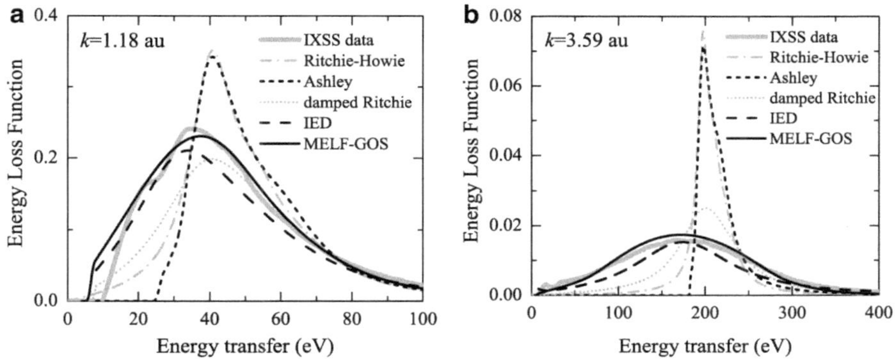

Fig. 15.3 Comparison of the experimental ELF of liquid water [42, 51] against the ELF predicted by different extension algorithms for the OELF. The labels in each figure identify the data [42, 51] and the models [17, 18, 20–22]. Results for two momentum transfer are shown: **(a)** $k = 1.18$ a.u. and **(b)** $k = 3.59$ a.u.

In this section we check the reliability of these models to describe the ELF of liquid water at finite momentum transfer by comparing the calculated ELF(k, ω) with the more recent available experimental results [42, 51], which we take as a benchmark.

In figures 15.3a and 15.3b we show the experimental ELF of liquid water [42,51] for two values of the momentum transfer, $k = 1.18$ a.u. and 3.59 a.u., respectively, as well as the results obtained from the various dielectric models previously discussed. The experimental broadening (for liquid water [51], as well as other materials [63]), is consistent with the theoretical expectation that single-particle excitations should gradually prevail over collective excitations as the momentum transfer increases.

Although all models predict the shift of the ELF to high energy transfers, only the MELF model [19, 22, 61] and the improved extended-Drude model (IED) [20, 60] give the correct broadening and reduction in intensity as k increases.

The more consistent account of plasmon damping introduced by Mermin [56] provides a notable improvement over the original Lindhard dielectric function [12]. Besides, the absence of any adjustable dispersion coefficients in the MELF method [19, 22, 61], makes it very suitable for providing a realistic extension scheme to the momentum space for any material.

The relatively simple dispersion formulae adopted in the IED model seem also very effective in reproducing the experimental data while retaining the convenience of working with Drude-type ELF. As noted by Kuhr and Fitting [50], the dispersion of the damping-coefficient provides the expected momentum broadening of the Bethe ridge and results in a notable improvement over earlier extended-Drude models. Moreover, the modified-quadratic dispersion used for the energy-coefficient, Eq. 15.14, shifts the position of the peak to lower energy transfers in better agreement with the experimental data than the pure-quadratic dispersion of the Ritchie-Howie [17] and Ashley [18] models.

The width and intensity of the ELF in the damped Ritchie model [21] broadens and diminishes, respectively, as k increases, but not enough to agree with the experimental data.

15.4 Results and discussions

We present in this section the main magnitudes characterising the electronic energy loss of proton beams through liquid water. Our framework is the plane-wave (first) Born approximation (PWBA), which provides accurate output over a substantial portion of the electronic regime. The most important advantage of the PWBA consists of the fact that, essentially, the scattering problem is reduced to find suitable descriptions for the target ELF over the complete range of energy- and momentum-transfer, i.e. the Bethe surface.

The influence of the experimental set of data for the OELF into the stopping power and the energy-loss straggling of liquid water is discussed, as well as, the relevance of the different methodologies used to extend the valence excitation spectrum to non-zero momentum transfers. All the results that follow use the GOS model (in the hydrogenic approach) to describe the oxygen K-shell electron excitations.

Finally, the depth dose distributions of protons in liquid water are evaluated for the different models of the Bethe surface, and the corresponding results are compared and discussed.

15.4.1 *Magnitudes characterizing the energy loss distribution: S and Ω^2*

A proton moving through a material experiences charge-exchange processes that modify its charge state. Therefore, the evaluation of the total stopping power, S, Eq. 15.1, requires the knowledge of the stopping power for each one of the projectile charge states, which are H^+ and H^0 for a proton beam.

In Fig. 15.4 we show the stopping power of H^+ and H^0 in liquid water obtained by the MELF-GOS model [19, 22] for a wide range of incident energies, where the IXSS data [16] for the OELF of liquid water are employed. The inset of Fig. 15.4 shows the energy dependence of the charge fractions of H^+ and H^0 in liquid water, as obtained from a parameterization to experimental data [25]. It can be seen that neutral hydrogen dominates at low projectile energies, whereas bare protons are most abundant at higher energies. In the intermediate region, both charge states are almost equally probable.

Although S_+ and S_0 look quite similar, the different behaviour of the charge state fractions ϕ_+ and ϕ_0 as a function of the projectile energy significantly

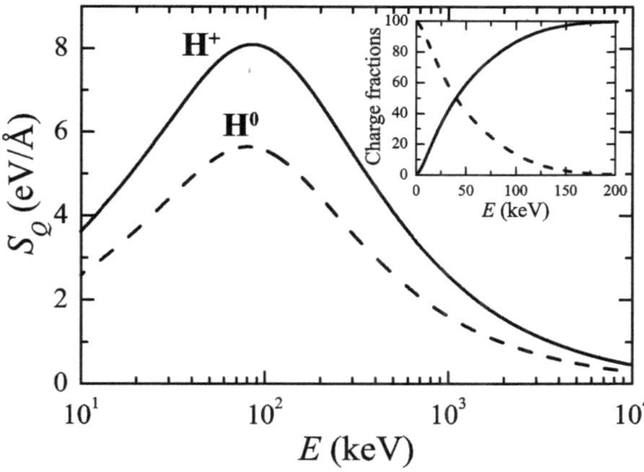

Fig. 15.4 Stopping power of H^+ (solid line) and H^0 (dashed line) in liquid water, S_+ and S_0 respectively, as a function of the incident energy. The calculations are done with the MELF-GOS method, starting from a fitting to IXSS experimental data in the optical limit [16]. The inset contains the charge fractions of H^+ (solid line) and H^0 (dashed line), ϕ_+ and ϕ_0 respectively, as a function of the projectile energy, obtained from [25]

affect the contribution of H^+ or H^0 to the total stopping power S, Eq. 15.1, at different energies. The latter contributes only at low projectile energy, having its maximum at around 25 keV and being insignificant at energies higher than 200 keV, whereas the maximum of the former appears at around 130 keV, being ~3 times larger than the contribution from H^0 and extending to higher energies. Therefore the stopping power S for energies larger than the maximum stopping power is mainly due to H^+, whereas at energies in the range of a few keV the stopping of H^0 becomes more significant.

In what follows we check the influence in the electronic energy loss of protons due to the input data used to construct the OELF of liquid water, from which the Bethe surface is obtained. In Fig. 15.5 we show the stopping power and the energy-loss straggling of protons in liquid water obtained through the MELF-GOS model, from Eqs. 15.1–15.3 after using the two different sets of experimental data for liquid water in the optical limit [15,16]. Solid lines derive from the IXSS data [16] whereas dashed lines come from the REF data [15]. These results depicted in Fig. 15.5 prove that, despite the strong differences (around 50% over the maximum energy transfer, as shown in Fig. 15.1) between the IXSS and the REF data of the OELF for liquid water, the discrepancies in the stopping power are lower than the 10%, mainly around the maximum stopping, whereas divergences in the energy-loss straggling are smaller in all the energy range.

After making clear the differences in the S and Ω^2 of liquid water due to the OELF used as input in the dielectric formalism, in what follows we discuss the significance on the proton stopping power due to the different schemes used to extend the valence excitation spectrum of liquid water over all the energy-momentum transfer. In Fig. 15.6 we show S, as a function of the proton incident energy, obtained from the different models described previously: the extended Drude model [17], the damped Ritchie model [21], the IED method [20, 60], the Ashley model [18], and the MELF-GOS method based in the Mermin ELF [19, 22, 52, 61]. All the calculations are based on the IXSS experimental data

Fig. 15.5 Stopping power and energy-loss straggling for a proton beam in liquid water, as a function of the incident energy. The MELF-GOS method was used, with a fitting to the IXSS data (solid lines) [16] or to the REF data (dashed lines) [15]. The contribution to S and Ω^2 from both, H^+ and H^0 are included in the calculation. The horizontal dashed line represents the Bohr straggling, $\Omega^2_{\mathrm{Bohr}} = 4\pi Z_1 Z_2^2 e^2 N$, which assumes that all the target electrons are free

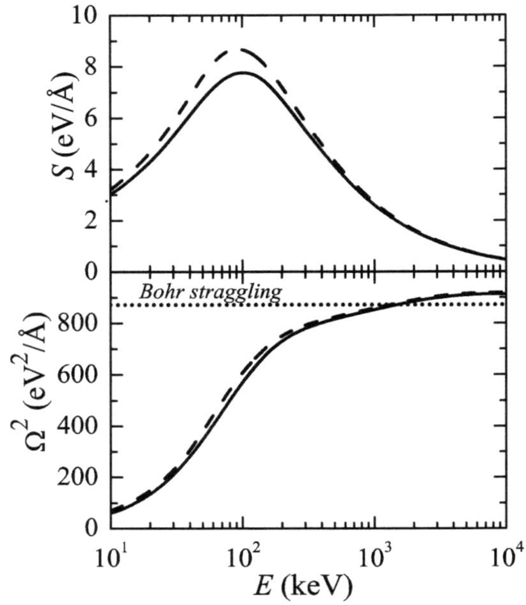

[16] for the OELF of liquid water, therefore, in the optical limit all the models are identical. Therefore, the only differences lying in the different scheme for the ELF extension to non-zero momentum transfer (that is, in the construction of the Bethe surface). As can be observed in Fig. 15.6, at high proton energies the results given by all models converge to the same stopping power, which is guaranteed by the fulfilment of the f-sum rule. However, at lower energies major discrepancies between the different analysed models appear.

Both extended-Drude models (by Ritchie and Howie [17] and by Ashley [18], which are based in the Lindhard-ELF) establish a pure quadratic dispersion relation for the energy coefficient, and provide comparable proton stopping powers, the latter being a little bit smaller than the former. At energies larger than the maximum stopping their results are similar to the ones obtained by the MELF-GOS model, however at lower energies S goes very quickly to zero. In order to explain this behaviour, we must note (see Fig. 15.3), that both Ritchie-Howie's and Ashley's models exhibit a sharp peak in the ELF, which is preserved as the momentum transfer increases (since damping is not included). Considering that the stopping power integrates the ELF into the transferred momentum and energy, for high proton energy the integration covers all the ω range, and the sharp ELF of the Ritchie-Howie and Ashley models contributes practically the same than the broad MELF-GOS model. However, for low proton energies only small transferred energies contribute to the integration, where the ELFs in the Ritchie-Howie and Ashley models are very small, given consistently low values of the stopping power. Similar arguments apply for explaining the small values of the proton stopping

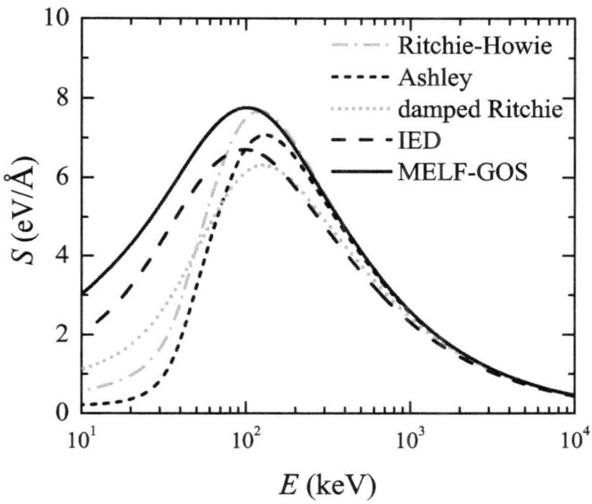

Fig. 15.6 Stopping power of liquid water for a proton beam, as a function of the incident energy, evaluated from different models to calculate the Bethe surface. In all cases the same fitting to the OELF from the IXSS data [16] was used. The contribution from both, H^+ and H^0 are included in the calculation. See the text for more details

power obtained by the damped Ritchie method [21] as compared with the MELF-GOS model in the whole proton energy range. Despite the fact that the empirical IED model [20, 60] includes the damping dispersion, providing a broad ELF and correctly predicts the peak position of the ELF, it provides a S value smaller than the one obtained by the MELF-GOS model, in all the range of energies evaluated. This is because the IED model underestimates the experimental ELF at all momentum transfer (see Fig. 15.3).

The results of the stopping power of liquid water for an incident proton beam, calculated by several studies in the literature [9, 19, 22, 64, 65], are depicted in Fig. 15.7 together with experimental data, the semiempirical code SRIM [66] and with the values provided by the ICRU report [7]. A short comment follows for the models that have not been discussed previously; all of them are based on the first Born approximation. The results by Dingfelder *et al.* [9] are based on cross-section data for several inelastic channels between protons and liquid water, using the optical measurements of Heller *et al.* [15] with a quadratic dispersion scheme. Emfietzoglou *et al.* [64] use the improved extended-Drude model to describe the Bethe surface of liquid water. The calculations by Akkerman *et al.* [65] provide the stopping power as a sum of valence electron excitation, core-electron ionization, and Barkas and Bloch terms, with Ashley's approximation for the ELF.

It is worth to notice that most of the experimental data correspond to measurements of protons on D_2O–ice [67, 68] and on H_2O–ice [69]. These data were used as input for obtaining the semiempirical [66] and tabulated [7] curves. The only experimental results available for liquid water [70, 71], covering the range from 0.3 to 2 MeV, were obtained with a thin liquid jet in vacuum, but its diameter was treated as a fitting parameter. The deviations for the stopping powers of the different phase states of water are most noticeable at low proton energies, but they dissapear as the energy increases [72].

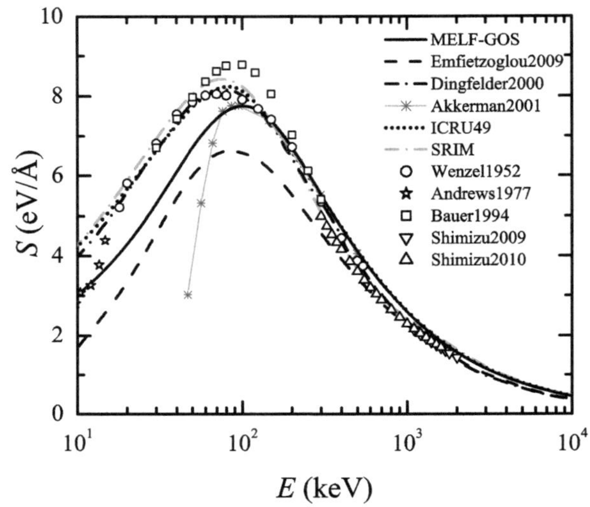

Fig. 15.7 Experimental, calculated and parameterized curves of the stopping power of liquid water, for proton beams. The corresponding sources are indicated in the inset. Notice that only the experiments by Shimizu *et al.* [70, 71] correspond to liquid water, the rest being done in ice water. See the text for more details

In Fig. 15.7 we observe that for energies larger than 300 keV, both the MELF-GOS and the ICRU49 curves are rather similar; however discrepancies appear for energies around and lower than the maximum stopping. The discrepancies between the results by Dingfelder *et al.* [9] and the MELF-GOS [19, 22] model could be attributed to the different OELF used as input by these models.

Given the discrepancies predicted by the different models at low proton energies, new experimental data of the stopping power of liquid water for protons are necessary to elucidate its behaviour for projectile energies around and lower the maximum stopping.

15.4.2 Depth-dose distributions

In what follows we use the simulation code SEICS (Simulation of Energetic Ions and Clusters through Solids) [73–75] to calculate the depth dose distribution of a proton beam travelling through a liquid water target. This code is based on a combination of the Monte Carlo and Molecular Dynamics methods to follow dynamically the trajectories of the protons incident on the target until they are stopped. Thus, knowing the coordinates, velocities and charges of the projectiles at each time it is possible to find the deposited energy by the projectile as a function of the depth into the irradiated target. The SEICS code includes the electronic force on the projectile, which is mainly responsible for the energy loss in the energy range of keV-MeV; this force is given by the stopping power S_Q, but taking into account the statistical fluctuation around that mean value, which is provided through the energy-loss straggling Ω_Q^2. The SEICS code also includes the interaction of the

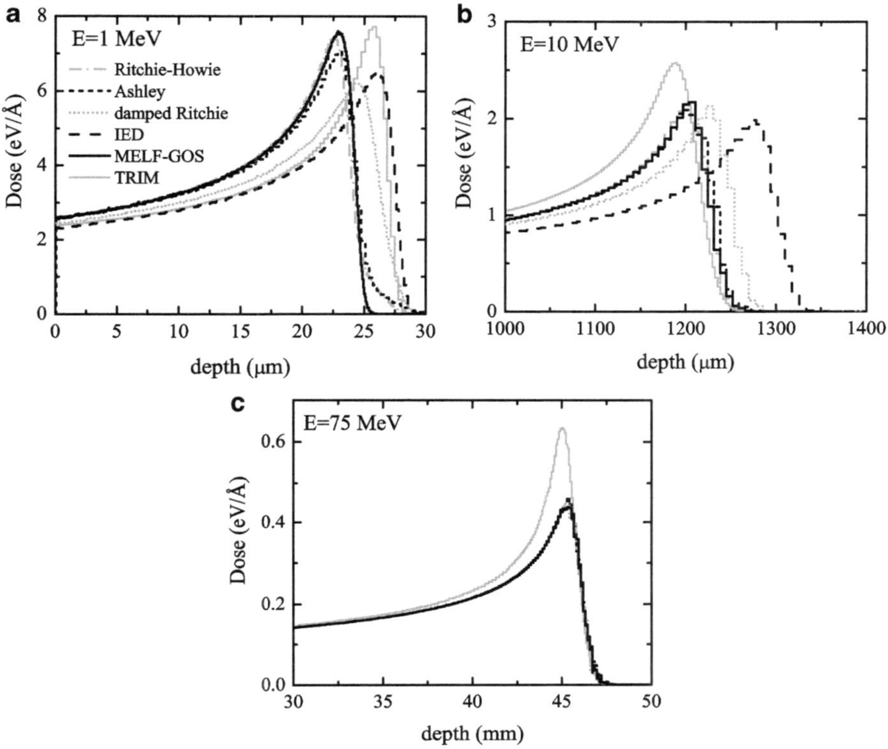

Fig. 15.8 Depth-dose distribution of proton beams in liquid water obtained by the SEICS code using different models for the stopping power and the energy-loss straggling (see more details in the text). The TRIM calculation [66] is also included for comparison. The proton incident energy corresponds to **(a)** $E = 1\,\text{MeV}$, **(b)** $E = 10\,\text{MeV}$, and **(c)** $E = 75\,\text{MeV}$. Notice that different scales are used in the abscissas of each figure

projectile with the target nuclei through elastic collisions, which contribute mainly to the angular deflection of the projectile and to the nuclear energy loss, the latter being more significant when the projectile energy is small and the projectile is near to stop at the end of its trajectory. The processes of electron capture and loss by the projectile are also included into the simulation, and they become especially significant at the Bragg peak.

In Figs. 15.8 we show the depth dose distribution, i. e. the Bragg curve, of proton beams passing through liquid water, for several incident energies (1 MeV, 10 MeV and 75 MeV). As the key input into the SEICS simulation code are the target stopping power and energy-loss straggling, we will use the values for S_Q and Ω_Q^2 obtained from the different models previously discussed, namely the extended-Drude model (grey dash-dotted line) [17], the Ashley model (short dashed line) [18], the damped Ritchie model [21] (grey dotted line), the improved extended-Drude model (IED) (dashed line) [20, 60], and the MELF-GOS model (solid line) [19, 22].

The curves from the TRIM code [66] are also shown for comparison purposes. Differences into the depth-dose distributions come mainly from the different values of the stopping power provided by each model (shown in Fig. 15.6); if the energy-loss straggling were not included in the simulations, the Bragg peak would be sharper and deeper.

All the models we have presented here (except the results from SRIM) predict the same mean ionization energy, $I = 79.4$ eV, because their common starting point was the OELF derived from the IXSS data [16]. For energies larger than 10 MeV the SEICS code uses the Bethe stopping power. However there are still discrepancies in the Bragg peak position predicted by the different models.

The depth-dose distribution obtained from the Ritchie-Howie [17], the Ashley [18] and the MELF-GOS [19, 22] models are rather similar, since their stopping power is comparable at proton energies larger than 200 keV. However, the Bragg peak calculated from the damped Ritchie [21] and the IED [20, 60] for 1 MeV and 10 MeV proton beams are shifted deeper as compared to the previous ones, since these models provide smaller values for the stopping power at energies lower than several MeV (see Fig. 15.6). However for 75 MeV proton beams, the depth-dose distributions obtained from all the models are quite similar since a large portion of the energy loss (as the projectile energy decreases from 75 MeV to 10 MeV) is evaluated with the same stopping power, namely the one provided by the Bethe formula with $I = 79.4$ eV (which is common to all models).

Therefore, it is important to notice that for high incident energies the position of the Bragg peak is mostly determined at the millimeter scale by the value of the mean excitation energy into the Bethe formula. But differences in the stopping power values provided by the different extension algorithms at proton energies less than a few MeV imply shifts in the Bragg peak of the order of micrometers ($\sim$4 μm when $E = 1$ MeV, $\sim$100 μm when $E = 10$ MeV and $\sim$300 μm when $E = 75$ MeV), which could have microdosimetric implications.

15.5 Conclusions

The electronic energy deposited by a proton beam in liquid water has been evaluated for several extended optical-data models currently used in the literature. We describe diverse methodologies based on Drude's, Lindhard's and Mermin's ELF. The choice of the current OELF data (either REF [15] or IXSS [16]) for liquid water has a significance of $\sim$10% around the maximum stopping power. Nonetheless the procedure to extend the OELF to non-zero momentum transfer is crucial to obtain ELF values consistent with available experimental data at finite momentum transfer [42, 43]. Only the MELF-GOS [19, 22] and the IED [20, 60] models satisfactorily reproduce the experimental Bethe surface.

We want to call the attention on the influence into the stopping magnitudes (especially at proton energies around and lower than the maximum stopping) of the different methodologies used to extend the OELF at finite momentum transfer.

While at high proton energies all the models predict rather similar stopping power values, at energies smaller than the maximum stopping most of the stopping curves go very quickly to zero, except the MELF-GOS model and in a less extent the IED model.

The depth dose distributions of proton beams in liquid water have been also calculated through the simulation code SEICS, using as input data the different stopping power and energy loss straggling provided by the extended optical-data models. As a result, we conclude that the position of the Bragg peak is sensitive to the values of the stopping power at energies between several hundreds of keV and several tens of MeV.

Acknowledgments R.G.M. and I.A. acknowledge financial support from the Spanish Ministerio de Ciencia e Innovación (Project FIS2010-17225). Financial support for I.K. and D.E. by the European Union FP7 ANTICARB (HEALTH-F2-2008-201587) is recognized. This work has benefited from the collaboration within COST Action MP 1002, Nanoscale Insights into Ion Beam Cancer Therapy.

References

1. Wilson R. R., Radiology 47, 487–491 (1946)
2. Tobias C. A., Lawrence J. H., Born J. L., McCombs R. K., Roberts J. E., Anger H. O., Low-Beer B. V. A., Huggings C. B., Cancer Research 18, 121–134 (1958)
3. Sisterson J., Nuclear Instruments and Methods in Physics Research B 241, 713–716 (2005)
4. Nikjoo H., Uehara S., Emfietzoglou D., Brahme A., New Journal of Physics 10, 075006-1–075006-28 (2008)
5. Solov'yov A. V., Surdutovich E., Scifoni E., Mishustin I., Greiner W., Physical Review E 79, 011909-1–011909-7 (2009)
6. Inokuti M., Reviews of Modern Physics 43, 297–347 (1971); Addenda, Reviews of Modern Physics 50, 23–35 (1978)
7. ICRU Report 49, Stopping Powers and Ranges for Protons and Alpha Particles (International Commission on Radiation Units and Measurements, Bethesda, Maryland, 1992)
8. Uehara S., Toburen L. H., Wilson W. E., Goodhead D. T., Nikjoo H., Radiation Physics and Chemistry 59, 1–11 (2000)
9. Dingfelder M., Inokuti M., Paretzke H. G., Radiation Physics and Chemistry 59, 255–275 (2000)
10. Date H., Sutherland K. L., Hayashi T., Matsuzaki Y., Kiyanagi Y., Radiation Physics and Chemistry 75, 179–187 (2006)
11. Fermi E., Physical Review 57, 485–493 (1940)
12. Lindhard J., Det Kongelige Danske Videnskabernes Selskab. Matematisk-fysiske Meddelelser 28 (8) (1954)
13. Ritchie R. H., Physical Review 106, 874–881 (1957)
14. Ritchie R. H., Physical Review 114, 644–654 (1959)
15. Heller J. M., Hamm R. N., Birkhoff R. D., Painter L. R., Journal of Chemical Physics 60, 3483–3486 (1974)
16. Hayashi H., Watanabe N., Udagawa Y., Kao C. C., Proceedings of the National Academy of Sciences of the USA 97, 6264–6266 (2000)
17. Ritchie R. H., Howie A., Philosophical Magazine 36, 463–481 (1977)
18. Ashley J. C., Journal of Electron Spectroscopy and Related Phenomena 46, 199–214 (1988)

19. Abril I., Garcia-Molina R., Denton C. D., Pérez-Pérez J. F., Arista N., Physical Review A 58, 357–366 (1998)
20. Emfietzoglou D., Cucinotta F. A., Nikjoo H., Radiation Research 164, 202–211 (2005)
21. Dingfelder M., Ritchie R. H., Turner J. E., Friedland W., Paretzke H. G., Hamm R. N., Radiation Research 169, 584–594 (2008)
22. Heredia-Avalos S., Garcia-Molina R., Fernández-Varea J. M., Abril I., Physical Review A 72, 052902-1-052902-9 (2005)
23. Sigmund P., Particle penetration and radiation effects. General aspects and stopping of swift point charges (Springer-Verlag, Berlin, 2006)
24. Pines D., Elementary excitations in solids, 2nd printing (Benjamin, New York, 1964)
25. Schiwietz G., Grande P. L., Nuclear Instruments and Methods in Physics Research B 175–177, 125–131 (2001)
26. Brandt W., Kitagawa M., Physical Review B 25, 5631–5637 (1982); corrections to the BK model appear in Brandt W., Nuclear Instruments and Methods in Physics Research 194, 13–19 (1982)
27. Garcia-Molina R., Abril I., Denton C. D., Heredia-Avalos S., Nuclear Instruments and Methods in Physics Research B 249, 6–12 (2006)
28. Sigmund P., Nuclear Instruments and Methods in Physics Research B 85, 541–550 (1994)
29. Sigmund P., Nuclear Instruments and Methods in Physics Research B 135, 1–15 (1998)
30. Segui S., Dingfelder M., Fernández-Varea J. M., Salvat F., Journal of Physics B: Atomic, Molecular and Optical Physics 35, 33–53 (2002)
31. Daniels J., Optics Communications 3, 240–243 (1971)
32. Seki M., Kobayashi K., Nakahara J., Journal of the Physical Society of Japan 50, 2643–2648 (1981)
33. Warren S. G., Applied Optics 23, 1206–1255 (1984)
34. Toburen L. H., McLawhorn S. L., McLawhorn R. A., Carnes K. D., Dingfelder M., Shinpaugh J. L., Radiation Research 174, 107–118 (2010)
35. Egerton R. F., Electron Energy-Loss Spectroscopy in the Electron Microscope (Plenum Press, New York, 1989)
36. Fano U., Annual Reviews of Nuclear Science 13, 1–66 (1963)
37. Dressel M., Grüner G., Electrodynamics of Solids. Optical Properties of Electrons in Matter (Cambridge University Press, Cambridge, 2003)
38. Tanuma S., Powell C. J., Penn D. R., Journal of Electron Spectroscopy and Related Phenomena 62, 95–109 (1993)
39. Smith D. Y., Inokuti M., Karstens W., Shiles E., Nuclear Instruments and Methods in Physics Research B 259, 1–5 (2006)
40. Chantler C. T., Olsen K., Dragoset R. A., Kishore A. R., Kotochigova S. A., Zucker D. S., X-ray form factor, attenuation and scattering tables (version 2.1). National Institute of Standards and Technology, Gaithersburg, MD, 2005. Available at http://www.nist.gov/pml/ffast/index.cfm
41. Henke B. L., Gullikson E. M., Davis J. C., Atomic Data and Nuclear Data Tables 54, 181–342 (1993)
42. Watanabe N., Hayashi H., Udagawa Y., Bulletin of the Chemical Society of Japan 70, 719–726 (1997)
43. Hayashi H., Watanabe N., Udagawa Y., Kao C.-C., Journal of Chemical Physics 108, 823–825 (1998)
44. Fernández-Varea J. M., Mayol R., Salvat F., Liljequist D., Journal of Physics: Condensed Matter 4, 2879–2890 (1992)
45. Fernández-Varea J. M., Mayol R., Liljequist D., Salvat F., Journal of Physics: Condensed Matter 5, 3593–3610 (1993)
46. Dingfelder M., Inokuti M., Radiation and Environmental Biophysics 38, 93–96 (1999)
47. Planes D. J., Garcia-Molina R., Abril I., Arista N. R., Journal of Electron Spectroscopy and Related Phenomema 82, 23–29 (1996)
48. Ding Z.-J., Shimizu R., Surface Science 222, 313–331 (1989)
49. Ding Z. J., Shimizu R., Scanning 18, 92–113 (1996)

50. Kuhr J.-Ch., Fitting H.-J., Journal of Electron Spectroscopy and Related Phenomena 105, 257–273 (1999)
51. Watanabe N., Hayashi H., Udagawa Y., Journal of Physics and Chemistry of Solids 61, 407–409 (2000)
52. Abril I., Denton C. D., de Vera P., Kyriakou I., Emfietzoglou D., Garcia-Molina R., Nuclear Instruments and Methods in Physics Research B 268, 1763–1767 (2010)
53. Mahan G. D., Many-Particle Physics, 3rd ed. (Kluwer Academic-Plenum Publishers, New York, 2000)
54. Penn D. R., Physical Review B 35, 482–486 (1987)
55. Sturm K., Advances in Physics 31, 1–64 (1982)
56. Mermin N. D., Physical Review B 1, 2362–2363 (1970)
57. Moreno-Marín J. C., Abril I., Garcia-Molina R., Nuclear Instruments and Methods in Physics Research B 193, 30–35 (2002)
58. Behar M., Fadanelli R. C., Abril I., Garcia-Molina R., Denton C. D., Nagamine L. C. C. M., Arista N. R., Physical Review A 80, 062901-1–062901-8 (2009)
59. Behar M., Denton C. D., Fadanelli R. C., Abril I., Cantero E. D., Garcia-Molina R., Nagamine L. C. C. M., European Physical Journal D 59, 209–213 (2010)
60. Emfietzoglou D., Abril I., Garcia-Molina R., Petsalakis I. D., Nikjoo H., Kyriakou I., Pathak A., Nuclear Instruments and Methods in Physics Research B 266, 1154–1161 (2008)
61. Garcia-Molina R., Abril I., Denton C. D., Heredia-Avalos S., Kyriakou I., Emfietzoglou D., Nuclear Instruments and Methods in Physics Research B 267, 2647–2652 (2009)
62. Abril I., Garcia-Molina R., Denton C. D., Kyriakou I., Emfietzoglou D., Radiation Research 175, 247–255 (2011)
63. Tirao G., Stutz G., Silkin V. M., Chulkov E. V., Cusatis C., Journal of Physics: Condensed Matter 19, 046207-1–046207-11 (2007)
64. Emfietzoglou D., Garcia-Molina R., Kyriakou I., Abril I., Nikjoo H., Physics in Medicine and Biology 54, 3451–3472 (2009)
65. Akkerman A., Breskin A., Chechik R., Lifshitz Y., Radiation Physics and Chemistry 61, 333–335 (2001)
66. Ziegler J. F., Biersack J. P., Ziegler M. D., SRIM. The Stopping and Range of Ions in Matter, (SRIM Co., Chester, Maryland, 2008)
67. Wenzel W. A., Whaling W., Physical Review 87, 499–503 (1952)
68. Andrews D. A., Newton G., Journal of Physics D: Applied Physics 10, 845–850 (1977)
69. Bauer P., Kaferbock W., Necas V., Nuclear Instruments and Methods in Physics Research B 93, 132–136 (1994)
70. Shimizu M., Kaneda M., Hayakawa T., Tsuchida H., Itoh A., Nuclear Instruments and Methods in Physics Research B 267, 2667–2670 (2009)
71. Shimizu M., Hayakawa T., Kaneda M., Tsuchida H., Itoh A., Vacuum 84, 1002–1004 (2010)
72. Liamsuwan T., Uehara S., Emfietzoglou D., Nikjoo H., International Journal of Radiation Biology 87, 141–160 (2011)
73. Garcia-Molina R., Denton C. D., Abril I., Arista N. R., Physical Review A 62, 012901-1–012901-5 (2000)
74. Heredia-Avalos S., Garcia-Molina R., Abril I., Physical Review A 76, 012901-1–012901-12 (2007)
75. Garcia-Molina R., Abril I., Heredia-Avalos S., Kyriakou I., Emfietzoglou D., Physics in Medicine and Biology 56, 6475–6493 (2011)

Chapter 16
Quantum-Mechanical Contributions to Numerical Simulations of Charged Particle Transport at the DNA Scale

Christophe Champion, Mariel E. Galassi, Philippe F. Weck, Omar Fojón, Jocelyn Hanssen, and Roberto D. Rivarola

Abstract Two quantum mechanical models (CB1 and CDW-EIS) are here presented to provide accurate multiple differential and total cross sections for describing the two most important ionizing processes, namely, ionization and capture induced by heavy charged particles in targets of biological interest. Water and DNA bases are then successively investigated by reporting in particular a detailed study of the influence of the target description on the cross section calculations.

16.1 Introduction

Numerical models and codes based on Monte Carlo (MC) techniques represent powerful tools for simulating 'event-by-event' radiation track structure at the nanometer level. It is worth noting that the success of MC energy transport codes essentially depends on the accuracy of both the theoretical model assumptions and

C. Champion (✉)
Laboratoire de Physique Moléculaire et des Collisions, ICPMB (FR CNRS 2843), Institut de Physique, Université Paul Verlaine-Metz, 57078 Metz Cedex 3, France

Université Bordeaux 1, CNRS/IN2P3, Centre d'Etudes Nucléaires de Bordeaux-Gradignan, CENBG, Chemin du Solarium, BP 120, 33175 Gradignan, France
e-mail: champion@univ-metz.fr

M.E. Galassi • O. Fojón • R.D. Rivarola
Instituto de Física Rosario, CONICET, Universidad Nacional de Rosario, 2000 Rosario, Argentina

P.F. Weck
Department of Chemistry, University of Nevada Las Vegas, Las Vegas, NV 89154, USA

J. Hanssen
Laboratoire de Physique Moléculaire et des Collisions, ICPMB (FR CNRS 2843), Institut de Physique, Université Paul Verlaine-Metz, 57078 Metz Cedex 3, France

G.G. Gómez-Tejedor and M.C. Fuss (eds.), *Radiation Damage in Biomolecular Systems*, Biological and Medical Physics, Biomedical Engineering,
DOI 10.1007/978-94-007-2564-5_16, © Springer Science+Business Media B.V. 2012

the physical input data used, *i.e.*, the cross sections implemented into the code for describing at the finest scale the charged particle induced collisions. Thus, in view of their potential applications in radioprotection, radiobiology, medical imaging and even in radiotherapy for treatment planning, most of the existing MC codes are based on cross sections in water, this molecule being considered as a good surrogate for biological targets. Furthermore, to overcome the lack of existing data in liquid water, they have traditionally relied on theoretical descriptions of condensed-phase interaction probabilities for model systems combined with parameters extrapolated from gas phase studies, or entirely on gas phase data. However, let us mention that significant advances have been achieved in the last few years for considering water in liquid phase and then evaluating the expected differences in terms of spatial patterns of energy deposition between liquid and vapor water [1].

Besides, MC track structure simulations play an important role to provide a quantitative understanding of the mechanisms of radio-induced damages. On this subject, numerous Monte Carlo codes have been developed among which we distinguish the specialized Monte Carlo codes - usually called "track structure codes" - which have been developed for microdosimetry simulations (see for example [2] and references therein). These codes are able to simulate precisely particle-matter interactions, the so-called "physical stage", some of them including also additional features, e.g. taking into account the "physico-chemical" and "chemical" stages which take place after the "physical" stage and allow in particular the simulation of oxidative radical species. With the use of sophisticated geometry models, some of the available MC codes are even able to predict - with a reasonable precision - direct and non-direct biological damages to the DNA molecule. This is the case of the PARTRAC software which is nowadays the most advanced Monte Carlo simulation package for modeling the biological effects of radiation. On the other hand, several general-purpose MC codes are already accessible to scientists for the simulation of particle transport. Among them we can cite the most commonly used, namely, EGS, FLUKA and MCNP with their different available versions. Some of them are limited to the simulation of electron and photon interactions, while others include a comprehensive description of hadronic interactions for a large variety of ions. However, in the major part these codes limit their lower energy range applicability down to 1 keV, which is not compatible with functionalities specific to microdosimetry. Such codes should indeed be able to simulate particle track structures (incident particles and the full consequent shower of secondary particles) over lengths at the nanometer scale, compatible with the DNA molecular size and sub-cellular scale. However, most of them are based on a semi-empirical description of the main ionizing process by means of least-squares fittings of experimental measurements of differential as well as total cross sections.

In this context, in the past we have developed a Monte Carlo code called TILDA for tracking heavy charged particles in liquid water [3] in which all the ion- and electron-induced interactions are described in details [4], liquid water being first used for modelling the biological medium. However, DNA lesions - and more particularly those involved in clustered damages - are nowadays considered of prime importance for understanding the radio-induced cellular death process (see for example [5]). Thus, further theoretical models as well as experimental data on ion-induced collisions at the DNA level remain crucial to go beyond the simple approximation which consists in modelling the biological matter by water as usually done in the existing track-structure numerical simulations. However, ionization and fragmentation of isolated nucleobases have until now received only little interest and have been essentially focused on the cross-section determination for electron-induced collisions. Indeed, ion-induced collisions have rarely been reported in the literature and to the best of our knowledge only few works exist (see for example [6]). On the theoretical side, we essentially find two recent approaches, namely, a first (semi)-classical one based on the CTMC-COB approach (previously tested for water [7]) and a second quantum-mechanical one [8] providing doubly and singly differential as well as total cross sections for proton, α-particle and bare carbon ion beams impacting on adenine, cytosine, thymine and guanine bases.

The present chapter deals with the theoretical models we have recently developed - within the quantum mechanical framework - for describing the ionization and the electronic capture processes induced by heavy charged particles in both water and DNA components. The obtained results will be reported in terms of multi-differential and total cross sections by pointing out the relative importance of the target description.

In the following sections, atomic units (a.u.) are used throughout unless indicated otherwise.

16.2 Theoretical approach

Let us first consider the processes of electronic capture and ionization induced by bare ion beam impact on mono-electronic atoms. This description may be then extended and applied to the case of multi-electronic atomic and molecular targets. The involved particles being charged, let us remind that the interactions are governed by long-range Coulomb forces, which might be small but never equal to zero even at infinite long distances. As a consequence, the wave function representing this situation can not be written as a product of free-particle (plane-waves) wave functions. The long-range nature of the potential results in the appearance of a Coulomb phase or distortion. Thus, the initial and final state wave functions in the three-particle system considered are chosen in such a way as to represent the physical problem and the correct asymptotic conditions at very large distances. This is of fundamental importance to avoid the presence of divergences in the scattering matrix elements associated with the population of elastic intermediate channels [9].

According with the studied reactions, different approximations which verify correct boundary conditions are employed. We will present them into the straight line version of the impact parameter approximation, where the internuclear vector $\vec{R}$, the impact $\vec{\rho}$ and the collision velocity $\vec{v}$ are related by the expression

$$\vec{R} = \vec{\rho} + \vec{v}t, \tag{16.1}$$

where t is the evolution time ($t = 0$ corresponding to the distance of closest internuclear separation). Finally, let us note that the collision is described in a reference frame fixed to the target nucleus.

16.2.1 Electron ionization

Two different models are here used to investigate this reaction: the first-order Born approximation with correct boundary conditions (CB1) [10] and the continuum distorted wave-eikonal initial state one (CDW-EIS) [11, 12], the first one being considered as an extension - to the case of ionization - of the model originally introduced for describing the electron capture [9, 13–15]. The main difference between both models resides in the fact that in CB1, one-center wave functions (target and continuum states in the only presence of the residual target) are chosen whereas in CDW-EIS, two-center wave functions (target and continuum states in the simultaneous presence of the projectile and residual target fields) are selected.

Thus, in the CB1 approach the initial and final wave functions are respectively given by the expressions

$$\begin{aligned}\chi_\alpha^+ &= \varphi_\alpha(\vec{x})\exp(-i\varepsilon_\alpha t)\exp\left[i\frac{Z_P(Z_T-1)}{v}\ln(vR-\vec{v}\cdot\vec{R})\right]\\ &= \psi_\alpha\exp\left[i\frac{Z_PZ_T}{v}\ln(vR-\vec{v}\cdot\vec{R})\right]\end{aligned} \tag{16.2}$$

and

$$\begin{aligned}\chi_\beta^{-(i)} &= (2\pi)^{-3/2}\exp\left(i\vec{k}\cdot\vec{x}-i\frac{k^2}{2}t\right)N^*(\xi)\,{}_1F_1(-i\xi;1;-ikx-i\vec{k}\cdot\vec{x})\\ &\quad\times\exp\left[-i\frac{Z_P(Z_T-1)}{v}\ln(vR+\vec{v}\cdot\vec{R})\right]\\ &= \psi_\beta\exp\left[-i\frac{Z_PZ_T}{v}\ln(vR+\vec{v}\cdot\vec{R})\right]\end{aligned} \tag{16.3}$$

where Z_P and Z_T denote the projectile and target nuclear charge, respectively.

In (16.2), $\varphi_\alpha(\vec{x})$ represents a target bound state with corresponding orbital energy ε_α whereas in (16.3), $\vec{k}$ is the linear momentum of the emitted electron as measured from the target nucleus, $\xi = Z_T/k$, and $\vec{x}$ is the position vector of the electron with respect to that nucleus. The terms contained in the first line of (16.3) represent the electron moving in a continuum state of the residual target field, being $N(a) = \exp(\pi a/2)\Gamma(1+ia)$ the normalization constant of the continuum factor ${}_1F_1(-i\xi;1;-ikx-i\vec{k}\cdot\vec{x})$. The last exponential factors in both (16.2) and (16.3) correspond to the asymptotic limits $t\to-\infty$ and $t\to+\infty$ of the interaction of the projectile with the electron and target nucleus in the entry and exit channels,

$$V_{\alpha,\beta} = -\frac{Z_P}{s}+\frac{Z_PZ_T}{R}\xrightarrow[t\to\pm\infty]{}\frac{Z_P(Z_T-1)}{R}, \tag{16.4}$$

respectively and where $\vec{s}$ denotes the position of the ejected electron with respect to the projectile nucleus. These exponential factors differ in form because they posses either correct outgoing or incoming conditions associated to the asymptotic scattering of the projectile in the field of the full target. The wave functions χ_α^+ and $\chi_\beta^{-(i)}$ include the long range character of the perturbation potential in the entry and exit channels respectively and the supra-indexes ($\pm$) indicate the outgoing or incoming character of these wave functions. It is immediate to observe that for ionization the asymptotic limit in the exit channel is formally only valid in the region defined by $x \ll R$, what leads to a validity condition for the CB1 approximation restricted to the cases where $k \ll v$. CB1 is also expected to describe the binary encounter process which is dominated by a binary projectile-electron interaction. In fact, as the ionized electron can move in all regions of the coordinate space, the wave functions representing the exit channel must verify the asymptotic boundary condition

$$\chi_\beta^{-(i)} \xrightarrow[t,x,s\to\infty]{} (2\pi)^{-3/2} \exp(i\vec{k}\cdot\vec{x}) \exp\left[i\frac{Z_T}{k}\ln(kx+\vec{k}\cdot\vec{x}) + i\frac{Z_P}{p}\ln(ps+\vec{p}\cdot\vec{s})\right] \exp\left[-i\frac{Z_P Z_T}{v}\ln(vR+\vec{v}\cdot\vec{R})\right] \tag{16.5}$$

being $\vec{p} = |\vec{k}-\vec{v}|$ the linear momentum of the ejected electron as seen from the projectile. This condition presents a two-center character, considering that both the projectile and residual target fields appear as acting simultaneously on the electron [16, 17]. Thus, a two-center approximation which verifies this limit, namely, the CDW-EIS approach, has been introduced by Crothers and McCann [11].

In the CDW-EIS framework, initial and final *distorted* wave functions are chosen as

$$\begin{aligned} \chi_\alpha^+ &= \varphi_\alpha(\vec{x}) \exp(-i\varepsilon_\alpha t) \exp\left[\frac{Z_P}{v}\ln(vs+\vec{v}\cdot\vec{s})\right] \exp\left[i\frac{Z_P Z_T}{v}\ln(vR-\vec{v}\cdot\vec{R})\right] \\ &= \psi_\alpha \exp\left[i\frac{Z_P Z_T}{v}\ln(vR-\vec{v}\cdot\vec{R})\right] \end{aligned} \tag{16.6}$$

and

$$\begin{aligned} \chi_\beta^{-(i)} &= (2\pi)^{-3/2} \exp\left(i\vec{k}\cdot\vec{x} - i\frac{k^2}{2}t\right) N(\xi)\, {}_1F_1(-i\xi;1;-ikx-i\vec{k}\cdot\vec{x}) \\ &\quad N(\zeta)\, {}_1F_1(-i\zeta;1;-ips-i\vec{p}\cdot\vec{s}) \exp\left[-i\frac{Z_P Z_T}{v}\ln(vR+\vec{v}\cdot\vec{R})\right] \\ &= \psi_\beta \exp\left[-i\frac{Z_P Z_T}{v}\ln(vR+\vec{v}\cdot\vec{R})\right] \end{aligned} \tag{16.7}$$

with $\zeta = Z_P/p$. It is easy to verify that χ_α^+ satisfies also the asymptotic limit in the entry channel (see (16.2)). According to the definitions given in (16.6)–(16.7), CDW-EIS presents a two-center character in both the initial and final channels.

A further model was introduced [18, 19] for the particular case of proton impact. Within this model, a CB1 type wave function was chosen in the entry channel whereas a continuum projectile factor with a dynamical projectile charge $\zeta_d = 1 - \left|\vec{v} - \vec{k}\right| / v$ was chosen in the exit channel instead of the one defined in (16.7). The continuum factor in this case is given by the expression $N(\zeta_d/p)_1F_1(-i\zeta_d/p; 1; -i(ps + \vec{p}\cdot\vec{s}))$. It is easy to show that for slow ejected electrons the behavior of the model recovers the CB1 approximation while electrons moving with a velocity close to the proton one feel the projectile field with an effective charge $\zeta_d = 1$.

16.2.2 Electron capture

It is well known that to describe adequately the electron capture process, where one electron is promoted from a bound state of the target to a bound state of the projectile, second and higher orders of the Born series are necessary [20]. This condition is crucial, for example to describe capture through two-step mechanisms [21, 22]. Thus, this reaction is analyzed using two different two-center approximations, the CDW-EIS [23] and the continuum distorted wave (CDW) ones [24], which in fact implicitly contain higher orders of the Born series. The difference between CDW-EIS and CDW resides in the choice of the initial distorted wave function. While in CDW-EIS, the initial wave function is chosen as for electron ionization (see (16.6)), in CDW it is taken as

$$\chi_\alpha^+ = \varphi_\alpha(\vec{x})\exp(-i\varepsilon_\alpha t)N^*(\lambda)_1F_1(i\lambda; 1; ivs + i\vec{v}\cdot\vec{s})\exp\left[i\frac{Z_P Z_T}{v}\ln(vR - \vec{v}\cdot\vec{R})\right]$$
$$= \psi_\alpha \exp\left[i\frac{Z_P Z_T}{v}\ln(vR - \vec{v}\cdot\vec{R})\right] \tag{16.8}$$

where $\lambda = Z_P/v$. Both approximations verify correct boundary conditions in the entry channel. The final wave function in both CDW-EIS and CDW is proposed as

$$\chi_\beta^{-(c)} = \varphi_\beta(\vec{s})\exp\left(-i\varepsilon_\beta t + i\vec{v}\cdot\vec{x} - i\frac{v^2}{2}t\right)N(\gamma)_1F_1(-i\gamma; 1; -ivx - i\vec{v}\cdot\vec{x})$$
$$\exp\left[-i\frac{Z_P Z_T}{v}\ln(vR + \vec{v}\cdot\vec{R})\right] = \psi_\beta \exp\left[-i\frac{Z_P Z_T}{v}\ln(vR + \vec{v}\cdot\vec{R})\right] \tag{16.9}$$

where $\varphi_\beta(\vec{s})$ is a projectile bound state wave function with corresponding energy ε_β and $\gamma = Z_T/v$. The wave function given by (16.9) satisfies the correct asymptotic conditions for electron capture, namely,

$$\chi_\beta^{-(c)} \xrightarrow[t\to+\infty]{} \varphi_\beta(\vec{s}) \exp\left(-i\varepsilon_\beta t + i\vec{v}\cdot\vec{x} - i\frac{v^2}{2}t\right)$$
$$\exp\left[-i\frac{Z_P(Z_T-1)}{v}\ln(vR + \vec{v}\cdot\vec{R})\right] \tag{16.10}$$

16.2.3 Transition amplitudes and scattering matrix elements

The first-order of the transition amplitude for electron capture and ionization as a function of the impact parameter $\vec{\rho}$ is given by the expressions

$$A_{\alpha\beta}^{+}(\vec{\rho}) = -i\int_{-\infty}^{+\infty} dt \left\langle \chi_\beta^{-} \left| \left(H_{el} - i\frac{\partial}{\partial t}\right)^{\dagger} \right| \chi_\alpha^{+} \right\rangle = (\rho v)^{2iZ_PZ_T/v} a_{\alpha\beta}^{+}(\vec{\rho}) \tag{16.11}$$

and

$$A_{\alpha\beta}^{-}(\vec{\rho}) = -i\int_{-\infty}^{+\infty} dt \left\langle \chi_\beta^{-} \left| \left(H_{el} - i\frac{\partial}{\partial t}\right) \right| \chi_\alpha^{+} \right\rangle = (\rho v)^{2iZ_PZ_T/v} a_{\alpha\beta}^{-}(\vec{\rho}) \tag{16.12}$$

in the post and prior versions, respectively. In (16.11)–(16.12), H_{el} is the electronic Hamiltonian which can be obtained, according to an eikonal treatment, from the Hamiltonian of the total system by excluding the kinetic energy operators associated to the relative movement between the collision aggregates. The term $(\rho v)^{2iZ_PZ_T/v}$ appearing in both equations describes the Rutherford scattering of the projectile in the target nuclear field. Thus, if the internuclear interaction is completely excluded we can define the reduced transition amplitudes $a_{\alpha\beta}^{+}(\vec{\rho})$ and $a_{\alpha\beta}^{-}(\vec{\rho})$ as

$$a_{\alpha\beta}^{+}(\vec{\rho}) = -i\int_{-\infty}^{+\infty} dt \left\langle \psi_\beta \left| \left(\tilde{H}_{el} - i\frac{\partial}{\partial t}\right)^{\dagger} \right| \psi_\alpha \right\rangle \tag{16.13}$$

and

$$a_{\alpha\beta}^{-}(\vec{\rho}) = -i\int_{-\infty}^{+\infty} dt \left\langle \psi_\beta \left| \left(\tilde{H}_{el} - i\frac{\partial}{\partial t}\right) \right| \psi_\alpha \right\rangle, \tag{16.14}$$

where ψ_α and ψ_β have been previously defined for each one of the reactions and models analyzed and $\tilde{H}_{el}$ can be obtained from H_{el} by excluding the internuclear potential.

Introducing the two-dimensional Fourier transform

$$R_{\alpha\beta}(\vec{\eta}) = \frac{1}{2\pi} \int d\vec{\eta} \exp(i\vec{\eta} \cdot \vec{\rho}) a_{\alpha\beta}(\vec{\rho}), \tag{16.15}$$

where $\vec{\eta}$ is the transverse momentum transfer and where $R_{\alpha\beta}(\vec{\eta})$ is related to the scattering matrix element $T_{\alpha\beta}(\vec{\eta})$ through the expression given by

$$R_{\alpha\beta}(\vec{\eta}) = \frac{T_{\alpha\beta}(\vec{\eta})}{2\pi v}, \tag{16.16}$$

the doubly differential cross section (DDCS) for electron ionization, as a function of the energy $E_k = k^2/2$ and the solid angle subtended by the ejected electron, can be written [25] as

$$\frac{d\sigma}{dE_k d\Omega_k} = k \int d\vec{\rho} \left|a_{\alpha\beta}(\vec{\rho})\right|^2 = k \int d\vec{\eta} \left|R_{\alpha\beta}(\vec{\eta})\right|^2, \tag{16.17}$$

where the Parseval's theorem has been used [11]. It results from (16.11)–(16.12) and (16.17) that the internuclear potential does not play any role when differential cross sections are integrated over the scattering angle subtended by the projectile. Single differential cross sections $d\sigma/dE_k$ and $d\sigma/d\Omega_k$ as a function of the final energy of the ejected electron or of the solid angle subtended by this electron, can be obtained by integration of (16.17) over Ω_k and E_k, respectively.

It must be noted that when the variable charge model for ionization is used and in accordance to the relationship obtained with the traditional first-Born approximation [18, 19], this relationship is here extended to the CB1 approximation, such that it is written now as

$$\left|R^{S}_{\alpha\beta}(\vec{\eta})\right|^2 = \left|N\left(\frac{\zeta_d}{p}\right)\right|^2 \left|R^{CB1}_{\alpha\beta}(\vec{\eta})\right|^2. \tag{16.18}$$

The first multiplicative factor in the r.h.s. of (16.18) is known as the Salin's factor. Finally, note that for electron capture, the expression (16.17) gives the total cross section.

16.2.4 The case of multi-electronic atomic targets

The reduction of a single electron reaction for multi-electronic atomic targets to a three-body one, composed by the projectile, the residual target and an *active* electron, has been done by Rivarola *et al.* [26] for electron capture and then extended by Fainstein *et al.* [12] for ionization. In such reduction the independent electron model is employed and the electrons in the residual target are supposed to remain as frozen in their initial orbitals. This approximation was then applied for molecular targets for both electronic capture [27] and ionization [28] processes.

As we do not deal in this work with cross sections as a function of the projectile scattering angle, the application of the theoretical models above described for mono-electronic atomic targets to the case of multi-electronic ones can be immediately done by considering $\varphi_\alpha(\vec{x})$ and $\varphi_\beta(\vec{s})$ as the orbital wave functions of the active electron, ε_α and ε_β the corresponding orbital energies, and setting $\xi = Z_T^*/k$ and $\gamma = Z_T^*/v$ (with Z_T^* an effective target charge given by $Z_T^* = \sqrt{-2n_\alpha\varepsilon_\alpha}$ where n_α is the principal quantum number of the initial orbital).

16.3 Molecular description of targets of biological interest

Describing ionizing collisions in molecular systems by a quantum-mechanical approach remains a difficult task essentially due to the multi-centered nature of the target. To overcome this difficulty, many attempts were proposed: *i)* a first one which consists in representing the molecular cross sections as a weighted sum of the cross sections of the different atomic components of the molecule, namely, the well-known Bragg's additivity rule, *ii)* a second technique - called complete neglect of differential overlap (CNDO) - where the molecular orbitals are expressed in terms of atomic orbitals of the atomic constituents, and finally *iii)* a third method which describes the populations of the target by means of molecular orbitals constructed from a linear combination of atomic orbitals in a self-consistent field approximation (MO-LCAO-SCF). These three approaches have been here tested for describing at the nanometric scale the ion-induced ionization and capture processes in water.

16.3.1 The case of the isolated water molecule

In the simple Bragg's additivity rule the different cross sections (differential as well as total) are calculated as the sum of the ones corresponding to each atom of the molecule weighted by the number of atoms in the molecule. Thus, for water, we can write

$$\sigma(H_2O) = \sigma(O) + 2\sigma(H), \tag{16.19}$$

where the atomic initial bound states of H and O are described by Roothaan-Hartree-Fock (RHF) [29] or Slater-type [30] wave functions. This approach has been already used in the past for vapor water (see for example [31]) and rather good agreement was reported with experimental data.

The second approach (CNDO) consists in writing each of the five molecular orbitals of the water target as a linear combination of atomic orbitals of its atomic constituents, namely, the orbitals H(1s), O(1s), O(2s) and O(2p) orbitals. Senger and co-workers [32, 33] have then proposed a series of weighting factors for numerous molecular targets including the water molecule whose coefficients are reported in Table 16.1.

Table 16.1 Population of the molecular orbitals used in the CNDO description of the H_2O molecule

Molecular orbital	Population
$1a_1$	2 O(1s)
$2a_1$	1.48 O(2s) + 0.52 H(1s)
$1b_2$	1.18 O(2p) + 0.82 H(1s)
$3a_1$	1.44 O(2p) + 0.34 H(1s) + 0.22 O(2s)
$1b_1$	2 O(2p)

Furthermore, let us note that this description was also used in the binary-encounter-dipole (BED) model developed by Kim and Rudd [34, 35] for providing ionization cross sections for a large set of molecules impacted by electrons.

Finally, we propose here a third description of the water molecule which has been successfully applied for treating the ionization of simple molecules like CH_4, NH_3 and H_2O by electrons [36] as well as by light-ion impact, namely, H^+, He^{2+} and C^{6+} ions [2, 37–39]. In these works, we have used the molecular description provided by Moccia who reported one-center ground state wave functions for molecules of the type XH_n, namely, for HF, CH_4 and SiH_4 [40], for NH_3, NH_4^+, PH_3 and PH_4^+ [41], and for H_2O, H_2S and HCl [42]. The molecular orbitals were expressed in terms of Slater-like functions all centered at a common origin coinciding with the X nucleus since the electronic density was - for these molecules - mainly governed by a "central" atom. Thus, providing suitable analytical wave functions was quite similar to the atomic case. Furthermore, note that the problem of evaluation of multi-center integrals depends on the type of basis functions used. Indeed, although it appears that there are no convenient and practical ways to evaluate such integrals for more than two non-aligned centers when Slater-type functions are used, it is worth noting that the use of Gaussian functions for the radial part decreases the difficulties even if it is clear that the Gaussian basis set needs probably 40% more such functions to achieve comparable results [40]. Under these conditions, the ten bound electrons of the water molecule were distributed among five one-center molecular wave functions corresponding to the five molecular orbitals of the water molecule (for more details concerning the coefficients needed for this kind of molecular description, we refer the reader to our previous works [2,4,36,38] and to the supplementary material online at http://www.aip.org/pubservs/epaps.html [43]).

Finally, it is important to note that these molecular wave functions refer to the calculated equilibrium configurations, *i.e.* to the geometrical configurations which, among many others considered, give the minimum of the total energy and agree with the experimental data in terms of HOH angle, bound O–H length, 1st ionization potential and electric dipole moment as reported in [42].

Table 16.2 Binding energies (in eV) for the different sub-shells of the water molecule in gaseous and liquid phases [49]

Water phase	$1b_1$	$3a_1$	$1b_2$	$2a_1$	$1a_1$
Vapor	12.61	14.73	18.55	32.20	539.70
Liquid	10.79	13.39	16.05	32.30	539.00

16.3.2 From vapor to liquid water

As highlighted above, numerical track-structure MC simulations were successfully developed for modeling the charged particle transport in biological medium and then providing a detailed description of the three-dimensional energetic deposit cartography. To that end, the modeling of the ion-induced ionizing processes in water and more particularly of ionization is of prime importance. Water ionization by charged particle impact (electrons as well as heavy charged particles) has been a matter of active research since the 70's particularly in the field of radiobiology for modeling the radio-induced damages (see for example [44] and [45]), the biological matter being commonly simulated by water. However, due to the scarcity of experimental measurements in its liquid phase, water was essentially studied in its vapor phase by assuming that describing the particle track-structure in liquid matter could be done, in a first step, either by applying the well-known "gas-phase approximation" *i.e.* via a simple linear extrapolation to unit density environment of the liquid or by converting the highly excited Rydberg states occurring in gaseous water into ionization (see for example [46]).

Another approach has consisted in implementing into the cross section calculations the binding energies of the liquid water phase whose values differ from gaseous water by about 2–4 eV essentially for the three outermost subshells see Table 16.2. Thus, we find an abundant literature dedicated to heavy charged particles-transport numerical simulations in gaseous and liquid water: see for example Nikjoo *et al.* [45] and González-Muñoz *et al.* [47] and more particularly the interesting work of Emfietzoglou and co-workers (see for example [48]) where the influence of the water phase on the singly differential and total ionization cross sections for protons was analyzed.

16.3.3 The DNA target

As stated above, ion-induced collisions on DNA bases have been rarely experimentally investigated. On the theoretical side, only few attempts were proposed for predicting total ionization cross sections. Among these, we essentially find two approaches: a first (semi)-classical one generally based on a classical-trajectory Monte Carlo (CTMC) type description and a second one developed in the quantum-mechanical framework and limited - for the major part of the existing studies - to the use of the first Born approximation. The "semi-classical group" may be illustrated by the study of Bacchus-Montabonel *et al.* [50] on C^{q+} ($q = 2$–4) induced collisions with Uracil and recently by our home-made CTMC code coupled to a classical over-barrier (COB) criteria [7, 51] where total cross sections of single electron loss processes (capture and ionization) were reported for collisions between multiply charged ions, namely, H^+, He^{2+} and C^{6+} (with impact energies ranging from 10 keV/amu to 10 MeV/amu) and DNA bases. To the best of our knowledge, the "quantum group" is represented by only two works, namely, the recent work of Dal Cappello *et al.* [52] where differential and total ionization cross sections have been reported for protons impinging on cytosine molecules and by our recent first Born description [8] in which differential as well as total cross sections for ions impinging on DNA bases were reported and compared to the rare existing experimental data. However, in our first work, we have described the biomolecules of interest via a CNDO approach similar to that reported by Bernhardt and Paretzke [53] and by considering only the 5 highest occupied MOs whereas the twenty highest occupied MOs are here taken into account, which represents a significant improvement over the original work. Total-energy calculations for all nucleobases were performed in the gas phase with the Gaussian 09 software at the RHF/3-21G level of theory [54]. The computed binding energies of the occupied molecular orbitals of the nucleobases were scaled, so that their theoretical first ionization potential coincides with the experimental value measured by Hush and Cheung [55]. The effective number of electrons in each atomic subshell was derived from a standard Mulliken population analysis. In this analysis only atomic sub-shells that contribute with an occupation number larger than 0.1 to each MO have been considered. Then, these atomic occupation numbers were normalized in order to account for a full 2-electron occupancy of each MO. The input parameters obtained for a target of Adenine are reported in Table 16.3.

16.4 Multiple ionization of biological molecules

In order to study the multiple ionization of molecular targets of biological interest, an independent electron model has been recently proposed [56]. The corresponding total cross sections were calculated by employing a binomial distribution of single particle probabilities [57]. Within this model, the cross section corresponding to ionization of q electrons from a particular molecular orbital composed of N equivalent electrons may be expressed as

$$\sigma_q^N = 2\pi \int_0^{+\infty} d\rho\rho P_q(\rho) = 2\pi \frac{N!}{q!(N-q)!} \int_0^{+\infty} d\rho\rho p(\rho)^q (1-p(\rho))^{N-q}, \quad (16.20)$$

where $p(\rho)$ is the probability per electron - for a given impact parameter ρ - for having a single ionization and $P_q(\rho)$ the probability to ionize q electrons, both

Table 16.3 Population and binding energies of the molecular orbitals used in the CNDO description of the Adenine molecule

Molecular orbital	Binding energies (eV)	Population
1	8.44	1.06 N(2p) + 0.94 C(2p)
2	9.98	1.23 N(2p) + 0.77 C(2p)
3	10.55	2.00 N(2p)
4	11.39	2.00 N(2p)
5	11.71	2.00 N(2p)
6	12.88	2.00 N(2p)
7	13.50	1.50 N(2p) + 0.50 C(2p)
8	15.23	1.46 N(2p) + 0.54 C(2p)
9	16.34	1.38 C(2p) + 0.62 H(1s)
10	16.85	0.56 N(2p) + 1.44 C(2p)
11	17.29	1.33 N(2p) + 0.67C(2p)
12	17.5	0.69 N(2p) + 1.31 C(2p)
13	18.42	2.00 N(2p)
14	18.99	1.42 N(2p) + 0.56C(2p)
15	20.10	2.00 N(2p)
16	21.32	2.00 N(2p)
17	22.86	0.87 N(2p) + 0.39 C(2p) + 0.39 N(2s) + 0.35 C(2s)
18	23.89	0.51 N(2p) + 1.51 C(2p) + 0.98 C(2s)
19	24.4	0.96 N(2p) + 1.04 C(2s)
20	28.35	2.00 C(2s)

from the same molecular orbital. Thus, for the case of a molecule composed by M orbitals, each one of them with an occupancy number $N_i (i = 1, \ldots, M)$, the probability for ionization of a total number of q electrons from this target is given by

$$P_q(\rho) = \sum_{q_1,\ldots,q_M=0}^{N_1,\ldots,N_M} \prod_{i=1}^{M} \frac{N_i!}{q_i!(N_i - q_i)!} [p_i(\rho)]^{q_i} [1 - p_i(\rho)]^{N_i - q_i}, \qquad (16.21)$$

with $q = \sum_{i=1}^{M} q_i$, being q_i the ionization degree of the *ith*-orbital. Then, the q-fold ionization cross section is obtained as

$$\sigma_q = 2\pi \int_0^{+\infty} d\rho \rho P_q(\rho). \qquad (16.22)$$

According to (16.21) only q_i electrons can be ionized from each *ith*-orbital, excluding the possibility of the others electrons to be ionized. Thus, $P_q(\rho)$ is known as an *exclusive* probability. This denomination is also extended to the cross section σ_q.

Then, the total or net ionization cross section reads,

$$\sigma_T = \sum_{i=1}^{M} 2\pi N_i \int_0^{+\infty} d\rho\rho p_i(\rho) = 2\pi \int_0^{+\infty} d\rho\rho \left(\sum_{q=1}^{N_i \times M} q P_q(\rho) \right) = \sum_{q=1}^{N_i \times M} q\sigma_q. \tag{16.23}$$

16.5 Results

16.5.1 Ion-induced ionization process in water

Figure 16.1a shows a comparison between the experimental DDCS data taken from Toburen and Wilson [31] and the CB1 and CDW-EIS results (solid and dashed line, respectively) for 0.5 MeV-protons and ejected energies E_e ranging from 12 eV to 750 eV. In both cases the water target is described within the CNDO approach. We observe that the CB1 model reproduces with a good agreement the experimental observations except for small angles, in particular at some fixed ejected electron energies, namely, 100 eV, 250 eV and 750 eV. Indeed, when the ejected electron is leaving the target in the forward direction with a speed comparable to that of the projectile, the interaction between the two charged particles is not necessarily weak and then the use of the 1st Born approximation is inappropriate. Thus, as already observed and reported by Rudd and Macek [58], the ejected electron angular distributions (DDCS) exhibit a big rise at low ejected angles *i.e.* in the forward direction. This process - called electron capture to the continuum (ECC) or charge transfer to the continuum - is all the more conspicuous that the velocity of the ejected electron is close to that of the scattered proton (in our case, it corresponds to electrons ejected with an energy of approximately 270 eV) and may be though in a simple image as a capture of a bound electron from the target molecule into a continuum state of the proton. However, this image is not completely true, considering the fact that for forward emission the electron travels in the combined field of the projectile and residual target [16, 17, 59]. It explains why this two-center effect is also experimentally observed for other electron energies different of the corresponding to ECC. Thus, the use of perturbative models that do not include a two-center representation in the initial and final wave functions results in a substantial underestimation of the doubly differential ionization cross sections. On the contrary, when distorted-wave approaches are used (such as is the case within the CDW-EIS approximation), the strong interaction between the projectile and the ejected electron is considered: the theoretical predictions exhibit then a better agreement with the experimental observations at an electron energy of 250 eV. However, in the backward directions CDW-EIS results clearly underestimate the experiments.

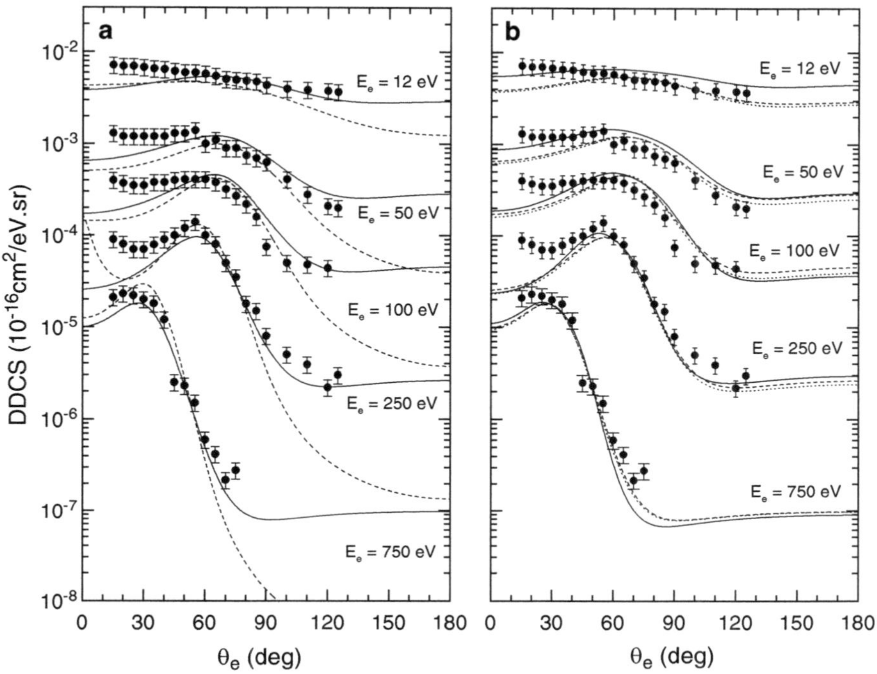

Fig. 16.1 DDCS for 0.5 MeV protons in water vapor for various ejected electron energies, namely, $E_e = 12\,\text{eV}$, 50 eV, 100 eV, 250 eV and 750 eV. Panel (**a**): CB1 and CDW-EIS results (solid and dashed line, respectively) with a water molecule described within the CNDO approach. Panel (**b**): CB1 results performed by describing the water molecule with a MO LCAO-SCF wave function (solid line), a CNDO description (dashed line) and via the Bragg's additivity rule (dotted line). The experimental data (circles) are taken from Toburen and Wilson [31]

This behavior may be partially attributed to the influence of the dynamic screening produced by the electrons remaining bound to the target on the evolution of the ionized one. It was recently shown that this dynamic screening which is in part neglected in the present CDW-EIS calculations, plays an important role on DDCS for the case of He ionization [60, 61]. In Fig. 16.1b, we compare the experimental data of Toburen and Wilson to additional CB1 results obtained by using different approaches to describe the water target (see §3.1), namely, the Bragg's additivity rule (dotted line), the previous CNDO method (dashed line) and the MO-LCAO-SCF description provided by Moccia [42] (solid line). Slight discrepancies may be observed in particular at low ejected energies where the molecular description of Moccia clearly improves the agreement with the experimental observations in particular in the forward ejection direction. Furthermore, let us note that the Bragg's and the CNDO approximations give DDCS which are very close to each other.

To improve the agreement between the experimental and the theoretical CB1-DDCS for protons, we have introduced the well-known Salin's factor [18] into the

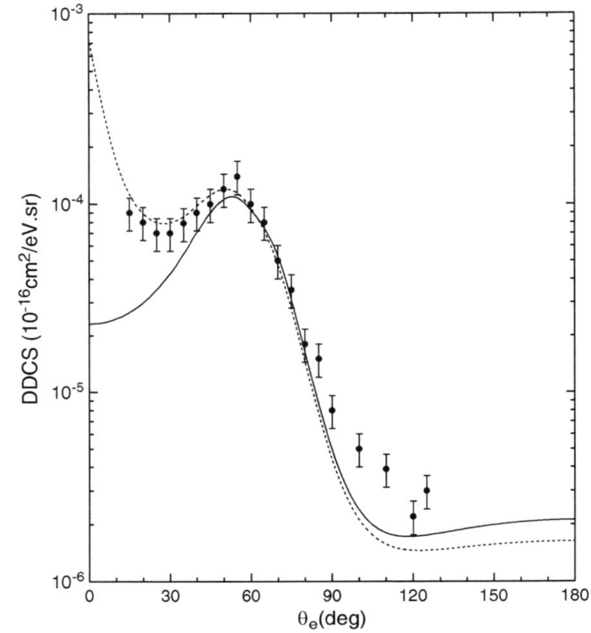

Fig. 16.2 DDCS for 0.5 MeV protons ejecting an electron of $E_e = 250$ eV in water vapor. CB1 results performed by describing the water molecule with a MO-LCAO-SCF wave function and by taking into account the Salin's factor (dashed line) or not (solid line). The experimental data (circles) are taken from Toburen and Wilson [31]

previously described CB1 model (see §16.2.1). Thus, the mechanism of electron transfer to the continuum is introduced through this multiplicative factor. Under these conditions, the obtained DDCS (see Fig. 16.2) present the ECC peak and clearly improve the agreement with the experimental observations at small ejected angles. Let us note nevertheless that the agreement remains quite unsatisfactory at large angles as already reported by Madison [62] for helium targets impacted by 100 keV and 200 keV protons.

Similarly, we compare in Fig. 16.3a the CB1 and CDW-EIS DDCS (solid and dashed line, respectively) for a water molecule (described within the CNDO approach) impacted by 6 MeV/u C^{6+} ions and for ejected electron energies ranging from 19.2 eV to 384 eV. The recent experimental measurements taken from [39] are also reported for comparison. For this system, the ejection energies are relatively low considering that the ECC mechanism must be preferably present at much higher electron energies (approximately 3300 eV). The qualitative behavior of CB1 and CDW-EIS is similar to that of Fig. 16.1a, showing in general that CDW-EIS provides a better description of experiments in the binary encounter peak region. At higher ejection energies considered experimental data present an unexpected behavior in this angular domain. In Fig. 16.3b, CB1 DDCS obtained by using the Bragg's rule, the CNDO and the Moccia's molecular representations are shown. It can be observed again that the best agreement with experimental data is found when the more complete Moccia's description is employed.

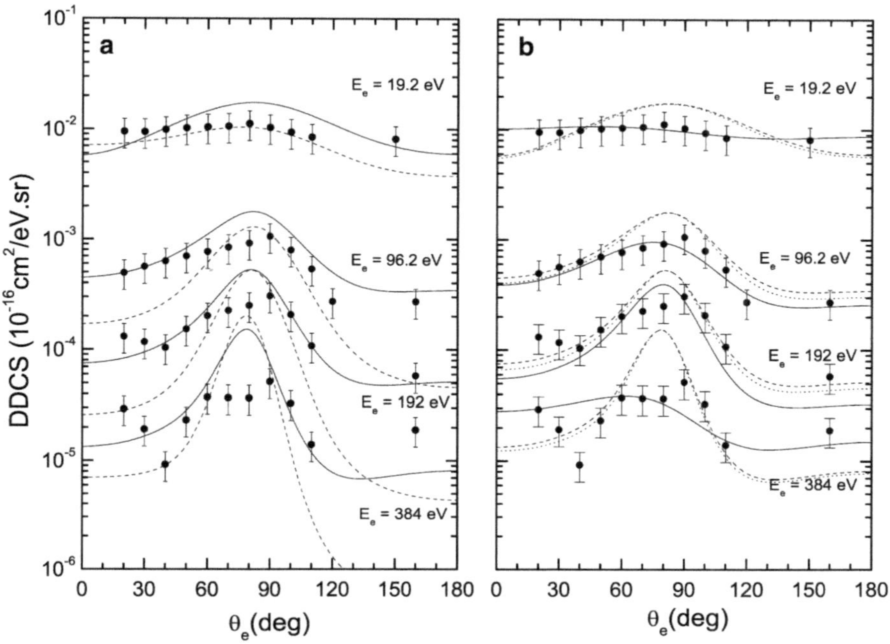

Fig. 16.3 DDCS for 6 MeV/amu C^{6+} ions in water vapor for various ejected electron energies, namely, E_e = 19.2 eV, 96.2 eV, 192 eV and 394 eV. Panel (**a**): CB1 and CDW-EIS results (solid and dashed line, respectively) with a water molecule described within the CNDO approach. Panel (**b**): CB1 results performed by describing the water molecule with a MO-LCAO-SCF wave function (solid line), a CNDO description (dashed line) and via the Bragg's additivity rule (dotted line). The experimental data (circles) are taken from [39]

By integrating the DDCS over the ejection solid angle, we obtain the singly differential cross (SDCS) whose example is reported in Fig. 16.4 where a comparison between the experimental data for protons taken from [31] and [63] and our present calculations is reported. Like previously, the left panel reports CB1 and CDW-EIS predictions provided by using a CNDO approach for describing the water molecule. A reasonably good agreement between the experience and both present theoretical models may be observed for ejected electron energies greater than 10 eV, the Auger electron peak being obviously not reproduced by the CB1 and CDW-EIS calculations. For lower ejected energies ($E_e < 10$ eV), the agreement is obviously less satisfactory, the kinematics being far from the domain of applicability of the theoretical approximations used. On the right panel of Fig. 16.4, surprisingly the better representation of experiments is obtained when the Bragg's rule and CNDO representations of the target are employed when compared with the Moccia's one.

In Fig. 16.5a, CB1- and CDW-EIS-SDCS for 6 MeV/u C^{6+} ions are compared to the recent experimental measurements reported by Dal Cappello *et al.* [39].

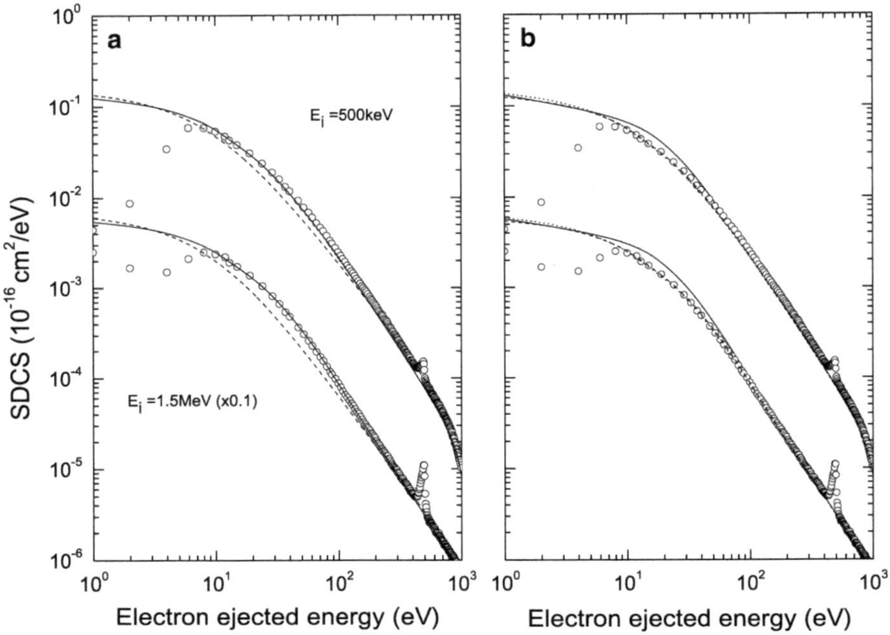

Fig. 16.4 SDCS for 500 keV and 1.5 MeV protons in water vapor. Panel (**a**): CB1 and CDW-EIS results (solid and dashed line, respectively) with a water molecule described within the CNDO approach. Panel (**b**): CB1 results performed by describing the water molecule with a MO-LCAO-SCF wave function (solid line), a CNDO description (dashed line) and via the Bragg's additivity rule (dotted line). The experimental data (circles) are taken from [31] and [63]

A good agreement is generally observed for both the results with nevertheless a regular overestimation for the CB1 predictions whereas the CDW-EIS ones show a very good accord. Furthermore, note that the CNDO approach used in Fig. 16.5a gives SDCS in close agreement with those obtained by using either the simple Bragg's additivity rule or the more sophisticated MO-LCAO-SCF approach (dotted and solid line in Fig. 16.5b, respectively) in particular for high ejected energies (E_e > 100 eV). In the low ejected energy regime, the best agreement is again surprisingly observed for the CNDO and the Bragg's description.

Figure 16.6a depicts an extensive comparison between the theoretical CB1-TCS (solid line) and CDW-EIS-TCS (dashed line) for protons, α-particles and C^{6+} ions and the available experimental measurements. In both cases the water molecule is described within the CNDO approach. Evident discrepancies may be observed at intermediate and low incident energy regimes, *i.e.*, where the validity criteria of the 1st Born approximation is no more respected. On the contrary, the distorted-wave approach shows a very good agreement with the experiments at intermediate energies. At high impact energy, the two approaches provide close predictions with nevertheless a slight overestimation for the CB1 model. In Fig. 16.6b we report

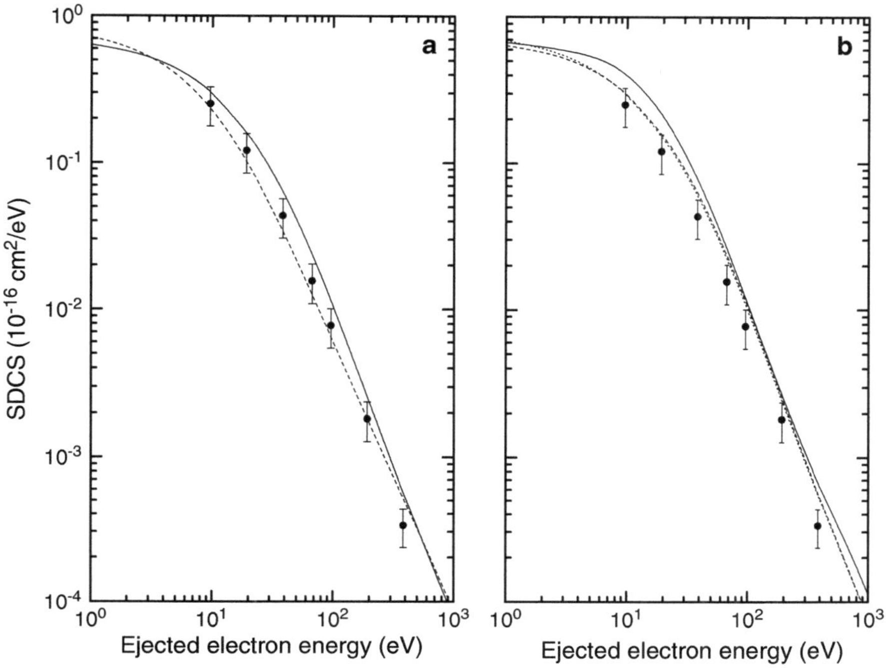

Fig. 16.5 SDCS for 6 MeV/amu C^{6+} ions in water vapor. Panel (**a**): CB1 and CDW-EIS results (solid and dashed line, respectively) with a water molecule described within the CNDO approach. Panel (**b**): CB1 results performed by describing the water molecule with a MO-LCAO-SCF wave function (solid line), a CNDO description (dashed line) and via the Bragg's additivity rule (dotted line). The experimental data (circles) are taken from [39]

the CB1 results provided by using different ways for describing the water target, namely, the Bragg's additivity rule (dotted line), the CNDO approach (the dashed line) and finally the MO-LCAO-SCF method (solid line). In the low impact energy domain, huge discrepancies may be observed between the MO-LCAO-SCF TCS and both the Bragg's and CNDO ones. As it could be expected, these differences tend to disappear when the incident energy increases.

16.5.2 Ion-induced capture process in water

In Fig. 16.7a, we report the total electron capture cross sections calculated within the CDW and the CDW-EIS framework (solid and dashed line, respectively), using for both approximations a CNDO description of the target. For CDW-EIS, we

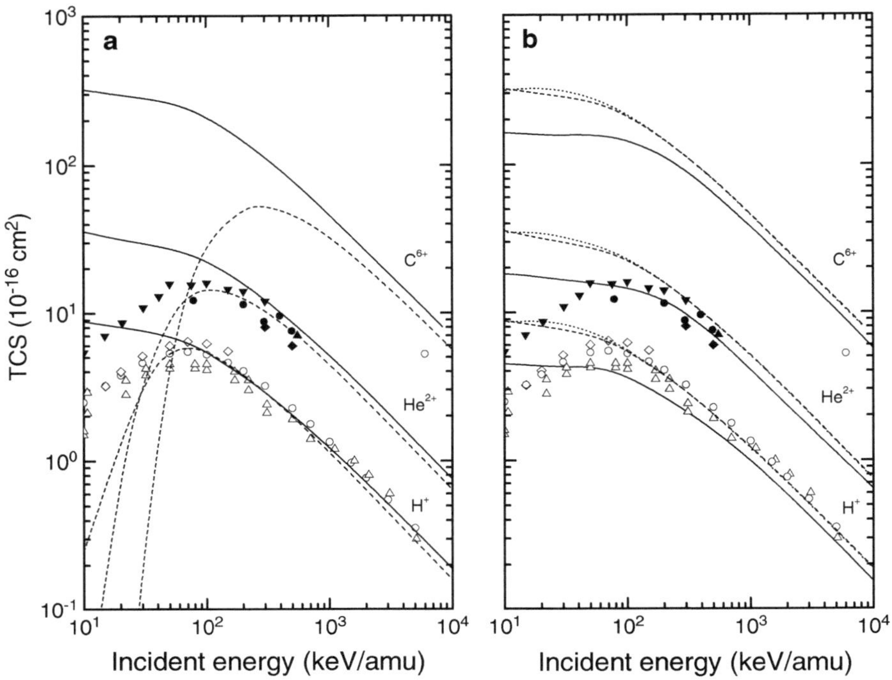

Fig. 16.6 Total ionization cross sections for H^+, He^{2+} and C^{6+} ions in water vapor. Panel (**a**): CB1 and CDW-EIS results (solid and dashed line, respectively) with a water molecule described within the CNDO approach. Panel (**b**): CB1 results performed by describing the water molecule with a MO-LCAO-SCF wave function (solid line), a CNDO description (dashed line) and via the Bragg's additivity rule (dotted line). The experimental measurements taken from various sources are represented by symbols (for more details we refer the reader to [2, 37, 39])

clearly observe a very good agreement between the theoretical predictions and the reported experimental data for all collision energies. Let us note nevertheless that both models converge when the incident energy increases.

In Fig. 16.7b, we note that the three target descriptions before employed, give cross sections very close to each other over the whole incident energy range, showing that the capture total cross sections are not much sensitive to the target representation.

16.5.3 From vapor water to liquid water

The influence of the thermodynamical phase of water on total cross sections for the processes of electron ionization (Fig. 16.8a) and electron capture (Fig. 16.8b) for impact of proton beams is analyzed. The case of liquid water is treated by considering the corresponding orbital energies given in Table 16.2. It can be observed that

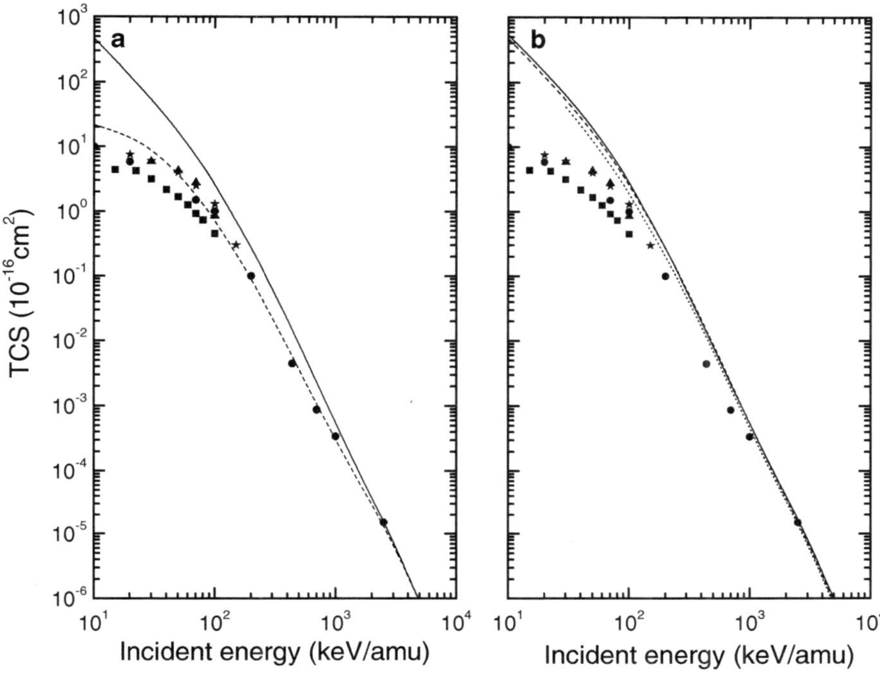

Fig. 16.7 Total cross sections for proton-induced capture in vapor water. Panel (**a**): CDW and CDW-EIS results (solid and dashed line, respectively) with a water molecule described within the CNDO approach. Panel (**b**): CDW results performed by describing the water molecule with a MO-LCAO-SCF wave function (solid line), a CNDO description (dashed line) and via the Bragg's additivity rule (dotted line). The experimental measurements taken from various sources are represented by symbols (for more details we refer the reader to [64])

the target phase plays a minor role on the determination of total cross sections both for ionization calculations obtained by employing the CDW-EIS and CB1 approximations as well as for capture ones where CDW-EIS and CDW models were used.

16.5.4 Multiple processes in water

In Fig. 16.9, q-fold ionization cross sections σ_q are presented as a function of the q-ionization degree of liquid water for the cases of equal-velocity C^{6+} and protons with energy of 1 MeV/amu. The target is described within the CNDO approximation while the CDW-EIS model is employed to calculate the corresponding cross sections. It can be observed that σ_q decreases in a steeper way for the case of protons than for C^{6+}. This behavior can be roughly explained considering that CB1 predicts cross sections scaling as Z_P^{2q}. Deviations from this scaling law must be attributed to two-center effects included in CDW-EIS.

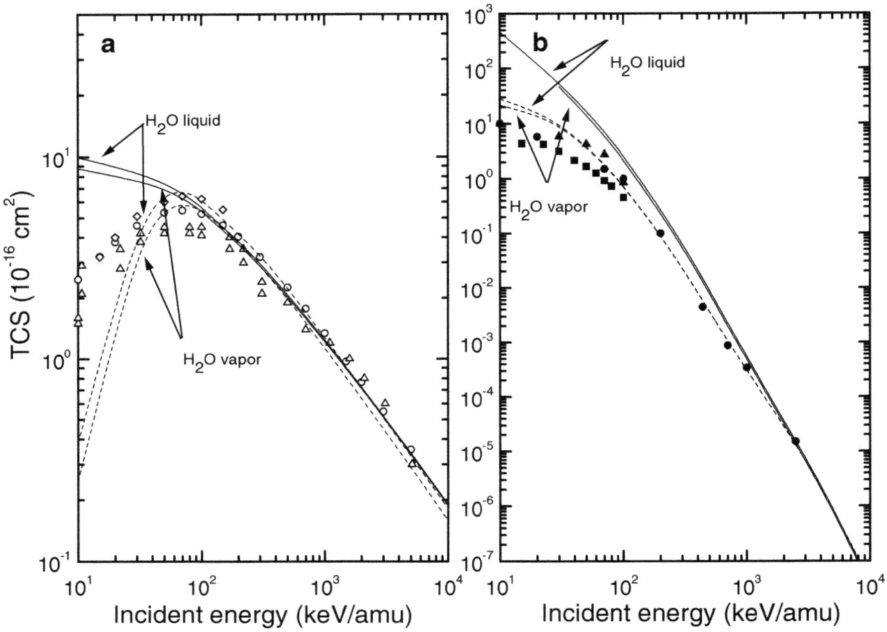

Fig. 16.8 Ionization and capture total cross sections for H^+ in liquid and gaseous water both described within the CNDO approach. Panel (**a**): Ionization: CB1 and CDW-EIS results (solid and dashed line, respectively). Panel (**b**): Capture: CDW and CDW-EIS results (solid and dashed line, respectively)

It has been shown [65, 66] that multiple ionization of water irradiated by high linear energy transfer-ion beams is responsible for the creation of a large amount of HO^2/O_2^- radicals and O_2 molecules in liquid water radiolysis in agreement with experiments [67, 68]. These radicals interacting with DNA can provoke damage to biological matter.

16.5.5 From water to DNA

Total cross sections for electron ionization and electron capture from Adenine by proton beams are presented in Figs. 16.10a and 16.10b, respectively. The target is described by using the CNDO representation presented in Table 16.3 within the CB1 and CDW-EIS approximations for ionization and the CDW-EIS and CDW ones for capture. The only available experimental points [69, 70] are also reported for comparison. Thus, for ionization, we first observe in Fig. 16.10a that the experimental data at 80 keV (taken from [69]) is largely underestimated by both theories whereas the recent measurement reported by Iriki et al. [70] at 1 MeV is well reproduced by the two models. We observe also that the CB1 and the CDW-EIS calculations are in reasonable agreement for impact energies larger than 100 keV.

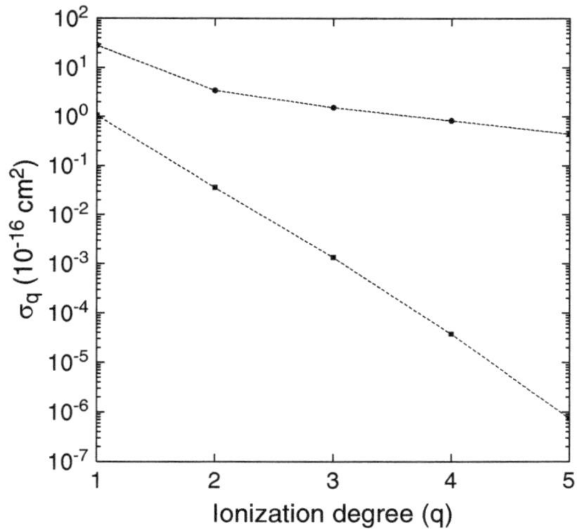

Fig. 16.9 Theoretical CDW-EIS q-fold ionization cross sections as a function of the q ionization degree for impact of 1 MeV/amu-C^{6+} ions (circles) and 1 MeV-protons (squares) in liquid water

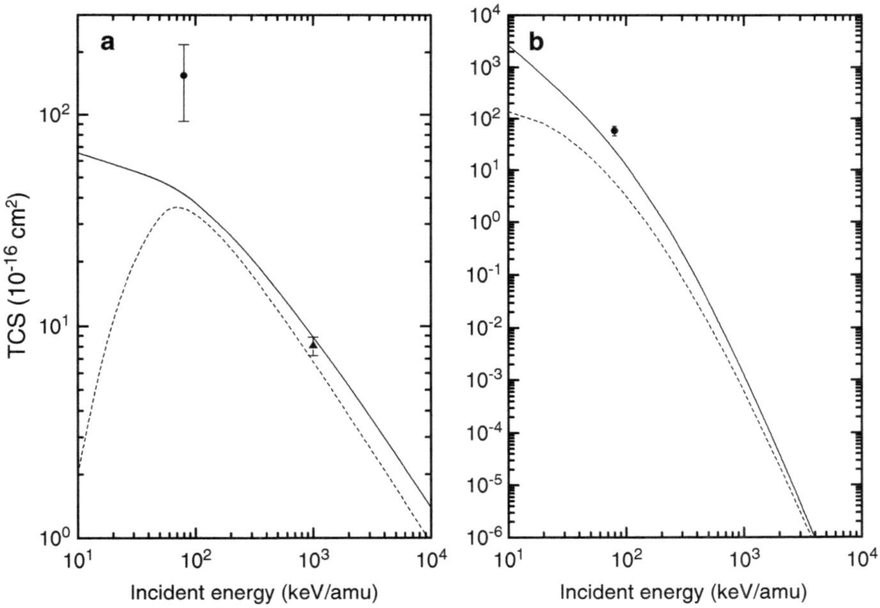

Fig. 16.10 Panel (**a**): CB1 and CDW-EIS ionization cross sections for protons in Adenine (solid and dashed line, respectively). Panel (**b**): CDW and CDW-EIS capture cross sections for protons in Adenine (solid and dashed line, respectively). The experimental data are taken from [69] (solid cicles) and from [70] (solid triangle)

Similarly, from Fig. 16.10b, it clearly appears that the experimental capture cross section at 80 keV is quite well reproduced by the CDW model whereas an underestimation is noted when the CDW-EIS model is used. Let us note that we here also observe a good convergence between the two models at high impact energies.

16.6 Monte Carlo simulation of charged particle transport in biological medium

Modeling of DNA damages from ionizing radiation is today an active and intense field of research. As stated above, Monte Carlo methods have been widely adopted in the radiobiology community since they can reproduce the stochastic nature of interactions between elementary particles and matter. In this context, we aim to develop a Monte Carlo code able to provide a full description of proton tracks in water and DNA components over a wide impact energy range (from several hundreds of MeV down to the Bragg peak region) including all the secondary particle histories.

In brief - and as commonly performed in the major part of the existing step-by-step Monte Carlo codes - the transport simulation will comprise series of random samplings which determine *i)* the distance to the next interaction (related to the mean free path, this latter being calculated from the total cross section), *ii)* the type of interaction which occurs at the point selected in *i)* and *iii)* the energy and direction of the resultant particles according to the type of interaction selected in *ii)*. These latter will be successively determined via random samplings among the pre-tabulated singly and doubly differential cross sections, respectively. Particular ionization or excitation potential will be assumed as locally deposited and the incident energy is reduced from the corresponding energy (including as well as potential and secondary kinetic energy transfers). All these steps will be consecutively followed for all resultant particles until their kinetic energy falls below the predetermined cut-off value (here 10 keV for incident protons and 7.4 eV for secondary electrons what corresponds to the water excitation threshold). Note that sub-threshold electrons will be assumed to deposit their energy where they are created. In these conditions, the code will provide by way of row data the coordinates of all the interaction events as well as the type of collision together with the energy loss, the energy deposited at each interaction point and the kinetic energy of the resultant particle(s) in the case of inelastic collision.

In this context, all the above-reported quantum-mechanical cross sections will be used as input data for describing the ion-induced interactions (ionization and capture) at the multi-differential scale. Furthermore, a detailed analysis of the influence of the target description as well as that of the quantum-mechanical model used for describing the ionizing processes will be done.

16.7 Conclusion

Single ionization and capture of water and DNA bases impacted by heavy charged particles of medical interest (protons and carbon ions) have been here theoretically studied by employing two different quantum mechanical models based on the CDW-EIS and CB1 approaches.

Considering the ionization process in water, the comparison of the CB1 to the CDW-EIS doubly differential cross sections has clearly shown that the extreme angle regions were not well described, with in particular a systematic underestimation of the experimental DDCS by the second model. Besides, singly differential cross sections have shown a very good agreement with the experimental data whereas the total cross sections have pointed out the well-know overestimation of the CB1 model at low incident energies.

Furthermore, the proton-induced capture in water has been studied within the continuum distorted wave model and good agreement with available measurements was reported.

Additionally, for both ionization and capture, we have reported a detailed study of the influence of the target description on the cross section calculations.

Finally, ionization and capture in DNA components have been studied and the comparison of our theoretical results with the scarce experimental data points available suggests that new experimental data are needed for a better understanding of the reactions investigated.

All the reported cross sections will be implemented into a new home-made Monte Carlo code devoted to a fine description of the track-structure of charged particles in biological medium.

Acknowledgments We acknowlegde Dr. Paula Abufager for providing us with the computer program to calculate CDW-EIS electron capture total cross sections.

References

1. Nikjoo, H., Goodhead, D. T., Charlton, D. E., and Paretzke, H. G., 1991, *Int. J. Radiat. Biol.* **60**, 739–756.
2. Champion, C., Boudrioua, O., Dal Cappello, C., Sato, Y., and Ohsawa, D., 2007, *Phys. Rev. A.* **75**, 032724.
3. Champion, C., L'Hoir, A., Politis, M. -F., Fainstein, P. D., Rivarola, R. D., and Chetioui, A., 2005, *Radiat. Res.* **163**, 222–231.
4. Champion, C., 2003, *Phys. Med. Biol.* **48**, 2147–2168.
5. Friedland, W., Jacob, P., Paretzke, H. G., Ottolenghi, A., Ballarini, F., and Liotta, M., 2006, *Radiat. Prot. Dosim.* **122**, 116120.
6. Le Padellec, A., Moretto-Capelle, P., Richard-Viard, M., Champeaux, J. -P., and Cafarelli, P., 2008, *J. Phys.: Conf. Ser.* **101**, 012007.
7. Abbas, I., Lasri, B., Champion, C., and Hanssen, J., 2008, *Phys. Med. Biol.* **53**, N1–N11.
8. Champion, C., Lekadir, H., Galassi, M. E., Fojón, O., Rivarola, R. D., and Hanssen, J., 2010, *Phys. Med. Biol.* **55**, 6053–6067.
9. Dewangan, D. P., and Eichler, J., 1985, *J. Phys. B* **18**, L65–L69.
10. Champion, C., Hanssen, J., and Rivarola, R. D., in the Book Series *Advances in Quantum Chemistry*, to appear (2011).
11. Crothers, D. S. F., and McCann, J. F., 1983, *J. Phys. B* **16**, 3229–3242.
12. Fainstein, P. D., Ponce, V. H., and Rivarola, R. D., 1988, *J. Phys. B* **21**, 287–299.
13. Belkić, Dž., Gayet, R., and Salin, A., 1979, *Phys. Rep.* **56**, 279–369.
14. Belkić, Dž., Gayet, R., Hanssen, J., and Salin, A., 1986, *J. Phys. B* **19**, 2945–2953.

15. Dewangan, D. P., and Eichler, J., 1986, *J. Phys. B* **19**, 2939–2944.
16. Stolterfoht, N. *et al.*, 1987, *Euro. Phys. Lett.* **4**, 899–904.
17. Fainstein, P. D., Ponce, V. H., and Rivarola, R. D., 1991, *J. Phys. B* **24**, 3091–3119.
18. Salin, A., 1969, *J. Phys. B* **2**, 631–639.
19. Salin, A., 1972, *J. Phys. B* **5**, 979–985.
20. Miraglia, J. E., Piacentini, R. D., Rivarola, R. D., and Salin, A., 1981, *J. Phys. B* **14**, L197–L202.
21. Thomas, L. H., 1927, *Proc. R. Soc. A* **114**, 561–576.
22. Shakeshaft, R., and Spruch, L., 1979, *Rev. Mod. Phys.* **51**, 369–381.
23. Martínez, A. E., Deco, G. R., Rivarola, R. D., and Fainstein, P. D., 1989, *Nucl. Instrum. Methods Phys. Res. B* **43**, 24–28.
24. Cheshire, I. M., 1964, *Proc. Phys. Soc. (London)* **84**, 89–98.
25. Belkić, Dž., 1978, *J. Phys. B* **11**, 3529–3552.
26. Rivarola, R. D., Piacentini, R. D., Salin, A., and Belkić, Dž., 1980, *J. Phys. B* **13**, 2601–2609.
27. Corchs, S. E., Rivarola, R. D., and McGuire, J. H., 1993, *Phys. Rev. A* **47**, 3937–3944.
28. Galassi, M. E., Rivarola, R. D., and Fainstein, P. D., 2004, *Phys. Rev. A* **70**, 032721.
29. Clementi, E. and Roetti, C., 1974, *At. Data Nucl. Data Tables* **14**, 177–478.
30. Slater, J. C., 1930, *Phys. Rev.* **36**, 57–64.
31. Toburen, L. H., and Wilson, W. E., 1977, *J. Chem. Phys.* **66**, 5202–5213.
32. Senger, B., and Rechenmann, R. V., 1984, *Nucl. Instrum. Methods Phys. Res. B* **2**, 204–207.
33. Senger, B., 1988, *Z. Phys. D: At., Mol. Clusters* **9**, 79–89.
34. Kim, Y. -K., and Rudd, M. E., 1994, *Phys. Rev. A* **50**, 3954–3967.
35. Hwang, W., Kim, Y-K., and Rudd, M. E., *J. Chem. Phys.* **104**, 2956–2966.
36. Champion, C., Hanssen, J., and Hervieux, P. -A., 2004, *J. Chem. Phys.* **121**, 9423–9429.
37. Boudrioua, O., Champion, C., Dal Cappello, C., and Popov, Y. V., 2007, *Phys. Rev. A.* **75**, 022720.
38. Champion, C., Boudrioua, O., and Dal Cappello, C., 2008, *J. Phys.: Conf. Ser.* **101**, 012010.
39. Dal Cappello, C., Champion, C., Boudrioua, O., Lekadir, H., Sato, Y., and Ohsawa, D., 2009, *Nucl. Instrum. Methods Phys. Res. B* **267**, 781–790.
40. Moccia, R., 1964, *J. Chem. Phys.* **40**, 2164–2176.
41. Moccia, R., 1964, *J. Chem. Phys.* **40**, 2176–2186.
42. Moccia, R., 1964, *J. Chem. Phys.* **40**, 2186–2192.
43. See EPAPS Document No. E-JCPSA6-121-011442. A direct link to this document may be found in the online article's HTML reference section. The document may also be reached via the EPAPS homepage (http://www.aip.org/pubservs/epaps.html) or from ftp.aip.org in the directory/epaps/. See the EPAPS homepage for more information.
44. Friedland, W., Dingfelder, M., Jacob, P., and Paretzke, H. G., 2005, *Radiat. Phys. Chem.* **72**, 279–286.
45. Nikjoo, H., Uehara, S., Emfietzoglou, D., and Brahme, A., 2008, *New J. Phys.* **10**, 075006.
46. Turner, J. E., Paretzke, H. G., Hamm, R. N., Wright, H. A., and Ritchie, R. H., 1982, *Radiat. Res.* **92**, 47–60.
47. González-Muñoz, G., Tilly, N., Fernández-Varea, J. M., and Ahnesjö, A., 2008, *Phys. Med. Biol.* **53**, 2857–2875.
48. Emfietzoglou, D., and Moscovitch, M., 2003, *Nucl. Instrum. Methods Phys. Res. B* **209**, 239–245.
49. Dingfelder, M., Inokuti, M., and Paretzke, H. G., 2000, *Radiat. Phys. Chem.* **59**, 255–275.
50. Bacchus-Montabonel, M. C., Labuda, M., Tergiman, Y. S., and Sienkiewicz, J. E., 2005, *Phys. Rev. A* **72** 052706.
51. Lekadir, H., Abbas, I., Champion, C., Fojón, O., Rivarola, R. D., and Hanssen, J., 2009, *Phys. Rev. A* **79**, 062710.
52. Dal Cappello, C., Hervieux, P. -A., Charpentier, I., and Ruiz-López, M. F., 2008, *Phys. Rev. A* **78**, 042702.
53. Bernhardt, Ph., and Paretzke, H. G., 2003, *Int. J. Mass. Spectrom.* **223–224**, 599–611.
54. Frisch M. J., *et al.*, Gaussian 09, Revision A.02, Gaussian, Inc., Wallingford CT, 2009.

55. Hush, N. S., and Cheung, A. S., 1975, *Chem. Phys. Lett.* **34**, 11–13.
56. Galassi, M. E., and Rivarola, R. D., to be submitted (2011).
57. Lüdde, H. J., and Dreizler, R., 1985, *J. Phys. B* **18**, 107–112.
58. Rudd, M. E., and Macek, J. H., in *Case Studies in Atomic Physics*, edited by E. W. McDaniels and M. C. McDowell (North-Holland, Amsterdam, 1972), Vol3., pp. 47–136.
59. Stolterfoht, N., DuBois, R. D., and Rivarola, R. D., 1997, Electron emission in heavy ion-atom collisions, Springer Series on Atoms and Plasmas, Ed. Springer-Verlag, ISSN 0177–6495.
60. Monti, J. M., Fojón, O. A., Hanssen, J., and Rivarola, R. D., 2010, *J. Atom. Mol. Opt. Phys.* **2010**, 1–7.
61. Monti, J. M., Fojón, O. A., Hanssen, J., and Rivarola, R. D., 2010, *J. Phys. B* **43**, 205203.
62. Madison, D. H., 1973, *Phys. Rev. A* **8**, 2449–2455.
63. Bolorizadeh, M. A., and Rudd, M. E., 1986, *Phys. Rev. A* **33**, 888–892.
64. Lekadir, H., Abbas, I., Champion, C., and Hanssen, J., 2009, *Nucl. Instrum. Methods Phys. Res. B* **267**, 1011–1014.
65. Gervais, B., Beuve, M., Olivera, G. H., Galassi, M. E., and Rivarola, R. D., 2005, *Chem. Phys. Lett.* **410**, 330–334.
66. Gervais, B., Beuve, M., Olivera, G. H., and Galassi, M. E., 2006, *Radiat. Phys. Chem.* **75**, 493–513.
67. Laverne, J. A., and Schuler, R. H., 1992, *Radiat. Phys. Chem.* **96**, 7376–7378.
68. Baldacchino, G., *et al.*, 1998, *Nucl. Instrum. Methods Phys. Res. B* **146**, 528–532.
69. Tabet, J., *et al.*, 2010, *Phys. Rev. A* **82**, 022703.
70. Iriki, Y., Kikuchi, Y., Imai, M., and Itoh, A., 2011, *Phys. Rev. A* **84**, 032704.

Chapter 17
Multiscale Approach to Radiation Damage Induced by Ions

Andrey V. Solov'yov and Eugene Surdutovich

Abstract We present the overview of the multiscale approach to radiation damage by ions. This approach has been designed to understand the effects happening on different scales in time, space, and energy, which influence the cell damage following irradiation by ions. The multiscale approach includes the topics starting from the energy loss by projectiles and production of secondaries to the pathways of cell damage. This particular paper is devoted to the analysis of complex damage of DNA. The complex damage is important because cells in which it occurs are less likely to survive. We consider several possible approaches to the calculation of complex damage and suggest different developments involving comparisons with experiments.

17.1 Introduction: multiscale approach to radiation damage

The multiscale approach to the radiation damage induced by irradiation with ions is aimed to the phenomenon-based quantitative understanding of the scenario from incidence of an energetic ion on tissue to the cell death [1, 2]. This approach joins together many spatial, temporal, and energetic scales involved in this scenario. The success of this approach will provide the phenomenon-based foundation for ion-beam cancer therapy, radiation protection in space, and other applications of ion beams. Main issues addressed by the multiscale approach are ion stopping in the medium [3], production and transport of secondary particles produced as a

A.V. Solov'yov (✉)
Frankfurt Institute for Advanced Studies, 60438 Frankfurt am Main, Germany
e-mail: solovyov@fias.uni-frankfurt.de

E. Surdutovich
Department of Physics, Oakland University, Rochester, Michigan 48309, USA
e-mail: surdutov@oakland.edu

G.G. Gómez-Tejedor and M.C. Fuss (eds.), *Radiation Damage in Biomolecular Systems*, Biological and Medical Physics, Biomedical Engineering,
DOI 10.1007/978-94-007-2564-5_17, © Springer Science+Business Media B.V. 2012

result of ionization and excitation of the medium [3–5], interaction of secondary particles with biological molecules, most important with DNA [1,2], the analysis of induced damage, and evaluation of probabilities of subsequent cell survival or death. Evidently, this approach is interdisciplinary, since it is based on physics, chemistry, and biology. Moreover, it spans several areas within each of these disciplines.

The multiscale approach started with the analysis of ion propagation, which resulted in the description of the Bragg peak and the energy spectrum of secondary electrons [3–6]. The practical goal of these works was providing a recipe for economic calculation of the Bragg peak position and shape. Theoretically, they concluded that the cross section of ionization of molecules of the medium singly-differentiated with respect to the energies of secondary electrons is the most important physical input on this scale, the longest in distance and highest in energy. The relativistic effects play an important role in describing the position of the Bragg peak as well as the excitation channel in inelastic interactions [3]. The effect of charge transfer and projectile scattering affect its shape [3]. The effects of nuclear fragmentation happening in the events of projectile collisions with the nuclei of the medium are also important on this scale. The next scale in energy and space is related to the transport of the secondaries, which has been considered in Refs. [1, 7], but it may still be revisited. The results of this analysis will give the spatial distributions of secondary particles as well as the accurate radial dose distribution.

The goal of the analysis of DNA damage mechanisms is the obtaining of the effective cross sections for the dominant processes, which should be taken into account in order to calculate the probability of different lesions caused by different agents. The above three stages of processes, represent not only different spatial scales, but also different time scales slowing from 10^{-21} to 10^{-5} seconds. We would like to calculate the spatial distribution of primary DNA damage, defined by the longest biochemical time, including the degree of complexity of this damage. Then, the repair and other biological effects can be included and thus the relative biological effectiveness (RBE) can be calculated. The RBE is one of the key integral characteristics of the effect of ions compared to that of photons. This ratio compares the doses of different projectiles leading to the same biological result. The calculation of RBE using the multiscale approach will be a result of a constructive quantitative analysis to physical, chemical, and biological phenomena and its predictive power will be scientifically sound. Conditions or environment related to the radiation damage may vary, if, e.g., the dose deposition is fast as in laser-driven beams, chemically active components increasing the number of active agents are present, or biological factors are more important, etc. The multiscale approach capable of including these variations will be more versatile than the existing approaches to calculating the RBE.

In [7], the radial dose distribution has been addressed. Traditionally, the radial dose is related to the radial distribution of damage. However, this does not include the complexity of damage, which may not be directly related to the dose. It is still not clear how to relate the dose with complexity of damage. This work is a step in this direction.

Finally, the analysis of possibility thermo-mechanical damage pathways has started in Refs. [3, 6] and has further advanced in Refs. [8, 9]. This idea stems from the fact that the energy lost by an ion is transferred to the tissue and it is then thermalized. We analyzed this transition in [8] and used it as an initial condition for hydrodynamic expansion described by a cylindrical shock wave in ref. [9]. These works predict a rapid rise of temperature and pressure in the vicinity of the track. Then when the expansion starts, the pressure is high on the wave front, but quickly drops in the wake of the wave causing large pressure gradients, and, therefore, strong forces which may rupture bonds of biomolecules, which may be located within several nm from the track. It was shown that these forces can be strong enough, more than 10 nN, but act only for a very short time. An estimate of work done by this force, based on [9], is several eV, but still more research is needed in order to investigate whether this does represent a separate mechanism of damage.

17.2 Distributions of the complex damage

Complex damage is defined as a number of DNA lesions, such as double strand breaks (DSB), single strand breaks, abasic sites, damaged bases, etc., that occur within two consecutive DNA convolutions so, that when repair mechanisms are engaged, they treat all these cluster of lesions as a single damage site [10]. In [11], the complexity of DNA damage has been quantified by defining a cluster of damage as a damaged portion of DNA molecule by several independent agents, such as secondary electrons, holes, or radicals. The calculation of damage complexity and its distribution is a very important stage in the multiscale approach, since it is closely related to the probability of the cell death as a result of damage. It is one of the defining factors in calculating of RBE.

17.2.1 Damage complexity distribution from the random walk approach

In Refs. [1, 2], we considered the targeting of DNA molecules. This direction led to the calculation of the radial distribution of DSBs with respect to the ion track. This calculation was limited by only considering secondary electrons to be the agents of DNA lesions. Nevertheless, this allowed to make an estimation of a number of DSBs produced by ions per unit length of track in the vicinity of the Bragg peak. The results obtained in these works were comparable with the experimental data. The approach of Refs. [1, 2] can be used for calculating the radial distribution of damage complexity. In [11], we have quantified the complexity having related it to the number of agents causing different lesions in a given volume. Then, N, the average number of lesions per this volume, has been estimated as a product of the volume, the number density of agents, and the probability of inducing damage by impact. The probability of complex damage is then a Poisson distribution $P(N, \nu)$, where ν is the degree of complexity.

Now we can make the next step and calculate the radial distribution of complex damage with respect to the ion track. In the simplest case, when all agents are equivalent, this distribution will be given by

$$P(r, \nu) = \exp\left(-N(r)\right) \frac{N(r)^{\nu}}{\nu!}, \tag{17.1}$$

where $N(r)$ is the average number of interactions per volume of a cluster at a given distance r from the track. Besides the volume of the cluster, taken to be a volume of two consecutive DNA convolutions, $N(r)$ contains two more parameters; these are the number density of agents at r and the probability of a lesion. They always appear as a product and therefore they can be considered as a single parameter. Still, it is very much desired to calculate both of them. The number density can be obtained from the analysis of the transport of the secondaries. The first step in this direction is to consider the diffusion of the secondaries from the place of their origin as was done in [1]. The further development will include the chemical reactions including these secondaries and the transport with the account of the angular distributions.

Let us assume that the diffusion of the secondaries is radial and calculate the number of particles which penetrate a patch of $A \approx 16$ nm^2, representing the effective surface of the target. For simplicity, let us assume that the patch is parallel to the track and that the patch is perpendicular to the direction of diffusion. Then the fluence through this patch is given by the following expression,

$$N_a(r) = \int_k dkAD \frac{\partial P_2(k,r)}{\partial r} \epsilon(k) \frac{dN}{d\zeta}, \tag{17.2}$$

where

$$P_2(k,r) = \frac{1}{\pi k l^2} \exp\left(-\frac{r^2}{kl^2}\right), \tag{17.3}$$

is the probability density for a particle to diffuse by r from the track in 2-D, k is the number of steps in a random walk, l is the mean free path, $D = l^2/4$ is the diffusion coefficient multiplied by the average time of between consecutive collisions, $\epsilon(k)$ is the attenuation factor, $\frac{dN_s}{d\zeta}$ is the number of secondary particles produced per one nm of the ion's track. For the estimation, we can take $\frac{dN_s}{d\zeta} = 20$, which corresponds to the doubled average number of ionizations per one nm of ion's track in the vicinity of the Bragg peak [1, 3]. We doubled this number to include further ionizations and the holes, also playing the role in damage.

The result of this distribution is plotted in Fig. 17.1. As expected, N_a decreases as the secondary particles diffuse out.

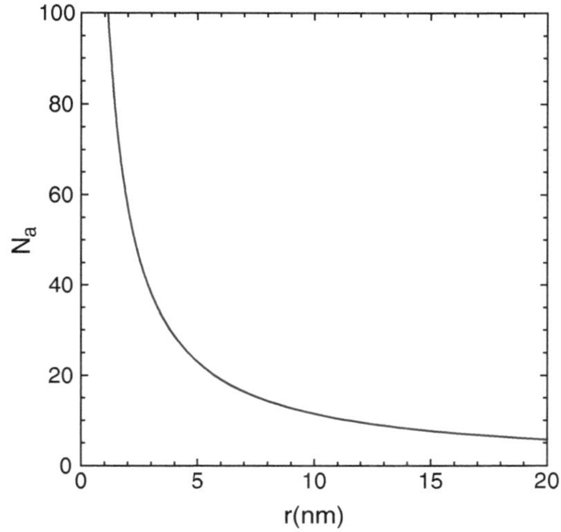

Fig. 17.1 Distribution of secondaries incident on a nucleosome in the vicinity of the ion's track

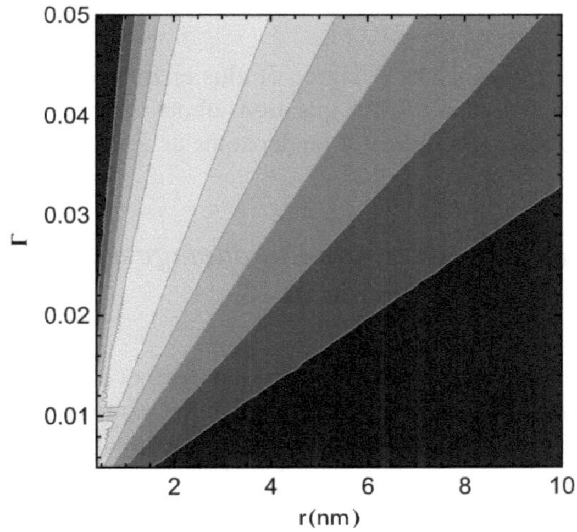

Fig. 17.2 Radial distribution of clusters of three lesions depending on different values of Γ. Color map from dark to light: $P(3) < 0.05$, $0.05 < P(3) < 0.10$, $0.10 < P(3) < 0.15$, $0.15 < P(3) < 0.20$, $0.20 < P(3) < 0.25$, and $P(3) > 0.25$

The next step is multiplying the number of agents $N_a(r)$ by the probability of inducing a lesion Γ to obtain $N(r)$. Then we can apply (17.1) and obtain the distribution of probability of observing of clustered damage of a given degree. The results for different values of Γ are plotted in Fig. 17.2. The shape can be explained by the properties of the Poisson distribution. A similar calculation can be done for any degree of complexity ν. If Fig. 17.3 the distributions of clusters of two and clusters of three lesions are compared for $\Gamma = 0.02$.

These distributions can be compared with the nano-dosimetric data [12] and the cell-survival dependences on the radius. Such comparisons may clarify the

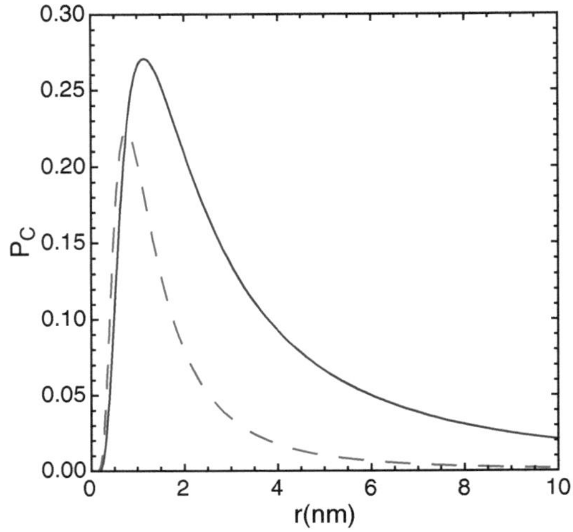

Fig. 17.3 Radial distribution of clusters of two (solid line) and clusters of three lesions (dashed line) for $\Gamma = 0.02$

significance of a degree of clusterization and its relation with lethality of damage. It may also resolve questions about the probability of lesions upon impact and the choice of volume of nucleosome as a unit.

17.2.2 Derivation of damage complexity from the radial dose distribution

As was shown in [7], the radial number density distribution is related with the radial dose. This relation involves another parameter, the average energy loss per collision. This parameter, however, can be estimated using the data on a variety of inelastic collisions. Then the number density of at least some agents can be compared to the experiments on the radial dose.

Let us now relate the dose distribution around a single ion's track to the distribution of clusters of DNA damage. The former can be measured using the contemporary nano-dosimetry. The latter can be observed by studying DNA repair foci. If any DNA lesion "consumes" the average energy of ΔW, we can calculate the absorbed energy and its density, i.e., dose. This absorbed energy (per unit of length of the track) should sum up to the linear energy transfer (LET) by the projectile. More precisely, it should sum up only to a portion of LET, which is left from the energy lost in the medium, some of which was spent on formation of agent of damage. From the other side, if we know the dose, which went to the biological damage, we can calculate the complexity, not necessarily knowing the agent of damage and its distribution. If ΔW, still quantifies the average energy per lesion, then we can calculate the number density of lesions directly from the biological dose $\mathcal{D}$ as $n = \mathcal{D}/\Delta W$, Then the above ν-fold cluster density is given by (17.1).

For example, if the LET is 0.9 keV/nm, as in the vicinity of the Bragg peak for C ions, 0.2% of this energy goes to biodamage, and the average energy loss per lesion is 3 eV, then the lesion density is of the order of 2×10^{-3} nm^{-3} and then the average number of lesions per nucleosome is 2. Then we can calculate the probabilities of ν-fold cluster damage in this nucleosome as $p(\nu, 2)$. There are two parameters in this estimate, the average energy per lesion and the percent of energy going to biodamage. The average energy per lesion can be found from quantum chemistry as the weighted average between the thresholds of all included processes. The energetics of these processes is known even if the cross sections are not. The second number includes the probability or cross sections of these processes. These values are available and for now, they can be taken from available experiments, e.g., on interactions of electrons with DNA [13].

Which of the two ways is preferable for the calculation of the distribution of cluster damage? If the number densities of agents at any distance to from the track are known, i.e., if it is possible to compare their calculations with nano-dosimetric experiments, along with the probabilities of inducing certain lesions, the first route leads to the radial distribution of clustered damage. This radial distribution potentially can be compared to the distribution of damaged cells. Then the first way is definitely preferable. If only the survival rate dependence on the distance of the track and the radial dose distribution are known, then the second route is the only resort and one can calculate the unknown parameter assuming the relation between a certain degree of clustering and cell death.

17.2.3 Integral damage complexity, distribution along the track

Still another path can be taken if we decide to ignore the radial distribution of dose or number density and just consider the longitudinal distribution of clusters. This may be relevant for the current experimental state of the art. When experimentalists study foci, which reveal the efforts of the proteins to fix the damaged DNA, they observe that the foci are very large, compared to the scale of radial distribution of the dose. The experimentalists can measure the linear density of clusters along the track and they hypothesize about the number of certain lesions, such as DSB, per unit length [14].

In order to obtain the longitudinal distributions, we can integrate the radial distribution of the complex damage starting from (17.2) substituted to (17.1) with a chosen Γ over the radius and thus present the longitudinal distribution of the complex damage.

$$P_{\zeta}(\nu) = \int_0^{\infty} \exp(-N(r)) \frac{N(r)^{\nu}}{\nu!} 2\pi r dr, \tag{17.4}$$

This gives the numbers for complex damages per μm. These numbers can be compared with experiments and this can give still another relation for the unknown parameters such as Γ.

17.3 Conclusions

We presented a brief description of the current status of the multiscale approach to the radiation damage by ions. There are many aspects in this approach in different areas of physics and other sciences.

The essential difference of this approach from other techniques is in approach to complex damage. This approach, presented in this paper is an invitation to experimentalists in biophysics and nano-dosimetry to think about possible ways of finding patterns corresponding to complexity of damage in the observed data. The main point in our approach to damage complexity is that it can be described by a spatial distribution, which is related to distribution of the cell survival rate and to the radial dose distribution.

The multiscale approach was designed in order to understand the mechanisms, which make the ion-beam therapy work. This includes the understanding of what is truly different between different therapies. Does the concentration of dose account for everything? Perhaps no, otherwise there would not be a question about the damage complexity. However, how different are the dose and complexity distributions? The answers to these questions are important and they will help to improve radiation therapy.

Acknowledgements We are grateful to the support of the authors' collaboration by the Deutsche Forschungsgemeinschaft.

References

1. A. Solov'yov, E. Surdutovich, E. Scifoni, I. Mishustin, W. Greiner, Phys. Rev. **E79**, 011909 (2009)
2. E. Surdutovich, A. Solov'yov, Europhys. News **40/2**, 21 (2009)
3. E. Surdutovich, O. Obolensky, E. Scifoni, I. Pshenichnov, I. Mishustin, A. Solov'yov, W. Greiner, Eur. Phys. J. D **51**, 63 (2009)
4. E. Surdutovich, E. Scifoni, , A. Solov'yov, Mutat. Res. **704**, 206 (2010)
5. E. Scifoni, E. Surdutovich, A. Solovyov, Phys Rev. E **81**, 021903 (2010)
6. O. Obolensky, E. Surdutovich, I. Pshenichnov, I. Mishustin, A. Solov'yov, W. Greiner, Nucl. Inst. Meth. B **266**, 1623 (2008)
7. E. Scifoni, E. Surdutovich, A. Solov'yov, Eur. Phys. J. D **60**, 115 (2010)
8. M. Toulemonde, E. Surdutovich, A. Solov'yov, Phys. Rev. E **80**, 031913 (2009)
9. E. Surdutovich, A. Solov'yov, Phys. Rev. E **82**, 051915 (2010)
10. S. Malyarchuk, R. Castore, L. Harrison, DNA Repair **8**, 1343 (2009)

11. E. Surdutovich, A. Yakubovich, A. Solov'yov, Eur. Phys. J. D **60**, 101 (2010)
12. B. Cassie, A. Wroe, H. Kooy, N. Depauw, J. Flanz, H. Paganetti, A. Rosenfeld, Med. Phys. **37**, 311 (2010)
13. L. Sanche, Eur. Phys. J. D **35**, 367 (2005)
14. F. Tobias, M. Durante, G. Taucher-Scholz, B. Jakob, Mutat. Res. **704**, 54 (2010)

Chapter 18
Track-Structure Monte Carlo Modelling in X-ray and Megavoltage Photon Radiotherapy

Richard P. Hugtenburg

Abstract The use of track structure calculations in radiotherapy using conventional low-LET radiation sources is discussed. Microdosimetry and emergent nanodosimetry methods are considered in explaining variations in quality factors associated with clinical practice and in vitro data. Transformation rate in the human derived for the in vitro system CGL1 is presented as a model for the induction of secondary cancer, a late effect associated with radiotherapy treatment.

18.1 Introduction

The use of radiation in cancer therapy is highly developed and has been shown to be successful only with high orders of accuracy [1]. The principal measure of radiation efficacy is the absorbed dose, however an increasingly wide variety of radiation sources are now used in radiotherapy and the assumption that the clinical response is dependent only on the absorbed dose is inaccurate in many cases. Despite the advantages of new modalities, including better normal tissue sparing, and less sensitivity to oxygenation effects, e.g. from ion-beam therapy or biological targeting, it is likely that similar, high orders of accuracy are needed in the determination of effective dose.

Physical models that go beyond the fundamental quantity of dose are increasingly being used in radiation therapy in order to better understand and to optimally design treatments. This has been driven by the adoption of new modalities, such as ion-beam therapy and boron neutron-capture therapy, where corrections to the dose to determine clinical efficacy are substantial. These models utilise models based on calculated linear energy transfer (LET) or microdosimetric quantities such as the

---

R.P. Hugtenburg (✉)
College of Medicine, Swansea University, Swansea, SA2 8PP United Kingdom
e-mail: r.p.hugtenburg@swansea.ac.uk

G.G. Gómez-Tejedor and M.C. Fuss (eds.), *Radiation Damage in Biomolecular Systems*, Biological and Medical Physics, Biomedical Engineering,
DOI 10.1007/978-94-007-2564-5_18, © Springer Science+Business Media B.V. 2012

lineal energy. Both measures have been shown to be highly valuable in characterizing the damage from high-LET sources such as neutrons and ion-beams. As well as being measurable with a tissue-equivalent proportional counter (TEPC), lineal energy is a more useful quantity for low-LET radiation as tracks tend to be highly contorted.

Tumour control in radiotherapy is largely dictated by biological processes that occur in the short-term, such as radiation-induced apoptosis of tumour cells, and in general the models describing these processes for a wide range of radiation types are reasonable. Lindborg and Grindborg [2] utilised microdosimetry measurements and the experience gained in clinical practice in moving from 250 kV X-rays to Co-60 sources with an associated decrease in the quality factor of around 25%, and to neutron radiotherapy, was used to estimate the size of the critical structures relevant in a microdosimetric approach. The study indicates a sensitive structure in the range 6-9 nm, which are comparable with the diameter of individual strands of DNA, these results contrast with models that are based on micron-sized volumes, i.e. critical structures that are on the scale of the entire genome.

In the case of long-term effects such as cancer induction, radiation damage is necessarily complex, depending on the ablation of spatially distant regions of the genome, e.g. of multiply-redundant tumour suppression and repair mechanisms. Microdosimetry is of considerable value in this context, in predicting results in a wide variety of in vitro experimental studies. In particular it has been shown that for a wide range of radiation types and biological endpoints, the relative biological response is closely proportional to the dose-weighted average lineal energy, y_D. This has been explained in terms of the production of sub-lesions, such as DNA strand breaks, through energy deposition and their subsequent chemical interaction to form a critical lesion. The probability of two sub-lesions interacting is proportional to the square of the concentration of sublesions, i.e. y^2.

This theory of dual radiation action [6] potentially provides a mechanistic basis for the linear-quadratic model, which has its most important utility in radiotherapy practice. A debate, primarily aired in the American journal Medical Physics [7–11] concluded that although the mechanisms believed to underpin biological response at high doses are still uncertain, and perhaps beyond physical models, the linear-quadratic model is, at least, valuable in its utility in radiotherapy.

There is perhaps a begrudging acceptance of the utility of these models and several authors have presented microdosimetric arguments for determining the effectiveness of kilovoltage X-ray interoperative radiotherapy systems [12, 13] in comparison with conventional therapeutic sources, including evaluating variation in effective dose with depth. Verhaegen *et al.* have examined the effective dose associated with brachytherapy seeds and intensity modulated radiotherapy (IMRT) with a linear accelerator [14–16]. Recent commentaries remind us of the many problems that still remain with physical modelling of therapeutic effect [17–21].

The ICRU recommends that a quality factor of 1 be employed for radiation up to an LET of 10 keV/micron. Recently it has been shown that the quality factor for photons and Auger electrons varies substantially from unity in certain circumstances. Data from studies utilising in vitro assays that are dependant on complex damage, including micronuclei induction, dicentric formation and transformation, show greater variation in radiobiological effects than predicted by LET and microdosimetry models. Meta-studies of the available data for dicentric formation and micronuclei induction suggest that the quality factor for cancer induction may be significantly increased for radiation of LET of 3 keV/micron or more [22, 23] with the associated public health consequences for low-energy diagnostic X-rays, such as used in mammographic screening for breast cancer.

Part of the complexity in these studies relates to the lack of consensus over what is the appropriate reference source. The references typically chosen are Co-60 gamma-rays and 250 kVp X-rays. Pattison *et al.* [24, 25] have suggested that a linear accelerator-based source can achieve close correspondence to the gamma component of the atomic bomb detonations at Hiroshima and Nagasaki, thereby linking the effective dose to this important epidemiological dataset in the Life-Span Study (LSS). Other workers confirm the importance of geometry and scattering in vivo and organ type related to the choice of reference [26–28].

In vitro transformation has an important relationship to cancer induction, in that it is believed that the cell transforms from normal rate of cell division to an abnormal rate, characterising the transformed state, when specific tumour suppressor complexes are ablated by radiation. The degeneracy of tumour suppression systems in cells is regarded as an important threshold in the progression of normal cells to a cancerous state. The X-rays from mammography X-rays systems yield a particularly high relative biological response (RBE) to high-energy X-ray references in transforming in vitro systems such as CGL1, suggested to be in excess of 4.

In extrapolating these results to real-world risk factors there are a number considerations. Firstly, CGL1 differs from normal cells, in that it is a tetraploid cell, containing two copies of the human genome. CGL1 consists of the HeLa cell-line combined with a normal human fibroblast, and possesses multiply redundant cellular mechanics. Secondly the doses from mammography examinations are notably smaller than the doses needed to observe significant changes transformation from the background transformation rate in CGL1. There is however considerable value in developing reliable physical models, they are likely to be valuable in the high dose context of radiotherapy, and can be modified according to known biological structures and redundancy.

Attempts at modelling effective dose for mammographic X-rays by Verhaegen *et al.*, utilising microdosimetric principles, determined RBE that were somewhat smaller and in contradiction with the data from recent studies, including transformation and dicentric formation. Does this data from low-LET, i.e. photon and electron, sources suggest further problems for microdosimetric theory in circumstances where is has been considered to be robust in the past This work, in focusing on the mechanistic basis for secondary cancer induction in radiotherapy, examines these findings and possible modifications, in the light of recent progress in predicting and measuring radiation damage at the nanoscale.

18.2 Microdosimetry principles

The lineal energy spectrum of a radiation source refers to the probability that a certain amount of energy will be deposited in a specified volume. The lineal energy quantity has an advantage over LET in that it can be measured directly with a TEPC, however difficulties arise in measuring effective volumes smaller than a few hundred nanometres, i.e. on the scale of individual strands of DNA. Recently workers have tended to concentrate on calculating the quantity via Monte Carlo track structure modelling.

Track structure calculations have mostly been performed with the use of dedicated microdosimetry codes designed utilising uniform media and a selected range of materials, often just liquid and gaseous water. In recent years the use of general purpose Monte Carlo codes has been demonstrated [29–31] and these are particularly valuable in circumstances where inhomogeneities are modelled or when a wider range of materials are need, e.g. in the case of binary therapies [32–35].

The lineal energy, y, is defined as the energy deposited, ϵ divided by the mean cord-length, $\bar{l}$, for the volume of interest, i.e.,

$$y = \frac{\epsilon}{\bar{l}}. \tag{18.1}$$

For convex objects $\bar{l} = 4V/S$ where V is the volume and S the surface area, reducing to $2/3d$ for spheres of diameter d.

The probability that an event occurring in the volume of interest has lineal energy in the range, $[y, y+dy]$, is given by the probability density function (PDF) $f(y)dy$. Two moments of this distribution, the frequency, y_F and dose, y_D averaged lineal energies are useful;

$$y_F = \int y f(y) dy,$$
$$y_D = \int y^2 f(y) dy. \tag{18.2}$$

In particular y_F relates to the dose, D, per single event in a volume of mass, m, according to the relation,

$$D = \frac{n y_F \bar{l}}{m}, \tag{18.3}$$

where n, the average number of events associated with dose D, is used to determine the Poisson distribution of events occuring in the the specific energy distribution, $z(D)$. n also characterises the likelihood of multiple track events and therefore higher order processes, such as quadratic effects.

The work of Zaider and Brenner (1985) [36] and refined by an ICRU committee [37] showed that quality dependence on specific energy, $Q(z)$, and lineal energy, $Q(y)$ could be calculated from in vitro data. The quality factor is given as,

$$\overline{Q} = \frac{1}{D}\int yf(y)Q(y)dy, \tag{18.4}$$

where y is the lineal energy, the energy deposited in a volume of mean cord-length, $\overline{l}$. At high doses Q should be generalised to utilise the specific energy quantity, i.e.,

$$\overline{Q} = \frac{1}{D}\int zf(z)Q(z)dz. \tag{18.5}$$

Many in vitro studies have shown that $Q(z)$ is essentially linear in z, until z approaches 100 keV/micron where saturation effects become important [38]. That is $\overline{Q} \approx z_D$, and in the limit of zero dose $\overline{Q} \approx y_D$. These studies lend support to the theory of dual radiation action and the value of y_D in predicting relative biological effect (RBE) at low dose limits. The theory states that there is a quadratic dependence on z on the formation of critical lesions, where z is considered to the concentration of molecular sublesions in the relevant volume, i.e. a chemical process in the nucleus. The spectrum of sublesions is $E(z) = \beta z^2$. It can be shown that the average yield of lesions as a function of dose is,

$$\overline{E(z)} = \beta(z_D D + D^2) = \alpha D + \beta D^2. \tag{18.6}$$

where α and β are the coefficients usually associated with the linear-quadratic theory. Accordingly $\alpha/\beta = z_D$, and in the limit of zero dose $\alpha/\beta = y_D\,\overline{l}/m$. The theory says nothing about the value of the β coefficient and is therefore independent of radiation type within the microdosimetric framework. y_D is often said to be proportional to α, however this may not be appropriate given the larger uncertainties asssociated with determining β, and its influence on the calculated value of α for the high-doses typically needed for in vitro studies.

18.3 Calculating high-dose response: Specific energy

Lineal energy spectra and moments calculated in water from in-air spectra for a range of X-ray energies are published in the PhD thesis of Verhaegen (Ghent U.) who utilised the TRION track-structure code [39]. This data has been used as the basic input data for single event, lineal energy spectra. TRION utilises gaseous water cross-sections. More recently TRION has been been updated to include liquid water cross-sections [40] however these cross-sections are expected to be less reliable.

Specific energy has been calculated from lineal energy distributions with a Monte Carlo sampling method that is easily implemented in Matlab code. The Poisson distribution of events traversing the volume of interest is determined by calculating the mean number of traversing events, n, as a function of dose. The average number of independent tracks per Gy is shown in Table 18.1 for 10 and 1000 nm spherical volumes and for X-ray energies of 10,100 and 1250 keV.

Table 18.1 Average number of events per Gy, n, y_F and y_D in 10 and 1000 nm spheres for a range of photon energies

	10 nm			1000 nm		
Energy (keV)	n	y_F	y_D	n	y_F	y_D
10	8.6×10^{-5}	7.6	18.4	16.8	3.5	5.5
100	7.4×10^{-5}	6.7	16.7	2.7	1.79	4.1
1250	6.4×10^{-5}	5.7	14.8	1.42	0.29	1.79

Table 18.2 Predicted α/β ratios (Gy) for a range of X-ray photon beams in a 1000 nm and 500 nm spheres, the later providing better agreement with measured α/β ratios for transformation in CGL1

Energy (keV)	10	100	1250	28 keV Mo/Mo	6 MV linac
1000 nm	1.22	0.88	0.39	1.09	0.63
500 nm	6.6	5.1	3.3	5.9	4.5

The frequency distribution of y, $f(y)$, is sampled n times from a cumulative probability density function (CDF) and summed. The process is repeated in order to obtain adequate statistical accuracy and approximately 10^5 samples were made per dose point. This is a straightforward method of calculating the specific energy distribution, where $f(y)$ distribution can cover several orders of magnitude. Various numerical methods are given elsewhere [42, 43]. Table 18.2 gives the fitted α/β ratios for these sources. These are comparable with α/β ratios in the range 0.5 - 3 Gy experienced clinically [17].

18.4 Modelling CGL1 transformation with microdosimetric principles

Using published dose-weighted mean lineal energy y_D published for a mammography unit [15] and a linear accelerator [16], the frequency and dose weighted lineal energy y_f and y_D were obtained as shown in Table 18.2.

The standard microdosimetric model says that the value of β is constant for all radiation types. The published data on CGL1 transformation reports that β varies widely and is close to zero in the case of the mammography X-rays, however error ranges in the published data do support a constant value of β. Strong linear dependence is also a feature of high-LET radiation such as heavy ions and neutron sources, however it is reasonable that a β component may be difficult to determine accurately when α is large relative to β.

The linear quadratic components, α and β, have been published for several studies of CGL1 transformationas summarised in Table 18.3. It can be seen that the values of α/β predicted from the 1 micron diameter microdosimetric volume are somewhat lower and somewhat less dispersed than determined from these experiments. Larger dispersal will occur for a larger microdosimetric volume, however this is in contradiction with a smaller sensitive volume required to give the

Table 18.3 Refitted α/β ratios (Gy) for transformation in CGL1 assuming a constant value of β. The value of β is the average of the values measured across the sources considered, in accord with the assumption that β should not vary with radiation type in a microdosimetric model. Ranges for the β value determined by the authors are given in parentheses, indicating the typical level of uncertainty associated with the approach

Study	α/β (Gy)	β (Gy^{-2})	Source used
Heyes and Mill	7.2	0.055 (0.012 - 0.079)	MoMo 29 kVp X-rays
Heyes and Mill	4.5		$^{90}Sr/^{90}Y$ β-rays
Heyes and Mill	4.3		A-bomb simulation
Göggelmann *et al.*	15.8	0.070 (-0.012 - 0.071)	29 kVp MoMo X-rays
Göggelmann *et al.*	11.5		220 kVp X-rays
Frankenberg *et al.*	13	0.049 (0.026 - 0.071)	29 kVp MoMo X-rays
Frankenberg *et al.*	0.2		220 kVp X-rays

higher α/β ratios reported in this experiment. None-the-less the strong dependence of the α/β ratio on the mean chord length, $\bar{l}$, implied by equation 18.3, means that only a small change in the diameter of the sensitive volume (to 500 nm) gives better overall agreement with measured α/β ratios of the reference sources.

18.5 Discussion

It is difficult to discount the microdosimetric basis for the quadratic effect in these systems. In contrast the likelihood of multiple crossings occurring in nanoscopic volumes becomes vanishingly small in radiation effects that are dictated by mechanics on the nanoscale. The solution to larger dispersion in the response to low LET radiation of certain in vitro systems probably lies at nanoscopic dimensions. A clue relates to the breakdown in the linear dependence of the quality factor Q(y) in high LET beams, usually referred to as saturation. This is explained with the observation that above 100 keV/micron (100 eV/nm) there will be typically more than the two ionisations needed to create a double-strand break hence decrease in the efficiency of ionisation. The quality factor dependence used to correct for this is on one hand accounting for chemical effects on the micron scale, and on the other, ionisation on the nanoscale.

Much more of a focus has been made on ionisation in recent years. Firstly track-structure is now more confidently being calculated on the nanoscale, and there are emergent experimental methods able to count and quantify ionisation for a variety of materials nanoscopic gaseous volumes, including DNA constituents and analogues [44]. In these methods where the numbers of ionisation are relatively small the energy deposited is less apprpriate and instead the number of damaging ionisation events are counted. Table 18.1 also gives microdosimetric data for 10 nm diameter spheres. $y_F\bar{l}$ gives the average energy deposited per event and ranges from 38-51 eV for the sources considered. While y_D shows us that the variation in RBE calculated using conventional microdosimetric methods would be smaller, the energy deposition is very close to that of the W-value for water, i.e. the typical

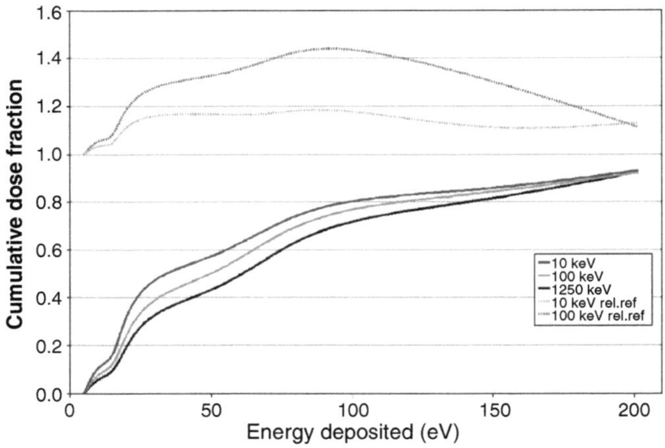

Fig. 18.1 Energy deposition cumulative with dose for 10, 100 and 1250 keV photons in a 10 nm spherical volume. A significant proportion of the energy deposited in nanoscale volumes by low-LET radiation occurs below the W-value for water and hence there is reduced likelihood of the small number of ionisations required, e.g. for SSB and DSB. The ratio of the amount of dose associated with energy deposition above a certain energy threshold is also plotted for 10 keV and 100 keV relative to 1250 keV photons as a reference radiation

amount of energy required to generate an ionisation event. It is therefore likely that many events in this energy-range fail to generate an ionisation event.

A correction for this effect can be envisaged by examining the proportion of dose occurring above a certain energy threshold. Figure 18.1 shows the proportion of dose for 10, 100 and 1250 keV beams occurring below energy that could be interpreted a certain ionisation threshold. Whether a single ionisation is required or two ionisations in order to form a double-strand break, differences in the rate of ionisation between these three radiation sources of 30 to 40% are present. This is an additional contribution to the RBE observed for soft x-rays relative to the X-ray reference radiation sources.

It is important to note that as these processes are occurring on the nanoscale, there cannot be multiple-track events and the correction will be independent of dose. Therefore a model that takes nanoscale effects into account must account for this separately to micron-scale processes that are quadratic in dose. This principle is also true for corrections made for saturation effects in heavy ions; that there is a microdosimetic model, describing the the overall rates of critical lesion formation and a nanodosimetric model defining the efficiency of ionisation. Strickly speaking it is not possible to decouple nanoscale processes from microdosimetric effects as track-ends can differ widely in their energy deposition. Track structure codes that provide coupled statistical distributions of energy and ionisation on the micron and nanoscale are needed.

18.6 Conclusion

Current work examines an important constraining aspect of radiotherapy, that of secondary cancer formation. Future direction in radiotherapy will depend the importance of high conformality versus low dose 'splash' with an increasingly younger population of cancer patients to consider, and increasing survival times.

Improvements in track-structure modelling in recent years and the failure of microdosimetric models to explain a variety of biological end-points, including those arising from new methodologies such as electrophoresis focussing on fragmentation of DNA, have meant the theory has gone out of favour. At the same time the need for bioeffect modelling in radiotherapy has grown and has continued to embrace methods that have arisen from microdosimetry, including the widespread utilisation of linear-quadratic theory, and the use of quality factors moving from conventional photon sources to new soft X-ray sources, ion and neutron beams. This is driven in part by an acknowledgement that radiotherapy needs to be performed accurately in order to achieve therapeutic gain, irrespective of whether the source of uncertainty is physical or biological in origin.

Data from transformation studies in low-LET sources show variation in the RBE that is somewhat larger than traditionally expected for photon sources, and predicting using traditional microdosimetric paradigms. While in vitro system like CGL1 have many differences to cancer induction in vivo, there are several valuable aspects to their use. Radiobiological effects in these studies are noted in a dose range that closely coincides to that of radiotherapy doses. Transformation is closely similar to carcinogenesis, in that in both cases the inactivation of tumour suppressing genes within the genome is an important, perhaps rate limiting step. Finally such models can be scaled geometrically to match the size and distribution of known tumour-suppressor complexes utilising, for example, recent findings relating to chromosome distribution [45].

Transformation assays provide further evidence of the shortcomings in the microdosimetry model, they also suggest the manner in which a more complete theory will be arrived at; that there is value in the microdosimetric paradigm. Microdosimetry is needed to explain quadratic effects observed in vitro and clinically, but often processes of damage on the scale of the DNA, i.e. nanometres, also require consideration. A fact observed by H.H. Rossi nearly 20 years ago [46].

Acknowledgements Frank Verhaegen is acknowledged for providing his thesis and the basic data on lineal energy spectra therein. Geoff Heyes, Alan Beddoe and Ihsan Al Affan are thanked for their helpful discussion.

References

1. B.J. Mijnheer, J.J. Battermann, A. Wambersie, Radiotherapy and Oncology, **8**, 237–252 (1987)
2. C. Lindborg, J.-E. Grindborg, Radiat. Prot. Dosim. **70**, 541–6 (1997)

3. W. Friedland, P. Jacob, H.G. Paretzke, M. Merzagora, A. Ottolenghi, Radiat. Environ. Biophys. **38** 39–47 (1999)
4. H. Nikjoo, P. ONeill, W.E. Wilson, D.T. Goodhead, Radiat. Res. **156** 577–83 (2001)
5. Y. Hsiao, R.D. Stewart, Phys. Med. Biol. **53**, 233–244 (2008)
6. A.M. Kellerer, H.H. Rossi, Radiat. Res. **75**, 471–488 (1978)
7. M. Zaider, Med. Phys. **25**, 791–792 (1998)
8. R.K. Sachs, D.J. Brenner, Med. Phys. **25**, 2071–2073 (1998)
9. M. Zaider, Med. Phys. **25**, 2074–5 (1998)
10. D.J. Brenner, M. Zaider, Med. Phys. **25**, 1055–1057 (1998)
11. P. Pihet, H.G. Menzel, Med. Phys. **26**, 848–852 (1999)
12. D.J. Brenner, C.-S. Leu, J.F. Beatty, R.E. Shefer, Phys. Med. Biol. **44**, 323–333 (1999)
13. M. Zaider, L. Hanin, Med. Phys. **38**, 574–583 (2011)
14. F. Verhaegen, I.A. Castellano, Radiat. Prot. Dosim. **99**, 393–6 (2002)
15. B. Reniers, F. Verhaegen, Radiat. Prot. Dosim. **122**, 401–3 (2007)
16. H.H. Liu, F. Verhaegen, Radiat. Prot. Dosim. **99**, 425–7 (2002)
17. D.J. Carlson, R.D. Stewart, X.A. Li, K. Jennings, J.Z. Wang, M. Guerrero, Phys. Med. Biol. **49**, 4477–4491 (2004)
18. R.D. Stewart, X.A. Li, Med. Phys. **34**, 3739–3751 (2007)
19. D.J. Brenner, Semin. Radiat. Oncol. **18**, 234-239 (2008)
20. J.P. Kirkpatrick, J.J. Meyer, L.B. Marks, Semin. Radiat. Oncol. **18**, 240-243 (2008)
21. H.T.Jr. Peterson, J. Radiol. Prot. **30**, 801–811 (2010)
22. G.J. Heyes, A.J. Mill, M.W. Charles, J. Radiol. Prot. **29**, A123–A132 (2009)
23. N. Hunter, C.R. Muirhead, J. Radiol. Prot. **29**, 5–21 (2009)
24. J.E. Pattison, L.C. Payne, R.P. Hugtenburg, A.H. Beddoe, M.W. Charles, Radiat. Prot. Dosim. **109**, 175–180 (2004)
25. J.E. Pattison, R.P. Hugtenburg, A.H. Beddoe, M.W. Charles, Radiat. Prot. Dosim. **95**, 125–135 (2001)
26. L. Walsh, W. Rühm, A.M. Kellerer, Radiat. Environ. Biophys. **43**, 145–151 (2004)
27. D. Harder, N. Petoussi-Henß, D. Regulla, M. Zankl, G. Dietze, Radiat. Prot. Dosim. **109**, 291–295 (2004)
28. A.M. Kellerer, H. Roos, Radiat. Prot. Dosim. **113**, 245–250 (2005)
29. R.D. Stewart, W.E. Wilson, J.C. McDonald, D.J. Strom, Phys. Med. Biol. **47**, 79–88 (2002)
30. C.S. Liu, C.-J. Tung, Y.H. Hu, C.M. Chou, T.C. Chao, C.C. Lee, Nucl. Instrum. Meths. B. **267**, 1823–9 (2009)
31. T.C. Chao, Y.S. Huang, F.Y. Hsu, Y. Hsiao, C.C. Lee, C.J. Tung, Radiat. Prot. Dosim. **143**, 248–252 (2011)
32. R.P. Hugtenburg, Z. Chaoui, J.E. Pattison, Nucl Instrum. Meths. A **580**, 157–160 (2007)
33. R.P. Hugtenburg, Eu. J. Radiol. **68**, S126–8 (2008)
34. R.P. Hugtenburg, A.E.R. Baker, S. Green, Appl. Radiat. Isotop. **67**, 433–5 (2009)
35. R.P. Hugtenburg, A.E.R. Baker, S. Green, Appl. Radiat. Isotop. **67**, S168-170 (2009)
36. M. Zaider, D.J. Brenner, Radiat. Res. **103**, 302–316 (1985)
37. ICRU: Report 40. The quality factor in radiation protection. International Commission of Radiation Units and Measurements, Bethesda, MA, (1986)
38. R.J. Cannell, D.E. Watt, Phys. Med. Biol. **30**, 255-258 (1985)
39. A.V. Lappa, E.A. Bigildeev, D.S. Burmistrov, O.N. Vasilyev, Radiat. Environ. Biophys. **32**, 1–19 (1993)
40. E.A. Bigildeev, V. Michalik, Radiat. Phys. Chem. **47**, 197–207 (1996)
41. W. Friedland, M. Dingfelder, P. Jacob, H.G. Paretzke, Radiat. Phys. Chem. **72**, 279–86 (2005)
42. H.H. Rossi, M. Zaider, Microdosimetry and its applications, Springer-Verlag, Berlin (1996)
43. A. Mozumder, Y. Hatano, (Eds.), Charged Particle and Photon Interactions with Matter. Chemical, Physicochemical and Biological Consequences with Applications, Marcel Dekker, New York, (2004)

44. G. Garty, S. Shchemelinin, A. Breskin, R. Chechik, G. Assaf, I. Orion, V. Bashkirov, R. Schulte, B. Grosswendt, B.: Nucl. Instrum. Meth. A, **492** 212–235 (2002)
45. M.N. Cornforth, K.M. Greulich-Bode, B.D. Loucas, J. Arsuaga, M. Vázquez, Sachs, M. Brückner, M. Molls, P. Hahnfeldt, L. Hlatky, D.J. Brenner, J. Cell Biol. **159** 237-244
46. H.H. Rossi, Radiation Protection Dosimetry **52**, 9–12 (1994)

Chapter 19
Simulation of Medical Linear Accelerators with PENELOPE

Lorenzo Brualla

Abstract Monte Carlo simulation of clinical linear accelerators (linac) allows accurate estimation of the absorbed dose in a patient. However, its routine use in radiotherapy units has been hindered by the difficulties related to efficient programming of the simulation files and the usually long computation times required. PENELOPE is a Monte Carlo general-purpose radiation transport code that describes the coupled transport of photons, electrons and positrons in arbitrary materials and complex geometries. Although PENELOPE by itself is perfectly suited for the simulation of linacs, it nevertheless imposes a programming effort on the end users wishing to do so. In this chapter a brief review is given on several programs that facilitate the simulation of linacs and computerised tomographies using PENELOPE as the Monte Carlo engine. Variance-reduction techniques implemented in these codes, which allow an efficient simulation of linacs, including multileaf collimators, are also described. The chapter ends with an example of a simulation with PENELOPE of a linac irradiating a highly conformed small electron field used for the treatment of the conjunctival lymphoma of the eye. The example shows the simulation of a linac and a computerised tomography of a segmented eye.

19.1 Brief description of a linac

Medical linear accelerators (linacs) are routinely used in radiotherapy units for the treatment of cancer. The principle on which all linacs are based is the same: to accelerate electrons through resonant cavities to energies on the order of a few MeV [1]. The pencil beam leaving the accelerating structure is nearly monoenergetic with a diameter of about 1 mm. In general, Monte Carlo simulations

---

L. Brualla (✉)
Strahlenklinik, Universitätsklinikum Essen, Germany
e-mail: lorenzo.brualla@uni-due.de

G.G. Gómez-Tejedor and M.C. Fuss (eds.), *Radiation Damage in Biomolecular Systems*, Biological and Medical Physics, Biomedical Engineering,
DOI 10.1007/978-94-007-2564-5_19,

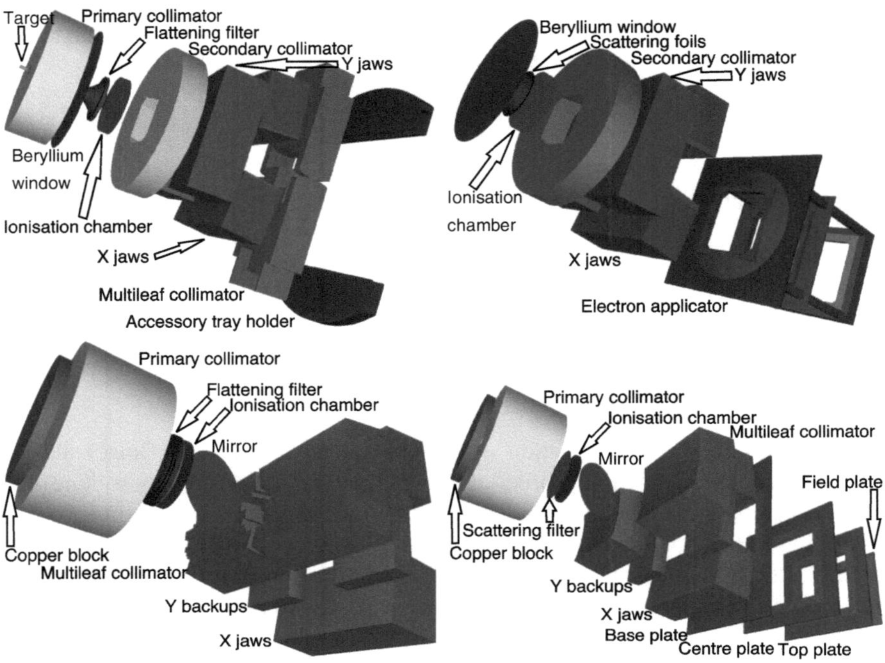

Fig. 19.1 Drawings of: Varian Clinac 2100 C/D operating in photon mode at 18 MV (upper left) and electron mode at 6 MeV (upper right), Elekta MLCi operating in photon mode at 10 MV (lower left) and electron mode at 4 MeV (lower right)

start from that position in the linac head, assuming as primary electron source a pencil beam with given spatial and energy distributions. Particles are then simulated downstream of the linac head. Therefore, from a Monte Carlo simulation point of view, the relevant constructive elements of the linac are those found downstream of the primary electron source.

Some linacs operate only with electron beams (e.g., Siemens Mevatron ME), others with photon beams (e.g., Varian Clinac 600 C/D), while others can operate either with electron or photon beams (e.g., Varian Clinac 2100 C/D). Those irradiating with electron beams usually include some thin material layers downstream of the primary source, called scattering foils, whose purpose is to spread the pencil beam and hence to cover a large field. Linacs irradiating with photon beams have a thick material target, usually made of tungsten, just downstream of the primary electron source. This target produces photons by bremsstrahlung emission. In many cases a flattening filter is placed in the position of the scattering foils in order to homogenize the energy distribution of the emitted photons. Downstream of the aforementioned constructive elements a series of collimating structures are found whose purpose is to conform the beam to the required field shape. Figure 19.1 shows four drawings of the constructive elements of the Varian Clinac 2100 C/D and Elekta MLCi operating in photon and electron modes.

The first Monte Carlo simulation of a medical linac was done by M. Udale in 1988 [2]. A few years later D.W.O. Rogers and co-workers introduced the code BEAM for simulation of medical linacs [3]. After these pioneering works, the field has observed an exponential growth in the number of publications related to Monte Carlo simulation of linacs. Such growth has derived in efforts to implement Monte Carlo simulations in the clinical routine within radiotherapy units [4, 5]. The first simulation of a linac with PENELOPE appeared in 2001 [6]. After that work, the PENELOPE code system has been increasingly used for linac simulation (e.g., [7–11]).

19.2 The PENELOPE code system

PENELOPE is a set of FORTRAN subroutines written for performing general-purpose Monte Carlo simulations of radiation transport in arbitrary materials and complex geometries [12, 13][1]. The code can simulate the coupled transport of electrons, photons and positrons in the energy range from 50 eV up to 1 GeV. PENELOPE includes a material database with the first 99 elements of the periodic table as well as 181 compounds. Should a compound not present in the database be required, this can be created by means of its stoichiometric formula.

Geometries in PENELOPE can be programed with the code PENGEOM included in the distribution. With PENGEOM it is possible to define geometries by grouping quadric surfaces to form bodies. Bodies, in its turn, can also be grouped to form more complex structures. When dealing with elaborate geometries, such as a linac, the way bodies are defined and grouped is critical in terms of simulation speed.

PENELOPE is a subroutine package, therefore, end users are responsible for writing a steering main program. Examples of main programs are provided with the distribution package. Although this approach is general enough for dealing with a wide range of applications, it imposes a burden on users that only need to simulate some specific linacs. These users are required to write their own geometry file and to code or adapt a main program. These tasks are error-prone and require some knowledge of physics, Fortran programing and PENGEOM syntax. In order to ease these tasks several codes have been published. The following codes are not part of the PENELOPE distribution, however they are useful for users interested in simulating linacs.

[1]Distributed at the OECD Nuclear Energy Agency Data Bank (http://www.nea.fr) and the Radiation Safety Information Computational Center (http://www-rsicc.ornl.gov).

19.3 PENEASY: a modular main program

PENEASY [26] is a modular, general-purpose main program for PENELOPE that includes various source models and tallies. The rationale is to provide a tool suitable for a wide spectrum of applications so that users do not need to develop a specific code for each new case. For those cases where these models are insufficient and some additional coding or adaptation needs to be done, its modular structure is designed to reduce programming effort to a minimum. The code is free, open software[2] mostly written in FORTRAN 77, although it has recourse to some extensions included in the Fortran 95 standard.

A convenient feature is the possibility of initialising the pseudo-random number sequence with seeds read from an external file. This can be used in conjunction with the package of Linux scripts CLONEASY [14] to parallelise execution in a straightforward way, without altering the code or the input file—except for the field defining the seeds. It is also interesting to remark that some parameters, such as the maximum number of histories or the allotted simulation time, can be changed by the user during execution by sending messages via an external file that is read at regular time intervals.

PENEASY frees the end user of PENELOPE from the task of writing a main program, however, the code still needs for its execution a configuration file, a material file and a geometry file.

The configuration file contains the following information: (i) the maximum number of histories to simulate and the allotted time; (ii) the integer numbers (seeds) to initialise PENELOPE's pseudo-random number generator; (iii) the parameters for the source models (e.g., a binned energy spectrum); (iv) the names of the geometry files; (v) the name of the materials data file and the transport parameters for each material; (vi) the configuration parameters for each tally; and (vii) the parameters for the application of variance-reduction techniques. Although the code provides a layout of the configuration file which only requires editing work in order to adapt it to a particular simulation, it is necessary to have some knowledge on the physics of radiation transport and the intricacies of Monte Carlo simulation in order to produce a simulation that computes meaningful results in a reasonable time.

The material file can be generated by means of the `material.f` program distributed with PENELOPE and its execution poses no major problem. However, writing the geometry file is a delicate task.

While simulating geometries by grouping quadric surfaces is a general enough approach for a large number of applications, it nevertheless imposes a limitation on the shape of the objects that can be represented. A field in which quadric surfaces are not well suited is medical physics problems that require the simulation of realistic anatomic structures. In these cases the geometry is usually represented in terms of a uniform grid of homogeneous parallelepipedic volume elements, or voxels.

[2]The code can be downloaded from http://www.upc.es/inte/downloads/penEasy.htm.

Voxelised models of the patient's anatomy can be obtained from computerised tomography (CT) scans, which can be processed to obtain the approximate chemical composition and mass density of each voxel [5, 15]. PENEASY, includes a set of geometry subroutines that allow the simulation of objects formed by combinations of quadric surfaces and voxels, or by voxels alone.

19.4 PENLINAC: a geometry pre-processor for PENGEOM

PENLINAC [27][3] is a code package intended to customise PENELOPE for the simulation of treatment machines in radiotherapy. The package is basically a set of subroutines that allows defining radiation sources and creating binary phase-space files. Phase spaces are evaluated via a program which exploits the abundant particle data saved, to permit filtering analyses by diverse criteria, such as nature of the particle, energy, region of impact on the tally plane, material and site of origin, material and site of last interaction, among others. Nevertheless, the PENLINAC probably most distinctive feature relies on simplification of the creation of geometries.

PENLINAC introduces a method to construct the geometry targeted on the machine head assembly components. Components are reproduced by shaping predefined structures. There is a set of such basic structures; each one designed to better suit one particular machine piece e.g. a stack of concentric truncated cones is meant to construct flattening filters. So, building a component is a matter of first determining the most appropriated structure and then defining its shape. This is done by writing—in a formatted text file—the real physical dimensions, location and materials of the component. The idea is to avoid dealing directly with the model used by PENGEOM, as it becomes quite intricate for a treatment machine. However, compatibility is maintained by providing a program to transform the definition of components to its representation in the form of bodies delimited by quadric surfaces of PENGEOM. Thus, the final geometry can be visually debugged and used for PENELOPE simulations.

The package also contains a main simulation program incorporating the set of subroutines to handle sources and phase spaces and a fast program to calculate dose distributions in homogeneous media. The latter optimises the transport by saving the time spent by other simulation programs in solving quadric equations to transport particles through the geometry.

[3] The code can be downloaded from http://sites.google.com/site/penlinacusers/.

19.5 PENEASYLINAC: a code for automatic simulation of linacs

PENEASYLINAC [26] is a Fortran program that automatically generates the configuration, material and geometry files required for simulating a given linac with PENELOPE/PENEASY. The code does not require the user to know the geometrical details and the most adequate simulation parameters (initial beam, absorption energies, variance-reduction techniques, etc.) A library of preprogrammed parts of linacs with movable accessories such as multileaf collimators, electron applicators and wedges is coded inside PENEASYLINAC, together with a set of adaptive variance-reduction techniques. Currently a wide range of Varian and Elekta linacs are programed in PENEASYLINAC. The code is an evolution of AUTOLINAC [16].

19.6 Variance-reduction techniques

Results obtained with Monte Carlo simulations are essentially exact. Their quality only depends on the interaction models employed, i.e., differential cross sections, and the transport algorithms. However, any Monte Carlo estimated quantity has an associated statistical uncertainty. The statistical uncertainty depends on the number of histories that have been simulated. The statistical uncertainty can be reduced to zero using an infinite sequence of particle histories. Statistical uncertainty is the major drawback of Monte Carlo methods. However, it is possible to reduce to some extent the uncertainty of a Monte Carlo estimated quantity without recurring to a larger sequence of particle histories, and therefore larger simulation times. The techniques that allow this uncertainty reduction are known as variance-reduction techniques [17, 18].

In order to apply variance-reduction techniques in a given simulation, each particle must be associated with an adimensional number called the statistical weight. This is not a physical quantity, but indicates the contribution of that particle to a quantity being estimated. In case of not using variance-reduction techniques all particles have a statistical weight equal to 1, that is, all particles contribute equally to the quantity being tallied.

In this section only the variance-reduction techniques that are already implemented in the aforementioned codes will be discussed. That is, the variance-reduction techniques discussed here are those that do not require programming effort from the user.

Interaction forcing: It consists of artificially increasing the interaction cross section for a given material and interaction mechanism [12]. In that way particles being transported through that material will interact more often than they would normally do. In order to avoid a bias in the estimated results, the statistical weight of the descendent particles produced in a forced interaction have their statistical weight reduced.

Interaction forcing is particularly useful when a better statistics is required on particles travelling through thin material layers or low density materials. For example, water is nearly transparent to photons on the order of a few MeV, so lower uncertainties are obtained when Compton interactions are forced.

Range rejection: Charged particles that have travelled far away from the zone of interest and whose chance of contributing to the tallied quantities is negligible are discarded so to avoid loosing computation time while simulating them. The technique can be combined with Russian roulette (see below) in order to keep the simulation unbiased.

Russian roulette: It is about discarding particles whose probability to contributing to the tallied quantities is low, or whose contribution to the tallied quantities will be low due to a low statistical weight. Particles are eliminated from the simulation with probability $K<1$. In order to keep the simulation unbiased, particles that survive a Russian roulette have their weight increased by a factor $1/(1-K)$.

Splitting: A particle which has a large probability of contributing to the final scored quantity is split, that is, n copies of it are made and then each of this copies is independently simulated. The statistical weight of each split particle is multiplied by a factor $1/n$ in order to keep the simulation unbiased. Figure 19.2 shows a graphical representation of splitting applied to the simulation of a Varian Clinac 2100 C/D.

Rotational splitting: It is applied to particles travelling through geometries with cylindrical symmetry when the primary source has also the same kind of symmetry. This is the case of Varian Clinacs from the primary source downstream to the ionisation chamber. It is a kind of splitting in which each split particle is rotated about the central beam axis an azimuthal angle [19]. The azimuthal angle between two neighbouring split particles is given by $\Delta\phi = 2\pi/n$. The particle direction cosines of each replica are transformed in order to keep the direction of the original particle relative to the central beam axis. See Fig. 19.2.

Fan splitting: It is a particular case of rotational splitting in which the particles that have travelled through a cylindrically symmetric geometry are split in a preferred direction. Fan splitting is suited for off-axis fields [26]. The rationale is to split particles in the direction of the off-axis field and hence increase the speed at which the final score quantity is tallied. Figure 19.2 shows how particles are split in a preferred direction covering the off-axis field.

Movable skins: The movable-skin method consists of defining relevant zones of the geometry in which an accurate transport of radiation is performed, whereas in less relevant zones the transport of some particles is discontinued [16]. By dividing collimating structures, such as, collimators, jaws, multileaf collimators, etc. in zones located near and away from the beam axis it is possible to simulate particles that will contribute to the penumbra and, at the same time, simplify the transport of particles that have a small chance of leaving the corresponding body. The thickness of the zone in which accurate simulation is performed, that is the skin, depends on the kind of particle and the energy. The surfaces dividing skin from non-skin regions can be moved in order to adapt the thickness to a given configuration of the linac.

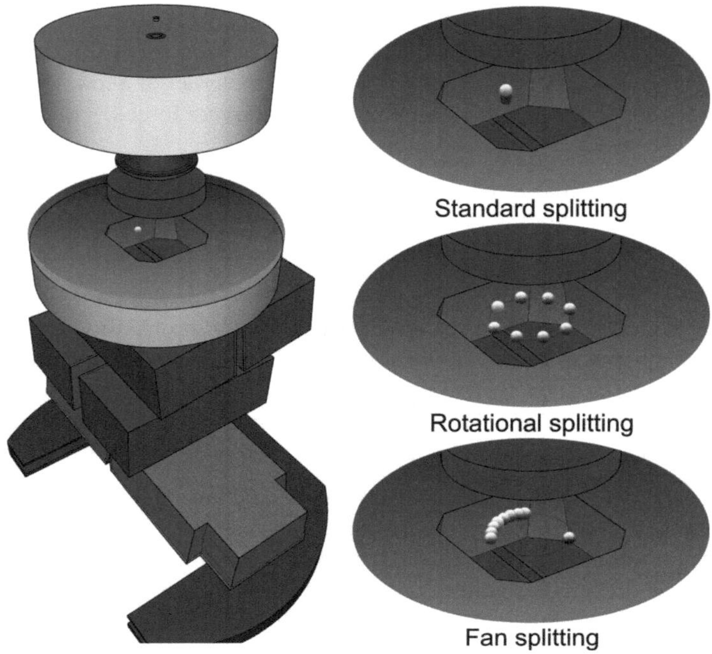

Fig. 19.2 Graphical representation of the three kinds of splitting techniques (right panel). The plane where splitting is applied is depicted as a translucent surface upstream the secondary collimator of the Varian Clinac 2100 C/D shown on the left panel

19.7 Efficient coding of geometries with PENGEOM

Although coding the geometry is not a variance-reduction technique by itself, the efficiency of a simulation largely depends on the way a geometry is coded. Depending on the logical structure used for a given geometry its simulation can be feasible or insurmountable. This is particularly true for the case of multileaf collimators (MLC). A MLC is a device made up of individual 'leaves' of a high atomic number material, usually tungsten, that can move independently in and out of the beam in order to block and form complex shapes. Typically a MLC consists of several tens of leaves, sometimes more than hundred. The complexity of a MLC in terms of the number of simulated parts as well as due to the shape of the leaves, demands for special coding strategies.

In PENGEOM quadric surfaces are grouped to form volumes. These volumes can be defined as BODY or MODULE. A BODY can be limited by surfaces and other BODYs, whereas MODULEs can only be limited by surfaces, but they can contain other BODYs and MODULEs inside. Faster simulations are achieved when only MODULEs are used and when every MODULE contains the smaller number of MODULEs inside, ideally only two. A geometry written that way is said to be hierarchically very ramified.

In Brualla et al. [16] the coding of the Varian MLC with 52 leaves is described. In the geometry of that MLC movable-skins are included. Usage of movable-skins renders the simulation of the MLC 12 times more efficient with respect to a simulation without them.

19.8 Measuring the performance of a simulation

Comparing the overall performance of two MC simulations is a non-straightforward task. They can be compared based on a number of quantities. However, conclusions reached on comparing a single figure could be misleading regarding the overall performance of the two simulations under scrutiny.

The simulation time, the number of particles that reach a scoring plane, the number of simulated histories per second and the efficiency of the requested estimators are figures that are automatically given by PENEASY. The efficiency ϵ of an estimated quantity is computed in PENEASY using the definition

$$\epsilon = \left(\frac{\bar{Q}}{2\sigma_Q}\right)^2 \frac{1}{T},$$

where $\bar{Q}$ is the average of the estimated quantity, σ_Q the standard deviation of $\bar{Q}$ and T the CPU time of the simulation. A 'history' refers to a primary electron entering the linac head and all the secondary particles generated by it.

In simulations where the absorbed dose is estimated in gridded or voxelised structures, twice the standard statistical uncertainty of the dose at each bin or voxel is also automatically obtained. For the overall comparison of two dose distributions the gamma index can be used [20, 21].

In PENELOPE doses are expressed in units of eV/g per primary electron. One eV/g equals to one Gy/(mA·s), so knowing the current exiting the exiting the vacuum window in mA and the irradiation time in s, the dose in Gy is obtained.

Comparing two simulations in terms of the tallied phase-space files (PSF) poses additional problems, in particular when variance-reduction techniques are used. Comparing the simulations either in terms of particles that have reached the scoring PSF, or the simulation time, or the number of simulated histories could be meaningless. The latent variance of a PSF provides a figure of merit useful for comparing PSFs [6]. For obtaining the latent variance several simulations must be run for a given PSF, using for each simulation a larger splitting number.

19.9 Simulating a linac and a CT geometry

This section shows the simulation of a Varian Clinac 2100 C/D used for the irradiation of a conjunctival lymphoma of the eye with a 6 MeV electron beam. Radiotherapy of ocular tumors involves very small target volumes and regions

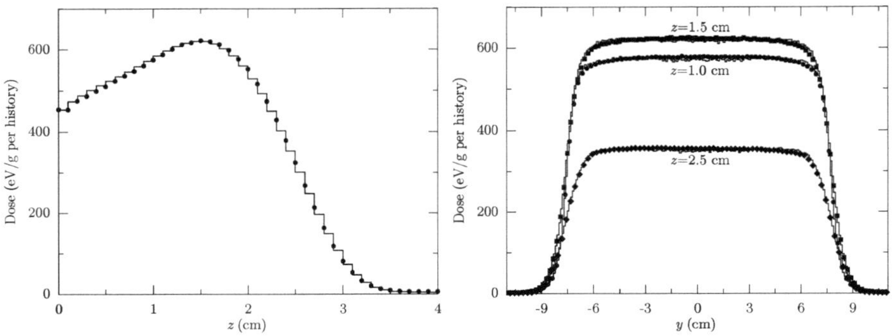

Fig. 19.3 Absorbed depth dose (left) and lateral dose profiles (right) for an electron beam nominal energy of 6 MeV and a field size of 15 × 15 cm^2. Experimental measurements are shown with symbols. Simulation results obtained with PENELOPE are shown with histogram. Standard deviation bars are not shown being smaller than the symbol size

of interest, and therefore needs dedicated treatment techniques. This is especially challenging in the treatment of stage I Non-Hodgkin lymphomas of the conjunctiva where external beam radiotherapy of the conjunctiva is the sole conservative and curative approach [22]. The conjunctiva is a complex target volume. It covers the anterior part of the eyeball excluding the cornea, reflects close to the equator to form the fornix and covers the inner side of the lids. This topography results in a form similar to a hemispherical shell that is extended from the surface up to a depth of approximately 1.5 cm.

Before simulating the specific field used for the irradiation of the conjunctiva, it is necessary to adjust the primary electron beam parameters so to reproduce experimental absorbed dose data in water from a reference field. For that purpose, a simulation of the Varian Clinac 2100 C/D at 6 MeV nominal energy with the standard 15 × 15 cm^2 electron applicator is run. A drawing of the the simulated geometry is shown in the upper right panel of Fig. 19.1.

The simulated results shown in Fig. 19.3 are obtained from a monoenergetic pencil-beam point-like primary electron source of 7.184 MeV. The agreement between experimental and simulated data is better than 1% of the maximum dose, with a distance-to-agreement of less than 1 mm. A PSF is tallied just upstream of the third scraper of the electron applicator.

For the external beam radiotherapy treatment of the conjunctival lymphoma a dedicated collimator inserted in the third scraper of the electron applicator is used [23]. This collimator consists of a cerrobend block, 1.6 cm thick, with a central cylindrical hole of diameter equal to 3.0 cm. A PMMA slab 0.2 cm thick is placed on top of the insert. Aligned with the central beam axis, and attached from below the slab, there is a hanging PMMA rod whose purpose is to shield the eye lens of the patient. The length of the rod is 6.0 cm and its diameter equals to 1.0 cm (see Fig. 19.4). The distance from the upstream surface of the slab to the source is 93.2 cm.

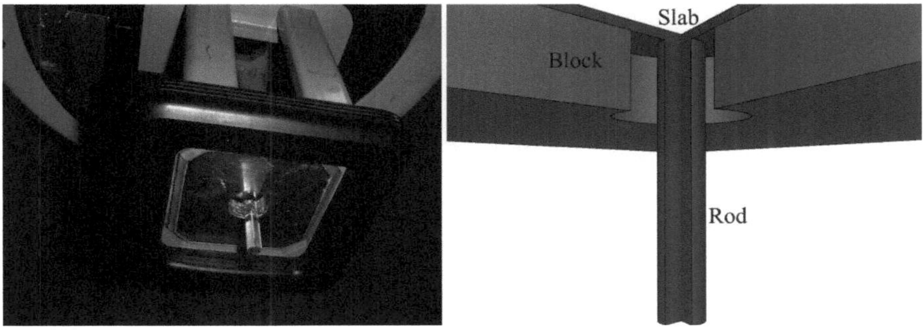

Fig. 19.4 Picture of the collimator used for the treatment of the conjunctival lymphoma inserted in the third scraper of the electron applicator, viewed from the patient's perspective in head-first supine position (left). Drawing of the simulated collimator where a wedge has been removed to show the inner structure (right)

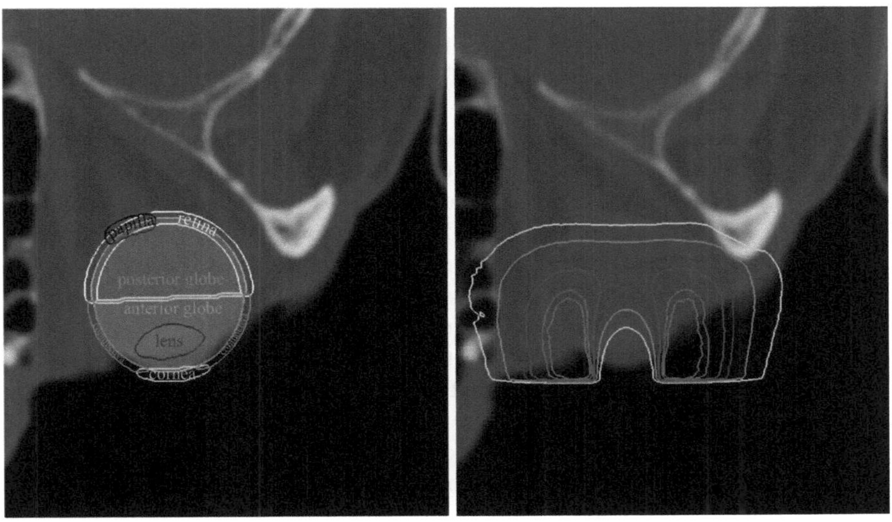

Fig. 19.5 Segmentation of the evaluation anatomical structures on the isocenter slice. The contour of the lacrimal gland is not shown since it does not intersect this slice (left). Isodose lines at the isocenter slice. Inner to outer relative isodose lines correspond to 100%, 95%, 80%, 70%, 50% and 30% of the maximum absorbed dose (right)

From the tallied PSF upstream of the dedicated collimator, particles are transported through the collimator itself and a subsequent PSF is tallied 0.2 cm above the tip of the patient's eye. A later simulation transports particles into a computerised tomography (CT) study of a patient. The CT study has $256 \times 256 \times 59$ voxels of size $0.03125 \times 0.03125 \times 0.1$ cm^3. Hounsfield units of the original CT are converted to mass density values via the calibration curve of the CT scanner. For material

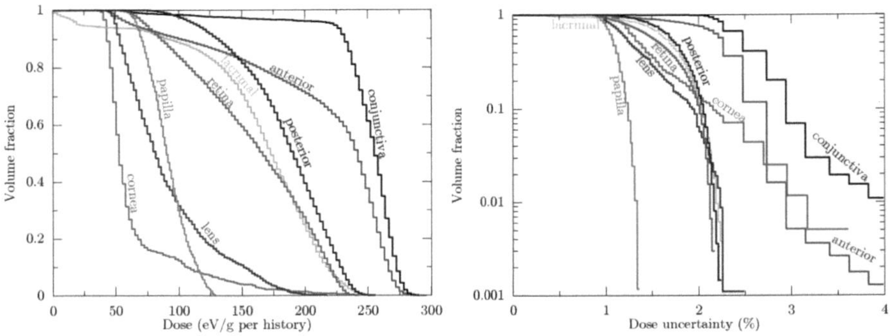

Fig. 19.6 Cummulative DVH (left) and UVH (right) for the considered anatomical structures.

assignment, three different media are considered namely water, air and bone. Since in the treatment a flat water bolus covers the eye, this bolus is added to the CT study used for the simulation.

The CT study is previously segmented in order to study the absorbed dose to different anatomical structures. The following structures of the eye have been considered in the segmentation: cornea, bulbar conjunctiva, retina and papilla. In addition, the eye is divided into two hemispheres, anterior and posterior globe, which can be evaluated separately. In the orbit, the lacrimal gland is also segmented. Figure 19.5 (left) shows all the aforementioned structures on a horizontal slice crossing the isocenter, except the lacrimal gland that does not intersect the isocenter slice. Figure 19.5 (right) shows the isodose lines superimposed to the isocenter horizontal slice. The plotted lines are constructed from the tallied dose in each voxel without applying any smoothing algorithm.

Segmentation of the CT study allows the calculation of dose volume histograms (DVH) [24]. Figure 19.6 shows cumulative DVHs, as well as uncertainty volume histograms (UVH) [25] for the segmented structures. Two standard deviations have been used for statistical uncertainty of the UVHs. The UVH plot of the conjunctiva shows that less than 1% of the voxels of the conjunctiva have a statistical uncertainty higher than 4% (2σ).

Acknowledgements The author is grateful to F. Salvat (Universitat de Barcelona), J. Sempau and M. Rodríguez (Universitat Politècnica de Catalunya) for their suggestions in relation to the codes described herein. W. Sauerwein (Universitätsklinikum Essen) and F.J. Zaragoza (Universitat Politècnica de Catalunya) are thanked for their collaboration in the example presented in section 19.9.

References

1. C.J. Karzmark, C.S. Nunan, E. Tababe, *Medical Electron Accelerators* (McGraw-Hill, 1993)
2. M. Udale, Phys. Med. Biol. **33**, 939 (1988)

3. D. Rogers, B. Faddegon, G. Ding, C. Ma, J. Wei, et al., Med. Phys. **22**, 503 (1995)
4. I.J. Chetty, B. Curran, J.E. Cygler, J.J. Demarco, G. Ezzell, B.A. Faddegon, I. Kawrakow, P.J. Keall, H. Liu, C.M.C. Ma, D.W.O. Rogers, J. Seuntjens, D. Sheikh-Bagheri, J.V. Siebers, Med. Phys. **34**, 4818 (2007)
5. N. Reynaert, S.C. van der Marck, D.R. Schaart, W.V. der Zee, C.V. Vliet-Vroegindeweij, M. Tomsej, J. Jansen, B. Heijmen, M. Coghe, C.D. Wagter, Radiat. Phys. Chem. **76**, 643 (2007)
6. J. Sempau, A. Sanchez-Reyes, F. Salvat, H. ben Tahar, Phys. Med. Biol. **46**, 1163 (2001)
7. V. Moskvin, C. DesRosiers, L. Papiez, R. Timmerman, M. Randall, P. DesRosiers, Phys. Med. Biol. **47**, 1995 (2002)
8. P. Carrasco, N. Jornet, M.A. Duch, L. Weber, M. Ginjaume, T. Eudaldo, D. Jurado, A. Ruiz, M. Ribas, Med. Phys. **31**, 2899 (2004)
9. L. Blazy, D. Baltes, J.M. Bordy, D. Cutarella, F. Delaunay, J. Gouriou, E. Leroy, A. Ostrowsky, S. Beaumont, Phys. Med. Biol. **51**, 5951 (2006)
10. E. Sterpin, F. Salvat, R. Cravens, K. Ruchala, Phys. Med. Biol. **53**, 2161 (2008)
11. S. García-Pareja, P. Galán, F. Manzano, L. Brualla, A.M. Lallena, Med. Phys. **37**, 3782 (2010)
12. F. Salvat, J.M. Fernández-Varea, J. Sempau, PENELOPE—*A code system for Monte Carlo simulation of electron and photon transport* (OECD Nuclear Energy Agency, Issy-les-Moulineaux, France, 2008)
13. J. Sempau, E. Acosta, J. Baró, J.M. Fernández-Varea, F. Salvat, Nucl. Instrum. Meth. B **132**, 377 (1997)
14. A. Badal, J. Sempau, Comput. Phys. Commun. **175**, 440 (2006)
15. W. Schneider, T. Bortfeld, W. Schlegel, Phys. Med. Biol. **45**, 459 (2000)
16. L. Brualla, F. Salvat, R. Palanco-Zamora, Phys. Med. Biol. **54**, 4131 (2009)
17. R.Y. Rubinstein, *Simulation and the Monte Carlo method* (Wiley, New York, 1981)
18. A.F. Bielajew, D.W.O. Rogers, in *Monte Carlo tansport of electrons and photons*, ed. by T.M. Jenkins, W.R. Nelson, A. Rindi (Plenum, New York, 1988), pp. 407–419
19. L. Brualla, W. Sauerwein, Radiat. Phys. Chem. **79**, 929 (2010)
20. D. Low, W. Harms, S. Mutic, J. Purdy, Med. Phys. **25**, 656 (1998)
21. A. Bakai, M. Alber, F. Nüsslin, Phys. Med. Biol. **48**, 3543 (2003)
22. S.E. Coupland, M. Hummel, H. Stein, Surv. Ophthalmol. **47**, 470 (2002)
23. L. Brualla, R. Palanco-Zamora, A. Wittig, J. Sempau, W. Sauerwein, Phys. Med. Biol. **54**, 5469 (2009)
24. R.E. Drzymala, R. Mohan, L. Brewster, J. Chu, M. Goitein, W. Harms, M. Urie, Int. J. Radiat. Oncol. **21**, 71 (1991)
25. I. Chetty, M. Rosu, M. Kessler, B. Fraass, R.T. Haken, F. Kong, D. McShan, Int. J. Radiat. Oncol. **65**, 1249 (2006)
26. J. Sempau, A. Badal, L. Brualla. Med. Phys. **38**, 5887 (2011)
27. M.L. Rodriguez. Phys. Med. Biol. **53**, 4573 (2008)

Part III
Biomedical Aspects of Radiation Effects

Chapter 20
Repair of DNA Double-Strand Breaks

Biochemical and Spatio-Temporal Aspects

Martin Falk, Emilie Lukasova, and Stanislav Kozubek

Abstract The genetic information of cells continuously undergoes damage induced by intracellular processes including energy metabolism, DNA replication and transcription, and by environmental factors such as mutagenic chemicals and UV and ionizing radiation. This causes numerous DNA lesions, including double strand breaks (DSBs). Since cells cannot escape this damage or normally function with a damaged genome, several DNA repair mechanisms have evolved. Although most "single-stranded" DNA lesions are rapidly removed from DNA without permanent damage, DSBs completely break the DNA molecule, presenting a real challenge for repair mechanisms, with the highest risk among DNA lesions of incorrect repair. Hence, DSBs can have serious consequences for human health. Therefore, in this chapter, we will refer only to this type of DNA damage. In addition to the biochemical aspects of DSB repair, which have been extensively studied over a long period of time, the spatio-temporal organization of DSB induction and repair, the importance of which was recognized only recently, will be considered in terms of current knowledge and remaining questions.

20.1 Introduction – DSB repair

It is estimated that as many as 2×10^4 to 1 million DNA lesions occur in every cell of the human body each day [1,2] as a consequence of intracellular [3] and extracellular processes [reviewed, *e.g.*, in 4–6]. Thus, about 800 DNA lesions appear per cell per hour [7]. These include base alterations, abasic sites, intra- and inter-strand base cross links, DNA adducts and single- (SSB) or double- (DSB) strand DNA breaks. Among these lesions, DSBs, produced by ionizing radiation (IR) [5, 6, 8],

M. Falk (✉) • E. Lukasova • S. Kozubek
Institute of Biophysics, Academy of Sciences of CR, Brno, Czech Republic
e-mail: mfalk@seznam.cz; lukasova@ibp.cz; kozubek@ibp.cz

G.G. Gómez-Tejedor and M.C. Fuss (eds.), *Radiation Damage in Biomolecular Systems*, Biological and Medical Physics, Biomedical Engineering,
DOI 10.1007/978-94-007-2564-5_20,

some chemicals [9] and reactive free radicals that arise from cellular oxidative energy metabolism [7], represent unique damage that poses the most serious threat to human health [4, 10]. Indeed, even a single DSB can lead to cell death if left unrepaired or, more seriously, it can cause mutagenesis and cancer if repaired improperly [reviews 4, 10–12]. The accumulation of unrepaired or misrepaired DSBs is also responsible for chronic inflammation [13], aging, and some non-malignant neurodegenerative diseases [review 10, 14, 15]. DSB repair thus allows cell survival while the integrity of genetic information is under continuous pressure. For instance, minimally 5000 SSBs occur in DNA during a single cell cycle because of reactive oxygen species (ROS) production and about 1% of these are converted to DSBs; thus, approximately 50 endogenous DSBs are generated in a single cell during one replication round [11].

Since DSBs disrupt the DNA molecule, they are a challenge for DNA repair pathways. Consequently, aberrations in chromatin that have potentially serious deleterious effects on health, including cancer, may occur because of misrepair [*e.g.*, 10]. On the other hand, radiotherapy and many kinds of chemotherapy [review 16] rely on the induction of DSBs to kill tumor cells – since transformed cells divide quickly and are often defective in some of the DNA repair pathways, they may be more sensitive to DSB-induced cell killing than normal cells [reviews 17, 18]. Therefore, when it comes to cancer, DSB repair is a double edged sword; on the one side it is a causative agent of cancer and the other side it is an important mechanism for cancer treatment. However, despite the risk of cancerogenic mutations, DSBs mediate important physiological processes including immunoglobulin V(D)J rearrangements [reviews 19, 20] and class switch [21, 22], and may be the only chance to resolve stalled replication or transcription forks [reviews 23–25]. In addition, DSBs allow mitotic recombination [26] and, in whole populations, enable molecular evolution and life adaptation [25, 27]. Thus, DSB repair is important for fundamental life processes; hence, it has been studied in basic and medical research for more than five decades.

Since DSBs are the most deleterious lesions caused by ionizing radiation and radiomimetics [5–9], DSB repair is of much interest also in industrial and military applications and space exploration. Initiated by the discovery of radioactivity and its biological effects, interest in DSB repair has again increased for two reasons: 1) many current problems of mankind are somehow connected with DSB repair. For example, the development and application of a revolutionary ion-beam cancer therapy [28] require building of a much deeper scientific background in addition to existing empirical knowledge. As another example, plans for long interplanetary missions are not possible without knowledge of the health risks posed by cosmic radiation exposure to spacecraft crews. Finally, there is the potential for the misuse of radioactive materials by terrorists. 2) Technological and methodological progress in recent years enable much more sophisticated studies of DSB repair that are necessary to address these issues.

20.2 DNA repair pathways, guardians of the genome integrity: An overview

DNA repair is a complicated multistep process that is executed by precisely regulated sequences of spatio-temporally organized biochemical processes, initiated by DNA damage induction and aimed at the processing of a specific type of DNA lesion. Several specific repair pathways, interconnected in complicated networks [29, 30], participate in DNA repair, which mirrors the broad spectrum of lesions potentially induced (mentioned earlier). For example, base-excision repair (BER) focuses on damage to single bases. Nucleotide-excision repair (NER) recognizes distortions in the DNA helix, especially those caused by UV-light, including pyrimidine dimers and 6, 4-photoproducts [31]. Mismatch repair (MMR) recognizes mispaired but undamaged regions in newly synthesized DNA strands that result from polymerase errors or slippage during replication or recombination [32]. For surveys and descriptions of these pathways, refer to the particular literature reviews [33–37]; Nevertheless, it should be emphasised that most DNA repair processes use the complementary DNA strand as an undamaged template for the synthesis of removed damaged fragments, such that lesions are usually processed quickly without any persistent consequences for the genome.

In DSBs, however, the disruption of the DNA molecule causes two challenging problems. First, free DNA ends must be stabilized spatially to be rejoined correctly, and, second, a homologous template must be found elsewhere in the genome, usually at the sister chromatid (because of space limitations); thus, the search for a homologous DNA sequence can be successful almost exclusively in S- or G_2-phase of the cell cycle [38, 39]. Otherwise, DSBs must be simply rejoined without the "knowledge" of the original DNA sequence, which may lead to mutations. Indeed, despite the fact that some alternative mechanisms exist [reviewed in 40], two dominant pathways conserved through evolution, homologous recombination (HR) and nonhomologous end-joining (NHEJ) were described in yeasts, mammals, and other organisms, including prokaryotes [reviews 41, 42].

When a DSB occurs, cells activate a complex network of DNA damage response (DDR) [29] in order to decide whether to reversibly stop the cell cycle and repair the damage [10], or stop it irreversibly and enter senescence [43] or programmed cell death (apoptosis) [review 44]. Alternatively, the cell may proceed to "adaptation," the process that enables it to enter mitosis even in the presence of unrepaired DSBs [reviewed in 45]. DDR proteins can be divided according to their functions: These include "sensors" that specifically recognize the lesion, "effectors" that participate in processing the lesion, and "mediators" that interact with signaling proteins to connect these two steps and individual DDR pathways. This organization of DDR enables precise regulation and high flexibility of cellular response as well as amplification of very weak damage signals [46]. However, while it is useful for educational purposes, this categorization of DDR proteins is a considerable simplification, since many of them have multiple activities in different repair steps. For example, the MRN complex (Mre11-Rad50-Nbs1) acts as one of the putative

DSB sensors, a co-activator of DSB checkpoint signaling (*e.g.*, by activating ATM), and as an effector protein both in the NHEJ and HR [46–50]. In addition, multiple regulatory feedback loops complicate our understanding of the temporal order of protein-protein and protein-DNA interactions involved in DDR. For instance, the identity of the first sensor necessary to recognize DSB lesions is in dispute.

20.3 "Biochemistry" of DSB repair pathways: What we think we know

20.3.1 Non-homologous end-joining (NHEJ)

ATM kinase [51], Ku proteins, and MRN complex [47–50] are among the candidates to be the "initiators" of DDR. ATM kinase is evidently the capellmeister, with a central role in regulating both the NHEJ and HR repair pathways, cell cycle checkpoints in G_1/S, intra-S and G_2/M, apoptosis and induction of specific transcriptional programmes in response to DSB [reviewed in 52–54]. Within intact cells, ATM exists as an inactive dimer. After DSB formation, ATM is autophosphorylated and dissociates into active monomers [55], which also results in the release of protein phosphatase 2A from ATM, so the activation of ATM is not further inhibited [56]. Full activation and amplification of the ATM signal is stimulated by MRN complex and DNA-PK bound to the sites of DSB [57, 58], where "semi-activated" ATM interacts with a broad spectrum of proteins. One of the ATM targets is the MRN complex that in a feedback loop activates ATM and enables its additional accumulation at the sites of DSB. The MRN complex is therefore another important and versatile protein that potentially participates in DSB recognition and initiation of DSB repair [12, 50]. It has multiple additional roles in both NHEJ and HR, especially in DSB-end processing [59–62], spatial stabilization of ends [63,64], and cellular signaling [50]. One of the first events in the NHEJ pathway is also binding of Ku heterodimer (Ku70/Ku80) to free DNA ends, where it stabilizes damaged chromatin, prevents DNA ends resection and potentially recognizes the lesion [65]. The most probable scenario therefore includes several parallel processes participating in sensing DSB, DDR activation, and signaling, with mutually stimulating effects.

Despite the complicated initiation step outlined, recent results [66–78] showed that most DSBs (about 85%, see Sect. 20.3.4, 20.4.4) can be repaired in principle with the participation of only five "core" proteins including Ku heterodimer, DNA-PKcs, XRCC4, and DNA ligase 3. Therefore, the NHEJ pathway *per se* (absent cellular signaling) may begin by the binding of Ku proteins to DSB ends, which enables their interaction with DNA-PK catalytic subunit (DNA-PKcs, a member of PIKK subfamily of PI-3 kinases [72]) and its activation [73,74]. Once active, DNA-PK holoenzyme phosphorylates several targets at the site of the DSB, including itself, which results in the dissociation of DNA-PKcs subunits from DSBs [75] and their exchange for NHEJ or HR proteins. DNA-PK (and Ku) thus probably tethers free DNA ends together, until DNA ligase and XRCC4 (LC complex) bound to DSBs and rejoin the ends [76, 77]. Refer to comprehensive reviews for a detailed description of the biochemistry of NHEJ and HR [78–85], and for repair of DSBs associated with replication or transcription [24] as well as the alternative repair pathways [40, 86].

20.3.2 Additional NHEJ components

The simple scenario described above is only possible when euchromatic DSBs with undamaged DNA ends [66–71] are being repaired. In that case, DSBs are rejoined quickly, often without changes in genetic information. However, additional steps that include modifications of chromatin structure and DNA-end processing are necessary to repair heterochromatic breaks and most DSBs induced by ionizing radiation (IR) that contain "dirty" ends with single-stranded overhangs and damaged bases [87]. This section will focus on the biochemical functions of "additional" NHEJ proteins and their cooperation, since the relationship between DSB repair and modifications of chromatin structure is the subject of Sec. 20.4.

Nuclease activities required to clear DNA ends were recognized for the Mre11 component of the MRN complex [59–61, 88] and, are, with its affinity to DNA, stimulated by another MRN member, Rad50 [89]. The third MRN member, Nbs1, mediates nuclear localization of the MRN complex [90] and further stimulates its activities [89–91]. However, the physiological importance of the exonuclease function of MRN is uncertain, since only 15% of DSBs induced by IR (that contain damaged DNA ends) require this protein for repair [13, 66–71, etc.]. Therefore, other endonucleases such as Artemis probably clear DNA ends [reviewed in 92], whereas MRN complex serves rather as an activator of ATM signaling[1] [25, 88, 93]. Since Artemis is activated by ATM, this interaction represents one of the contact points between NHEJ and ATM signaling that demonstrates participation of both of these pathways in the repair of more complex DSBs or DSBs located in more complex (dense) chromatin.

Another important early step of DSB repair is ATM-mediated phosphorylation of the histone H2AX [11, 94, 95], and the consequent formation of γH2AX foci that cover megabase regions on either side of DSB [96] (see Sec. 20.4.3 devoted to γH2AX foci), and create a platform for binding of "non-core" repair proteins (like MDC1 [97, 98], 53BP1 [98, 99], BRCA1 [100, 101], SMC1 and others [*e.g.*, 100, 102]) to the site of damage. These proteins are also phosphorylated by ATM and, once activated, bind to γH2AX foci. Even though the activity of these "mediators" is dispensable for initial binding of DSB sensors to the lesion sites [103] and also for repair of most DSBs [66–71], they are necessary for proper DDR response [98, 101, 104] and repair of a subgroup of DSBs that are repaired only

[1] the function of MRN as a DSB sensor [53, 66, 93] is also hardly conceivable if only 15% of DSBs require MRN for their repair (even thought MRN colocalizes with DSBs in general)

with difficulty [66–71] (discussed later). Although the data are contradictory, it is possible that MDC regulates checkpoint activation [98, 100, 105] and binding of ATM, 53BP1, BRCA1, and other proteins to DSBs, by way of recruiting RNF8 to the sites of lesions and by ubiquitination of H2AX [106, 107, summarized in 98]. 53BP1 and BRCA1 then push DDR to either NHEJ or HR [101, 108–110], which suggests interconnections not only between NHEJ and ATM signaling, but also between NHEJ and HR, and shows how individual repair pathways cooperate and, at the same time, "compete" for DSB substrates. Therefore, the main function of these mediators that exhibit pleiotropic functions with many partners, is probably to serve as a scaffold for protein-protein interactions at the sites of DSBs [101]. The spectrum of "non-core" proteins recognized to operate in DSB repair is broad and continues to grow. Many of these proteins participate in reorganization of chromatin structure at the sites of DSBs to allow their repair, which is discussed in Section 20.4.3.

20.3.3 Homologous recombination (HR)

HR [reviewed, *e.g.*, in 85,111,112] is enabled by extensive $5' \rightarrow 3'$ resection of DNA ends by MRN complex (MRX in yeasts) and other exonucleases [60–62, 112]. This processing generates long 3′-single-stranded DNA (ssDNA) overhangs that mediate recombination, which are immediately bound by RPA proteins with a high affinity for ssDNA. RPAs protect DNA from degradation and formation of secondary DNA structures, as well as from interaction with other HR proteins until they are replaced with Rad51 in a process mediated by Rad52, Rad55 and Rad57 proteins. Rad51, the key HR protein, forms a nucleoprotein filament with ssDNA and initiates the recombination step of HR by invading an undamaged homologous DNA duplex (stimulated by Rad54, Srs2 helicase, etc.). Dissociation of Rad51 allows strand pairing and extension of the invading strand by DNA polymerase. The second 3′-DNA overhang at the opposite side of DSB can be captured by the displaced DNA strand of the donor chromatid (to form a double-Holiday junction) or the extended invading strand can be released from the complex with the donor chromosome and anneal with the complementary overhang on the opposite DSB end of the original chromatid. Holiday junctions are resolved (by resolvase) to yield crossovers or non-crossovers, and remaining DNA gaps are filled by DNA polymerase while DNA ligation is accomplished by DNA ligase [85]. In synthesis-dependent strand annealing (SDSA), DSB repair is accomplished as described above; however, only non-crossovers are produced.

Importantly, while repair fidelity tends to be simplified as "error-prone" NHEJ and "error-free" HR, HR may in fact yield loss of heterozygosity or mutated repair products such as translocations and other chromosomal aberrations [25, 113], because of crossovers between imperfectly homologous (mutated or allelic) loci or repeated sequences. Thus, repair fidelity is modulated by a precise NHEJ that rejoins "clear" DSB ends as well as the less precise HR [see 25 for the review, 113].

20.3.4 Interplay between DSB repair pathways

The relative contribution of the individual pathways important for DSB repair depends upon many factors that vary among organisms [reviews 25, 112]. In yeasts, homologous recombination (HR) is the most common pathway [reviewed in 25,85]. In higher eukaryotes, however, cells spend most of their lives in the G_0/G_1 phase of the cell cycle and the homologous sequences of the paired chromosomes are not easily accessible, because homologous chromosomal territories are usually distant from each other [114, 115]. Most of DSBs in higher eukaryotes are thus simply rejoined by non-homologous end-joining (NHEJ) [41, reviewed in 116, 117] that allows sufficiently rapid removal of DSBs ($t_{1/2}$: 5–30 min [40]) from the large complex genomes of mammals without the requirement for a homologous template [review 118]. However, the risk of repair errors is greater for NHEJ than HR. Because of the resection of DNA ends required to resolve complex DSBs (frequently generated by IR [119–121]), small terminal nucleotides deletions or insertions occur even in the case of a "successful" repair process [86]. NHEJ thus produces more potentially carcinogenic chromosomal mutations than HR [122]. However, considering the variety of possible DSB substrates of NHEJ [86], and that only about 3% of the mammalian genome contains gene coding sequences (as compared to 70% in yeasts) [25], NHEJ represents a very efficient and versatile repair pathway in higher eukaryotes even with the acceptable risk of genetic damage [123–125]. Especially under physiological conditions, about 25–50% of endogenous, nuclease-induced DSBs are repaired without errors in yeast and mammalian cells [25, 126]. However, it should be noted that in addition to gene coding sequences, at least 50% of mammalian DNA encodes for functional RNAs, *e.g.* miRNAs that play important roles in regulation of gene transcription [25]. Moreover, other non-coding DNA sequences may contain "epigenetic" information that is manifested in various ways (*i.e.*, chromatin structure).

Homologous recombination probably begins to dominate repair processes when the cell enters S-phase and the homologous DNA becomes accessible on the sister chromatid [41]. Theoretically, HR can proceed also between repeated, ectopic and allelic loci [25], but proof of significant HR activity in G_1-phase is lacking [42,117]. Recently published results of Mao et al. [42, 117] show that NHEJ predominates over other repair pathways throughout the mammalian cell cycle, and that HR is limited to the S-phase, where it nevertheless still occurs less frequently than NHEJ and also less frequently than previously thought. These conclusions may be in line with a new observation that only heterochromatic DSBs (representing about 15% of all DSBs) are repaired by HR in S/G_2 [70]. Taken together, it seems that NHEJ and HR cooperate in DSB repair as they process different DSB targets [25,66,127]. The trigger for NHEJ or HR at the particular DSB is however unknown and how the two pathways interact is not understood, thought it is obvious that structural characteristics of the DSB lesion and surrounding chromatin, cell type, proliferation rate, and other factors evidently play an important role, in addition to the phase of the cell cycle. For a detailed review on this topic see Shrivastav et al. [25, and citations therein]. For example, it was shown that HR inhibits NHEJ: in G_1-cells, the resection of DSB ends is inefficient, since it is dependent on CDK1, the expression of which is carefully regulated through the cell cycle. Hence, the formation of the invading nucleoprotein filament that is necessary to initiate HR is prevented. Moreover, the expression of many repair proteins is also genetically regulated with a respect to the cell cycle. For instance, the levels of Rad51 and Rad52 that participate in HR, are not enhanced until the S-phase [128]. However, since NHEJ also functions in S- and G_2-phases, it is evident that the presence of a homologous DNA template *per se* is not sufficient to switch off NHEJ and initiate HR. Instead, it seems that HR and NHEJ compete for DSB targets to initiate DSB repair, proteins common to both repair pathways including MRN complex, ATM, PARP-1, RAD18, DNA-PKcs and BRCA1 bind to DSBs [129] and commit the cell to one pathway or the other. With complex local chromatin and global cell conditions, additional proteins specific for a particular pathway bind to the DSB and the entire process becomes irreversible. Therefore, a complex network of interconnected repair mechanisms operates in the cell nucleus, rather than separate repair pathways.

In support to this, also alternative end-joining repair mechanisms that are "hybrids" of NHEJ and HR were identified [reviews 40, 86, 130]. These include backup non-homologous end joining (B-NHEJ) [reviewed in 130–132], single-strand annealing (SSA) [133, 134], and microhomology-mediated end joining (MMEJ) [reviewed in 86], although it is not clear whether they represent variants of the same repair mechanisms or "separate" repair pathways. Similar to HR, SSA requires relatively long homologous sequences at DSB ends for successful rejoining. Whereas, unlike HR, this homology is located within the same DNA molecule and, unlike NHEJ, extensive resection of DNA ends on both sides of DSB is required [133,134]. Since homology is provided by repetitive sequences, one repeat and DNA sequence between the two interacting repeats are usually lost and SSA is therefore always error-prone [review 85,112]. Thus, these alternative pathways probably serve only as life-saving backup repair mechanisms that are activated when "classic" DNA-PK-dependent NHEJ (D-NHEJ) [review 130] is not functional; nevertheless, their existence and character suggest a high flexibility and interconnection of repair mechanisms.

20.4 DSB repair and chromatin structure: Difficult answers to simple questions

It was well proved during the last few years that higher-order chromatin structure influences DSB induction, repair, and the formation of chromosomal translocations (see [135] and citations therein for review). However, the multidisciplinary nature of the research on radiation damage (that covers physical, chemical and biological process over a time scale ranging from 10^{-22} s to years and spatial scale from picometers to meters) makes experiments and their complex interpretation difficult. Therefore, while the biochemistry of DSB repair is largely understood, the spatio-temporal organization of DSB repair is less clear.

20.4.1 Chromatin structure, function and dynamics in interphase nuclei

In previous chapters, the temporal order of interactions between damaged DNA and the proteins involved in DDR were described. Nevertheless, in eukaryotes, DNA exists as chromatin, a complex of DNA, histones, RNA, and non-histone proteins [136–138]. These interactions, in turn, result to formation of hierarchical higher-order chromatin structures and nonrandom organization of chromatin in the cell nucleus. At the lowest level, DNA wraps around histone octamers and forms nucleosomes, the basic units of chromatin [139, 140]. Consequently, nucleosomes form the nucleohistone fibre that is hierarchically wrapped to yield mitotic chromosomes with the maximum DNA compaction [141–143]. Such extreme compaction allows DNA (about 2–3 m long in humans [144, 145]) to fit into a cell nucleus of about 10 μm in diameter and is necessary to allow cell division; however, it precludes "physiological" functions of DNA such as transcription, replication, and DNA repair [reviewed in 146–148]. In interphase nuclei, therefore, DNA must adopt a less compact conformation that lies between the above mentioned extremes [142, 143]. This decondensation of individual mitotic chromosomes results in formation of spatially separated "chromosomal territories" (CHT) [review 149] that exhibit some intermingling [150–152] and, in some aspects, a specific nuclear localization [153–155, and many others]. This organization has been described as "order in randomness" [153], since it seems to be more probabilistic than deterministic; however, this notion continues to be under dispute [114, 115, 156–158, etc.]. For instance, there is no unequivocal opinion on the level of mutual positioning of individual CHTs and transmission of this order to daughter cells. On the other hand, it is well proofed that highly expressed CHTs [159, 160] are preferentially localized closer to the nuclear center in spherical cycling cells than less expressed chromosomes that usually appear closer to the nuclear membrane [reviewed, *e.g.*, in 153, 161, 162]. In flat cells (e.g. normal human fibroblasts), however, CHTs might be distributed according to their size [review 156, 162, 163] and this positioning of individual CHTs can change as the cells enter quiescence, senescence, or as found in some diseases [164–168]. Nuclear organization of CHTs therefore seems to be plastic, depending on the organism, cell type, cell cycle, level of development, and degree of differentiation [164–168]. This variability complicates our recognition to the problem in its whole complexity, which is reflected in a broad spectrum of models that were suggested to (solely) highlight different aspects of chromatin organization in CHTs and cell nuclei [153–158, 161, 162, 169–172, *etc.*]. Important questions in the context of DSB repair therefore are, how can chromatin organization and its dynamics influence the probability of translocations between the particular chromosomes [reviewed 135] and how can nuclear environment modify expression profiles of the translocated chromosomal parts, *e.g.* as a result of their heterochromatinization after translocation to heterochromatic nuclear domains.

Similar rules of organization apply also for chromatin inside the chromosomal territories [153, 154, 171, 172]. Genetic information is not distributed homogeneously along the DNA molecule; highly expressed genes frequently form clusters named RIDGEs (Regions of Increased Gene Expression) and silent genes form clusters named antiRIDGEs [159, 160]. Both RIDGEs and antiRIDGEs are spread nonrandomly through the human genome, forming chromosomes with high, middle and low levels of overall transcription. To exert their functions, chromatin in RIDGEs is more decondensed than that in antiRIDGEs, which allows better access of transcription machinery (and other proteins) to DNA. Conversely, the condensed chromatin of antiRIDGEs contributes to gene silencing [reviewed, *e.g.*, in 173].

These structural differences of chromatin are dictated by covalent posttranslational modifications of histones (methylation, acetylation, ubiquitination, phosphorylation, etc.) and DNA methylation [148, 174–176]. Precise positioning of different combinations of histone modifications throughout the genome constitutes an epigenetic "histone code" [reviews 148, 176–178] that strictly regulates binding of additional regulatory and effector proteins to the particular DNA loci. For example, heterochromatin binding protein 1 (HP1) is required for heterochromatin formation and specifically binds, by its chromodomain, to the histone H3 dimethylated on the lysine 9 (dimetH3K9) in DNA loci that should be genetically silenced [178, 179]. Conversely, trimethylated histone H3 on the lysine 4 (trimetH3K4) attracts proteins associated with gene activation (see reviews, [180–182]). Histone modifications are hereditary and can be reversed by specific enzymes or histone replacement. This histone code can thus facilitate regulation of local chromatin structure and expression profiles during development, cell differentiation, replication, and DNA damage responses or other cellular stresses [180, 181].

In three-dimensional space, the nonrandom distribution of active and inactive DNA loci within the genome, as evidenced by specific epigenetic modifications, results in the formation of structurally and functionally distinct subchromosomal domains. Active regions of the chromosome may protrude from their subdomains and even from CHTs [183, 184], which allows some dynamic intermingling of chromatin such as spatial colocalization of several (coregulated) genes in transcription factories [185–188] or interactions between transcription enhancers and gene promoters [189]. On the other hand, long-range pan-nuclear movement of undamaged chromatin has been rarely reported [190–193]. Therefore, the nucleus does not have either random structure (formerly liken to a cap of soup with randomly swimming DNA "noodles") or rigid, deterministic organization, but it has a dynamic higher-order chromatin architecture necessary for nuclear functions.

20.4.2 Chromatin structure and DSBs: Questions, questions

DSBs cause extensive damage not only to primary DNA but also to chromatin structure at all levels. Therefore, it is necessary not only to understand the biochemical aspects of DSB repair, but also to reveal how DSB repair is organized in time and in the space of the cell nucleus, to determine how chromatin structure influences its sensitivity to DSB and the efficiency of repair processes [66–71, 135, 194–205, and others]. In fact, we lack the generally accepted answers even to very basic questions, as: Are DSBs induced with the same efficiency in decondensed chromatin with active genes as they are in condensed heterochromatin that is genetically silent? And how is this influenced by the linear energy transfer (LET) of IR, dose rate, presence of radical scavengers and other conditions? Once DSBs are induced, are they repaired at their sites of origin or do they migrate into nuclear subdomains that are more permissive for repair or perhaps to specialized "repair factories", where several DSBs are repaired simultaneously? Hence, do chromosomal translocations between the particular loci form because of their predetermined physical proximity in the nonrandom higher-order chromatin structure (nuclear architecture) or because of their mutual approach evoked by DSB repair processes? If the latter case is true, do these dynamics reflect chromatin damage, changes of local chromatin structure at the sites of DSB that are required to enable their repair, or directed movement of DSBs to repair competent subdomains of the cell nucleus? If they exist, can repair factories form *de novo* anywhere in the cell nucleus or only in special nuclear subcompartments?

20.4.3 γH2AX foci and changes of higher-order chromatin structure associated with DSB repair

As already mentioned (Sec. 20.3.2), one of the first and most prominent changes made to histones, and thereby chromatin structure, at the sites of DSBs is the phosphorylation of the histone H2AX by ATM kinase (or ATR, DNA-PK-Cs). H2AX accounts for 2–25% of the H2A histone pool in mammalian cells [206, 207] and is phosphorylated on serine 139 [reviewed in 11, 94] in a megabase region on either side of the lesion [96]. Using anti-γH2AX antibody, phosphorylated H2AX (termed γH2AX) can be detected microscopically as nuclear foci that appear within minutes of formation of DSBs and is a specific marker of these lesions [reviewed in 11, 94, 96]. In present, this is the most versatile and sensitive method for detection of DSBs formed even at cGy or mGy IR doses [reviewed in 127, 208–210]. In combination with flow cytometry, immunodetection of γH2AX allows high-throughput analysis of cell populations, whereas in combination with high-resolution confocal microscopy, it provides an unusual tool to study DSB repair in individual spatially fixed cells. This becomes useful, for example, when different cell types in heterogeneous tissues like blood must be studied. A "Top-Tech" method available allows visualization of fluorescently labeled repair proteins that colocalize with γH2AX foci[2] in living cells, so DSB repair can be continuously monitored under physiological conditions [197, 199, 211]. Detection of γH2AX foci thus fundamentally contributed to the research of many aspects of DSB repair, with important conclusions made on the basis of their presence in the nucleus; however, an exact nature of γH2AX foci still remains in dispute.

For example, it is not known if one γH2AX focus is equal to one DSB, since there are discrepancies between the numbers of DSBs and repair kinetics as determined by comparison of results from pulse field gel electrophoresis (PFGE) and γH2AX immunostaining. While most DSBs were repaired in about 30 min to 1 hour, as determined by PFGE, γH2AX foci persist in nuclei much longer, usually several hours or even days post-irradiation (PI) [review 135]. However, it should be noted that PFGE requires extremely high IR doses to produce detectable amount of DSBs, since its sensitivity is much lower than that of γH2AX immunostaining. In addition, differences were observed especially with tumor cells, frequently under different experimental conditions (type of radiation, the irradiation dose, dose rate, cell type, time PI of DSB quantification etc.) [reviews 135, 212–214]. Therefore, it is not evident whether these results are experimental artefacts or if they have a biological meaning. Nevertheless, if we ignore the additional problem of clustered DSBs[3], the prevailing belief is that the number of DSBs correlates 1:1 with the number of γH2AX foci [reviewed in 135].

The most interesting questions are those associated with late γH2AX foci that persist in nuclei for a long time. It is unclear if these persistent foci are unrejoined DSBs that can only be repaired with difficulty (as they usually colocalize with repair proteins like Mre11, Nbs1, or 53BP1), irreparable DSBs, sites of misrepaired DSBs (chromosomal translocations), or DSBs where, despite successful rejoining of DNA ends, the original chromatin structure (and/or epigenetic code) has not been restored or was irreversibly damaged [reviews 135, 194]. The last alternative is supported by results suggesting that not only genetic mutations, but also epigenetic alterations that persist at the DSB sites can lead to the development of cancer or other diseases associated with "DSBs" (neurodegenerative diseases, aging etc.) [194]. If so, deregulation of gene expression caused by epigenetic alterations may affect much larger DNA domains than genetic mutations usually introduced by NHEJ[4], *e.g.* point mutations or small insertions and deletions[5] [194]. Indeed, epigenetic mutations such as hypermethylation of CpG islands in gene promoters (that is frequently responsible for silencing tumor suppressors) are well known causes of cancer. The equivalence of persistent γH2AX foci and unrestored chromatin structure at the sites

---

[2]however it should be kept on mind that not all individual foci of repair proteins necessarily colocalize with γH2AX and represent DSBs [199]

[3]frequently induced by (densely) ionizing IR

[4]since epigenetic "mutations" (like persisting γH2AX phosphorylation) may affect megabase-large chromatin domains

[5]except translocations

of DSBs may be supported also by the colocalization of heterochromatin protein 1 (HP1, [reviews 215, 216]) with γH2AX foci at later times PI [201, 217–220], since HP1 might function therein to re-condense chromatin structure previously relaxed by repair processes.

On the other hand, many authors identify γH2AX foci with unrepaired or even unreparable DSBs (with unrejoined DNA). Together with the observation that some γH2AX foci persist in nuclei throughout several cell divisions, this presumption lead to the postulation of a so called "adaptation to DSB damage", which enables the cell to enter mitosis even in the presence of unrepaired DSBs. If confirmed to be true, this process may be the last chance to eliminate damaged cells once other mechanisms have failed. The problem is that even though most cells die because of their inability to progress through mitosis, some cells might survive with a damaged genome. Therefore, whatever the persistent γH2AX foci are, they seem to signal an increased risk of the development of cancer (although it may be caused by various pathogenetic mechanisms).

In addition to γH2AX phosphorylation, a lot of other repair-associated changes of the epigenetic code at DSBs or perhaps at the pan-nuclear level (*e.g.*, associated with altered transcription programs after formation of DSBs) serve as specific platforms for protein-DNA and protein-protein interactions in the context of DSB repair and complex DDR signaling. For an overview of the "epigenetic memory" associated with DSB repair; see the review by Orlowski et al., [194].

20.4.4 Chromatin structure and the mechanism of DSB repair

DSB repair includes rapid- and slow-kinetic components [review 221], where the latter was supposed to represent the repair of complex DSBs that are processed only with difficulty [reviews 67, 92]. Nevertheless, concerning the structurally and functionally distinct chromatin domains, DSBs were thought to be introduced homogeneously and repaired by the same molecular mechanism with the same efficiency. Recent results however indicate that both DSB induction [review 135, 200] and repair [66–71, 135, 151, 194–200, *etc.*] are markedly influenced by chromatin compaction and structure. It was revealed that about 85% of DSBs introduced by sparsely ionizing IR do not require ATM activity for their repair [66, 67, 92]; surprisingly, DSBs that persisted unrepaired were almost exclusively localized in condensed heterochromatin [66–71]. Therefore, it seems that different "NHEJ" mechanisms with different efficiencies and fidelities act on the open and condensed chromatin substrates. In this model, "euchromatic" DSBs are rapidly rejoined by the "core" NHEJ members (Sec. 20.3.1), whereas more complicated processes associated with ATM-signaling pathway (Sec. 20.3.2) are required for repair of DSBs in dense heterochromatin [66–71]. This is in accordance with other results [135, 200, 222] that show DSB repair proceeds faster in euchromatin than in heterochromatin. By contrast, the proportion of DSBs in euchromatin progressively increases up to about 60 min PI [200, 201]. This seeming paradox can, however, be easily explained by rapid chromatin decondensation around heterochromatic DSBs, which is necessary either to allow a repair competent environment (increase the accessibility of huge repair complexes to damaged DNA) at the original sites of DSB lesions or to enable movement of DSBs into nuclear subcompartments that are more convenient locations for repair processes to occur [discussed in 135].

Both these hypotheses are not mutually exclusive and are not unprecedented in cellular biology: The potential movement into repair factories may be similar to the dynamic associations of genes in transcription factories, as discussed in the next section (20.4.5). On the other hand, it is also generally accepted that the condensed nature of heterochromatin poses a barrier to enzymes and other proteins that operate on DNA [223] and it must relax to allow transcription and replication [224–226]; therefore, chromatin decondensation may also be expected at the sites of heterochromatic DSBs. Indeed, epigenetic histone modifications typical for open euchromatin like increased H4K5 acetylation, decreased H3K9 dimethylation, and other modifications were observed at the sites of the DSBs almost immediately after their induction [198]. In accordance, as already discussed, ATM activates repair proteins that mediate chromatin decondensation such as KAP-1 (KRAB-associated protein 1, also known as TIF1b, TRIM28 or KRIP-1 [68, 227, 228]); importantly, this ATM activity is necessary only for the repair of heterochromatic DSBs. In support of this, dissociation of Heterochromatin protein 1 (HP1 [215, 216]) from affected heterochromatin domains was observed soon after irradiation [201, 217–220]. Moreover, in addition to local chromatin decondensation observed at the sites of lesions, global pan-nuclear decondensation initiated by DSB damage was described by Ziv et al. [227]. Finally, ATM initiates a complex DNA damage response that includes cell cycle arrest, which provides additional time for repair of heterochromatic breaks that can only be processed slowly.

Chromatin structure is, therefore, an important determinant of the initiation phase of DSB repair. Since, in addition to ATM, a lot of other proteins like γH2AX, MDC1, 53BP1, RNF8, RNF168 and Artemis are specifically required only for processing of DSBs in heterochromatin [70, 228], the above results can be interpreted as adaptive modification of NHEJ to problematic chromatin structure[6]. All together, DSB repair in heterochromatin seems to be slower, less efficient, and is potentially associated with increased formation of chromosomal translocations as discussed in Sec. 20.4.5.

Higher-order chromatin structure, however, also plays an important role in the terminal phase of DSB repair, as already discussed in the Sec. 20.4.3 in the context of persistent γH2AX foci. Since gene transcription is regulated by chromatin structure that is determined by an epigenetic code (Sec. 20.4.2), the original chromatin status must be somehow reconstructed along megabase-sized

---

[6]Multiple complex DSBs introduced into DNA by densely ionizing radiation [229,230] also require additional extensive processing of damaged DNA ends as compared to repair of simple DSBs induced by low-LET IR [231–233]. Both higher-order chromatin structure and DSB characteristics thus may determine individual steps of the repair mechanism

chromosome sequences that encompass DSBs to restore normal gene transcription. This process logically requires removal of repair-specific histone modifications and *de novo* re-configuration of the original epigenetic code.

The most striking reversion is the disappearance of γH2AX from DSBs after repair has been accomplished. This process is mediated either by γH2AX dephosphorylation or by replacement with non-phosphorylated histone H2AX [234–237]. Since re-accumulation of HP1 protein at DSB sites was also reported at longer times PI (about 4 hours) [201], these processes may reflect re-condensation of chromatin domains that had been relaxed during DSB repair. The exact steps necessary to reconstitute chromatin after DNA ends have been rejoined are not understood, nevertheless it is reasonable to hypothesize that unrestored chromatin structure is dangerous for the cell: Since "epimutations" do affect chromatin structure in large, megabase-sized DNA regions and, on the other hand, do not affect DNA integrity, cells can theoretically proceed through mitosis without any obstacles and transmit the altered expression of genes to subsequent generations (as discussed in Sec. 20.4.3).

Chromatin structure thus significantly influences the mechanism, efficiency, and probably the fidelity of DSB repair at all its phases. Despite additional experiments are required to confirm these results, all together it seems that heterochromatin poses serious obstacles to repair since: 1) especially for densely ionizing radiation, a higher concentration of chromatin per unit nuclear volume as compared with euchromatin provides more DNA targets in close mutual proximity, so clustered, complex lesions can form easily (see Sec. 20.4.6); 2) at the same time, repair of heterochromatic DSBs requires extensive decondensation of chromatin at the sites of lesions that must be 3) followed by its re-condensation. Repair of heterochromatic DSBs therefore includes additional steps (as compared with euchromatin) that are mechanistically and topologically problematic. This not only complicates the repair but also induces a chromatin "movement" that might simplify formation of chromosomal translocations.

20.4.5 Mobility of DSBs and the mechanism of formation of chromosomal translocations

The question of DSB mobility is one of the mostly disputed questions in radiobiology [reviewed, *e.g.*, in 135]. The previously described chromatin decondensation at the sites of DSBs, relaxation of free DNA ends caused by DSB induction, activities of huge repair complexes, invasion of broken DNA chains into undamaged template (for HR): these and potentially other processes associated with DSB induction or repair result in reorganization of chromatin structure and might be accompanied by chromatin "movement." Indeed, after irradiating cells with densely ionizing α-particles, Aten et al. [195] observed clustering of γH2AX foci during the time PI, which was interpreted as migration of DSBs into the

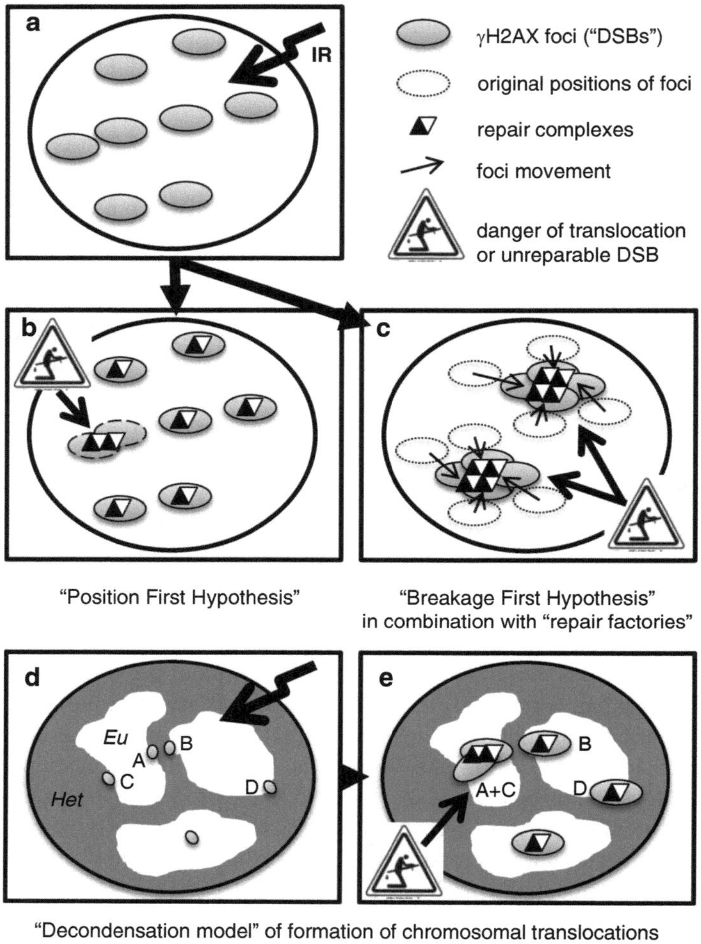

Fig. 20.1 The comparison of the "Position First Hypotheses" (**a, b**), "Breakage First Hypotheses" (**a, c**) and "Decondensation model" (**d, e**) of formation of chromosomal translocations.

In the case of the "Position First Hypothesis," DSBs (circles) induced by ionizing radiation (**a**) start to colocalize with repair proteins (white-black diamonds) and are repaired at the sites of their origin (**b**); translocations, therefore, form only between loci that have been close to each other already before their damage (**b**). The probability of a translocation between loci therefore correlates with their nuclear separation that is determined by the higher-order chromatin structure. On the other hand, the "Breakage First Hypothesis" presupposes migration (arrows) of DSBs (marked by γH2AX foci, ellipses) into specialized repair factories, where several DSBs are repaired in common (**c**). Chromosomal translocations can thus also form between previously distant loci (**c**) with an unspecified probability; specific loci might mutually associate more frequently because of the higher-order chromatin structure. The newest model of formation of chromosomal translocations, termed here the "Decondensation model" (**d, e**) [191, 271] counts with chromatin decondensation at the sites of DSB (**e**), which may result into the protrusion of damaged chromatin into low-dense chromatin subdomains ("chromatin holes," white) of the cell nucleus (**e**). Although most DSBs are repaired at their sites of origin, some can occasionally cluster "repair factories" where several DSBs are repaired simultaneously [195, 238, 239]. Markova et al. [209] arrived at similar conclusions when considering low-LET γ-rays. The mechanism proposed is not unprecedented in cellular biology. For example, "migration" of several co-transcribed genes into common transcription factories is well known, even for loci on different chromosomes [*e.g.* 152, 240]. Such compartmentalization of transcription processes provides many advantages to the cell including easier accumulation of proteins involved at transcription sites, better regulation, more intensive cellular signaling on weak initial stimuli, and lower energetic expenses, among others. Why can't a similar mechanism operate in DSB repair (Fig. 20.1a,c)? One reason could be the increased risk of misrejoining if multiple free DNA ends were captured in such a "factory". The hypothesis that counts with chromatin movement, potentially in combination with "repair factories", to explain the mechanism of formation of chromosomal translocations, is the "Breakage First Theory" (Fig. 20.1a,c); the release of free DNA ends caused by DSBs extends the movement of damaged chromatin, and consequently enables (complex) chromosomal translocations between originally distant DNA loci [135, 195, 241–243].

However, immunostaining of γH2AX foci in cells irradiated with low LET radiation (γ, X) showed relatively high spatial stability of DSBs during PI [199, 244, reviewed in 135]. In fact, the mobility of DSBs was equivalent to the Brownian movement of undamaged chromatin [135, 199, 244–247]. Similar results were also obtained for DSBs induced by other methods such as UV-lasers [197] or even heavy charged-particles [248–251]. Since individual DSBs colocalize with repair proteins already at their sites of origin, in the first few minutes PI, and later disappear without prior clustering with other lesions (marked as γH2AX foci), it is not probable that DSBs must migrate into repair factories to be processed [135, 199, 244–247]. DSBs are therefore usually repaired individually at the sites of their origin. This is related to the second widespread, and perhaps most accepted, hypothesis to describe the mechanism of formation of chromosomal translocations, the "Position First Theory" (Fig. 20.1a,b). If this were true, translocations would form only between DNA loci proximate in the cell nucleus before DSB induction [reviewed in 135, 242, 247].

Fig. 20.1 (continued) in a limited space of "chromatin holes" and produce translocations. Local higher-order chromatin structure determines to which "chromatin hole" a concrete DSB will protrude into, and thus also the probability of translocation between loci (**e**); nuclear distances between loci are therefore not the only parameter that influences the translocation probability. For example, DSBs "A" and "B" (**d, e**) that are localized at the opposite sides of a heterochromatic domain will protrude into different "chromatin holes" because of the heterochromatic "barrier" that separates them. The probability of translocation is therefore higher for loci "A" and "C" that protrude into the same chromatin hole, despite their original nuclear distance being larger as compared to that of "A" and "B." On the other hand, "B" and "D" are too distant to produce translocation, despite protruding to the same "chromatin hole"

So, why do some translocations appear much more frequently than others do? Is this because of the higher-order chromatin structure or DSB movement?[7] Moreover, in latter case, is there a higher chance that some loci will move to a particular nuclear domain and will mutually interact more frequently because of higher-order structure? Or, is a combination of both hypotheses required?

Since DSB movement was observed when cells were irradiated with densely [195], but usually not sparsely [197, 199], ionizing radiation, one explanation of the contradictory results regarding DSB mobility could be that high-LET IR fragments chromatin [247], due to high energy deposition along the particle track. It is indeed hardly conceivable that these different mechanisms (one processing DSBs at the sites of origin and one that requires their directed movement into repair factories) would operate depending on the quality of damaging IR or kind of DSBs produced. In support to this, our experiments in which DSBs were induced by low-LET γ-rays and immuno-detected as γH2AX, NBS1 or 53BP1 foci revealed generally little movement of DSBs, and, at the same time, showed that some foci had an exceptionally high mobility [199, reviewed in 135]. In agreement with the results of P. Jeggo's group [66–71] that showed a specific repair mechanism for heterochromatic DSBs (Sec. 20.4.4), mobile foci were almost exclusively located inside the condensed heterochromatin or at its border with decondensed euchromatin domains (called "chromatin holes" because of their low staining with DNA dyes in interphase nuclei) [199]. In addition, the seemingly chaotic movement of "mobile DSBs" was, in fact, directed from the condensed chromatin into chromatin holes, where it occasionally ended with the clustering of two or rarely more γH2AX foci [199; reviewed in 135]. Since heterochromatic DSBs largely colocalized with epigenetic markers suggesting chromatin decondensation (Sec. 20.4.4), their movement probably reflects the opening of dense heterochromatin domains that is initiated by DSB repair. The "directed" nature of this movement, although unexpected if it should result from "random" decondensation, can be also simply described since protrusion of damaged chromatin into chromatin holes is frequently easier than decondensation of the whole affected chromatin domain, especially when it comes to DSBs located close to the border with the chromatin hole [199; reviewed in 135] (Fig. 20.1d, e).

Importantly, the limited space within chromatin holes sometimes causes temporal or stable clustering of repair foci that protrude into the same hole [199]. So, what in fact represent these γH2AX clusters: complex multiple DSBs, repair factories or by-products of DSB repair with an increased risk of formation of chromosomal translocations? Since γH2AX clusters persist longer than single foci in cell nuclei, the last possibility seems to be most probable. A model that illustrates the potential relationships between higher-order chromatin structure, DSB repair, and formation of chromosomal translocations, is described in a detail in the review of Falk et al., 2010 [135] (Fig. 20.1d,e). Briefly, euchromatic DSBs are repaired at, or close to, their individual positions. Therefore, in accordance with the "Position First Theory,"

[7] provoked either by DSB induction or repair

the probability of their interactions and mis-rejoining is predetermined by a (global) higher-order chromatin structure that designates their initial nuclear positions and restricts their "movement" (Fig. 20.1d). However, heterochromatin must first decondense to allow DSB repair (Fig. 20.1e), and as a consequence, heterochromatic DSBs frequently protrude into the nearest chromatin hole where they can mutually interact with euchromatic or other heterochromatic DSBs that protruded into the same "hole" and produce chromosomal translocations (Fig. 20.1e). Since it enables interactions of DNA loci that were previously "distant", this resembles the "Breakage First Theory" (Fig. 20.1c). A critical concept in this model is that the initial nuclear distances of DSBs (the global chromatin architecture) are an important but not the only determinant of the probability of interactions between two or more particular loci. Indeed, the local higher-order chromatin structure finally determines whether concrete loci will protrude into the same or different chromatin holes (Fig. 20.1e). Thus, the highest probability of mutual interaction does not always correspond to the shortest nuclear distance. For example, if two loci A and B are located at the opposite sides of a small heterochromatin domain, they will protrude into different chromatin holes (Fig. 20.1e), which prevents their mutual interaction despite a very short nuclear distance between them. Alternatively, originally more distant loci A and C will protrude into the same chromatin hole, and, since they are not "too far away" from one another once they are induced[8], they pose the highest risk of forming chromosomal translocations (Fig. 20.1e). This model thus highlights the role of higher-order chromatin structure both in DSB repair and misrepair, and describes relationships between the factors and processes involved. Importantly, it combines aspects of both of the seemingly antagonistic "Breakage First" and "Position First" hypotheses (see Falk et al. [135]) (Fig. 20.1b, c).

20.4.6 Sensitivity of structurally and functionally distinct chromatin domains to DSB induction

As described in Sec. 20.4.1, genes are not homogeneously distributed along the DNA molecule, but form clusters of highly expressed and unexpressed genes [159, 160]. Since chromatin structure regulates and reflects the activity of genes (genetically silent regions are more condensed and associate with a large amount of heterochromatin binding proteins in contrast to the "open" structure of expressed chromatin), structurally and functionally distinct chromatin domains, and, at a higher level of organization also chromosomal territories, form in the three-dimensional space of the cell nucleus. As described earlier in Sections (20.4.4, 20.4.5), chromatin structure determines the mechanism of DSB repair and markedly influences its efficiency. Therefore, it is reasonable to ask, whether the higher-order

[8]due to the global higher-order chromatin structure

chromatin structure and nuclear architecture somehow also affect the sensitivity to DSB induction within structurally and functionally distinct chromatin domains. This is an important question as DSBs in a gene-dense chromatin cluster can damage many important genes including oncogenes and protooncogenes. Indeed, many experiments have shown preferential localization of γH2AX foci in decondensed euchromatin [66–71, 199–205, 250, 251] both after irradiating cells with low- and high-LET IR, but the explanation for this phenomenon is unknown.

Considering ionizing radiation, a particle that transits through the cell nucleus accompanied by a very high energy deposition along its path should theoretically damage chromatin without regard to its structure. If this is true, a high-density chromatin provides more DNA targets for high-LET IR. This is in accordance with early studies on the potentially different radiosensitivity of euchromatin and heterochromatin that revealed more chromosomal translocations in heterochromatin [252, and citations therein]. However, other authors [*e.g.*, 253] provided conflicting results and found that most chromosomal aberrations occurred in euchromatin. Indirect scoring of "DSBs" as the number of chromosomal translocations in eu- and hetero- chromatic bands of mitotic chromosomes, used in both studies, nevertheless considers not only the frequency of DSB induction but also processes of DSB repair and formation of chromosomal translocations, which also differ for these chromatin domains (Sec. 20.4.4, 20.4.5). Along with technological development, direct quantification of DSBs by a pulse-field gel electrophoresis (PFGE) showed a higher radiosensitivity of euchromatin, after exposure to both sparsely [254] and densely [255] ionizing radiation. Nevertheless, the results continue to be contradictory. For example, many authors reported equivalent sensitivities of different eu- and heterochromatin domains when using the same method [256–259] or they readdressed this question by analyzing complex genome rearrangements by newly available multicolor-FISH [260, 261].

As γH2AX foci were recognized as a specific marker of DSBs, their immunostaining in interphase nuclei [94–96, 208, 262, 263] promised to end this dispute, since it provides the highest sensitivity of DSB detection and, working on intact cells, eliminates problems with experiments on isolated chromatin [Sec. 20.4.3]. Using this method, γH2AX foci were found almost exclusively in decondensed "euchromatin". Unfortunately, it was then shown that heterochromatin is refractory to γH2AX foci formation [264], which can be alternatively explained by immediate heterochromatin decondensation at the sites of DSB induction that transforms heterochromatic DSBs to "euchromatic" [199, 200]. This could also explain the paradox where γH2AX foci are required only for the repair of heterochromatic DSBs [66–71] and, at the same time, heterochromatin structure does not allow their formation [264]. A simple detection of γH2AX foci nevertheless does not allow their scoring in the context of the original chromatin structure.

This obstacle was recently eliminated [200] by employing the ImmunoFISH method [265] to simultaneously label γH2AX foci with specific chromosomal (sub)domains. Thus, comparing the numbers of γH2AX foci introduced in individual chromosomal territories[9] or RIDGE and antiRIDGE gene clusters of equivalent molecular sizes [200], it was possible to score DSBs independently of changes of the original chromatin structure at the sites of DSBs (monitored *e.g.* by DNA dyes) [200]. In the case of DSBs formed in a condensed antiRIDGE domain, all lesions were correctly considered to be heterochromatic[10], despite chromatin decondensation at the sites. For sparsely ionizing γ-radiation, these experiments showed decondensed chromatin had a higher susceptibility to DSB damage [200]. Since most of the damage introduced with γ-rays is not caused by the photon *per se* but by the indirect effect of irradiation [266], these results were interpreted to mean that the structural features[11] of heterochromatin [267] might protect DNA against free radicals [135, 200, 268–270] generated by water radiolysis [271, 272]. Active DNA loci, containing important genes, therefore seem to be at a higher risk for radiation damage, with previously described potential consequences for human health [200; reviewed in 135]. Overall chromatin structure can therefore contribute to the different radiosensitivity seen in different cell types or in developmental stages. On the other hand, after an exposure to densely ionizing radiation, the number of DSBs generated could be a positive function of chromatin density, since a higher locally deposited energy results in a different mechanism of DNA damage (characterised by a prevalence of the "direct effect") than γ-radiation. Surprisingly, many authors showed that DSB distribution is nonrandom also in this case; however, it is unclear whether this observation can be attributed to chromatin structure [273, 274].

20.5 Conclusions and Perspectives

It is now generally accepted that, in addition to its biochemical aspects, the spatio-temporal organization of chromatin plays an important role in the formation of chromosomal translocations and DSB repair. Recent results show that higher-order chromatin structure can influence the sensitivity of chromatin to DSB formation, the complexity of DSBs produced, the speed, efficiency, and fidelity of DSB repair, the risk of formation of chromosomal translocations, and potentially other aspects of complex cellular response to DNA damage. Therefore, it is evident that an impact of higher-order chromatin structure on "DSB repair" is really complex. It seems that "opened" euchromatin that contains important genes is more sensitive to DSB induction; but, at the same time, it is more efficiently repaired and the risk of formation of chromosomal translocations in euchromatin is probably lower than that of heterochromatin. On the other hand, the complex structure of heterochromatin

---

[9] of similar molecular sizes or nuclear volumes

[10] and not misinterpreted as euchromatic

[11] but not simply its compactness

partially shields DNA from harmful free radicals produced by irradiation, but it also complicates repair processes because of the requirement for extensive chromatin decondensation at the sites of DSBs, which then increases the risk of repair errors. This decondensation may also make it easier for originally relatively "distant" DSBs to interact and elevates the risk of chromosomal translocations. Consequently, the probability of mutual interaction of two particular loci is determined not only by their original nuclear distances, but also by the local chromatin "texture."

Recent developments in microscopy and molecular methods, facilitate experiments in living cells that are necessary to continually monitor DSB repair under physiological conditions and in real-time. However, processes associated with DSB induction and repair, formation of chromosomal translocations and, finally, clinical manifestation of DNA damage, encompass a broad range of times and spatial dimensions. Additional research will therefore require a multidisciplinary approach with many benefits including a better understanding of the repair mechanisms that will allow specific inhibition or stimulation, either by influencing expression of genes involved or by modification of chromatin structure. This will enable developments to improve current radiotherapy as well as more advanced chemotherapeutics. For example, a new and very promising ion-beam cancer therapy could not be safely and fully introduced without additional physical, chemical, biological, medical and technological research. Many other benefits that can be anticipated were already discussed in Sec. 20.1.

Acknowledgments Supported by the IAA500040802 project of the Grant Agency of CR

References

1. Lodish H, Berk A, Matsudaira P, Kaiser CA, Krieger M, Scott MP, Zipursky SL, Darnell J (2004) Molecular Biology of the Cell, 5th edn. WH Freeman, New York
2. Mani RS, Chinnaiyan AM (2010) Nat Rev Genet 11(12), 819–29. Review
3. Vilenchik MM, Knudson AG (2003) Proc Natl Acad Sci U S A 28; 100(22), 12871–6.
4. Obe G, Johannes C, Schulte-Frohlinde D (1992) Mutagenesis 7(1), 3–12. Review
5. Olive PL (1998) Radiat Res 150(5 Suppl), S42–51. Review
6. Jeggo PA (1998) DNA breakage and Repair. In: Hall JC, Dunlap JC, Friedmann T, Giannelli F (ed) Advances in Genetics 38, Academic Press, San Diego.
7. Vilenchik MM, Knudson AG Jr (2000) Proc Natl Acad Sci U S A 97(10), 5381–6.
8. Harper JV, Anderson JA, O'Neill P (2010) DNA Repair 9(8), 907–913. DOI:10.1016/j.dnarep.2010.06.002
9. Povrik F and Austin MJF (1991) Mutat Res 257, 127–143.
10. Jackson SP, Bartek J (2009) Nature 461(7267), 1071–8. Review
11. Podhorecka M, Skladanowski A, Bozko P (2010) J Nucleic Acids 2011. doi:10.4061/2011/920161Review
12. Khanna KK, Jackson SP (2001) Nat Genet 27(3), 247–54. Review
13. Morio T, Kim H (2008) Int J Biochem Cell Biol 40(4), 598–603. DOI:10.1016/j.biocel.2007.12.007
14. Nijnik A, Woodbine L, Marchetti C, et al (2007) Nature 447(7145), 686–90.

15. Caldecott KW (2011) Eukaryotic DNA Damage Surveillance and Repair (Molecular Biology Intelligence Unit), 1st edn. Springer
16. Ljungman M (2009) Chem Rev 109(7), 2929–50. Review
17. Fan R, Kumaravel TS, Jalali F, Marrano P, Squire JA, Bristow RG (2004) Cancer Res 64(23), 8526–33.
18. Powell SN, Kachnic LA (2008) Anticancer Agents Med Chem 8(4), 448–60. Review
19. Jung D, Alt FW (2004) Unraveling V(D)J recombination; insights into gene regulation. Cell 116(2), 299–311. Review
20. Jung D, Giallourakis C, Mostoslavsky R, Alt FW (2006) Annu Rev Immunol 24, 541–70. Review
21. Franco S, Murphy MM, Li G, Borjeson T, Boboila C, Alt FW (2008) J Exp Med 205(3), 557–64.
22. Chaudhuri J, Basu U, Zarrin A, Yan C, Franco S, Perlot T, Vuong B, Wang J, Phan RT, Datta A, Manis J, Alt FW (2007) Adv Immunol 94,157–214. Review
23. Limoli CL, Giedzinski E, Bonner WM, Cleaver JE (2002) Proc Natl Acad Sci U S A 99(1), 233–8.
24. Rothstein R, Michel B, Gangloff S (2000) Genes Dev 14(1), 1–10. Review
25. Shrivastav M, De Haro LP, Nickoloff JA (2008) Cell Res 18(1), 134–47. Review
26. Keeney S, Neale MJ (2006) Biochem Soc Trans 34(Pt4), 523–5. Review
27. Maresca B, Schwartz JH (2006) Anat Rec B New Anat 289(1), 38–46. Review
28. Obolensky OI, Surdutovich E, Pshenichnov I, Mishustin I, Solov'yov AV, Greiner W (2008) Nucl Instrum Methods Phys Res Section B: Beam Interactions with Materials and Atoms 266(8), 1623–1628. DOI:10.1016/j.nimb.2007.11.054.
29. Harper JW, Elledge SJ (2007) Mol Cell 28(5), 739–45. Review
30. Pontier DB, Tijsterman M (2009) Cur Biol 19(16), 1384–1388. DOI:10.1016/j.cub.2009.06.045
31. Griffiths AJF, Wessler SR, Carroll SB Doebley J (2011) Introduction to genetic analysis, 10th edn. Macmillan.
32. Bertrand P, Tishkoff DX, Filosi N, Dasgupta R, Kolodner RD (1998) Proc Natl Acad Sci U S A 95(24), 14278–83.
33. Hoeijmakers JH (2001) Nature 411(6835), 366–74. Review
34. Hoeijmakers JH (2001). Maturitas 38(1), 17–22. Review
35. Peterson CL, Côté J (2004) Genes Dev 18(6), 602–16. Review
36. Spry M, Scott T, Pierce H, D'Orazio JA (2007) Front Biosci 12, 4191–207. Review
37. Neher TM, Turchi J (2011) Antioxid Redox Signal [in press]
38. Branzei D & Foiani M (2008) Nat Rev Mol Cell Biol 9, 297–308. DOI:10.1038/nrm2351
39. Lisby M, Barlow JH, Burgess RC, Rothstein R (2004) Cell 118(6), 699–713. DOI:10.1016/j.cell.2004.08.015
40. Iliakis G (2009) Radiother Oncol 92(3), 310–5. Review
41. Sonoda E, Hochegger H, Saberi A, Taniguchi Y, Takeda S (2006) DNA Repair 5(9–10), 1021–9. Review
42. Mao Z, Bozzella M, Seluanov A, Gorbunova V (2008) DNA Repair 7, 1765–1771.
43. Van Nguyen T, Puebla-Osorio N, Pang H, Dujka ME, Zhu C (2007) J Exp Med 204(6), 1453–61.
44. Roos WP, Kaina B (2006) Trends Mol Med 12(9), 440–50. Review
45. Bartek J, Lukas J (2007) Curr Opin Cell Biol 19(2), 238–45. Review
46. Sochanowicz B, Szumiel I (2005) Nukleonika 50(4), 129–138.
47. Assenmacher N, Hopfner KP (2004) Chromosoma 113(4), 157–66. Review
48. Williams RS, Williams JS, Tainer JA (2007) Biochem Cell Biol 85(4), 509–20. Review.
49. Iijima K, Ohara M, Seki R, Tauchi H (2008) J Radiat Res (Tokyo) 49(5), 451–64. Review
50. Lamarche BJ, Orazio NI, Weitzman MD (2010) FEBS Lett 584(17), 3682–95. Review
51. Yosef Shiloh (2003) Nat Rev Cancer 3, 155–168. DOI:10.1038/nrc1011
52. Abraham RT (2001) Genes Dev 15(17), 2177–96. Review
53. Lavin MF, Kozlov S (2007) Cell Cycle 6(8), 931–42. Review

54. Kastan MB & Lim D (2000) Nat Rev Mol Cell Biol 1, 179–186. DOI:10.1038/35043058
55. Bakkenist CJ, Kastan MB (2003) Nature 421(6922), 499–506.
56. Goodarzi AA, Jonnalagadda JC, Douglas P, Young D, Ye R, Moorhead GB, Lees-Miller SP, Khanna KK (2004) EMBO J 23(22), 4451–61.
57. Uziel T, Lerenthal Y, Moyal L, Andegeko Y, Mittelman L, Shiloh Y (2003) EMBO J 22(20), 5612–21.
58. Lee JH, Paull TT (2005) Science 308(5721), 551–4.
59. Zhuang J, Jiang G, Willers H, Xia F (2009) J Biol Chem 284(44), 30565–73.
60. Paull TT, Gellert M (1998) Mol Cell 1(7), 969–79.
61. Zhuang J, Jiang G, Willers H, Xia F (2009) J Biol Chem 284(44), 30565–73.
62. Usui T, Ohta T, Oshiumi H, Tomizawa J, Ogawa H, Ogawa T (1998) Cell 95(5):705–16.
63. Williams RS, Moncalian G, Williams JS, et al (2008) Cell 135(1), 97–109.
64. Hopfner KP, Craig L, Moncalian G, Zinkel RA, Usui T, Owen BA, Karcher A, Henderson B, Bodmer JL, McMurray CT, Carney JP, Petrini JH, Tainer JA (2002) Nature 418(6897), 562–6.
65. Tomimatsu N, Tahimic CG, Otsuki A, Burma S, Fukuhara A, Sato K, Shiota G, Oshimura M, Chen DJ, Kurimasa A (2007) J Biol Chem 282(14), 10138–45.
66. Löbrich M, Jeggo PA (2005) DNA Repair 4(7), 749–59. Review
67. Riballo E, Kühne M, Rief N, Doherty A, Smith GC, Recio MJ, Reis C, Dahm K, Fricke A, Krempler A, Parker AR, Jackson SP, Gennery A, Jeggo PA, Löbrich M (2004) Mol Cell 16(5), 715–24.
68. Goodarzi AA, Noon AT, Deckbar D, Ziv Y, Shiloh Y, Löbrich M, Jeggo PA (2008) Mol Cell 31(2), 167–77.
69. Goodarzi AA, Noon AT, Jeggo PA (2009) Biochem Soc Trans 37(Pt 3), 569–76.
70. Noon AT, Shibata A, Rief N, Löbrich M, Stewart GS, Jeggo PA, Goodarzi AA (2010) Nat Cell Biol 12(2), 177–84.
71. Goodarzi AA, Jeggo P, Lobrich M (2010) DNA Repair 9(12), 1273–82
72. Smith GCM and Jackson SP (1999) Genes Dev 13, 916–934.
73. Gottlieb TM, Jackson SP (1993) Cell 72(1), 131–42.
74. Meek K, Gupta S, Ramsden DA, Lees-Miller SP (2004) Immunol Rev 200, 132–41. Review
75. Chan DW, Chen BP, Prithivirajsingh S, Kurimasa A, Story MD, Qin J, Chen DJ (2002) Genes Dev 16(18), 2333–8.
76. Downs JA, Jackson SP (2004) Nat Rev Mol Cell Biol 5(5), 367–78. Review
77. Drouet J, Frit P, Delteil C, de Villartay JP, Salles B, Calsou P (2006) IJ Biol Chem 281(38), 27784–93.
78. Houtgraaf JH, Versmissen J, van der Giessen WJ (2006) Cardiovascular Revascularization Medicine 7(3), 165–172. DOI:10.1016/j.carrev.2006.02.002
79. Iliakis G, Wang H, Perrault AR, Boecker W, Rosidi B, Windhofer F, Wu W, Guan J, Terzoudi G, Pantelias G (2004) Cytogenet Genome Res 104(1–4):14–20. Review
80. Pastink A, Eeken JC, Lohman PH (2001) Mutat Res 480–481, 37–50. Review
81. Haber JE (2000) Trends Genet 16(6), 259–264. DOI:10.1016/S0168-9525(00)02022-9
82. Yano K, Morotomi-Yano K, Adachi N, Akiyama H (2009) J Radiat Res (Tokyo) 50(2), 97–108. Review
83. Weterings E, van Gent DC (2004) DNA Repair (Amst) 3(11), 1425–35. Review
84. Lieber MR, Ma Y, Pannicke U, Schwarz K (2003) Nat Rev Mol Cell Biol 4(9), 712–20. Review
85. Dudas A, Chovanec M (2004) Mutat Res 566(2), 131–67.
86. McVey M, Lee SE (2008) Trends Genet 24(11), 529–38. Review
87. Prise KM, Pinto M, Newman HC and Michael BD (2001) Rad Res 156(5), 572–576.
88. Stracker TH, Theunissen JW, Morales M, Petrini JH (2004) DNA Repair 3(8–9), 845–54. Review
89. Paull TT, Gellert M (1999) Genes Dev 13(10), 1276–88.
90. van den Bosch M, Bree RT, Lowndes NF (2003) EMBO Rep 4(9), 844–9. Review
91. Desai-Mehta A, Cerosaletti KM, Concannon P (2001) Mol Cell Biol 21(6), 2184–91.
92. Jeggo PA, Löbrich M (2005) Cell Cycle 4(3), 359–62. Review

93. Lee JH, Paull TT (2004) Science 304(5667), 93–6.
94. Rogakou EP, Pilch DR, Orr AH, Ivanova VS, Bonner WM (1998) J Biol Chem 273(10), 5858–68.
95. Sedelnikova OA, Pilch DR, Redon C, Bonner WM (2003) Cancer Biol Ther 2(3), 233–5. Review
96. Rogakou EP, Boon C, Redon C, Bonner WM (1999) J Cell Biol 146(5), 905–16.
97. Stucki M, Jackson SP (2006) DNA Repair 5(5), 534–43.
98. Minter-Dykhouse K, Ward I, Huen MS, Chen J, Lou Z (2008) J Cell Biol 181(5), 727–35.
99. Anderson L, Henderson C, Adachi Y (2001) Mol Cell Biol 21, 1719–1729.
100. Lou Z, Chini CC, Minter-Dykhouse K, Chen J (2003) J Biol Chem 278(16), 13599–602.
101. Lowndes NF (2010) DNA Repair 9(10),1112–6.
102. Paull TT, Rogakou EP, Yamazaki V, Kirchgessner CU, Gellert M, Bonner WM (2000) Curr Biol 10(15), 886–95.
103. Celeste A, Fernandez-Capetillo O, Kruhlak MJ, Pilch DR, Staudt DW, Lee A, Bonner RF, Bonner WM, Nussenzweig A (2003) Nat Cell Biol 5(7), 675–9.
104. Ward IM, Minn K, van Deursen J, Chen J (2003)Mol Cell Biol 23(7), 2556–63.
105. Goldberg M, Stucki M, Falck J, D'Amours D, Rahman D, Pappin D, Bartek J, Jackson SP (2003) Nature 421(6926), 952–6.
106. Huen MS, Grant R, Manke I, Minn K, Yu X, Yaffe MB, Chen J (2007) Cell 131(5), 901–14.
107. Kolas NK, Chapman JR, Nakada S, Ylanko J, Chahwan R, Sweeney FD, Panier S, Mendez M, Wildenhain J, Thomson TM, Pelletier L, Jackson SP, Durocher D (2007) Science 318(5856), 1637–40.
108. Venkitaraman AR (2004) Nat Rev Cancer 4(4), 266–76. Review
109. Bouwman P, Aly A, Escandell JM, et al (2010) Nat Struct Mol Biol 17(6), 688–95.
110. Bunting SF, Callén E, Wong N, et al (2010) Cell 141(2), 243–54.
111. Sung P, Klein H (2006) Nat Rev Mol Cell Biol 7(10), 739–50. Review
112. van den Bosch M, Lohman PH, Pastink A (2002) Biol Chem 383(6), 873–92. Review
113. Shcherbakov VP, Plugina L, Shcherbakova T, Sizova S, Kudryashova E (2011) DNA Repair 10(1), 16–23.
114. Heride C, Ricoul M, Kiêu K, von Hase J, Guillemot V, Cremer C, Dubrana K, Sabatier L (2010) J Cell Sci 123(Pt23), 4063–75.
115. Nagele RG, Freeman T, McMorrow L, Thomson Z, Kitson-Wind K, Lee H (1999) J Cell Sci 112(Pt4), 525–35.
116. Critchlow SE, Jackson SP (1998) Trends Biochem Sci 23(10), 394–8. Review
117. Mao Z, Bozzella M, Seluanov A, Gorbunova V (2008) Cell Cycle 7, 2902–2906.
118. Daley JM, Palmbos PL, Wu D, Wilson TE (2005) Annu Rev Genet 39, 431–51. Review
119. Goodhead DT (1994) Int J Radiat Biol 65, 7–17.
120. Ward JF (1994) Int J Radiat Biol 66, 427–432.
121. Pastwa E, Neumann RD, Mezhevaya K, Winters TA (2003) Radiat Res 159, 251–61.
122. Zhang Y, Rowley JD (2006) DNA Repair 5(9–10), 1282–97. Review
123. Karanjawala ZE, Grawunder U, Hsieh CL, Lieber MR (1999) Curr Biol 9(24), 1501–4.
124. Ferguson DO, Sekiguchi JM, Chang S, Frank KM, Gao Y, DePinho RA, Alt FW (2000) Proc Natl Acad Sci U S A 97(12), 6630–3.
125. Sharpless NE, Ferguson DO, O'Hagan RC et al (2001) Mol Cell 8, 1187–1196.
126. Lin Y, Lukacsovich T, Waldman AS (1999) Mol Cell Biol 19, 8353–8360.
127. Rothkamm K, Krüger I, Thompson LH, Löbrich M (2003) Mol Cell Biol 23(16), 5706–15.
128. Chen F, Nastasi A, Shen Z, Brenneman M, Crissman H, Chen DJ (1997) Mutat Res 384(3), 205–11.
129. Kim JS, Krasieva TB, Kurumizaka H, Chen DJ, Taylor AM, Yokomori K (2005) J Cell Biol 170(3), 341–7.
130. Iliakis G (2009) Radiother Oncol 92(3), 310–5. Review
131. Wu W, Wang M, Wu W, Singh SK, Mussfeldt T, Iliakis G (2008) DNA Repair 7(2), 329–38.
132. Singh SK, Wu W, Zhang L, Klammer H, Wang M, Iliakis G (2011) Int J Radiat Oncol Biol Phys 79(2), 540–8.

133. Ivanov EL, Sugawara N, Fishman-Lobell J, Haber JE (1996) Genetics 142(3), 693–704.
134. Frankenberg-Schwager M, Gebauer A, Koppe C, Wolf H, Pralle E and Frankenberg D (2009) Rad Res 171(3), 265–273.
135. Falk M, Lukasova E, Kozubek S (2010) Mutat Res 704(1–3), 88–100. Review
136. Bekhor I, Feldman B (1976) Biochemistry 15(22), 4771–4777. DOI:10.1021/bi00667a004
137. Wolffe AP (1999) Chromatin, Structure and Function, 3rd ed. Academic, San Diego.
138. Chen P, Li G (2010) Protein Cell 1(11), 967–71.
139. Xu M, Zhu B (2010) Protein Cell 1(9), 820–9.
140. Ray-Gallet D, Almouzni G (2010) Essays Biochem 48(1), 75–87. Review
141. Berezney R (2002) Adv Enzyme Regul 42, 39–52. Review
142. Woodcock CL, Ghosh RP (2010) Cold Spring Harb Perspect Biol 2(5), a000596. Review
143. Klug WS & Cummings MR & Spencer CA (2008) Concepts of Genetics, 9th edn. Prentice Hall.
144. Matthews HR (1997) DNA Structure Prerequisite Information. In: Campbell, Reece, Mitchell, Taylor (ed) Biology Concept and Connections. Benjamin Cummings, California.
145. McGraw (1997) Hill Encyclopedia of Science and Technology. McGraw Hill, New York.
146. McNally JG. (2009) J Cell Biol 185(1), 7–9.
147. Dillon N (2004) Biol Cell 96(8), 631–7. Review
148. Jenuwein T, Allis CD (2001) Science 293(5532), 1074–80. Review
149. Cremer T, Cremer C (2001) Nat Rev Genet. 2001 Apr;2(4), 292–301. Review
150. Branco MR, Pombo A (2006) PLoS Biol 4(5), e138.
151. Aten JA, Kanaar R (2006) PLoS Biol 4(5), e155.
152. Simonis M, Klous P, Splinter E, Moshkin Y, Willemsen R, de Wit E, van Steensel B, de Laat W (2006) Nat Genet 38(11), 1348–54.
153. Kozubek S, Lukásová E, Jirsová P, Koutná I, Kozubek M, Ganová A, Bártová E, Falk M, Paseková R (2002) Chromosoma 111(5), 321–31.
154. Lukásová E, Kozubek S, Kozubek M, Falk M, Amrichová J (2002) Chromosome Res 10(7), 535–48.
155. Falk M, Lukásová E, Kozubek S, Kozubek M (2002) Gene. 2002 Jun 12;292(1–2), 13–24.
156. Bolzer A, Kreth G, Solovei I, Koehler D, Saracoglu K, Fauth C, Müller S, Eils R, Cremer C, Speicher MR, Cremer T (2005) PLoS Biol 3(5), e157
157. Parada LA, McQueen PG, Misteli T (2004). Genome Biol 5(7), R44.
158. Neusser M, Schubel V, Koch A, Cremer T, Müller S (2007) Chromosoma 116(3), 307–20.
159. Caron H, van Schaik B, van der Mee M, et al (2001) Science 291(5507), 1289–92.
160. Versteeg R, van Schaik BD, van Batenburg MF, Roos M, Monajemi R, Caron H, Bussemaker HJ, van Kampen AH (2003) Genome Res 13(9), 1998–2004.
161. Misteli T (2005) Bioessays 27(5), 477–87. Review
162. Cremer T, Cremer M (2010) Cold Spring Harb Perspect Biol 2(3), a003889. Review
163. Sun HB, Shen J, Yokota H (2000). Biophys J 79(1), 184–90.
164. Boyle S, Gilchrist S, Bridger JM, Mahy NL, Ellis JA, Bickmore WA (2001) Hum Mol Genet 10(3), 211–9.
165. Bridger JM, Boyle S, Kill IR, Bickmore WA (2000) Curr Biol 10(3), 149–52.
166. Croft JA, Bridger JM, Boyle S, Perry P, Teague P, Bickmore WA (1999) J Cell Biol 145(6), 1119–31.
167. Mehta IS, Amira M, Harvey AJ, Bridger JM (2010) Genome Biol 11(1), R5.
168. Brown KE, Baxter J, Graf D, Merkenschlager M, Fisher AG (1999) Mol Cell 3, 207–217.
169. Cremer C, Münkel C, Granzow M, Jauch A, Dietzel S, Eils R, Guan XY, Meltzer PS, Trent JM, Langowski J, Cremer T (1996). Mutat Res 366(2), 97–116. Review
170. Razin SV, Gromova II (1995) BioEssays 17, 443–450.
171. Kosak ST, Groudine M (2002) Dev Cell 2(6), 690–2.
172. Visser AE, Aten JA (1999) J Cell Sci 112(Pt19), 3353–60.
173. Francastel C, Schübeler D, Martin DI, Groudine M (2000) Nat Rev Mol Cell Biol 1(2), 137–43. Review
174. Sinčić N, Herceg Z (2010) Curr Opin Oncol [Epub ahead of print].

175. Strahl BD, Allis CD (2000) Nature 403(6765), 41–5.
176. Nightingale KP, O'Neill LP, Turner BM (2006) Curr Opin Genet Dev 16(2), 125–36. Review
177. Turner BM (2007) Nat Cell Biol 9(1), 2–6.
178. Lachner M, O'Carroll D, Rea S, Mechtler K, Jenuwein T (2001) Nature 410(6824), 116–20.
179. Nakayama J, Rice JC, Strahl BD, Allis CD, Grewal SI (2001) Science 292(5514), 110–3.
180. Allis D, Jenuwein T, Reinberg D (2007) Epigenetics. Cold Spring Harbor Laboratory Press, New York.
181. Rossetto D, Truman AW, Kron SJ, Côté J (2010) Clin Cancer Res 16(18), 4543–52.
182. Ruthenburg AJ, Allis CD, Wysocka J (2007) Mol Cell 25, 15–30.
183. Volpi EV, Chevret E, Jones T, et al. (2000) J Cell Sci 113, 1565–1576.
184. Williams RRE, Broad S, Sheer D, Ragoussis J (2002) Exp Cell Res 272, 163–175.
185. Osborne CS, Chakalova L, Brown KE, Carter D, Horton A, et al. (2004) Nat Genet 36, 1065–1071.
186. Spilianakis CG, Lalioti MD, Town T, Lee GR, Flavell RA (2005) Nature 435, 637–645.
187. Sutherland H, Bickmore WA (2009) Nat Rev Genet 10(7), 457–66. Review
188. Dorier J, Stasiak A (2010) Nucleic Acids Res 38(21), 7410–21.
189. Nolis IK, McKay DJ, Mantouvalou E, Lomvardas S, Merika M, Thanos D (2009) Proc Natl Acad Sci U S A 106(48), 20222–7.
190. Woodcock CL, Ghosh RP (2010) Cold Spring Harb Perspect Biol 2(5), a000596. Review
191. Chubb JR, Boyle S, Perry P, Bickmore WA (2002) Curr Biol 12(6), 439–45.
192. Bridger JM and Mehta IS (2011) Nuclear Molecular Motors for Active, Directed Chromatin Movement in Interphase Nuclei. In: Adams NM and Freemont PS (ed) Advances in Nuclear Architecture, 1st edn. Springer, DOI:10.1007/978-90-481-9899-3_5
193. Belmont A (2003) Curr Opin Cell Biol 15(3), 304–10. Review
194. Orlowski C, Mah LJ, Vasireddy RS, El-Osta A, Karagiannis TC (2010) Chromosoma [Epub ahead of print]. Review
195. Aten JA, Stap J, Krawczyk PM, van Oven CH, Hoebe RA, Essers J, Kanaar R (2004) Science 303(5654), 92–5.
196. Lukas C, Bartek J, Lukas J (2005) Chromosoma 114(3), 146–54. Review
197. Kruhlak MJ, Celeste A, Nussenzweig A (2006) Cell Cycle 5(17), 1910–2.
198. Loizou JI, Murr R, Finkbeiner MG, Sawan C, Wang ZQ, Herceg Z (2006) Cell Cycle 5(7), 696–701. Review
199. Falk M, Lukasova E, Gabrielova B, Ondrej V, Kozubek S (2007) Biochim Biophys Acta 1773(10), 1534–45.
200. Falk M, Lukásová E, Kozubek S (2008) Biochim Biophys Acta 1783(12), 2398–414.
201. Falk M, Lukasova E, Gabrielova B, Ondrej V, Kozubek S (2008) J Phys Conf Ser 101
202. Cowell IG, Sunter NJ, Singh PB, Austin CA, Durkacz BW, Tilby MJ (2007) PLoS One 2(10), e1057.
203. Kinner A, Wu W, Staudt C, Iliakis G (2008)Nucleic Acids Res. 36(17), 5678–94. Review
204. Vasireddy RS, Karagiannis TC, El-Osta A (2010) Cell Mol Life Sci. 67(2), 291–4.
205. Karagiannis TC, Harikrishnan KN, El-Osta A (2007) Oncogene 26(27), 3963–71.
206. Wang H, Wang M, Wang H, Böcker W and Iliakis G (2005) J Cell Physiol 202(2), 492–502.
207. Redon C, Pilch D, Rogakou E, Sedelnikova O, Newrock K, Bonner W (2002) Curr Opin Genet Dev 12(2), 162–9. Review
208. Rothkamm K, Lobrich M (2003) Proc Natl Acad Sci USA 100(9), 5057–62.
209. Markova E, Schultz N, Belyaev IY (2007) Int J Radiat Biol 83(5), 319–29.
210. Szumiel I (2007) Analysis of DNA Double-Strand Breaks by Means of γ-H2AX Foci. In: Obe, Günter, Vijayalaxmi (ed) Chromosomal Alterations 1st eds. Springer, DOI:10.1007/978-3-540-71414-9_9
211. Wang Y, Shyy JY, Chien S (2008) Annu Rev Biomed Eng 10, 1–38. Review
212. Foray N, Arlett CF, Malaise EP (1997) Biochimie 79(9–10), 567–75. Review
213. Foray N, Monroco C, Marples B, Hendry JH, Fertil B, Goodhead DT, Arlett CF, Malaise EP (1998) Int J Radiat Biol 74(5),551–60.

214. Foray N, Charvet AM, Duchemin D, Favaudon V, Lavalette D (2005) J Theor Biol 236(4), 448–58.
215. Zeng W, Ball AR Jr, Yokomori K (2010) Epigenetics 5(4), 287–92. Review
216. Shimada A, Murakami Y (2010)Epigenetics 5(1), 30–3. Review
217. Dinant C, Luijsterburg MS (2009) Mol Cell Biol 29(24), 6335–40. Review
218. Ayoub N, Jeyasekharan AD, Bernal JA, Venkitaraman AR (2008) Nature 453(7195), 682–6.
219. Ayoub N, Jeyasekharan AD, Venkitaraman AR (2009) Cell Cycle 8(18), 2945–50. Review
220. Ball AR Jr, Yokomori K (2009) J Cell Biol 185(4), 573–5. Review
221. Frankenberg-Schwager M, Frankenberg D (1994) Radiat Res 138(1 Suppl), S97–100. Review
222. Surrallés J, Sebastian S, Natarajan AT (1997) Mutagenesis 12(6), 437–42.
223. Osley MA, Tsukuda T, Nickoloff JA (2007) Mutat Res 618(1–2), 65–80. Review
224. Müller WG, Walker D, Hager GL, McNally JG (2001) J Cell Biol 154(1), 33–48.
225. Nye AC, Rajendran RR, Stenoien DL, Mancini MA, Katzenellenbogen BS, Belmont AS (2002) Mol Cell Biol 22(10), 3437–49.
226. Carpenter AE, Memedula S, Plutz MJ, Belmont AS (2005) Mol Cell Biol 25(3), 958–68.
227. Ziv Y, Bielopolski D, Galanty Y, Lukas C, Taya Y, Schultz DC, Lukas J, Bekker-Jensen S, Bartek J, Shiloh Y (2006) Nat Cell Biol 8(8), 870–6.
228. Cann KL and Dellaire G (2011) Cell Biol 89(1), 45–60. DOI:10.1139/O10--113
229. Sage E, Harrison L (2010) Mutat Res. 2010 [Epub ahead of print].
230. Shikazono N, Noguchi M, Fujii K, Urushibara A, Yokoya A (2009) J Radiat Res (Tokyo) 50(1), 27–36. Review
231. Tsuchida Y, Tsuboi K, Ohyama H, Ohno T, Nose T, Ando K (1998) Brain Tumor Pathol 15(2), 71–6.
232. De Laney TF, Kooy HM (2007) Proton and Charged Particle Radiotherapy. Lippincott Williams & Wilkins.
233. Iwadate Y, Mizoe JE, Osaka Y, Yamaura A, Tsujii H (2001) Int J Radiat Oncol Biol Phys 50(3), 803–808.
234. van Attikum H, Gasser SM (2009) Trends Cell Biol 19(5), 207–17. Review
235. Ikura T, Tashiro S, Kakino A, et al (2007) Mol Cell Biol 27(20), 7028–40.
236. Keogh MC, Kim JA, Downey M, et al. (2006) Nature 439, 497–501.
237. Chowdhury D, Keogh MC, Ishii H, Peterson CL, Buratowski S, Lieberman J (2005) Mol Cell 20(5), 801–9.
238. Krawczyk PM, Stap J, van Oven C, Hoebe R, Aten JA (2006) Radiat Prot Dosimetry 122(1–4), 150–3.
239. Lisby M, Mortensen UH, Rothstein R (2003) Nat Cell Biol 5(6), 572–7.
240. Sutherland H, Bickmore WA (2009) Nat Rev Genet 10(7), 457–66. Review
241. Han W and Yu KN (2010) Ionizing Radiation, DNA Double Strand Break and Mutation. In: Urbano KV (ed) Advances in Genetics Research, Volume 4, Nova Science Publishers, Inc.
242. Natarajan AT (2002) Mutat Res 504(1–2), 3–16. Review
243. Nelms BE, Maser RS, MacKay JF, Lagally MG, Petrini JH (1998) Science 280(5363), 590–2.
244. Soutoglou E, Dorn JF, Sengupta K, Jasin M, Nussenzweig A, Ried T, Danuser G, Misteli T (2007) Nat Cell Biol 9(6), 675–82.
245. Meaburn KJ, Misteli T, Soutoglou E (2007) Semin Cancer Biol 17(1), 80–90. Review
246. Soutoglou E, Misteli T (2007). Curr Opin Genet Dev 17(5), 435–42. Review
247. Tobias F, Durante M, Taucher-Scholz G, Jakob B (2010) Mutat Res 704(1–3), 54–60. Review
248. Jakob B, Splinter J, Durante M, Taucher-Scholz G (2009) Proc Natl Acad Sci USA 106(9), 3172–7.
249. Jakob B, Splinter J, Taucher-Scholz G (2009) Radiat Res 171(4), 405–18.
250. Jakob B, Rudolph JH, Gueven N, Lavin MF, Taucher-Scholz G (2005) Radiat Res 163(6), 681–90.
251. Costes SV, Ponomarev A, Chen JL, Nguyen D, Cucinotta FA, Barcellos-Hoff MH (2007) PLoS Comput Biol 3(8), e155.
252. Natarajan AT, Ahnström G (1969) Chromosoma 28(1), 48–61.
253. Folle GA, Martínez-López W, Boccardo E, Obe G (1998) Mutat Res 404(1–2), 17–26.

254. Elia MC, Bradley MO (1992) Cancer Res 52(6), 1580–6.
255. Löbrich M, Cooper PK, Rydberg B (1996) Int J Radiat Biol 70(5), 493–503.
256. Rothkamm K, Löbrich M (1999) Mutat Res 433(3), 193–205.
257. Sak A, Stuschke M, Stapper N, Streffer C (1996) Int J Radiat Biol 69(6), 679–85.
258. Löbrich M, Kühne M, Wetzel J, Rothkamm K (2000) Gen Chromos Cancer 27(1), 59–68.
259. Fouladi B, Waldren CA, Rydberg B, Cooper PK (2000) Radiat Res 153(6), 795–804.
260. Puerto S, Ramirez MJ, Marcos R, Creus A, Surralles J (2001) Mutagenesis 16, 291–296.
261. Martínez-López W, Folle GA, Cassina G, Méndez-Acuña L, Di-Tomaso MV, Obe G, Palitti F (2004) Cytogenet Genome Res 104(1–4), 182–7.
262. Sedelnikova OA, Rogakou EP, Panyutin IG, Bonner WM (2002) Radiat Res 158(4), 486–92.
263. Löbrich M, Shibata A, Beucher A, Fisher A, Ensminger M, Goodarzi AA, Barton O, Jeggo PA (2010) Cell Cycle 9(4), 662–9.
264. Kim JA, Kruhlak M, Dotiwala F, Nussenzweig A, Haber JE (2007) J Cell Biol 178(2), 209–18.
265. Pombo A, Cuello P, Schul W, Yoon JB, Roeder RG, Cook PR, Murphy S (1998) EMBO J 17(6), 1768–78.
266. Degani N and Pickholz D (1973) Radiation Botany 13(6), 381. DOI:10.1016/S0033-7560(73)90193--2
267. Dietzel S, Zolghadr K, Hepperger C, Belmont AS (2004) J Cell Sci 117, 4603–4614.
268. Warters RL, Lyons BW (1992) Radiat Res 130, 309–318.
269. Begusova M, Gillard N, Sy D, Castaing B, Charlier M, Spotheim-Maurizo M (2005) Radiat Phys Chem 72, 265–270.
270. Davidkova M, Stisová V, Goffinont S, Gillard N, Castaing B, Spotheim-Maurizot M (2006) Rad Protect Dosim 10, 1–6.
271. Haygarth K, Bartels DM (2010) J Phys Chem A 114(28), 7479–84.
272. Burns WG (1989) Nature 339, 515–516. DOI:10.1038/339515c0
273. Radulescu I, Elmroth K, Stenerlöw B (2004) Radiat Res 161(1), 1–8
274. Newman HC, Prise KM, Michael BD (2000) Int J Radiat Biol 76(8), 1085–93.

Chapter 21
Differentially Expressed Genes Associated with Low-Dose Gamma Radiation

Growth Differentiation Factor (GDF-15) as a radiation response gene and radiosensitizing target

Hargita Hegyesi, Nikolett Sándor, Boglárka Schilling, Enikő Kis, Katalin Lumniczky, and Géza Sáfrány

Abstract We have studied low dose radiation induced gene expression alterations in a primary human fibroblast cell line using Agilent's whole human genome microarray. Cells were irradiated with ^{60}Co γ-rays (0; 0.1; 0.5 Gy) and 2 hours later total cellular RNA was isolated. We observed differential regulation of approximately 300–500 genes represented on the microarray. Of these, 126 were differentially expressed at both doses, among them significant elevation of GDF-15 and KITLG was confirmed by qRT-PCR. Based on the transcriptional studies we selected GDF-15 to assess its role in radiation response, since GDF-15 is one of the p53 gene targets and is believed to participate in mediating p53 activities. First we confirmed gamma-radiation induced dose-dependent changes in GDF-15 expression by qRT-PCR. Next we determined the effect of GDF-15 silencing on radiosensitivity. Four GDF-15 targeting shRNA expressing lentiviral vectors were transfected into immortalized human fibroblast cells. We obtained efficient GDF-15 silencing in one of the four constructs. RNA interference inhibited GDF-15 gene expression and enhanced the radiosensitivity of the cells. Our studies proved that GDF-15 plays an essential role in radiation response and may serve as a promising target in radiation therapy.

21.1 Introduction

Cellular responses to direct ionizing radiation exposure are mediated in part through modulation of gene expression. Although translational and post-translational effects are also important, much can be learned from global gene expression studies that

H. Hegyesi (✉) • N. Sándor • B. Schilling • E. Kis • K. Lumniczky • G. Sáfrány
Department of Molecular and Tumor Radiobiology, F. Joliot-Curie National Research Institute for Radiobiology and Radiohygiene, Anna 5, Budapest, 1221 Hungary
e-mail: hegyesi.hargita@osski.hu; nikisan@hp.osski.hu; schilling.boglarka@osski.hu; kis.eniko@osski.hu; lumniczky.katalin@osski.hu; safrany.geza@osski.hu

G.G. Gómez-Tejedor and M.C. Fuss (eds.), *Radiation Damage in Biomolecular Systems*, Biological and Medical Physics, Biomedical Engineering,
DOI 10.1007/978-94-007-2564-5_21,

compare transcript levels across the entire genome. Accordingly, gene expression profiling has been used to address many questions in radiation biology, including the prediction of radiation sensitivity of tumor cell lines [1, 2], or normal tissue [3] and estimating the absorbed dose for biodosimetry [4]. More sophisticated network analyses of transcriptomic data are also starting to provide insight into signaling pathways and key transcription factors involved in radiation responses [5]. Gene expression studies on fibroblasts have shown that several genes are regulated by low LET radiation [6, 7], and some of these transcriptional responses are predictive of radiation induced fibrosis [8]. Fibrosis is known to develop through the action of various proinflamatory cytokines (TNFα, IL-1, and IL-6), ROS, nitrogen species (NS) and growth factors [9]. Several reports suggest that, TGF-β also plays a role in radiation induced fibrosis [10].

In the present study we have used whole human genome microarrays to examine the response of human skin fibroblasts to low dose gamma radiation. Our aim was to find low dose specific genes and potential marker genes to predict radiosensitivity. Normal human fibroblasts previously have been shown to elicit radiation response in microbeam experiments and findings from these studies have implicated nitric oxide (NO) and TGF-β1 in a ROS related pathway [11, 12]. In our study several genes were differentially expressed in irradiated cells both after 0.1 Gy or 0.5 Gy irradiation. We selected the GDF-15 gene to asses its role in radiation response. GDF-15 is belongs to the TGF-β superfamily and it is an important downstream mediator of the response to DNA damage response. It is well known that TGF-β1 is activated by ionizing radiation-induced free radicals. It is a prototype of the multifunctional regulators of cell growth and differentiation, which stimulate connective tissue formation and decreases collagen degradation resulting in fibrosis [13].

Lentiviral vectors can infect non-dividing cells, and so have a broad cell tropism, are non-toxic, and have stable gene expression due to viral genome integration into cell chromosomes [14]. During RNA interference, the introduction of shRNA by transfection into a diverse range of organisms and cell types causes degradation of the complementary mRNA, thereby silencing gene expression.

The current study used a cell culture model of transfected normal fibroblast to compare the radiosensitivity and the durability of expression of a lentiviral-delivered shRNA to GDF-15.

21.2 Materials and Methods

21.2.1 Cell lines

Primary human fibroblast cell line (F11) was established from skin biopsies taken from foreskin samples of children undergoing circumcision for medical indications, as described previously [5]. To create immortalized cell lines, we used a retroviral plasmid vector construct kindly provided by Dr. Lansdorp (Terry-Fox Laboratory, BC Cancer Research Centre, Vancouver, British Columbia, Canada V5Z 1L3). The construct contained the human telomerase (hTERT) gene in the pMIG vector that also encoded the green fluorescence protein (GFP) marker gene. The plasmid construct was introduced into FlyA13 retroviral packaging cell line (bought from ECACC, Porton Down, Salisbury, UK) to produce infectious retroviral particles, which were used to transduce primary human fibroblast cells. After transduction, the GFP positive cells were selected by flow cytometry and hTERT expression was validated by qRT-PCR.

21.2.2 Radiation treatment, colony forming assay

Cells were exposed with 2Gy of ^{60}Co γ-rays (Gammatron-3; Siemens, Erlangen, Germany; dose rate; was 0.37 Gy/min) at ambient temperature. To measure radiation sensitivity, cells were seeded on 100 mm culture dishes at a density of 500–1500 cells per dish, and 24 hours later they were irradiated with 2 Gy (SF2). After incubation for 14 days, primary colonies in the dishes were fixed in methanol and stained with 1% methylene blue. Colonies consisting of more than 50 cells were scored as survivors.

21.2.3 RNA isolation and reverse transcription

Total RNA was isolated from irradiated and mock-irradiated cells using RNeasy Mini kit (Qiagen, Hilden, Germany) according to the manufacturer's instructions. Total RNA concentrations were determined by spectrometric analysis, and RNA quality was checked by formaldehyde agarose gel electrophoresis. 1 μg total RNA was reverse transcribed into cDNA using the *High Capacity* cDNA Reverse Transcription Kit according to the manufacturer's instructions (Life Technologies Corporation, Carlsbad, CA USA). The synthesized cDNA was stored at −80 °C or analyzed directly by real-time PCR.

21.2.4 Microarray Hybridization and Analysis

RNA concentration was measured using a NanoDrop-1000 spectrophotometer. All RNA samples had RNA integrity numbers >9.0 and 260nm/280nm absorbance ratios >2. Cyanine-3 (Cy3) and Cyanine-5 (Cy5) labeled cRNA was prepared from 0.1 μg RNA. 1.0 μg of cRNA hybridized to Agilent Whole Human Genome

Table 21.1 Oligonucleotide primers used in qRT-PCR experiments

GENE SYMBOL	FORWARD	REVERSE
TP53INP1	tcagcagaagaagaagaagaagag	agcaggaatcacttgtatcagc
CDKN1A	cctcatcccgtgttctccttt	gtaccacccagcggacaagt
GDF-15	tcacgccagaagtgcggctg	cgtcccacgaccttgacgcc
KITLG	cccttaggaatgacagcagtagca	gcccttgtaagacttggctgtctc
CXCL-12	aacgccaaggtcgtggtc	tggcttgttgtgcttacttgttt
CXCL-2	tgccagtgcttgcagac	tcttaaccatgggcgatgc
RAD54L	ccaaaacagtccttgccatt	cagccatcactttagcacga
DLGAP4	gtagaggacgactggcgaag	taggagaggttgcgcttgat
GAPDH	cgaccactttgtcaagctca	aggggtctacatggcaactg
Actin-β	ttgccgacaggatgcagaagga	aggtggacagcgaggccaggat

Oligo Microarrays (Agilent 251485025000-002, 4x44K) using the Gene Expression Hybridization Kit, and washed following Agilent's recommendations. Slides were scanned with the Agilent DNA Microarray Scanner and default parameters of Feature Extraction Software 9.1 (Agilent, Santa Clara, CA) were used for image analysis, data extraction, background correction, and flagging of non-uniform features. Gene ontology (GO) groups of genes whose expression was differentially regulated following 0.1 and 0.5 Gy gamma radiation exposures were identified. Analysis of GO groups, rather than individual genes, made it possible to reduce the number of tests conducted, and enabled findings among biologically related genes. eGOon software (version V2.0) was used to test for representation of annotation classes.

21.2.5 Real-time PCR (qRT-PCR)

To quantify mRNA levels, quantitative real-time PCR was performed using a Rotor-Gene, Corbett real-time PCR System (Invitrogen, Carlsbad, CA). The reaction mixture was composed of Maxima™ SYBR® Green qPCR Master Mix, (Fermentas Vilnius, Lithuania), 12,5 pM of each primer, and 2 μl cDNA template (obtained after RT as described above) in a volume of 25 μl. The cycle number at which the fluorescent signal crossed the detection threshold was denoted as the threshold cycle (CT). CT values obtained for the detected genes were normalized using β-actin and Glyceraldehyde 3-phosphate dehydrogenase (GAPDH) as internal standard. Each PCR reaction was run in duplicate and at least three independent experiments were performed. Relative fold-inductions were calculated by the ΔΔCT method as recommended. The primer pairs used in PCR studies are shown in Table 21.1.

21.2.6 Establishment of GDF-15-silenced cell lines

To investigate the effect of GDF-15 in radiation response we tried to silence the GDF-15 gene in immortalized human fibroblast cells using the MISSION Lentiviral Transduction system bought from SIGMA (Sigma, St Louis, MO, USA). The system contains five different shRNA for GDF-15 gene. Immortalized F11-hTERT human fibroblasts were transduced with the lentiviral vectors and stable clones were selected on the base of their puromycin resistance. Expression of GDF-15 mRNA was measured by quantitative PCR. We named the stable shGDF-15 expressing F11-hTERT cells as "shGDF-15"-: #1, #2, #3 and #4. The stable transfected F11-hTERT cells were propagated in the presence of puromycin (0.75 $\mu g/ml$).

21.2.7 Data analysis

Except where otherwise specified, results are shown as mean and standard deviation of three separate experiments and P values are calculated using paired Student's T test.

One-way analysis of variance and two-tailed t tests were used to compare differences among groups. $P < 0.05$ was considered statistically significant. Data were presented as means $\pm$ SE.

21.3 Results

21.3.1 Differentially expressed genes following radiation exposure

Three hundred sixty six differentially expressed genes were identified after 0.1 Gy gamma radiation in F11 primary fibroblast cells [15]. Of these 366 genes, 347 genes were induced, and 19 were repressed. Five hundred two differentially expressed genes were found after irradiating the cells by 0.5 Gy gamma radiation [16]. Of these, 450 genes were up-regulated, and 52 genes were down-regulated (Table 21.2). Remarkably, 126 of differentially expressed genes responded both after 0.1 and 0.5 Gy exposures (Fig. 21.1).

To elucidate the involvement of specific biological processes in cellular responses to low dose exposures at the transcriptional level, the gene ontology (GO) annotations of the genes were performed. The results of GO analysis of the differentially expressed genes are summarized in Table 21.2. Several genes encoding the subunits of ribosomal complex were up-regulated in F11 cells both after 0.1 and 0.5 Gy radiations. Other significantly up-regulated biological

Table 21.2 Gene ontology analysis using eGOn

Gene Ontology categories	0.1 Gy	Up	Down	0.5 Gy	Up	Down
Response to DNA damage	8	8	–	5	5	–
Programmed cell death	13	13	–	17	16	1
Regulation of cell cycle	6	5	1	17	12	5
Regulation of cell proliferation and growth	3	3	–	6	5	1
Regulation of cell differentiation	2	2	–	7	6	1
Intracellular signaling cascade	28	26	2	38	34	4
Cell surface receptor linked signal transduction	12	11	1	28	25	3
Protein metabolism	16	15	1	24	24	–
Nucleotide, nucleic acid metabolism	1	1	–	5	4	1
Cell-cell adhesion	10	9	1	21	20	1
Others	102	96	6	88	78	10
Not-annotated	165	158	7	246	221	25
Sum	366	347	19	502	450	52
%		94.81	5.4		89.64	10.36

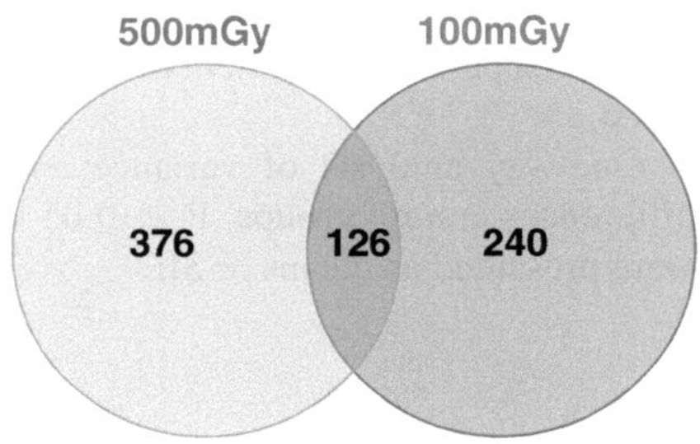

Fig. 21.1 Stylized Venn diagram depicting the changes in gene expression levels induced by 0.1 Gy or 0.5 Gy gamma-radiations

processes were the protein kinase cascade and apoptosis (Table 21.2). Of the 366 differentially expressed genes following 0.1 Gy irradiation, 13 genes had known roles in apoptosis, and all of them were induced. These included three members of the tumor necrosis factor (TNF)-receptor superfamily (TNFRSF16,) and other genes generally considered to be pro-apoptotic (BNIP2, BCLAF1) [1]. On the other hand, several anti-apoptotic genes were also up-regulated (GAS2L3 and CCAR1). Among repressed genes there were some known signal transduction targets (CNP and NBL1). Out of 502 differentially expressed genes after 0.5 Gy, 17 genes were involved in cell death/apoptosis, and 16 of them were induced. Four members of the tumor necrosis factor (TNF)-receptor superfamily (TNFRSF4, TNFRSF10C, TNFRSF11B and FAS) were induced following 0.5 Gy gamma-irradiation in comparison to 0.1 Gy exposures. The other difference regarding apoptosis-related gene expression changes following 0.1 and 0.5 Gy radiation was that few known genes involved in apoptotic mitochondrial pathway (BBC3) were differentially expressed in the latter case. In addition, cell cycle regulation, DNA damage response and intracellular signaling cascade related genes significantly dominated radiation responses. Eight DNA damage response/DNA repair genes

Table 21.3 Comparison of relative gene expression patterns obtained by microarray and qRT-PCR. ## Mean of treated to control ratios measured by microarray in three independent experiments. # Mean of qRT-PCR measurements of the same RNA used in microarray analyses. Each data point represents the mean ± SE,* = $P < 0.05$

	0.1 GY		0.5 GY	
GENE SYMBOL	Array ##	qRT-PCR #	Array ##	qRT-PCR #
TP53INP1	ND	1.33 ± 0.12	ND	1.738 ± 0.25*
CDKN1A	1.19	1.236 ± 0.19	1.31	1.43 ± 0.02*
GDF-15	1.33	1.39 ± 0.06*	1.44	1.67 ± 0.03*
DLGAP4	1.51	1.305 ± 0.13	1.55	1.28 ± 0.10
KITLG	1.51	1.72 ± 0.18*	1.33	1.58 ± 0.22*
CXCL-12	1.44	0.88 ± 0.079	1.24	1.2 ± 0.2
CXCL-2	1.02	1.37 ± 0.29	1.12	1.45 ± 0.049*
RAD54L	1.05	1.06 ± 0.15	0.92	1.01 ± 0.01

(CSNK1D, CRY1, MLH1, TRIB1, ANKRD17, DDIT3, TYMS and POLK) were up-regulated by 0.1 Gy radiation. Protein biosynthesis and positive regulation of cell surface receptor linked signal transduction involved genes represented the most notably up-regulated biological processes after radiation exposure. One of the key features of the gene expression profile was the high abundance of genes participating in ubiquitin cascade (UBA52, UBE2D, USP33, UBQLN2 and USP47). The most significantly down-regulated biological process was regulation of the cell cycle (Table 21.2).

QRT-PCR was used to confirm the differential expression of 8 of these genes (Table 21.3). Nearly one-third of the genes showed greater fold-changes by qRT-PCR than predicted by the arrays. Such ratio compression is often encountered in microarray experiments, and is thought to be probe or primer sequence dependent. In all, our microarray analysis was well supported by the qRT-PCR results.

21.3.2 Inhibition of GDF-15 gene expression by lentiviral shRNA constructs

Four lentiviral shRNA constructs were transduced into F11hTERT cells in an attempt to silence the GDF-15 gene. GDF-15 expression was evaluated by qRT-PCR. Using TRCN#1 (F11hTERT-shGDF-$15^{\#1}$) and TRCN#3 (F11hTERT-shGDF-$15^{\#3}$) lentiviral constructs relative expression of GDF-15 decreased to 0.448 ± 0.058 and 0.59 ± 0.106, respectively (Table 21.3). Using TRC#2 lentiviral particle construct (F11-hTERT-shGDF-$15^{\#2}$), the expression of GDF-15 was enhanced instead of inhibited. There was no significant success in gene silencing effect when we used the TRC#4 construct. The lentiviral shRNA construct, however, demonstrated significant advantages in posttranscriptional gene silencing in F11-hTERT-shGDF-$15^{\#1}$ cells.

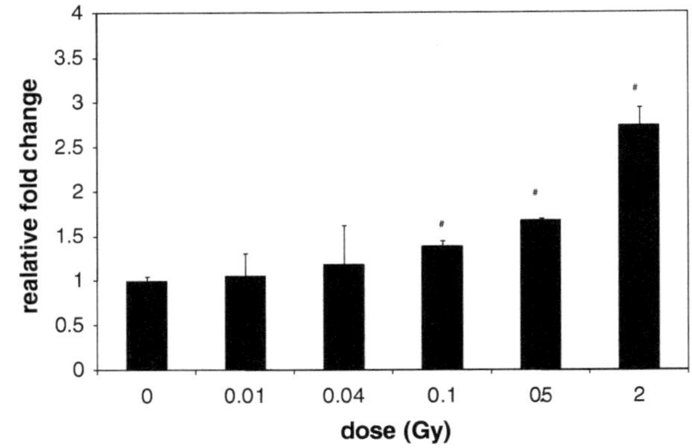

Fig. 21.2 Dose-dependent alterations in GDF-15 gene-expression after irradiation. Gene expression was measured 2 hours after exposure to gamma-radiations as described in Materials and Methods. Data are expressed as means ± SE from three independent experiments. Student's t test: as compared with the control sample, # = $P < 0.05$

Table 21.4 ShRNA-mediated suppression of the GDF-15 gene expression and its effect on the cellular radiosensitivity. a, Quantitative detection of shRNA-mediated suppression (%) of GDF-15 mRNA levels was analyzed in F11-hTERT-shGDF-15 cells by real-time RT-PCR. Data are means from at least three independent experiments. Each data point represents the mean ± SE of three experiments. * = $P < 0.05$. b, F11-hTERT and F11-hTERT-shGDF-15 cells were irradiated with 2 Gy gamma-radiations and RNA was isolated 2h later. Data are expressed as means ± standard error from three independent PCR runs. Each data point represents the mean ± SE of three experiments. * = $P < 0.05$. c, Cell survival after exposure of F11-hTERT and F11-hTERT-shGDF-15 cells (#1, #2, #3, #4), cells to gamma-ray with 2 Gy (SF2). Cell survival was measured using a colony forming assay

	[a]Relative expression of GDF-15	[b]2Gy-induced relative expression of GDF-15	[c]SF2
F11-hTERT	1	2.74 ± 0.21*	0.32
F11-hTERT-shGDF-15$^{\#1}$	0.44 ± 0.058*	0.845 ± 0.167	0.14*
F11-hTERT-shGDF-15$^{\#2}$	2.62 ± 0.28	ND	0.34
F11-hTERT-shGDF-15$^{\#3}$	0.59 ± 0.16*	1.25 ± 0.056	0.24
F11-hTERT-shGDF-15$^{\#4}$	1.09 ± 0.27	1.965 ± 0.29	0.28

21.3.3 The influence of GDF-15 on radiation resistance

GDF-15 gene expression showed a dose dependent pattern in irradiated F11-hTERT cells (Fig. 21.2). Next, we investigated radiation-induced GDF-15 alterations in shRNA silenced cells after irradiation with 2 Gy by qRT-PCR (Table 21.4). In wild type F11-hTERT cells GDF-15 expression increased about 2.74-fold (Table 21.4). GDF-15 expression also increased in F11-hTERT-shGDF-15$^{\#1}$ and F11-hTERT-shGDF-15$^{\#3}$ after 2 Gy exposure, but the level of GDF-15 was close to the level of GDF-15 in the unirradiated wild type cells (Table 21.4).

To assess the functional consequences of the GDF-15 gene silencing we compared the *in vitro* radiosensitivity of the parental F11-hTERT and F11-hTERT-shGDF-15 cells by colony forming assay. We found that efficient silencing of GDF-15 (F11-hTERT-shGDF-15$^{\#1}$) exhibited significantly increased radiation sensitivity after 2 Gy irradiation (Table 21.4).

21.4 Discussion

Previous studies have utilized microarrays to describe gene expression changes associated with ionizing radiation [17–19]. However, little is known about gene expression alterations following low dose ionizing radiation exposure in normal tissues and the dose-dependency of these transcriptional changes. In this study, we used whole genome human microarrays to conduct a genome-wide survey of transcriptional changes of the early response genes in normal fibroblast to low dose gamma-ray. We have identified genes that are induced or repressed following irradiation and have characterized their expression profiles. Furthermore, we have investigated the dose-dependency of the transcriptional responses. We found that 502 and 366 genes responded to 0.5 and 0.1 Gy radiations, respectively. 126 genes responded to both doses. Several of these genes including DDB2 and CDKN1A were already identified by us as radiation response gene after 2 Gy irradiations [5]. DDB2 and CDKN1A responded to radiation in peripheral white blood cells [20], and fibroblasts [7], as well.

Regarding low dose effects Warters et al. used Rank Products (RP) analysis and did not find genes that displayed differential gene expression compared to the control cells at 0.1 Gy. However, with the 1 Gy samples, they observed 13 genes that displayed increased expression 4 hours after irradiation. These were the annotated genes SESN1, CDKN1A, GDF-15, FDXR and HSPA4L as well as seven poorly annotated genes [21]. SESN1, CDKN1A, GDF-15, FDXR were also identified by us as consensus radiation response genes, [5]. Zhou et al. also analyzed the gene expression profile of normal human fibroblast and they found different patterns at different time points after exposure. Many of these genes were prototypical p53 target genes that mainly contribute to initiation and maintenance of G1 arrest through inhibition of cyclin-dependent kinases [22]. GDF-15 is a member of the transforming growth factor-β superfamily that was formerly identified by us as a radiation response gene [5]. GDF-15 is supposed to regulate tissue differentiation and maintenance. It has recently been shown to be induced by radiation in human colon cancer cell lines, as well [23].

Identifying the biological mechanisms that underlie intrinsic or acquired resistance of tumor cells to radiotherapy and to prevent normal tissues of radiation induced side effects is of critical importance for improving the morbidity and mortality associated with human malignancies. For reasons that have yet to be determined, fibroblasts seem to be particularly resistant to the cytotoxic effects of ionizing radiation. Most studies evaluating radiation sensitivity use cell culture models and little data are available translating the relevance of these findings to the complexity and variability of the clinical environment. Several features were used to reduce noise and artifacts and optimize our ability to identify relevant radio/chemotherapy related gene expression effects [24].

We used a cell type–specific analysis approach that focused specifically on normal diploid fibroblast avoiding heterogeneity in sample composition. Gene expression analyses determined that a group of qRT-PCR validated genes are significantly associated with radiation exposure to low dose gamma-ray in fibroblast cells. Further, many of these genes are components of complex cellular pathways of interacting proteins that were also statistically associated with radiotherapy exposure, supporting the concept that a network of responses contributes to cell survival and therapy resistance.

In this study, we found that stress response pathways, particularly those involving TGF-β related cytokines, might be involved in the mechanism of radiation resistance of fibroblast. Cytokines are released by many cells following radiation exposure, including endothelial cells, fibroblasts, immune cells and parenchymal cells. The interplay of these cytokines is thought to be responsible for the pathogenesis of many of the effects following radiation exposure [25]. Several cytokines and growth factors, including IL-1β, CXC-10 and GDF-15, are up-regulated in chemotherapy-resistant tumor cell lines, and GDF-15 can specifically modulate resistance [26]. Further experimental studies of chemo- or radioresistant cancer cells may clarify whether targeting these cytokines or their receptors could reduce tumor cell viability, and could protect surrounding normal fibroblast population following cytotoxic therapies. Although prior studies found that GDF-15 can exert pro-apoptotic effect in several cancer types [27] later others described growth promoting effects in malignant glioma [28]. Gene expression studies of GDF-15 in colon, prostate, and pancreas showed increased expression levels of GDF-15 in tumors compared to benign tissue [29]. In therapeutic context Modlich et al. [30] compared gene expression changes during the setting of neoadjuvant chemotherapy for patients with primary breast cancer. Expression profiles of paired tumor samples obtained before and 24 h after chemotherapy found increased expression of GDF-15 and several other genes following treatment. Shimizu et al. [31] reported an analysis of gene expression differences between 5-fluorouracil–chemoresistant and 5-fluorouracil–chemosensitive colon cancer cell lines and determined GDF-15 to be one of the most significantly up-regulated genes in resistant cells.

The results of our experiments with GDF-15 are consistent with these reports. The role of GDF-15 in radiation therapy has not been previously reported. A recent study using GDF-15 shRNA showed successful inhibition of radiation induced expression of GDF-15. In F11-hTERT-shGDF-15$^{\#1}$ cells, GDF-15 suppression was validated by qRT-PCR. However, the inhibition effect was weakened in the case of TCR#3 and TRCR#4 shRNA construct, respectively. This indicated that abundant gene silencing effect of GDF-15 could be achieved with only one shRNA expression construct (TCR#1), but later this effect was stable under long term cultivation.

In principle, there are two general siRNA delivery methods. One uses chemically synthesised 19-21nt siRNA with apparent short-lived effect [32]. The other approach is vector-based shRNA, including the use of plasmid vectors and viral vectors. Vector-based shRNA has a more prolonged effect and is relatively inexpensive. Our previous experience indicated that a retroviral transfer could facilitate better transfection than plasmid based transfection, and this was confirmed in human telomerase (hTERT) immortalized fibroblast cells. The advantages of lentivirus for gene transfer, especially to non-dividing cells, make the shRNA technique more accessible for specific *in vitro* and *in vivo* silencing in fibroblast populations. The lentiviral construct here showed significant silencing of GDF-15 in F11-hTERT shGDF-15$^{\#1}$ cells. There was 55.2% decrease of GDF-15 achieved at any time post transfection. Similarly Yang et al. presented that, complete transforming growth factor, beta receptor II (TBRII) silencing was not achieved, even though a lentiviral construct was employed, and was facilitated by a cationic transfection aid in renal fibrogenesis model [33].

In summary, in the present study, we show for the first time that in normal fibroblasts, GDF-15 is protective to the cytotoxic effects of radiation. Furthermore, we found that silencing of GDF-15 may confer a component of radiosensitivity to fibroblast cells, a finding that supports further studies designed to manipulate GDF-15 for therapeutic benefit. It will be important to determine if GDF-15 induces radiation resistance in human cancer cells as well. Because many tumors are radiation resistant, these findings would provide insight into novel therapeutic strategies to overcome radiation resistance through GDF-15 expression.

Acknowledgments This work was supported by the following grants: the European Union NOTE project (FP6-036465/2006), Hungarian OTKA K77766 and ETT 827-1/2009. The authors thank the expert technical assistance of Ms. Mária Frigyesi and Ms. Rita Lőkös.

References

1. Torres -Roca JF, Eschrich S, Zhao H, Bloom G, Sung J, McCarthy S, Cantor AB, Scuto A, Li C, Zhang S, Jove R, Yeatman T (2005) Cancer Res **65**(16), 7169–7176
2. Amundson SA, Do KT, Vinikoor LC, Lee RA, Koch-Paiz CA, Ahn J, Reimers M, Chen Y, Scudiero DA, Weinstein JN, Trent JM, Bittner ML, Meltzer PS, Fornace AJ Jr (2008) Cancer Res **68**(2), 415–424
3. Svensson JP, Stalpers LJ, Esveldt-van LRE, Franken NA, Haveman J, Klein B, Turesson I, Vrieling H, Giphart-Gassler M (2006) Plos Med **3**(10), 1904–1914
4. Paul S, Amundson SA (2008) Int J Radiat Oncol Biol Phys **71**(4), 1236–1244
5. Kumaraswamy S, Chinnaiyan P, Shankavaram UT, Lu X, Camphausen K, Tofilon PJ (2008) Cancer Res **68**(10), 3819–3826
6. Kis E, Szatmári T, Keszei M, Farkas R, Ésik O, Lumniczky K, Falus A, Sáfrány G (2006) Int J Radiat Oncol Biol Phys **66**(5), 1506–1514
7. Rodningen OK, Overgaard J, Alsner J, Hastie T, Borresen-Dale AL (2005) Radiother Oncol **77**(3), 231–240
8. Rodningen OK, Borresen-Dale AL, Alsner J, Hastie T, Overgaard J (2008) Radiother Oncol **86**(3), 314–320
9. Bentzen SM (2006) Nat Rev Cancer **6**(9), 702–713
10. Barcellos-Hoff MH (1996) J Mammary Gland Biol Neoplasia **1**(4), 353–363

11. Shao C, Prise KM, Folkard M (2008) Mutat Res **638**(1–2), 139–145
12. Sokolov MV, Smilenov LB, Hall EJ, Panyutin IG, Bonner WM, Sedelnikova OA (2005) Oncogene **24**(49), 7257–7265
13. Wynn TA (2008) Cellular and molecular mechanisms of fibrosis. J Pathol **214**(2), 199–210
14. Naldini L, Blomer U, Gage FH, Trono D, Verma IM (1996) Proc Natl Acad Sci USA **93**(21), 11382–11388
15. Supplementary material available at: http://www.osski.hu/osski/sbfo/mto/rdbs/radresgen100.xls
16. Supplementary material available at: http://www.osski.hu/osski/sbfo/mto/rdbs/radresgen500.xls
17. Amundson SA, Fornace AJ (2001) Radiat Prot Dosimetry **97**(1), 11–16
18. Amundson SA (2008) Radiat Environ Biophys **47**(1), 25–31
19. Khodarev NN, Park JO, Yu J, Gupta N, Nodzenski E, Roizman B, Weichselbaum RR (2001) Proc Natl Acad Sci USA **98**(22), 12665–12670
20. Ponten F, Lindman H, Bostrom A, Berne B, Bergh J (2001) J Natl Cancer Inst **93**(2), 128–133
21. Warters RL, Packard AT, Kramer GF, Gaffney DK, Moosb PJ (2009) Radiat Res **172**(1), 82–95
22. Zhou T, Chou JW, Simpson DA, Zhou Y, Mullen TE, Medeiros M, Bushel PR, Paules RS, Yang X, Hurban P, Lobenhofer EK, Kaufmann WK (2006) Environ Health Perspect **114**(4), 553–559
23. Okazaki R, Moon Y, Norimura T, Eling T. (2006) Radiat Res **165**(2), 125–130
24. Prisack HB, Karreman C, Modlich O, Audretsch W, Danae M, Rezai M, Bojar H (2005) Anticancer Res **25**(6C), 4615–4621
25. Rubin P, Johnston CJ, Williams JP, McDonald S, Finkelstein JN (1995) Int J Radiat Oncol Biol Phys **33**(1), 99–109
26. Huang CY, Beer TM, Higano CS, True LD, Vessella R, Lange PH, Garzotto M, Nelson PS (2007) Clin Cancer Res **13**(19), 5825–5833
27. Jutooru I, Chadalapaka G, Chintharlapalli S, Papineni S, Safe S (2009) Mol Carcinog **48**(8), 692–702
28. Roth P, Junker M, Tritschler I, Mittelbronn M, Dombrowski Y, Breit SN, Tabatabai G, Wick W, Weller M, Wischhusen J (2010) Clin Cancer Res **16**(15), 3851–3859
29. Mimeault M, Batra SK (2010) J Cell Physiol **224**(3), 626–635
30. Modlich O, Prisack HB, Munnes M, Audretsch W, Bojar H (2005) J Transl Med **3**, 32
31. Shimizu D, Ishikawa T, Ichikawa Y, Togo S, Hayasizaki Y, Okazaki Y, Danenberg PV, Shimada H (2005) Int J Oncol **27**(2), 371–376
32. Czauderna F, Fechtner M, Dames S, Aygün H, Klippel A, Pronk GJ, Giese K, Kaufmann J (2003) Nucleic Acids Res **31**(11), 2705–2716
33. Yang T, Zhang B, Pat BK, Wei MQ, Gobe GC (2010) J Biomed Biotechnol Journal of Biomedicine and Biotechnology, doi:10.1155/2010/859240

Chapter 22
Chromosome Aberrations by Heavy Ions

Francesca Ballarini and Andrea Ottolenghi

Abstract It is well known that mammalian cells exposed to ionizing radiation can show different types of chromosome aberrations (CAs) including dicentrics, translocations, rings, deletions and complex exchanges. Chromosome aberrations are a particularly relevant endpoint in radiobiology, because they play a fundamental role in the pathways leading either to cell death, or to cell conversion to malignancy. In particular, reciprocal translocations involving pairs of specific genes are strongly correlated (and probably also causally-related) with specific tumour types; a typical example is the BCR-ABL translocation for Chronic Myeloid Leukaemia. Furthermore, aberrations can be used for applications in biodosimetry and more generally as biomarkers of exposure and risk, that is the case for cancer patients monitored during Carbon-ion therapy and astronauts exposed to space radiation. Indeed hadron therapy and astronauts' exposure to space radiation represent two of the few scenarios where human beings can be exposed to heavy ions. After a brief introduction on the main general features of chromosome aberrations, in this work we will address key aspects of the current knowledge on chromosome aberration induction, both from an experimental and from a theoretical point of view. More specifically, *in vitro* data will be summarized and discussed, outlining important issues such as the role of interphase death/mitotic delay and that of complex-exchange scoring. Some available *in vivo* data on cancer patients and astronauts will be also reported, together with possible interpretation problems. Finally, two of the few available models of chromosome aberration induction by ionizing radiation (including heavy ions) will be described and compared, focusing on the different assumptions adopted by the authors and on how these models can deal with heavy ions.

F. Ballarini (✉) • A. Ottolenghi
University of Pavia - Department of Nuclear and Theoretical Physics, and INFN (National Institute of Nuclear Physics) – Sezione di Pavia, 27100 Pavia, Italy
e-mail: francesca.ballarini@pv.infn.it; andrea.ottolenghi@pv.infn.it

G.G. Gómez-Tejedor and M.C. Fuss (eds.), *Radiation Damage in Biomolecular Systems*, Biological and Medical Physics, Biomedical Engineering,
DOI 10.1007/978-94-007-2564-5_22,

22.1 Introduction

Mammalian cells exposed to ionising radiation during the G_0/G_1 phase of the cell cycle can show different types of chromosome aberrations (CAs) including dicentrics, translocations, rings, inversions, deletions and complex exchanges, the latter usually defined as chromatin rearrangements involving at least 3 breaks and 2 chromosomes. An exhaustive classification of the various aberration types was provided by Savage and Simpson [1]. In the majority of the experimental studies available in the literature, chromosome aberrations are observed at the first post-irradiation metaphase. However, Premature Chromosome Condensation (PCC) techniques, based either on fusion with mitotic cells or on treatments with chemicals such as Calyculin A [2], allow observation of aberrations during interphase at any time after irradiation. This can help minimising possible biases introduced by phenomena such as cell-cycle perturbations and interphase cell death, which occur with significant probability after exposure to high-dose and/or high-LET radiation including heavy ions. For a long time the experimental observations were based on Giemsa solid staining. In Giemsa all chromosomes are shown with the same colour, and thus only dicentrics, rings and acentric fragments (plus a few complex exchanges such as chromosomes with more than 2 centromeres) can be scored. The introduction of the Fluorescence *In Situ* Hybridisation (FISH) technique [3] represented a fundamental turn, allowing selective painting of one or more specific pairs of homologue chromosomes and thus detection of aberration types that are not visible with solid staining, such as translocations and many complex exchanges. The recent introduction of the so-called "multi-FISH" technique, which allows painting of each homologue pair with a different (pseudo-)colour, provided additional information, especially on the induction of very complex exchanges involving large numbers of chromosomes [4, 5, 13]. Site-specific probes and m-BAND techniques allow even more detailed investigations. For instance, the use of telomeric probes prevents mis-scoring of complete exchanges as incomplete ones, which are now considered to play a minor role.

22.1.1 Why should we care about chromosome aberrations?

Aberrations represent a fundamental step in the biological pathways leading either to cell death, or to cell conversion to malignancy. On one side dicentric chromosomes imply a decreased probability for the cell to be able to duplicate, whereas on the other side reciprocal translocations involving pairs of specific genes are strongly correlated with specific tumour types. Typical examples are the BCR-ABL translocation for Chronic Myeloid Leukaemia, which involves the ABL gene on chromosome 9 and the BCR gene on chromosome 22 [6], and the PML-RARα translocation for Acute Promyelocytic Leukaemia, which involves the PML and RARα genes on chromosomes 15 and 17, respectively [7, 8]. Causal relationships between chromosome aberrations and cancer were also proposed (e.g. [9]). Furthermore, aberration yields are used for biodosimetry applications, and more generally as biomarkers of exposure and risk [10]. In particular, the frequency of CAs in peripheral blood lymphocytes (PBL) was used to evaluate radiation exposure in victims of accidents including Chernobyl [11–19], in survivors of the Hiroshima and Nagasaki A-bombs [20, 21], and in astronauts exposed to the complex radiation environments encountered in space [22–26, 29].

22.1.2 Why should we care about heavy ions?

A possible scenario where human beings are exposed to heavy ions is tumour treatment with Carbon ions, which are now used by an increasing number of radiotherapy centres including Chiba (with about 3,000 patients treated until now) and Hyogo, in Japan, and Darmstadt in Germany. Other Carbon facilities started operating more recently in Pavia (Italy), Heidelberg (Germany) and other locations (see [27] for a review). Like protons, Carbon ions are characterized by a localization of energy deposition in the so-called "Bragg peak" region. This provides an improved dose conformation, also considering that their Relative Biological Effectiveness (RBE) in the plateau is sufficiently low (approximately 1, like for protons). Furthermore, Carbon beams are particularly suitable for treating radioresistant tumours because their RBE for clonogenic inactivation in the region of the (spread-out) Bragg peak can be up to 3 (also depending on the beam features, the considered cell line etc.), to be compared with the 1.1 value typically adopted for proton beams. However, treatment planning with heavy ions is particularly complex, also considering that at the energies of interest for hadrontherapy nuclear reactions of the primary particles with the beam-line constituents and with the various components of the human body play a non negligible role: projectile fragmentation gives rise to lighter fast particles which form a "tail" of dose beyond the SOBP. It is therefore of utmost importance to characterize hadrontherapy beams, and it is also desirable to monitor to what extent normal tissues are spared during the treatment. Blood is a normal tissue which is unavoidably exposed during radiotherapy, and the yield of chromosome aberrations in peripheral blood lymphocytes is considered as a reliable estimate of the equivalent whole-body dose [28]. Lymphocytes circulate in the blood vessels and are distributed throughout the body, mainly in lymph nodes, spleen, bone marrow, thymus and the gut lymphoid tissue. Damage to the haematopoietic tissue is therefore a major limiting factor with respect to the total dose delivered in a radiotherapy treatment, both for acute morbidity and for the risk of developing secondary cancers [29]. In this framework, Durante et al. [14] monitored the induction of chromosomal aberrations in PBL of cancer patients treated with X rays or Carbon ions at NIRS in Chiba, finding that the lymphocytes from C-ion patients carried less aberrations than those from X-ray patients. More details are reported in section 22.2.2.

The aforementioned exposure to space radiation represents another situation where human beings are exposed to heavy ions, though the scenario is made particularly complex by the fact that Galactic Cosmic Rays are a mixed field consisting of high-energy protons (about 87% in fluence), He ions (12%) and heavier ions (1%), the latter also called "HZE particles" (High "Z" and Energy). Astronauts are exposed to GCR continuously, with a dose rate of the order of 1 mSv/day in deep space. Despite the small contribution in fluence, the contribution of heavy ions to the total equivalent dose can be much higher, up to 50% [30]. Every astronaut wears a personal dosimeter and various dosimeters are located in different places of the International Space Station. However, due to the complexity of the exposure scenario and to the fact that some aspects of heavy-ion radiobiology are still not clear, the physical dose is not sufficient to estimate the corresponding damage and risk. Biological dosimetry with chromosome aberrations can be of particular help in case of exposure to space radiation since aberrations can take "automatically" into account peculiar aspects relative to the radiation field composition and modulation, as well as possible interactions with microgravity, stress etc. Comparison of post-flight aberration yields with *in vitro* gamma-ray calibration curves provides estimates of equivalent doses, as well as of space radiation quality, by taking into account measured absorbed doses. Monitoring of chromosome aberrations in astronauts' PBL has become routine in the last decade (see references quoted in section 22.1.1). An extensive study on CA induction in astronauts exposed to space radiation [12] is discussed in section 22.2.2.

Since space radiation is a low fluence rate scenario, when dealing with space research applications it is particularly important to characterize and quantify the effects of low fluences of heavy ions, down to single cell nucleus traversals. Sample calculations will be presented in section 22.3.2.

22.2 Heavy-ion-induced chromosome aberrations: *in vitro* and *in vivo* evidence

22.2.1 In vitro data

Most data on chromosome aberration induction by heavy ions come from *in vitro* experiments where living cells, quite often lymphocytes, were exposed to Carbon ions (mainly at NIRS in Japan) or heavier ions such as Iron, Silicon and other ions that are of interest for space radiation research (mainly in Brookhaven, USA). Due to their high LET, heavy ions have a higher RBE with respect to photons for most endpoints including chromosomal aberrations. Several works are available in the literature on CA induction in human cells exposed to heavy ions [e.g. 10, 12, 31–34, 37, 38]. The data indicate that the linear coefficient for dicentric induction increases with the radiation LET, peaking around 60–100 keV/μm, and decreases sharply at higher LET values. While Giemsa solid staining does not allow detailed scoring of complex exchanges, which are very effectively induced by heavy ions, more recent investigations employed FISH painting. The results indicated a much higher frequency of complex exchanges with respect to low-LET radiation, and the rearrangements were of greater complexity.

Most studies on chromosome aberrations induced by high-LET radiation measured the damage in metaphase cells. However, following exposure to high-LET radiation, the frequency of CA in interphase is significantly higher than in metaphase [35–37], most likely due to mitotic delay or interphase cell death. It is therefore desirable that the analysis of heavy-ion-induced CAs is performed by Premature Chromosome Condensation, which allows CA scoring during interphase thus reducing the cell population selection bias in metaphase and leading to higher RBE values.

A very complete and informative study of CA induction by heavy ions is that by George et al. [37], who irradiated *in vitro* human lymphocytes with several ions including 290 MeV/n C-12 (LET = 13.3 keV/μm), 490 MeV/n Si-28 (LET = 56 keV/μm), 550 MeV/nucleon Ar-40 (LET = 86 keV/μm), 1040 MeV/n Fe-56 (LET = 147 keV/μm), 500 MeV/n Fe-56 (LET = 200 keV/μm), 200 MeV/n Fe-56 (LET = 440 keV/μm) and 10,000 MeV/n Au-197 (LET = 1393 keV/μm). In most cases the doses were lower than 1 Gy, and simple and complex aberrations were scored by whole-chromosome FISH painting. For Carbon and Iron, the analysis was performed not only at the first post-irradiation mitosis, but also in interphase following Calyculin-A-induced PCC. PCC values were found to be considerably higher than metaphase values for all the considered ions. PCC dose-response curves were linear for simple exchanges, and linear-quadratic for complex exchanges. Table 22.1 summarizes observed whole-genome-equivalent yields of simple and complex exchanges (average number of exchanges per 100 cells) induced by similar doses of C-12 and Fe-56 of different energies and measured both at the first post-irradiation metaphase and in interphase with PCC. When reading Table 22.1, it has to be taken into account that the energies reported in the first column are in vacuum values, whereas the LET values in the second column are values on the samples. Therefore the energies on sample are lower than those reported in the table (e.g. 414 MeV/n instead of 500 MeV/n for 200 keV/μm Fe, and 115 MeV/n instead of 200 MeV/n for 440 keV/μm Fe). The table shows that interphase aberrations are higher than metaphase aberrations by a factor ranging between 2 and 3. The differences are particularly relevant for complex exchanges.

RBE estimates were also performed both for simple and for total (i.e. simple plus complex) exchanges. RBEmax values for metaphase simple exchanges ranged from 0.5 (for 200 MeV/n Fe-56) to 7.8 (for 500 MeV/n Ar-40). RBE values for simple exchanges derived from PCC data were all higher than the corresponding metaphase values. For instance, the RBEmax found for 1000 MeV/n Iron ions increased from 6.3 for metaphase samples to 18.1 for PCC samples. The values found for total exchanges were even higher, due to the significant contribution of complex exchanges. Also for total exchanges the RBEmax values derived from metaphase data were higher than the corresponding PCC values. The RBEmax calculated for 1040 MeV/n Iron ions increased from 9.6 for metaphase analysis to 26.1 for PCC analysis. The RBEmax-LET relationship, which reached a maximum

Table 22.1 Comparison between aberrations observed in metaphase and in interphase PCC samples following irradiation of human lymphocytes with Carbon and Iron ions (data from George et al. 2003)

Ion type and energy	LET on sample (keV/μm)	Dose (Gy)	Fluence (particles/cell)	Simple Exchanges/100 cells (metaphase)	Simple Exchanges/100 cells (interphase)	Complex Exchanges/100 cells (metaphase)	Complex Exchanges/100 cells (interphase)
C-12 290 MeV/n	13.3	1.2	16.9	23.3 ± 2.6	60.1 ± 8.0	3.5 ± 1.0	18.2 ± 4.4
Fe-56 1040 MeV/n	147	1.0	1.27	28.7 ± 3.0	76.3 ± 12.9	17.4 ± 2.3	58.8 ± 11.3
Fe-56 500 MeV/n	200	0.9 (metaphase) or 1.0 (interphase)	0.77 (metaphase) or 0.85 (interphase)	13.1 ± 2.2	48.2 ± 10.7	10.0 ± 2.0	21.7 ± 5.1
Fe-56 200 MeV/n	440	1.0	0.43	4.6 ± 1.1	10.1 ± 1.9	1.0 ± 0.5	12.0 ± 2.1

around 150 keV/μm, is consistent with other literature studies on CA induction. In particular, Testard et al. (1997) studied the induction of CAs by several ions in the LET range 21.7–1100 keV/μm, and FISH analysis indicated that CA induction increases with LET up to about 150 keV/μm. Wu et al. [38] investigated CA induction in metaphase cells irradiated with different ions including Fe-56 and C-12, with LET values up to 140 keV/μm. The RBE calculations for total exchanges scored with FISH showed values up to 2.5.

Besides outlining the question of possible biases due to interphase death/mitotic delay, which can be overcome thanks to the PCC technique, these studies clearly show that for heavy ions the scoring of complex exchanges, which is also strongly dependent on the chromosome painting technique, is a crucial issue. A recent multi-FISH study of CA induction in human lymphocytes exposed to 1000 MeV/n Iron ions and treated with chemically-induced PCC [39] shows that at 3 Gy, approximately 80% of exchanges are complex, compared with an average of 50% found by George et al. [40] with 2 FISH probes in the dose range 0.2–2 Gy.

22.2.2 *In vivo data: Carbon therapy patients and astronauts*

Therapeutic Carbon beams represent the only source of human exposure to heavy ions on Earth. As mentioned above, Durante et al. [41] monitored the induction of chromosome aberrations in PBL of patients exposed to a 290 or 350 MeV/n SOBP at NIRS, as well as of other patients exposed to 10 MV X-rays. For the 290 MeV/n Carbon beam, the LET was about 13 keV/μm in the plateau and raised from about 40 keV/μm in the proximal edge to 200 keV/μm at the distal fall-off, with clonogenic inactivation RBE values ranging between 2 and 3 along the SOBP. Similar values were found for the 350 MeV/n beam. Reciprocal exchanges (i.e. dicentrics plus translocations) were the most frequent aberration type scored during radiotherapy, but deletions and complex exchanges were observed as well. The fraction of aberrant PBL were found to increase with the number of delivered dose fractions, reaching a plateau at high doses. Interestingly, while C-ions were found to be more efficient than X rays at inducing chromosomal aberrations in PBL *in vitro* (showing a RBE of 1.43 ± 0.17 at 13 keV/μm and of 3.9 ± 0.4 at 83 keV/μm), for the patients considered in this study the fraction of aberrant PBL was lower after Carbon-ion treatments than after X-ray treatments. This result was interpreted as a proof of the improved dose distribution achieved with C-ions. Furthermore, the fraction of aberrant PBL was found to be well correlated with the lymphocyte loss during the treatment, suggesting that the reduced yield of C-ion-induced aberrations in lymphocytes implies a lower risk of acute bone marrow toxicity with respect to X-rays, as well as a lower risk of secondary cancers (the latter due to the correlation between aberrations in PBL and late cancer incidence, see e.g. [42]).

Monitoring of chromosome aberrations in astronauts' PBLs provides an example of chromosome aberration induction following *in vivo* exposure to a mixed field consisting of high-energy particles including heavy ions. Many studies are available in the literature obtained both by conventional Giemsa-staining and by FISH chromosome painting (see references quoted in section 22.1.1). Taken together, these studies show that biodosimetry estimates based on chromosome aberration measurements lie within the range expected from physical dosimetry and ICRP models.

A very interesting analysis was performed by Durante et al. [43], who measured chromosomal aberrations (with Giemsa or FISH) in PBL from 33 crew members involved in long-term missions on the Mir station or on the International Space Station (ISS), as well as on short-term taxi flights, spanning about ten years (from 1992 to 2003). Some cosmonauts were involved in up to five spaceflights, with a maximum total time in space of 748 days (in 3 flights), corresponding to a cumulative dose of 289 mGy. The average absorbed dose was about 4.3 mGy for short-term flights and 78 mGy for long-term missions. The total time spent by the same individual in Extra-Vehicular Activities (EVA) was also recorded, showing a maximum of 79 hours. While the dicentric yields observed after short-term missions (less than 3 months) were not significantly higher with respect to pre-flight levels, those measured following long-term missions (more than 3 months) in lymphocytes from cosmonauts at their first flight showed a highly significant increase, which was consistent with the values calculated from physical dosimetry data. The maximum post-flight dicentric yield (0.0075 ± 0.0028 dicentrics/cell) was observed for an astronaut who made a spaceflight of 189 days, receiving 81 mGy. Comparison of post-flight dicentric yields with pre-flight gamma-ray calibration curves indicated that the observed increase of dicentrics after long-term missions would correspond to an equivalent dose of 0.2 Sv, corresponding to a LEO space radiation quality factor of about 2.5. This is consistent with the 2.4 value calculated by Badhwar et al. [44] on the basis of LET spectra measurements on Mir and the ICRP model. Interestingly, for cosmonauts involved in two or more space flights, the yield of chromosomal inter-changes was not correlated to the total duration of the space sojourn, nor to the integral absorbed dose. The frequencies of dicentrics and translocations declined rapidly between two subsequent spaceflights, and the yields of stable translocations at the end of the last mission were generally of the same order as background aberration frequencies measured before the first mission. This suggested that the effects of repeated space flights are not simply additive for chromosome aberrations, that might be explained by taking into account changes in the immune system (and thus lymphocyte survival and repopulation) under microgravity conditions and/or other phenomena such as adaptive response to space radiation.

22.3 Heavy-ion-induced chromosome aberrations: theoretical models

Despite the recent significant advances in the experimental techniques and the large amount of available data, some aspects of the mechanisms underlying the induction of chromosome aberrations have not been fully elucidated yet. For example it is still not clear whether any DNA double-strand break (DSB) can participate in the formation of chromosome aberrations, or if more severe (i.e. clustered) breaks are required. Furthermore, while it is widely recognised that only breaks sufficiently close in space can interact and form exchanges, the relationship linking the initial distance between two breaks and their interaction probability is still not known. Both exponentially-decreasing functions and step functions were applied with equal success, though the latter seem to receive more support by the experimental evidence that DNA repair – including misrepair – mainly takes place in repair factories located in the small channels separating the various chromosome territories. Another object of debate is the possibility of having an exchange starting from a single radiation-induced chromosome break, which may lead to a (simple) exchange-type aberration mediated through subsequent induction of a second break by the enzymatic mechanisms involved in DNA repair [45].

Theoretical models and simulation codes can be of great help both as interpretative tools, for elucidating the underlying mechanisms, and as predictive tools, for performing extrapolations where experimental data are not available, typically at low doses and/or low dose rates. Various modelling approaches can be found in the literature; many of them are based on Lea's "Breakage-and-Reunion" theory [46]. Though the Revell's "Exchange Theory" was applied by various authors, the models based on Lea's approach better describe the induction of complex exchanges. Various reviews on chromosome aberration induction theories and models are available in the literature [47–50]. In the next section, we will present two of the few modelling approaches that can deal with heavy ions, since the vast majority of the available works are limited to photons and/or light ions.

22.3.1 A model based on interphase chromosomes and DSB production and rejoining

In 2002, Chatterjee and co-workers published a modelling work on chromosome aberration induction in human lymphocytes exposed to different radiation types including heavy ions [51]. The model explicitly takes into account interphase chromosome structure, intra-nuclear chromosome organization, and DSB production and rejoining in a faithful or unfaithful manner. More specifically, each of the 46 human chromosomes is modelled as a random polymer inside a spherical volume. The chromosome spheres are packed randomly within a spherical nucleus, with an allowed overlap degree controlled by a parameter Ω.

The induction of DSBs was modelled on the basis of radiation track-structure, and chromosome exchanges were assumed to arise from pairwise mis-rejoining of close DSB free-ends. Rejoining was modelled by a Monte Carlo procedure using a Gaussian proximity function controlled by an interaction range parameter σ. The parameters were fixed *a posteriori* by fitting the model predictions to experimental data. With an overlap parameter of 0.675 μm and an interaction range of 0.5 μm, the calculated yields of inter-chromosomal exchanges were found to be in good agreement with various experimental data sets relative to low-LET irradiation of human lymphocytes up to 4 Gy.

Inter-chromosome rejoining events induced by 1 GeV/n Fe ions were simulated in the dose range 0.22–2 Gy, corresponding to 1.0–9.12 average traversals per cell nucleus. The authors assumed that at high LET the rejoining process would be governed by the same proximity function with the same parameter values as for low-LET radiation. The calculated average number of DSBs per Gy per cell nucleus for Fe ions was found to be 80. The simulated dose-response curve for the inter-chromosome rejoining frequency was linear. No comparison with experimental data was performed. Since the simulation outcomes are in terms of (total) inter-chromosome rejoining events per cell, and not (total) chromosome exchanges per cell, it is not trivial to make direct comparisons with the results obtained with our model presented in section 22.3.2. However, it is interesting to note that the inter-chromosome rejoining yield predicted by Holley et al. for a single Fe-ion traversal (about 0.5 rejoining events per cell) is consistent with the total exchange yield (about 0.95 total exchanges per cell) predicted by our model. The numerical discrepancy can be due to different factors including the aberration scoring criteria: the fact that we take into account a very large number of complex exchanges might be an explanation for the higher number predicted by our model.

22.3.2 A mechanistic model and a Monte Carlo code based on radiation track structure

Starting from 1999 [52], we developed a mechanistic model and a Monte Carlo code based on radiation track structure at the nanometre level, which is now able to simulate the induction of the main aberration types (including dicentrics, translocations, rings, various complex exchanges and deletions) following irradiation of human lymphocytes with photons, light ions and heavy ions such as Carbon and Iron [53–59].

The main assumption of the model consists of considering chromosome aberrations as the "evolution" of clustered, and therefore severe, initial DNA breaks, that is the Complex Lesions mentioned above. This assumption relies on the fact that the dependence of CLs on radiation quality reflects that shown by mutation and inactivation data [60], whereas non-clustered DSB show a much weaker dependence on the radiation type and energy. Each CL is assumed to produce two independent chromosome free ends. Only free ends induced in neighbouring chromosomes or in the same chromosome are allowed to join and give rise to aberrations, reflecting the experimental evidence that DNA repair takes place within the channels separating the various chromosome "territories", which are basically non-overlapping intra-nuclear regions occupied by a single chromosome. Although the implementation of human fibroblast cell nuclei is in progress, the current version of the model mainly deals with human lymphocyte nuclei, which are modelled as 3-μm radius spheres. The 46 chromosome territories are described as (irregular) intra-nuclear domains with volume proportional to the chromosome DNA content, and each territory consists of the union of small adjacent cubic boxes. Repetition of chromosome territory construction with different chromosome positions provides different configurations for lymphocyte nuclei in the G_0 phase of the cell cycle.

The yield of induced $CL \cdot Gy^{-1} \cdot cell^{-1}$ is the starting point for dose-response-curve simulations. While for photons the lesions are randomly distributed in the cell nucleus, for light ions they are located along straight lines representing the cell nucleus traversals. Concerning heavy ions, which are still "work in progress", as a first approach a fraction of the lesions induced by a heavy ion are "shifted" radially to model the effects of the so-called "delta rays", which play a significant role in determining the features of heavy particle tracks. For a given dose D (in Gy), the average number of cell nucleus traversals n is calculated by $n = D\pi r^2/(0.16\,\text{LET})$ where the LET is expressed in keV/μm and r (in μm) represents either the cell nucleus radius (for light ions), or the nucleus radius plus the maximum range of delta rays (for heavier ions). An actual number is extracted from a Poisson distribution. For each cell nucleus traversal, random extraction of the point where the particle enters the nucleus provides the traversal length, being the direction fixed (parallel irradiation).

The average number of CLs per unit length along a cell nucleus traversal is calculated as $CL/\mu m = 0.16 \cdot CL \cdot Gy^{-1} \cdot cell^{-1} \cdot LET \cdot V^{-1}$, where V is the cell nucleus volume in μm^3. For each nucleus traversal, a Poisson distribution provides an actual number of lesions. Comparison of the CL positions to those of the boxes constituting the chromosome territories allows association of the lesions to the various chromosomes. Specific background (i.e. prior to irradiation) yields for different aberration types (typically 0.001 dicentrics/cell and 0.005 translocations/cell) can be included. Both Giemsa staining and whole-chromosome FISH painting can be simulated, and the implementation of multi-FISH is in progress. Small fragments, i.e. with size of about 10 Mbp, are not scored when the simulation outcomes are to be compared with experimental data, since these fragments can hardly be detected in experiments. Simulation of CL induction and rejoining for a sufficiently high number of times provides statistically significant aberration yields. Repetition of the process for different dose values allows obtaining dose-response curves for the main aberration types, directly comparable with experimental data.

In previous works the model has been tested for gamma rays, protons and He ions by comparing simulated dose-response curves with experimental data available in the literature, without performing any fit *a posteriori.* The good agreement between model prediction and experimental data for the induction of different aberration types allowed for model validation regarding both the adopted assumptions and the simulation techniques. Furthermore, the model has been applied to evaluate the induction of Chronic Myeloid Leukaemia [54] and to estimate dicentric chromosomes observed in lymphocytes of astronauts following long-term missions onboard the Mir space station and the International Space Station, on the basis of simulated gamma-ray dose response weighted by the space radiation quality factor

Table 22.2a Predicted and observed Simple Exchanges per 100 cells induced in human lymphocytes exposed to Iron ions (PCC data from George et al. 2003).

Dose (Gy)	Model prediction	PCC data
0.2	13.2	16.6 ± 2.5
0.5	31.1	46.5 ± 6.0
1.0	64.4	76.3 ± 12.9
1.5	94.4	94.5 ± 18.2
2.0	115.6	114.3 ± 19.9

Table 22.2b Predicted and observed Simple Exchanges per 100 cells induced in human lymphocytes exposed to Carbon ions (PCC data from George et al. 2003).

Dose (Gy)	Model prediction	PCC data
0.1	3.7	4.8 ± 0.8
1.2	65.6	60.1 ± 0.8

[55]. The extension of the model to heavy ions has started only recently, and the results are still preliminary. An example is reported in Tables 22.2a and 22.2b, which show calculated yields (average number per 100 cells) of whole-genome simple exchanges (i.e. dicentrics plus reciprocal translocations) induced by 1 GeV/n Fe ions (LET = 147 keV/micron) and 290 MeV/n Carbon ions (LET = 13.3 keV/micron), respectively. PCC data taken from the literature work discussed in section 22.2.1 [40] are also reported for comparison.

In the framework of space radiation research, we calculated that a single traversal by a high-energy (1 GeV/n) H- or He-ion does not give rise to aberration yields higher than the background levels (due to their high velocity combined with their low charge, which imply a low LET), whereas a single cell nucleus traversal by a 1 GeV/n Iron ion, which has the same velocity but much higher charge and thus LET, was found to induce 0.26 dicentrics (and 0.26 reciprocal translocations) per cell, and 0.45 complex exchanges per cell.

22.4 Conclusions

Some key aspects of the current knowledge on chromosome aberration induction by heavy ions were addressed, both from an experimental and from a theoretical point of view. More specifically, *in vitro* literature data were summarized and discussed, outlining the important role of interphase death/mitotic delay and that of complex-exchange scoring. *In vivo* data from cancer patients treated with Carbon ions and astronauts exposed to space radiation were also reported, confirming that chromosome aberrations in peripheral blood lymphocytes are reliable biodosimeters.

Two of the few available models of chromosome aberration induction that can deal with heavy ions were then described. In particular, the approach described in section 22.3.2 showed good agreement with *in vitro* PCC data on simple exchanges induced in human lymphocytes exposed to Carbon and Iron ions. This provided further support for the main assumption of the model, i.e. that aberrations arise from clustered, and thus severe, double-helix breaks. In progress for this model is also the implementation of chromosome aberration processing at mitosis, which determines whether the cell will fail duplication or it will be able to duplicate possibly giving rise to aberrated daughter cells. This is a key issue in radiobiology because on one side the duplication of aberrated cells implies an enhanced risk for normal tissue with possible consequences in terms of radiation protection, whereas on the other side the death of (tumour) cells is the main goal for radiotherapy.

Acknowledgements This work was partially supported by EU ("RISC-RAD" project, Contract no. FI6R-CT-2003-508842, and "NOTE" project, Contract no. FI6R-036465) and ASI (Italian Space Agency, "Mo-Ma/COUNT" project).

References

1. Savage J, Simpson P, Mutat Res **312**, 51–60 (1994)
2. Durante M, Furusawa Y, Gotoh E, Int J Radiat Biol **74**, 457–462 (1998)
3. Pinkel D, Straume T, Gray J., Proc Natl Acad Sci USA **83**, 2934–2938 (1986)
4. Cornforth M, Radiat Res **155**, 643–659 (2001)
5. Anderson R, Stevens D, Goodhead D, Proc Natl Acad Sci USA **99**, 12167–12172 (2002)
6. Mitelman F, Mutat Res **462**, 247–253 (2000)
7. de The H, Lavau C, Marchio A, Chomienne C, Degos L, Dejean A, Cell **66**, 675–684 (1991)
8. Faretta MR, Di Croce L, Pelicci PG, Seminars in Hematology **38**, 42–53 (2001)
9. Bonassi S, Hagmar L, Stromberg U, Huici Montagud A, Tinnerberg H, Forni A, Heikkila P, Wanders S, Norppa H, for the European Study Group on Cytogenetic Biomarkers and Health (ESCH), Cancer Res **60**, 1619–1625 (2000)
10. Durante M, Radiat Res **164**, 467–473 (2005)
11. Bauchinger M, Schmid E, Braselmann H, Int J Radiat Biol **77**, 553–557 (2001)
12. Sakamoto-Hojo E, Natarajan AT, Curado MP, Radiat Prot Dosim **86**, 25–32 (1999)
13. Kanda R, Minamihisamatsu M, Hayata I, Int J Radiat Biol **78**, 857–862 (2002)
14. Hsieh W, Lucas JN, Hwang JJ, Chan CC, Chang WP, Int J Radiat Biol **77**, 797–804 (2001)
15. Lindholm C, Tekkel M, Veidebaum T et al, Int J Radiat Biol **74**, 565–571 (1998)
16. Lloyd D, Moquet JE, Oram S, Edwards AA, Lucas JN, Int J Radiat Biol **73**, 543–547 (1998)
17. Johnson K, Nath J et al., Mutat Res **439**, 77–85 (1999)
18. Snigiryova G, Braselmann H, Salassidis K et al, Int J Radiat Biol **71**, 119–127 (1997)
19. Stephan G, Pressl S, Int J Radiat Biol **71**, 293–299 (1997)
20. Stram D, Sposto R et al, Radiat Res **136**, 29–36 (1993)
21. Nakano M, Kodama Y et al, Int J Radiat Biol **77**, 971–977 (2001)
22. George K, Durante M, Wu H, Willingham V, Badhwar G, Cucinotta F, Radiat Res **156**, 731–738 (2001)
23. Yang T, George K, Johnson AS, Durante M, Fedorenko BS, Biodosimetry results from space flight MIR-18 Radiat Res **148**, S17–S23 (1997)
24. Obe G, Johannes I, Johannes C, Hallman K, Reiz G, Facius R, Int J Radiat Biol **72**, 727–734 (1997)
25. Testard I, Ricoul M, Hoffschir F, Flury-Herard A, Dutrillaux B, Fedorenko B, Gerasimenko V, Sabatier L, Int J Radiat Biol **70**, 403–411 (1996)

26. Fedorenko B, Druzhinin S, Yudaeva L et al., Adv Space Res **27**, 355–359 (2001)
27. Amaldi U, Kraft G, J Radiat Res **48S**, A27–A41 (2007)
28. Edwards A, Radiat Res **148**, 39–44 (1997)
29. Kolb H, Bone marrow morbidity of radiotherapy. In: Plowman P, McElwain T, Meadows A (eds) Complications of cancer management, Oxford, pp 398–410 (1991)
30. Durante M, La Rivista del Nuovo Cimento **19**(12),1–44 (1996)
31. Durante M, Furusawa Y, George K, Gialanella G, Greco O, Grossi G, Matsufuji N, Pugliese M, Yang T, Radiat Res **149**, 446–454 (1998)
32. Kawata T, Durante M, Furusawa Y, George K, Takai N, Wu H, Cucinotta F, Int J Radiat Biol **77**, 165–174 (2001)
33. Ohara H, Okazaki N, Monobe M, Watanabe S, Kanayama M, Minamihisamastu M, Adv Space Res **22**, 1673–1682 (1998)
34. Testard I, Dutrillaux B, Sabatier L, Int J Radiat Biol **72**, 423–433 (1997)
35. George K, Wu H, Willingham V, Furusawa Y, Kawata T, Cucinotta F, Int J Radiat Biol **77**, 175–183 (2001)
36. Durante M, George K, Yang T, Radiat Res **148**, S45–S50 (1997)
37. Durante M, Furusawa Y, Majima H, Kawata T, Gotoh E, Radiat Res **151**, 670–676 (1999)
38. Wu H, Durante M, George K, Yang T, Radiat Res **148**, S102–S107 (1997)
39. Durante M, George K, Wu H, Cucinotta F, Radiat Res **158**, 581–590 (2002)
40. George K, Durante M, Willingham V, Wu H, Yang T, Cucinotta FA, Radiat Res **160**, 425–435 (2003)
41. Durante M, Yamada S, Ando K et al., Int J Radiat Oncol Biol Phys **47**, 793–798 (2000)
42. Rabbits T, Nature **372**, 143–149 (1994)
43. Durante M, Snigiryova G, Akaeva E et al., Cytogenet Genome Res **103**, 40–46 (2003)
44. Badhwar G, Atwell W, Reitz G, Beaujean R, Heinrich W, Radiat Meas **35**, 393–422 (2002)
45. Chadwick K, Leenhouts H, Int J Radiat Biol **33**, 517–529 (1978)
46. Lea DE, Actions of radiations on living cells. Cambridge University Press, Cambridge, UK (1946)
47. Savage J,, Mutat Res **404**, 139–147 (1998)
48. Ottolenghi A, Ballarini F, Merzagora M, Radiat Environ Biophys **38**, 1–13 (1999)
49. Edwards A, Int J Radiat Biol **78**, 551–558 (2002)
50. Hlatky L, Sachs R, Vazquez M, Cornforth M, Bioessays **24**, 714–723 (2002)
51. Holley W, Mian IS, Park S, Rydberg B, Chatterjee A, Radiat Res **158**, 568–580 (2002)
52. Ballarini F, Merzagora M, Monforti F, Durante M, Gialanella G, Grossi G, Pugliese M, Ottolenghi A, Int J Radiat Biol **75**, 35–46 (1999)
53. Ballarini F, Ottolenghi A, Adv Space Res **31**, 1557–1568 (2003)
54. Ballarini F, Ottolenghi A, Radiat Environ Biophys **43** (2004)
55. Ballarini F, Ottolenghi A, Radiat Res **164**, 567–70 (2005)
56. Ballarini F, Biaggi M, Ottolenghi A, Radiat Prot Dosim **99**, 175–182 (2002)
57. Ballarini F, Battistoni G, Cerutti F et al., Physics to understand biology: Monte Carlo approaches to investigate space radiation doses and their effects on DNA and chromosomes. In: Gadioli E (ed) Proc of the 11th International conference on nuclear reaction mechanisms, Varenna, Italy, June 12–16, 2006. Ricerca Scientifica ed Educazione Permanente suppl 126, pp 591–600 (2006)
58. Ballarini F, Alloni D, Facoetti A, Mairani A, Nano R, Ottolenghi A, Adv Space Res **40**, 1392–1400 (2007)
59. F. Ballarini, M.V. Garzelli, G. Givone, A. Mairani, A. Ottolenghi, D. Scannicchio, S. Trovati, A. Zanini (2008), Proc Int Conf on Nuclear Data for Science and Technology 2007, Nice, France, April 2007, vol. 2, pp 1337–1341 (pdf available at http://nd2007.edpsciences.org)
60. Belli M, Goodhead D, Ianzini F, Simone G, Tabocchini MA, Int J Radiat Biol **61**, 625–629 (1992)

Chapter 23
Spatial and Temporal Aspects of Radiation Response in Cell and Tissue Models

Kevin M. Prise and Giuseppe Schettino

Abstract Rapid advances in our understanding of radiation responses, at the subcellular, cellular, tissue and whole body levels have been driven by the advent of new technological approaches for radiation delivery. Ionising radiation microbeams allow precise doses of radiation to be delivered with high spatial accuracy. They have evolved through recent advances in imaging, software and beam delivery to be used in a range of experimental studies probing spatial, temporal and low dose aspects of radiation response. A range of microbeams have been developed worldwide which include ones capable of delivering charged particles, X-rays and electrons. The original rational for their development was as a precise means of measuring the effects of single radiation tracks. However, the ability to target radiation with microbeams at subcellular targets has been used to address fundamental questions related to radiosensitive sites within cells. Further developments include using microbeams to target more complex 3-D systems where the possibilities of utilizing the unique characteristics of microbeams in terms of their spatial and temporal delivery will make a major impact.

23.1 Introduction

Recently, there have been significant advances in our understanding of radiation responses, at the subcellular, cellular, tissue and whole body levels. This has been driven by the advent of new technological approaches for radiation delivery. In the clinic, the delivery of radiotherapy to cancer patients has become highly sophisticated using image guided approaches to deliver spatially shaped beams to maximize radiation dose to the tumour and minimize dose to the surrounding

K.M. Prise (✉) • G. Schettino
Centre for Cancer Research & Cell Biology, Queen's University Belfast, Belfast BT9 7BL, UK
e-mail: k.prise@qub.ac.uk; g.schettino@qub.ac.uk

G.G. Gómez-Tejedor and M.C. Fuss (eds.), *Radiation Damage in Biomolecular Systems*, Biological and Medical Physics, Biomedical Engineering,
DOI 10.1007/978-94-007-2564-5_23,

tissues. Similar advances have occurred in experimental studies. In particular, the development of microbeams allow precise doses of radiation to be delivered with high spatial accuracy into experimental models. They have evolved through recent advances in imaging, software and beam delivery to be used in a range of experimental studies probing spatial, temporal and low dose aspects of radiation response [1]. A range of microbeams have been developed worldwide which include ones capable of delivering charged particles, X-rays and electrons. Localised delivery of radiation at the subcellular level is proving a powerful tool. For example, localized production of radiation-induced damage in the nucleus allows probing of the key mechanisms of DNA damage sensing, signalling and repair. Crucially this can be done under conditions where cells retain viability and where the responses to relevant environmental, occupational or clinical doses can be tested. These approaches have started to unravel some of the early events which occur after localised DNA damage within cells.

The key rational for the development of modern microbeams originally came from the necessity to evaluate the biological effects of very low doses of radiation (down to exactly one particle track traversal) in order to evaluate environmental and occupational radiation risks. At these levels, only a few cells in the human body are exposed [2] separated by intervals of months or years. Due to the uncertainties of conventional irradiations and random Poisson distribution of tracks, such dose patterns cannot be simulated *in vitro* using conventional broad field techniques. Current excess cancer risks associated with exposure to very low doses of ionizing radiation are therefore estimated by extrapolating high dose data obtained from *in vitro* experiments or from epidemiological data from the atomic bomb survivors. This approach, however, suffers from limited statistical power and is unable to resolve uncertainties from confounding factors forcing the adoption of the precautionary linear non-threshold (LNT) model. Confounding this, there is experimental evidence of non-linear effects at low doses. These include genomic instability [3], low dose hypersensitivity [4] and the bystander effect [5, 6], which could potentially increase the initial radiation risk, while effects such as the adaptive response [7] may act as a protective mechanism reducing the overall risks at low doses. Microbeams allow accurate targeting of single cells and analysis of the induced damage on a cell-by-cell basis which is critical to assess the shape of the dose-response curve in the low dose region. Using microbeams, it has been possible to determine the effect of single particle track traversals for a range of biological endpoints including oncogenic transformation [8], micronuclei formation [9] and genetic instability [10].

23.2 Microbeam development

The development of microbeams is not new and has been an ongoing process over many years with the first UV microbeam being described by Chahotin back in 1912 [11]. However, it has been with the advances in imaging, computing and radiation

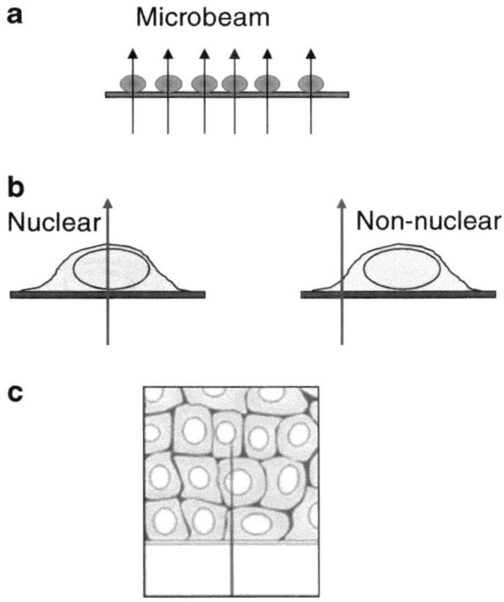

Fig. 23.1 Key uses for a microbeam as a means of (**a**) uniform irradiation of groups of cells with the same number of radiation tracks, (**b**) localized irradiation of a subcellular target, (**c**) localized irradiation of a tissue region

detection in recent years that ionising radiation-based microbeams have entered a phase of rapid development and application. It is currently estimated that there are around 30 radiation microbeams under development or operational world-wide [12]. Despite the advantages of deterministic irradiation achieved by targeting and analyzing cells individually having been recognized since the early 1950s [13], technology for developing sophisticated microirradiation facilities only became available in the late 1990s. Modern radiobiological microbeams are facilities able to deliver precise doses of radiation to preselected individual cells (or part of them) *in vitro* and assess their biological consequences on a single cell basis (Fig. 23.1). They are therefore uniquely powerful tools for addressing specific problems where very precise targeting accuracy and dose delivery are required.

Many of the current generation of microbeams are developed around particle accelerators in order to irradiated biological samples with an exact number of ions however X-ray and electron microbeams have also been developed and are routinely used. Many different experimental set-ups have been exploited in order to achieve a precise dose delivery to individual preselected cells (or part of them). In general, however, there are a few basic requirements that a microbeam facility has to fulfil in order to perform accurate radiobiological experiments (Fig. 23.2).

These are:

- Production of a stable radiation beam of micron or submicron size.
- Radiation detectors able to monitor, with high efficiency, the dose delivered to the samples and trigger the beam stop mechanism when the desired dose has been reached.
- Image system for sample localization. This will have to be supported by appropriate software for image analysis and co-ordinate recording.

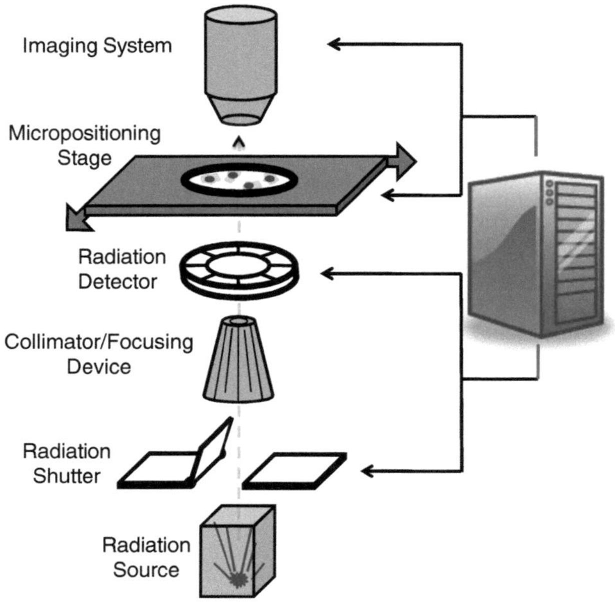

Fig. 23.2 Key elements of a microbeam using a charged particle microbeam as an example

– Micropositioning stage to align the samples with the radiation probe with high spatial resolution and reproducibility.

Charged particle microbeams can be grouped according to the approach used to reduce the radiation beam to sub-cellular dimensions. Many of the older microbeams used collimation approaches whereas more recently, electromagnetic focusing is currently the most popular approach reflecting both technological advances and the need for finer resolutions. The collimation approach is centred on the use of pinholes or collimators to physically obstruct the radiation beam allowing particles to emerge only through a narrow aperture. The advantages offered by the collimation approach include a relatively straightforward alignment and beam location, easy extraction into air (necessary for live cell irradiation) and reduction of particle flux to radiobiological relevant dose-rate ranges. On the other hand, particle scattering (which allows low energy particles to emerge with a wider lateral angle than the primary beam) represents a strong limitation to the final beam size and targeting accuracy. Collimators and set of apertures have been extensively used at the Gray Cancer Institute and Columbia University, pioneers of modern radiobiological microbeams. Using fused silica tubing with apertures as small as 1 μm in diameter, 90% protons and 99% of $^3He^{2+}$ ions were confined within a 2 μm spot [14] while using laser-drilled micro apertures (5 and 6 μm) a 5 μm beam with 91% of unscattered particles was achieved [15]. Collimation methods are still successfully used at JAERI (Takasaki, Japan) [16] and the INFN-LNL (Italy) [17].

The focusing approach has rapidly increased in popularity mainly driven by the availability of existing facilities (previously used for nuclear microscopy and elemental analysis) and the need to investigate the sensitivity of sub-cellular targets. Using a variety of electromagnetic quadrupole doublets [18, 19], extremely narrow charged particle beams (down to a few 10 s of nanometers) can be achieved in vacuum. Moreover existing nuclear microscopy microbeams can access a large range of ions and energies ranging from He to Au and U and LETs values up to 15000 keV/μm [20]. However, as the focused beam has to be extracted in air, significant scattering is introduced by the vacuum window, air gap and traversal of the cell support membrane. Focused spots of 1 μm or less on the samples are however achievable with a very low fraction of particles reaching the targets being scattered. Successful charged particle microbeams based on focus systems have been developed and are routinely used in Germany at the PTB [21], the GSI [22], the University of Munich [18] and Leipzig, [23], in Japan at the NIRS [24] and at Columbia University (USA) [19].

Another key feature of the modern microbeam facilities is the ability to deliver a precise number of particles. This requires a high efficiency detection system (which will trigger the signal when the pre-set number of events has been reached) coupled to a very fast beam shuttering system. Particle detection is probably the feature that differs most between microbeam facilities developed so far. They take two different approaches either detecting before or alternatively after the particles reach the biological sample. By placing the detector between the vacuum window and the samples, no further constraints are imposed on the sample holder or the cell environment while the inevitable detector-beam interaction reduces the quality and accuracy of the exposure. In order to minimize energy loss in the detector, only thin, transmission type detectors are appropriate. These detectors are generally thin films of plastic scintillators whose light flashes generated by the particles traversals are collected by a photomultiplier and then processed [25]. The alternative configuration consists of placing the detector behind the sample holder. Using this approach, no extra scattering is introduced by the detector and better targeting accuracy can in theory be reached. While conventional solid state detectors can be used [26], such configuration requires that the delivered particles have enough energy to pass through the samples setting a limit of the lowest energy usable. In many cases, it is also necessary to remove the culture medium requiring additional procedures (such as humidity control devices) to keep the cells viable during the irradiation process.

X-ray and electron microbeams have also been developed in order to provide quantitative and mechanistic radiobiological information that complement the charged particle studies. As photons do not suffer from scattering problems, X-ray microbeams are in theory capable of achieving radiation spots in air of an order of magnitude or smaller than those so far achieved with ion beams. Moreover, such high spatial resolution is maintained as the X-ray beam penetrates through cells making it possible to irradiate with micron and sub-micron precision, targets that are several tenths or hundreds of microns deep inside the samples. Current X-ray microbeams employ benchtop based electron bombardment X-ray sources [27] for energies up to a few keV or synchrotrons [28] for X-ray beams of a few 10s of keV. The X-ray focusing is generally achieved using diffraction gratings (Zone Plates) which are commonly used in X-ray microscopy to produce X-ray spots down to 50nm. For example, the soft X-ray microbeam at the Centre for Cancer Research & Cell Biology at Queen's University Belfast can deliver 278eV carbon-K, 1.49 keV Aluminium-K or 4.5 keV titanium K-X-rays using diffraction optics. Reflecting X-ray optics are also used [28] and current development [29] promise to significantly improve both spot size and dose rate.

As for X-ray, electron microbeams are generally self-contained units with relatively low development and maintenance costs. They rely on standard electron guns and electrostatic devices to produce and accelerate energetic electron beams which are subsequently reduced to micrometer size by the use of apertures or electromagnetic focusing [30]. As electrons greatly scatter as they interact with biological samples, it is impossible for electron microbeams to achieve targeting resolutions at the micron or submicron level despite the actual size of the focused beam. However, the great advantage of electron (and X-ray) microbeams concerns the ability to easily vary the energy (and therefore the LET) in order to investigate the relative biological importance of various parts of the electron track. In this respect, electron and X-ray microbeams complement the work done with charged particle facilities to investigate the LET dependence. Despite the main differences in radiation production and detection between charged particle, X-ray and electron microbeams, much of the requirements for biological sample imaging and processing remain similar.

23.3 Biological Studies with Microbeams

A major use of microbeams is as a means of probing the spatial and temporal evolution of radiation damage. In particular they can produce highly localised DNA damage under defined conditions complimenting laser approaches for time-resolved and spatial studies. It is widely accepted that the biological effectiveness of ionizing radiation is determined by the ionization pattern (i.e. track structure) produced inside cells or tissues [31]. Understanding the extent and pattern of DNA damage induced [32, 33] and their spatio-temporal evolution is therefore of critical importance for assessing biological risks of radiation exposure. Double stand breaks (DSBs) are considered the most critical DNA lesion induced by radiation due to the complexity of cellular mechanisms involved in the correct rejoining of physically separated DNA sections. The DNA damage caused by a charged particle traversal is the result of a complex clustering of ionizations which occurs along the particle path itself (core) and radially due to secondary electrons (penumbra). Track structures simulations [34] and experimental measurements performed in nanodosimetry detectors such as the Jet Chamber [35] and live cells [36] have determined how the spatial distribution of ionization critically depends on the mass and energy of the particle. As a consequence, charged particle beams are expected to induce clusters of DNA breaks which result in the formation of complex DNA-dsbs.

Despite the final end point being the physical separation of the DNA double helix, DNA-dsbs arising from cluster of DNA lesions present a more difficult challenge for the cellular DNA repair mechanisms. Nikjoo and colleagues [37] calculated that although the number of breaks per unit dose remains nearly constant with the LET, their complexity varies significantly. Whilst for low energy electrons only 20–30% of DNA-dsbs can be considered complex, this proportion increases to 70% for high LET α-particles and to 90% when base damages are included. In general, the complexity of the DNA breaks is rapidly enhanced by increasing the LET.

Crucially, DSBs resulting from multiple damage sites are often associated with loss of genetic material and high probability of incorrect rejoining which are responsible for late effects such as chromosomal aberrations and genetic mutations including carcinogenesis. Despite their clearly fundamental role in determining the fate of the irradiated cell, little is known about the spatio-temporal evolution of DSBs and their related repair events. There are currently two main aspects of great interest of the spatio-temporal evolution of DSBs: the first is related to breaks mobility within the cell nucleus while the second concerns the dynamic interaction and alternation of DNA repair proteins. Theoretical attempts to describe how ends from different DBSs meet to form chromosome aberrations have led to two conflicting theories. While the "contact first" theory proposes that interactions between chromosome breaks can only take place when DSBs are created in chromatin fibres that co-localize, the "breakage first" theory is based on DSBs moving over large distances before interacting. Extensive DSB migration and interaction is therefore the centre of open debates [38, 39]. Using microbeams, it is possible to induce DSBs in precise locations inside the cell nucleus (recent biological developments allow staining of chromosome domains in live cells [40]) at precise times and investigate their spatio-temporal evolution[1]. Being able to control the site and time of the damage induction allows investigations of the DSB mobility using conventional immunofluorescence techniques. Correlating this data to the extent of the effect induced can then provide critical information on how DSBs mobility affects DNA repair and subsequent cellular response.

Understanding the sequential steps in the processing of DNA damage by individual DNA repair proteins is key to understanding the spatial and temporal mechanisms of radiation response. The dynamic interaction and exchanges of DNA repair proteins at the site of damage is a critical aspect as it provides clues of the necessaries steps, functionality and requirements of the different enzymatic activities involved in the repair process. The current knowledge of repair/missrepair events that follow DSBs induction by ionizing radiation relies on immunofluorescence assays (i.e. using antibodies against modified histone proteins such as γ-H2AX or other damage response proteins) on fixed cells [36, 41] [42]. Despite some contradictory indications of chromatin movement and subsequent formation of repair clusters [39, 41], these data provide only a static view of a selected point in time from which it is very difficult to draw dynamic conclusions. Studies looking at the dynamics of DNA repair recruitment are currently being attempted [42, 43] using high atomic number charged particle irradiations (which form highly clustered ionizations) and high resolution microscopy. Modern microbeams are also equipped with state of the art imaging stations in order to accurately monitor the cell response to specific radiation insults. Moreover, the high precision in the delivery of radiation damage to sub-cellular sites using a wide range of LET radiations (from X-rays to heavy ions) and the single cell nature of the experiments represent a natural approach to follow cellular reactions to radiation insults in time. Using microbeam approaches, the spatio-temporal details of the irradiation of each sample within a population can be precisely controlled and the cellular response assessed on a cell by cell basis. Combined with the use of GFP-tagged proteins, these features make radiation microbeams a unique tool for the analysis of the spatio-temporal evolution of the DSBs repair processes in realtime.

23.4 Subcellular targeting and studies in tissue

Although radiation microbeams are playing an important role in studies of DNA damage and repair, a major advantage is the ability to target different regions within cells and tissues. This has been utilised by several groups interested in responses to low dose targeted irradiation. The standard paradigm for radiation effects has been based on direct energy deposition in nuclear DNA driving biological response [5]. Previous studies using radioisotope incorporation have shown that the DNA within the nucleus is a key target as ^{131}I-conconavalin A bound to cell membranes was very inefficient at cell killing, in contrast to ^{131}I-UdR incorporated into the nucleus [44]. These authors also found that dose delivered to the nucleus, rather than cytoplasm or membranes, determined the level of cell death. Recently it has been shown that irradiation of cytoplasm alone can induce an effect. Wu et al. [45] found increased levels of mutations in A_L cells after cytoplasmic irradiation using an α-particle microbeam. The types of mutations were similar to those that occurred spontaneously in unirradiated cells and were formed as a consequence of increased ROS species. Using a charged particle microbeam, it has been shown that bystander responses are induced in radioresistant glioma cells even when only the cell cytoplasm is irradiated, proving that direct damage to cellular DNA by radiation is not required to trigger the effect [46]. Under conditions of cytoplasmic-induced bystander signalling, disruption of membrane rafts also inhibits the response [47]. More recently several groups have reported an involvement of mitochondria in the signalling pathways involved in both cytoplasmically irradiated and bystander cells [48, 49]. This is an expanding area of research which is beginning to understand subcellular radiosensitivity.

Several groups have now extended studies from cell-culture models to more complex tissue models and *in vivo* systems. These are providing convincing evidence for a role for bystander responses of relevance to the *in vivo* situation. The original work done in this area used human and porcine ureter models. The ureter is highly organised with 4–5 layers of urothelium, extending from the fully differentiated uroepithelial cells at the lumen to the basal cells adjacent to the lamina propria or supporting tissue. Using a charged particle microbeam, it was possible to locally irradiate a single small section of ureter such that only 4–8 urothelial cells were targeted. The tissue was then cultured to allow an explant outgrowth of urothelial cells to form. When micronucleated or apoptotic cells were scored in this outgrowth, a significant bystander response was observed. Also, a significant elevation in the number of terminally differentiated urothelial cells was detected. Overall, this involves a much greater fraction of cells than those which were expressing damage. Typically in the explant outgrowth 50 – 60% of the cells are normally differentiated, but this increases by 10 – 20 % when a localised region of the original tissue fragment is irradiated with the microbeam [50]. Therefore, in this model, the major response of the tissue is to switch off cell division which may be a protective response where proliferation leading to additional damage propagation is prevented [51]. Further studies with microbeams have been done in other tissue models. In recent work in commercially available skin reconstruct models it has been possible to use localised irradiation with microbeam approaches and measure the range of bystander signalling. After localised irradiation of intact 3-D skin reconstructs, these can be incubated for up to 3 days before being sectioned for histological analysis of sections at different distances away from the irradiated area. With this approach it was observed that both micronucleated and apoptotic bystander cells could be detected up to 1mm away from the originally irradiated area [52]. Further studies have utilised other tissue reconstruct models including ones aiming to mimic radon exposure in the lung [53] and observed similar long-range effects. The role of cell to cell communication either directly via GJIC or indirectly via autocrine and paracrine factors may be highly tissue specific and unlikely to be exactly mimicked in an *in vitro* test system, so a combination of studies with both *in vitro* and *in vivo* models will need to be developed in the future.

Finally our developing views, on the response of cells and tissues to localised irradiation, is starting to impact on our understanding of responses to clinically relevant beams. For advanced radiotherapies, dose is delivered to the tumour, not as a single uniform exposure but, in a highly spatial and temporal manner. Specifically a series of beams are shaped and delivered from different angles into the patient to gradually "paint" dose into the tumour whilst minimising dose to the surrounding normal tissues. Although the aim is to give overall a uniform dose to the tumour, individual cells or regions within the tumours or surrounding normal tissues see highly variable radiation doses at any one time. Microbeams will be a valuable tool in mapping out these dose distributions, both spatially and temporally.

23.5 Concluding remarks

Microbeams are making a significant impact on our understanding of radiation responses in cells and tissue models. They are powerful probes alongside other approaches (such as laser-based systems) for following DNA damage and repair on an individual cell basis. They have a major advantage in that they allow these processes to be followed under conditions of direct physiological relevance to environmental, occupational and therapeutic doses. The technology for producing microbeam sources is rapidly changing allowing finer resolution beams and more optimal integration with cell imaging approaches. This will allow high throughput approaches where biological changes occur at low frequency in response to low dose exposures of relevance to radiation risk studies. Future advances in biology with the impact of live cell imaging approaches will allow DNA damage processing to be carefully mapped in real-time. A major challenge is to gain further insights into subcellular radiosensitivity mechanisms by probing at the nuclear and non-nuclear levels in both cell and tissue models. These studies need to consider both spatial and temporal aspects following responses in cells through to functional biological changes. A future advance will be to translate these approaches into *in vivo* models to understand the responses of these particularly to low dose exposure. These studies will also impact on our understanding of the effectiveness of advanced radiotherapy approaches where highly modulated spatial and temporal beams are delivered.

Acknowledgment The authors are grateful to Cancer Research UK [CUK] grant number C1513/A7047 and the European Union NOTE project (FI6R 036465) for funding their work.

References

1. G. Schettino, S.T. Al Rashid, K.M. Prise, Mutat Res **704** (2010) 68–77
2. D.J. Brenner, R.C. Miller, Y. Huang, E.J. Hall, Radiat. Res. **142** (1995) 61–69
3. M.A. Kadhim, S.A. Lorimore, K.M. Townsend, D.T. Goodhead, V.J. Buckle, E.G. Wright, International Journal of Radiation Biology **67** (1995) 287–293
4. B. Marples, M.C. Joiner, Radiation Research **133** (1993) 41–51
5. K.M. Prise, G. Schettino, M. Folkard, K.D. Held, Lancet Oncol **6** (2005) 520–528
6. K.M. Prise, J.M. O'Sullivan, Nat Rev Cancer **9** (2009) 351–360
7. D.R. Boreham, J.-A. Dolling, J. Broome, S.R. Maves, B.P. Smith, R.E.J. Mitchel, Radiation Research **153** (2000) 230–231
8. R.C. Miller, G. Randers-Pehrson, C.R. Geard, E.J. Hall, D.J. Brenner, Proc Natl Acad Sci USA **96** (1999) 19–22
9. K.M. Prise, M. Folkard, A.M. Malcolmson, C.H. Pullar, G. Schettino, A.G. Bowey, B.D. Michael, Adv Space Res **25** (2000) 2095–2101
10. M.A. Kadhim, S.J. Marsden, D.T. Goodhead, A.M. Malcolmson, M. Folkard, K.M. Prise, B.D. Michael, Radiat Res **155** (2001) 122–126
11. S. Chahotin, Biol Zent. **32** (1912) 623
12. S. Gerardi, J. Radiat Res **50** (2010) A15–A30
13. R.E. Zirkle, W. Bloom, Science **117** (1953) 487–493
14. M. Folkard, K.M. Prise, G. Schettino, C. Shao, S. Gilchrist, B. Vojnovic, Nuclear Instruments and Methods B **231** (2005) 189–194
15. G. Randers-Pehrson, C. Geard, G. Johnson, D. Brenner, Radiation Research **153** (2000) 221–223
16. Y. Kobayashi, T. Funayama, S. Wada, Y. Furusawa, M. Aoki, C. Shao, Y. Yokota, T. Sakashita, Y. Matsumoto, T. Kakizaki, N. Hamada, Biol Sci Space **18** (2004) 235–240
17. S. Gerardi, G. Galeazzi, R. Cherubini, Radiat Res **164** (2005) 586–590
18. A. Hauptner, S. Dietzel, G.A. Drexler, P. Reichart, R. Krucken, T. Cremer, A.A. Friedl, G. Dollinger, Radiat Environ Biophys **42** (2004) 237–245

19. A. Bigelow, G. Garty, T. Funayama, G. Randers-Pehrson, D. Brenner, C. Geard, J Radiat Res (Tokyo) **50** Suppl A (2009) A21–28
20. M. Heiss, B.E. Fischer, B. Jakob, C. Fournier, G. Becker, G. Taucher-Scholz, Radiat Res **165** (2006) 231–239
21. K. Greif, W. Beverung, F. Langner, D. Frankenberg, A. Gellhaus, F. Banaz-Yasar, Radiat Prot Dosimetry **122** (2006) 313–315
22. M. Heiss, B.E. Fischer, M. Cholewa, Radiation Research **161** (2004) 98–99
23. A. Fiedler, T. Reinert, J. Tanner, T. Butz, Nuclear Instruments and Methods in Physics Research B **260** (2007) 169–173
24. H. Imaseki, T. Ishikawaa, H. Isoa, T. Konishia, N. Suyaa, T. Hamanoa, X. Wanga, N. Yasudaa, M. Yukawaa, Nuclear Instruments and Methods in Physics Research Section B: Beam Interactions with Materials and Atoms **260** (2007) 81–84
25. M. Folkard, B. Vojnovic, K.J. Hollis, A.G. Bowey, S.J. Watts, G. Schettino, K.M. Prise, B.D. Michael, International Journal of Radiation Biology **72** (1997) 387–395
26. G. Randers-Pehrson, C.R. Geard, G. Johnson, C.D. Elliston, D.J. Brenner, Radiat Res **156** (2001) 210–214
27. M. Folkard, G. Schettino, B. Vojnovic, S. Gilchrist, A.G. Michette, S.J. Pfauntsch, K.M. Prise, B.D. Michael, Radiat Res **156** (2001) 796–804
28. Y. Kobayashi, T. Funayama, N. Hamada, T. Sakashita, T. Konishi, H. Imaseki, K. Yasuda, M. Hatashita, K. Takagi, S. Hatori, K. Suzuki, M. Yamauchi, S. Yamashita, M. Tomita, M. Maeda, K. Kobayashi, N. Usami, L. Wu, J Radiat Res (Tokyo) **50** Suppl A (2009) A29–47
29. P.D. Prewett, A.G. Michette, in: A.K. Freund, T. Ishikawa, A.M. Khounsary, D.C. Mancini, A.G. Michette, S. Oestreich (Eds.), Society of Photo-Optical Instrumentation Engineers (SPIE) Conference Series, 2001, pp. 180–187
30. M.B. Sowa, M.K. Murphy, J.H. Miller, J.C. McDonald, D.J. Strom, G.A. Kimmel, Radiat Res **164** (2005) 695–700
31. D.T. Goodhead, International Journal of Radiation Biology **56** (1989) 623–634
32. T.J. Jenner, C.M. deLara, P. O'Neill, D.L. Stevens, International Journal of Radiation Biology. **64** (1993) 265–273
33. K.M. Prise, M. Pinto, H.C. Newman, B.D. Michael, Radiat Res **156** (2001) 572–576
34. H.G. Paretzke, Adv Space Res **6** (1986) 67–73
35. B. Grosswendt, S. Pszona Radiat Environ Biophys **41** (2002) 91–102
36. B. Jakob, M. Scholz, G. Taucher-Scholz, Radiat Res **159** (2003) 676–684
37. H. Nikjoo, P. O'Neill, W.E. Wilson, D.T. Goodhead, Radiat Res **156** (2001) 577–583
38. R.M. Anderson, D.L. Stevens, D.T. Goodhead, Proc Natl Acad Sci USA **99** (2002) 12167–12172
39. B.E. Nelms, R.S. Maser, J.F. Mackay, M.G. Lagally, J.H.J. Petrini, Science **280** (1998) 590–592
40. J. Essers, A.B. Houtsmuller, L. van Veelen, C. Paulusma, A.L. Nigg, A. Pastink, W. Vermeulen, J.H. Hoeijmakers, R. Kanaar, Embo J **21** (2002) 2030–2037
41. J.A. Aten, J. Stap, P.M. Krawczyk, C.H. van Oven, R.A. Hoebe, J. Essers, R. Kanaar, Science **303** (2004) 92–95
42. A. Asaithamby, N. Uematsu, A. Chatterjee, M.D. Story, S. Burma, D.J. Chen, Radiat Res **169** (2008) 437–446
43. B. Jakob, J.H. Rudolph, N. Gueven, M.F. Lavin, G. Taucher-Scholz, Radiat Res **163** (2005) 681–690
44. R.L. Warters, K.G. Hofer, Radiat Res **69** (1977) 348–358
45. L.J. Wu, G. Randers-Pehrson, A. Xu, C.A. Waldren, C.R. Geard, Z. Yu, T.K. Hei, Proc Natl Acad Sci USA **96** (1999) 4959–4964
46. C. Shao, M. Folkard, B.D. Michael, K.M. Prise, Proc Natl Acad Sci USA **101** (2004) 13495–13500
47. H. Zhou, V.N. Ivanov, Y.C. Lien, M. Davidson, T.K. Hei, Cancer Res **68** (2008) 2233–2240
48. S. Chen, Y. Zhao, W. Han, G. Zhao, L. Zhu, J. Wang, L. Bao, E. Jiang, A. Xu, T.K. Hei, Z. Yu, L. Wu, Br J Cancer **98** (2008) 1839–1844

49. L. Tartier, S. Gilchrist, S. Burdak-Rothkamm, M. Folkard, K.M. Prise, Cancer Res **67** (2007) 5872–5879
50. O.V. Belyakov, M. Folkard, C. Mothersill, K.M. Prise, B.D. Michael, Radiation Protection Dosimetry **99** (2002) 249–251
51. O.V. Belyakov, M. Folkard, C. Mothersill, K.M. Prise, B.D. Michael, Mutat Res **597** (2006) 43–49
52. O.V. Belyakov, S.A. Mitchell, D. Parikh, G. Randers-Pehrson, S.A. Marino, S.A. Amundson, C.R. Geard, D.J. Brenner, Proc Natl Acad Sci USA **102** (2005) 14203–14208
53. O.A. Sedelnikova, A. Nakamura, O. Kovalchuk, I. Koturbash, S.A. Mitchell, S.A. Marino, D.J. Brenner, W.M. Bonner, Cancer Res **67** (2007) 4295–4302

Chapter 24
Therapeutic Applications of Ionizing Radiations

María Elena Sánchez-Santos

Abstract The aim of radiation therapy is to deliver a precisely measured dose of radiation to a defined tumour volume with minimal damage to the surrounding healthy tissue, resulting in the eradication of the tumour, a higher quality of life with palliation of symptoms of the disease, and the prolongation of survival at competitive cost. Together with surgery and pharmacology, radiotherapy is presently one of the most important therapeutical weapons against cancer. This chapter provides an overview of the clinical use of radiation, with emphasis on the optimisation of treatment planning and delivery, and a top level summary of state-of-the-art techniques in radiation therapy.

24.1 Radiation Oncology as a speciality in medicine

Radiation oncology is the medical speciality dealing with the use of ionizing radiations in the treatment of patients with malignant neoplasms (only occasionally those with benign conditions), alone or combined with other modalities. The aim of radiation therapy is to deliver a precisely measured dose of radiation to a defined tumour volume with minimal damage to the surrounding healthy tissue, resulting in the eradication of the tumour, a higher quality of life with palliation of symptoms of the disease, and the prolongation of survival at competitive cost.

M.E. Sánchez-Santos (✉)
Hospital Universitario La Paz, 28046 Madrid, Spain
e-mail: msanchezs.hulp@salud.madrid.org

G.G. Gómez-Tejedor and M.C. Fuss (eds.), *Radiation Damage in Biomolecular Systems*, Biological and Medical Physics, Biomedical Engineering,
DOI 10.1007/978-94-007-2564-5_24, © Springer Science+Business Media B.V. 2012

24.1.1 Historical development of Radiation Oncology

Shortly after Roentgen discovered x-rays in 1895 [1] and, three years later, Pierre and Marie Curie discovered Radium [2], it was observed that certain properties of ionizing radiations made them suitable for medical uses. Their ability to traverse the soft tissues of an organism was the beginning of diagnostic imaging methods and lead to a new specialty in medicine, Diagnostic Radiology. On the other hand, in 1922 the field of Radiation Therapy was founded at the International Congress of Oncology in Paris where evidence was presented showing that locally advanced larynx cancer could be cured with radiation [3].

From that moment on, advances in our knowledge about radiobiology and radiophysics and the technological development of equipment have allowed the development of both Diagnostic Radiology and Radiotherapy and the more recent appearance of Nuclear Medicine which uses radiopharmaceuticals to diagnose and treat different diseases.

Together with surgery and pharmacology, radiotherapy is presently one of the most important therapeutical weapons against cancer, which was estimated to affect more than 3 million new patients in 2006 in Europe [4]. Radiotherapy and surgery deal with cancer on a local/regional level, while pharmacology treats it at systemic level. The development of radiotherapy since the first x-ray and cobalt treatment equipments (see Fig. 24.1) has implied the disappearance of mutilating surgical techniques for achieving better results regarding both long-term tumour control and treatment morbidity. Amongst others, this is true for malignant breast, head and neck, prostata, and rectal tumours as well as sarcomas [5].

In 1990, the European Cancer Registry-Based Study of Survival and Care of Cancer Patients (EUROCARE) was created. In one of its studies, it was observed that patients diagnosed with cancer between 1995 and 1999 had an age-standardized five-year relative survival rate of 50.3% [6]. Nowadays, the mean optimum irradiation rate of cancer patients is estimated to be about 60% [7]. All this gives an idea about the importance that the therapeutic utilization of ionizing radiations has gained over the years.

In the last decades, advances have been achieved in cancer treatment. These advances are related to certain circumstances such as the improvement in diagnostic and screening tools, a better interdisciplinary communication among cancer surgeons, radiation oncologists, medical oncologists and pathologists, and a closer interaction among physicians and other sciences allowing the transfer of clinically useful biomedical discoveries and the emergence of cancer pharmacology. There is no doubt that the future of radiation therapy is very promising.

24.1.2 Radiobiological concepts in radiation therapy

The application of radiation therapy is based on a selective depopulation of tumour cells with the lowest possible damage to the surrounding normal tissues. In 1906,

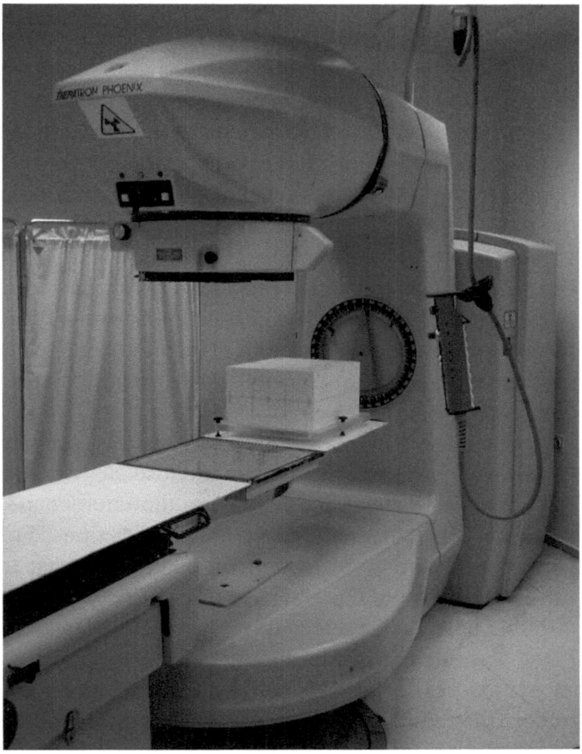

Fig. 24.1 Co-60 teletherapy unit formerly used for treating oncological patients. Utilization of this type of equipment for patient treatment has disappeared due to the availability of technologically more advanced ones

Bergonie and Tribondeau [8] formulated a law with data obtained experimentally. They concluded that the effectiveness of x-rays was greater on cells which have a greater reproductive activity, consequently they can destroy tumour cells preserving healthy tissues.

Radiosensitivity expresses the response of a tumour to irradiation. It reflects the degree and speed of regression of the tumour. *Radiocurability* refers to the eradication of the tumour at the primary or regional site. It reflects a direct effect of the irradiation which may not be parallel to the patient's ultimate outcome. There is no significant correlation between radiosensitivity and radiocurability. A tumour may be radiosensitive and yet incurable or relatively radioresistant and curable by irradiation. At least four factors have been considered to affect the different radiosensitivities of tumours [9]: the oxygen pressure in tumor cells (hypoxia is related to lower radiosensitivity), the proportion of clonogenic cells (proliferating cells are more radiosensitive), the inherent radiosensitivity of tumour cells and the capability of some cell lines to repair sublethal radiation damage.

Tumour control probability is higher the higher the radiation dose delivered. For every increment of radiation dose, a certain fraction of cells is killed. Numerous *dose response curves* from a variety of tumours have been published. Different irradiation levels will yield a different tumour control, depending on the extent of the lesion. A *boost* is a dose applied through small portals to residual disease. *Shrinking field technique* is the use of progressively size-reduced portals in order to apply higher radiation doses to the central portion of the tumour.

Ionizing radiations induce changes in normal tissues surrounding a tumour such as structural alterations, tissue destruction, severe dysfunction and death. They are related to the cell's radiation sensitivity and kinetic characteristics and depend on the total radiation dose delivered, the fractionation schedule used and the tissular volume treated. The *Minimal Tolerance Dose* TD5/5 is the radiation dose that will cause a severe complication rate of no more than 5% in normal tissues within 5 years of treatment [10]. There is a correlation between dose and both tumour control and the probability of complications. The *therapeutic ratio* is the optimal dose that will produce the maximum probability of tumour control with a minimal frequency of complications.

To increase the therapeutic ratio, various fractionation schedules are used in radiation therapy so that the total dose delivered is divided into a number of fractions. Dose–time factors express the interdependence of total dose, total time in which it is delivered and number of fractions. From a radiobiological point of view, the advantages of dose fractionation are that fractionation favours repair of sub-lethal damage, repopulation of cells between fractions, redistribution of cells throughout the cell cycle and reoxygenation. Dose fractionation allows a reduction in the absolute number of tumour cells by the initial fractions, reducing the number of hypoxic cells through cell killing and reoxygenation so that the amount of oxygen per remaining cell increases. This effect still increases because blood vessels previously compressed by a growing cancer are decompressed; this permits a better oxygenation reducing the distance that oxygen must diffuse trough tissue with each fraction. Fractionation exploits the difference in the recovery rate between normal tissues and neoplasic tumours and the patient's tolerance improves when applying fractionated irradiation [11].

It is important to remember that an inadequate fractionation schedule with a prolonged course of therapy and small daily fractions may allow the growth of rapidly proliferating tumours and may decrease early acute reactions but will not protect of serious late damage to normal tissue. The standard fractionation for radiation therapy is the delivery of five weekly fractions of 150–200 cGy. Altered fractionation schedules are:

Hyperfractionation A large number of dose fractions smaller than conventional are given daily. The total dose administered daily is 15–20% higher than for standard fractionation within a unchanged total period of time. The total dose delivered is higher than for standard fractionation [12].

Accelerated fractionation In multiple daily fractions, several conventional radiation dose fractions are delivered over a shorter total period reaching a similar total administered dose. With the concomitant boost, a standard dose fraction is delivered daily together with an additional dose to the final target volume (boost) during the schedule of general radiation therapy.

Split course Two conventional dose fractions are given daily, separated by a time brake between them. Total dose is unchanged or higher than for standard treatment and total treatment time is higher. I wonder if a larger line break is needed here in order to show the description of special items is finished?

Some principles are pertinent when selecting a fractionation schedule of radiation therapy to treat a patient [13]. Multiple daily fractions can be more effective in rapidly growing tumours with a cell line exhibiting a high growth fraction. Fractionation to less than one treatment per day and higher doses can be more efficacious for slowly-growing tumours or tumour cells with a large repair capacity for sublethal damages. Normal tissues behave as actively proliferating cells concerning acute reactions but as slowly proliferating cells concerning the tissue's manifest late injury; 4–8 hours should be allowed between fractions for a maximum repair of normal tissues. When applying accelerated fractionation over a shorter total period, some reduction in the total dose must be introduced. These schedules appear to be preferable for use with hypoxic cell sensitizers or other chemical modifiers of radiation response that require the presence of a high concentration of the compound in the tumour at the time of radiation exposure. Finally, the aim of hyperfractionation is to achieve the same incidence of late effects on normal tissue that is observed with a conventional regime, with an increase in tumour control probability.

The *dose rate* can significantly influence the biological response to a given dose. This effect is more evident for dose rates between 1–10 Gy/h. The biological effect achieved by a given irradiation dose decreases as the dose rate diminishes, allowing for an increase in cell repair. For high dose rates, the tumour dose must be decreased in comparison to that delivered at low dose rates because of the effect on normal tissues. The dose-rate-effect has special interest in brachytherapy and external cobalt units because its numerical value diminishes with time in these radiation therapy techniques [14].

24.1.3 Process of radiation therapy

The clinical use of radiation is a complex process that, from the moment when he/she is first referred to a radiation therapy unit until the moment they finish treatment, includes the following steps for a patient. It involves many health professionals with a variety of interrelated functions (radiation oncologists, physicists, technicians and nurses).

Clinical evaluation An initial evaluation of the patient is made and extent and nature of the tumour is determined by a complete physical examination and a review of all diagnostic studies. The full extent of the lesion should be determined and staging should be established accordingly. The radiation oncologist must be aware of the biological and pathological characteristics of the tumour, as well as any clinical manifestations, so that micro-extensions of the tumour can be included in the treated volume.

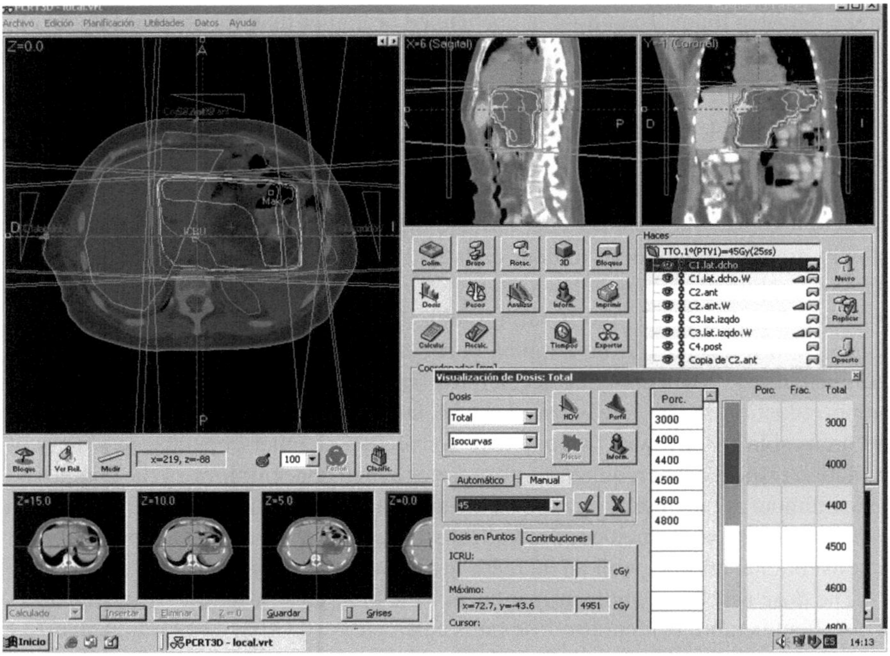

Fig. 24.2 3D planning screen. In each CT slice, treatment volume and healthy organs are drawn. This allows to obtain the exact radiation dose each of them receives

Therapeutic decision Choice of the therapeutic modalities that should be used for the patient. It is important to determine the treatment's objective—cure or palliation—and to evaluate the alternative therapeutic approaches.

Tumour localization Localization and definition of the tumour volume and the surrounding normal structures through a complete physical examination and pertinent radiographic (computed tomography, CT), magnetic resonance imaging (MRI) and/or radionuclide studies (positron emission tomography, PET).

Treatment planning in different steps:

- Localization-Simulation consists in the acquisition of radiological images of the patient under geometrical and anatomical treatment conditions. Additional anatomical data that may be necessary for dose planning and for the designation of immobilization and repositioning devices, shielding blocks, masks, etc. are usually obtained via CT, MRI, PET and ultrasound images.
- Determination of the treatment volumes, normal structures and organs at risk with a 3D planning system (Fig. 24.2). Three different tumour volumes are considered [15,16]: GTV—gross tumour volume (demonstrated tumour), CTV—clinical target volume (demonstrated and/or suspected tumour), and PTV—planning target volume (CTV plus a security margin to ensure delivery of the prescribed dose).

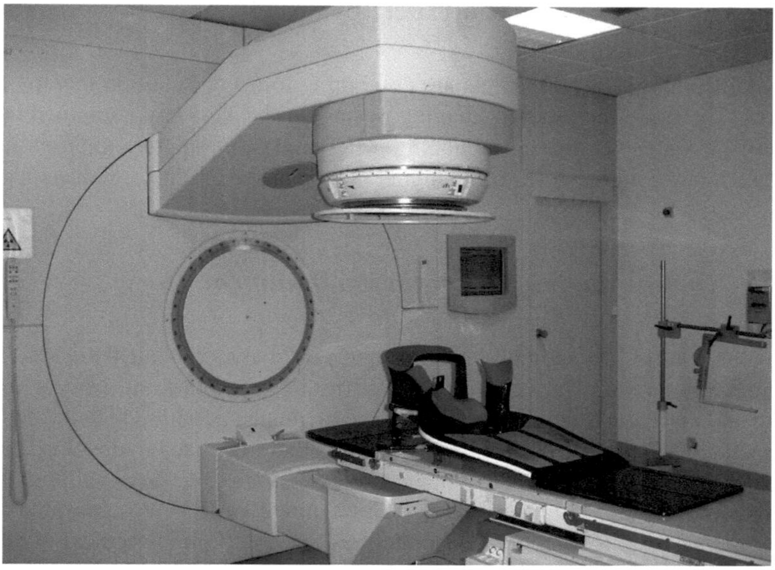

Fig. 24.3 Linear accelerator for clinical radiotherapy use. A positioning and immobilization device can be seen on the treatment couch. Below the couch, a system for obtaining the beam's eye view (see Sect. 24.2.3) of the treatment in course. At the right (background of the room), an alternative verification system using radiographic films can be seen

- Prescription of tumour dose and fractionation schedules.
- Prescription of dose limits for organs at risk. Organs at risk are normal tissues whose radiation sensitivity may significantly influence treatment planning and/or dose prescription. The dose limits for normal tissues have been tabulated [17] and organs at risk can be divided into three different classes according to the severity of the radio-induced lesions.
- Dosimetric procedures with designation of the treatment portals, dose calculation, beam selection and computation and isodose curve generation.
- Definitive dose prescription: selection of the treatment plan to be used for the patient.

Treatment delivery During treatment delivery, some verification procedures such as image-view or verification films (Fig. 24.3) and *in vitro* and/or *in vivo* dosimeters must be carried out. The patient's status must be evaluated periodically during the course of therapy in order to assess tumour response and the patient's tolerance of the treatment.

Follow-up after treatment After finishing treatment, periodic follow-up examinations of the patient are critical for evaluating the general condition of the patient and the tumour response and for a timely detection of recurrences or secondary effects on normal tissues.

24.2 Advances in radiation therapy

Several technologic developments are combining to bring radiation therapy into a new era. These are the modern imaging technologies like x-ray computed tomography (CT), magnetic resonance imaging (MRI) and positron emission tomography (PET) and new computers featuring higher power and reliability at reduced costs.

24.2.1 Optimization of the radiation treatment

Failure to control the tumour loco-regionally is a major contributing factor to a decreased survival. The effects predicted by dose-response curves for tumour control and normal tissue injury are based on the precision with which the dose and the irradiated volume are defined. Every effort should be made to develop accurate dose-response algorithms and to obtain the highest possible dose optimization in the irradiated volume. It is required to plan treatment as good as achievable and to deliver the selected treatment as accurately as possible. This is critical to achieve maximum tumour control probability and satisfactory results in normal tissues. The ICRU recommends a 5% accuracy for dose delivery computations [15, 16].

With advanced treatment delivery systems and new treatment planning systems, three dimensional conformal radiation therapy (3DCRT) has become possible. 3DCRT is an external beam radiation therapy in which the prescribed dose is conformed closely to the target volume. Its goal is to obtain an optimal dose distribution in patient tissues in order to achieve maximum tumour control probability and satisfactory results in normal tissues.

24.2.2 Optimization of treatment planning

Currently, computed tomography (CT) is the principal source of image simulation. CT allows more accurate definition of tumour volume and of the anatomy of normal structures and the generation of digitally reconstructed radiographs. Contiguous CT slices are used to define anatomical structures and target volumes by drawing contours slice by slice.

3D treatment planning allows calculating 3D dose distributions, optimizing dose distribution and permits the radiographic verification of the volume treated [18]. The dose optimization requires correction for inhomogeneities in tissue density and for the individual shape of the patient's body and the development of practical treatment planning capabilities.

These equipments allow the design of treatment portals and virtual simulation of therapy, so that external radiation beams of any possible orientation are simulated and spatial dose information is displayed. Optimization programs attempt to minimize the dose gradient across the target volume as well as the maximum dose to critical organs. They allow a critical evaluation of the treatment plan.

Dose-volume histograms depict the amount of target volume or critical normal structure receiving more than a specified dose level. This representation allows an evaluation of the plan for dose optimization.

There is a growing demand to incorporate the complementary information available from MRI, single-photon emission computed tomography (SPECT) and PET. MRI provides excellent soft tissue contrast with precise delineation of normal critical structures and treatment volumes. Single-photon emission computed tomography and positron emission tomography provide detailed functional information about tissue metabolism and radioisotope transport of the tumour.

The imaging study must be integrated with the treatment planning CT data set. This requires calculation of a 3D transformation that relates the coordinates of a particular imaging study to the planning CT coordinates. The 3D transformation is used to "fuse" information such as anatomic structure contours from the imaging study with the planning CT.

New trends in the integration of multimodality image data for 3D RTP are:

Radiobiologically optimized dose delivery using intensity and radiation quality modulation based on high-resolution PET-CT or magnetic resonance spectroscopic imaging (MRSI).

Inverse planning A verification of the intensity modulated radiation therapy (IMRT) plan and dose is performed before treatment delivery.

24.2.3 Optimization of treatment delivery

Advances in computer technology have allowed the development of 3DCRT to achieve an optimal dose distribution in patient tissues. Technical innovations should also be used to ensure higher accuracy in the delivery.

Multileaf collimators The new treatment machines can deliver multiple segments of the treatment with different beam apertures automatically under computer control. Computer-controlled conformal radiation therapy treatment machines allow performing any kind of conformal field shaping, with an individualized beam aperture.

Devices to reduce errors in patient positioning Immobilization devices like masks, alpha cradles, vacuum pillows and gate breathing (Fig. 24.4). Manoeuvres to exclude sensitive organs from the irradiated volume may be needed, too.

Beam's eye view The patient's contour is viewed as if the observer's eye were placed at the radiation source, looking along the radiation beam axis.

Clinical research in 3DCRT presents three categories:

- Technical innovations and improvements including target volume delineation, treatment planning, immobilization, and dose delivery.

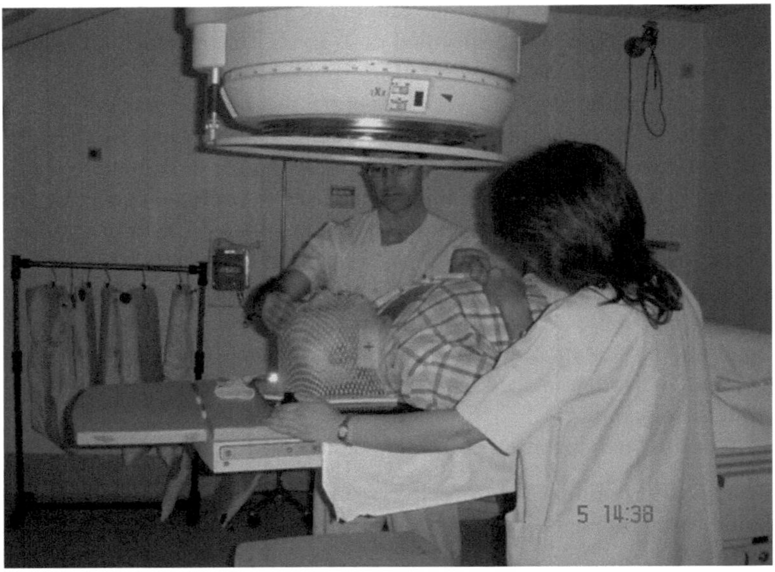

Fig. 24.4 Daily clinical practice: positioning of the patient before treatment using a head and neck immobilization mask. At the back, some vacuum pillows are visible

- Clinical trials aimed at diminishing dose to normal structures to reduce complication rates.
- Dose-escalation studies aimed at increasing dose to the target volume to improve local tumour control and survival.

24.3 State-of-the-art techniques in radiation therapy

The advent of 3DCRT and cross sectional imaging has developed the ability to modulate the radiation beam. Intensity modulation of the beam is by far the most important degree of freedom of dose delivery.

Intensity-modulated radiation therapy consists in a moving-beam computer-controlled conformal radiation therapy. To perform any kind of conformal field shaping, it is necessary to develop an individualized beam aperture. IMRT has developed an individualized beam aperture and intensity modulated three dimensional dose delivery technique by scanning high-energy narrow electron and photon beams. The most suitable accelerator will be in the range 6–15 MV for both superficially located and deep-seated targets. A narrow penumbra region of a photon beam ideally should contain low energy photons (4 MV), whereas the gross tumour volume, particularly when deep-seated targets are concerned, should be irradiated by high energy photons. This allows a better protection of organs at risk and escalation of dose inside the tumour.

Major research areas in IMRT are:

IMRT with mixed photon and electron beams Electron and photon beams are combined to create dose distributions that feature a steep dose fall-off at large depths similar to pure electron beams and flat beam profiles and sharp and depth-independent beam penumbras as in photon beams.

IMRT simultaneous integrated boost (SIB) Sophisticated technique where high precision treatment is required. The SIB allows a dose-escalation. That means to deliver higher total doses of radiation to smaller treatment volumes. A biologic effect of accelerated fractionation can occur within the tumour.

Light ion radiation therapy Intensity-modulated light ion beams are the ultimate tool in clinical practice. Intensity modulated proton therapy may be able to cure even the most advanced hypoxic and radiation-resistant tumours. Low ionization-density hydrogen ions and high ionization-density carbon ions are used. The high LET component is located only in the high-dose tumour volume. The low LET component is located in the surrounding normal tissues.

Tumour immobilization devices Tumour motion between and even during radiation treatments represents a major uncertainty. Efforts are required to ensure accurate conformal radiation therapy. The gate breathing system allows delivery of the radiation beam only in particular moments of the respiratory cycle and prevents it when the tumour is out of the beam.

Tomotherapy IMRT is an intensity modulated radiation therapy with daily serial or helical CT localization. It allows optimum target coverage and doing a frameless stereotaxis.

Stereotactic radiation therapy is a technique that delivers a large single fraction or multiple fractions of radiation to a number of small, stationary portals at different angles. Thus, usually small volumes are treated. Beams intersect at a common point within the body after entering through different points distributed over the skin surface. Extra precision regarding target localization and treatment geometry is required since high-dose gradients at field edges minimize dose deposition outside the target volume. In cranial stereotactic radiation therapy the patient's immobilization is done by fixing a semicircular stereotactic frame to his skull or using a mask. The collimator moves circumferentially along the frame. New stereotactic radiation therapy delivery machines allow the treatment of localizations different from intracranial tumours.

Intraoperative radiation therapy is a treatment technique that uses electron external beam irradiation for deep-seated cancers. The irradiation is concentrated to the tumour volume while the adjacent, surgically mobilized normal tissues, can be avoided.

High Dose-Rate (HDR) Brachytherapy is illustrated in Fig. 24.5. Brachytherapy consists of placing sealed radioactive sources very close to or in contact with the target tissue. With HDR, high doses can be safely delivered to a localized

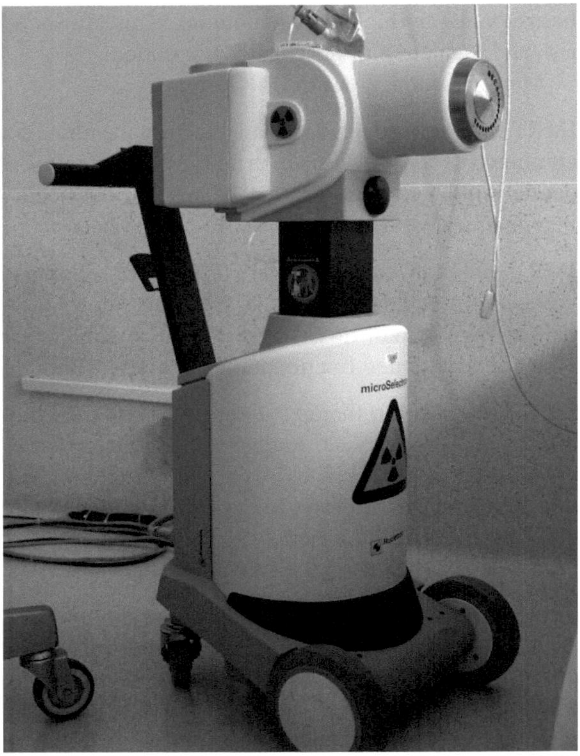

Fig. 24.5 HDR brachytherapy equipment. This treatment modality allows for the positioning of radioactive sources close to or within the tumour or volume treated. High doses are delivered in short time periods

target region over a short time. It represents a technological advance that improves physical dose delivery due to the short treatment time and negligible organ motion. It offers the dosimetric advantages of a remote-controlled oscillating source, radiation safety and protection with decreased personnel exposure and a reduced possibility of human error through computerized remote after-loading. General anaesthesia is not required in selected patients and complications associated with prolonged bed confinement are avoided, especially in elderly patients with comorbidities.

Adaptative radiotherapy is the latest development in radiation therapy. The dose delivery is biologically optimized using beam intensity and radiation quality modulation. The process is based both on three-dimensional conformal radiation therapy and dose delivery monitoring. This technique represents the maximal potential of percent accuracy in tumour response.

References

1. W.C. Roentgen, Br J Radiol **4**, 32 (1931)
2. P. Curie, M.P. Curie, G. Bemont, Compt Rend Acad Sci **127**, 1215 (1898)

3. H. Coutard, Am J Roentgenol **28**, 313 (1932)
4. International Agency on Research of Cancer. European Cancer Observatory. http://eu-cancer.iarac.fr (2009)
5. V.T.D. Vita, S. Hellmann, S.A. Rosenberg, *Cancer. Principles and Practice of Oncology*, 5th edn. (Lippincott, Williams and Wilkins Inc., USA, 1997)
6. M. Sant, C. Allemani, M. Santaquilani, Eur J Cancer **45**, 931 (2009)
7. R.C. of Radiologists, *Clinical Oncology Information Network Guideline for External Radiation Radiotherapy* (Pergamon Press, 2000)
8. J. Bergonie, L. Tribondeau, Rad Res **11**, 587 (1959)
9. W.T. Moss, W.N. Brand, *Therapeutic Radiology* (Mosby Company, Saint Louis, 1972)
10. P. Rubin, R. Cooper, *Radiation Biology and Radiation Pathology Syllabus* (American College of Radiology, Chicago, 1975)
11. W.T. Moss, W.N. Brand, *Radiation Oncology: Rationale. Techniques. Results.* (Mosby Company, Saint Louis, 1979)
12. J.C. Horiot, W. van den Bogaert, in *Frontiers in Radiation Therapy and Oncology*, vol. 22, ed. by J.M. Vaeth, J. Meyer (S. Karger AG, Basel, 1989), p. 149
13. J.D. Cox, Int. J. Radiat. Oncol. Biol. Phys. **13**, 1271 (1987)
14. E.J. Hall, D.J. Brenner, Int. J. Radiat. Oncol. Biol. Phys. **21**, 1403 (1991)
15. ICRU, Prescribing, recording and reporting photon beam therapy. Tech. Rep. 50, International Commission on Radiation Units and Measurements (1993)
16. ICRU, Prescribing, recording and reporting photon beam therapy (Supplement to ICRU report 50). Tech. Rep. 62, International Commission on Radiation Units and Measurements (1999)
17. B. Emami, J. Lyman, Int. J. Radiat. Oncol. **21**, 109 (1991)
18. Agencia de Evaluación de Tecnologías de Galicia, Efectividad y seguridad de los planificadores en 3D frente a los planificadores 2D y 2,5D en oncología radioterápica. Tech. rep., Consejo Interterritorial del Sistema Nacional de Seguridad (1999)

Chapter 25
Optimized Molecular Imaging through Magnetic Resonance for Improved Target Definition in Radiation Oncology

Dževad Belkić and Karen Belkić

Abstract Magnetic resonance spectroscopy (MRS) and spectroscopic imaging (MRSI) are a key modality in radiation oncology for brain and prostate tumors. Improved target definition for radiation therapy (RT) and distinction of changes due to RT from tumor recurrence have been greatly aided by MRSI. However, current applications of MRS/MRSI have limitations due to mainly the fast Fourier transform (FFT) and noise. Optimization of MRS/MRSI is possible by more advanced signal processing via the fast Padé transform (FPT). As a quotient of two polynomials, the FPT markedly improves the resolution of *in vivo* MR time signals encoded from the brain and reliably reconstructs all spectral parameters of metabolites. Due to high spectral density with numerous multiplet resonances, MRS/MRSI of the prostate is exceedingly difficult. The FPT applied to MRS data as encoded from normal and malignant prostate resolves all the genuine resonances, including multiplets and closely overlapping peaks. With synthesized time signals, the FPT fully retrieves all the input spectral parameters with machine accuracy. Such super-resolution is achieved without fitting or numerical integration of peak areas, thereby yielding the most accurate metabolite concentrations. This needs only short signal lengths that imply improved signal-to-noise ratios. These ratios are further enhanced by eliminating "noisy" Froissart doublets as confluent pole-zero pairs. Hence, only the true information is reconstructed by the FPT, as the prerequisite for clinically meaningful interpretations of *in vivo* time signals. With these long sought capabilities of advanced Padé-based signal processing, MRS and MRSI are poised to reach their full potential in radiation oncology.

Dž. Belkić (✉) • K. Belkić
Nobel Medical Institute, Karolinska Institute, Department of Oncology and Pathology, 171 76 Stockholm, Sweden
e-mail: Dzevad.Belkic@ki.se; Karen.Belkic@ki.se

G.G. Gómez-Tejedor and M.C. Fuss (eds.), *Radiation Damage in Biomolecular Systems*, Biological and Medical Physics, Biomedical Engineering,
DOI 10.1007/978-94-007-2564-5_25,

25.1 Introduction

Magnetic resonance imaging (MRI), a sensitive morphological imaging technique free from ionizing radiation is indispensable for timely cancer detection, but often has insufficient specificity. Magnetic resonance spectroscopy (MRS) can enhance specificity by elucidating the metabolic features of malignancy. Since molecular changes often precede morphologic alterations, sensitivity can be further improved by MRS. Molecular imaging can be accomplished by combining MRS and MRI yielding magnetic resonance spectroscopic imaging (MRSI). Rather than selecting a single voxel to encompass a specific volume, a spectrum is obtained at each point of selected grids thereby providing volumetric coverage. In this chapter, we highlight the achievements as well as current limitations of molecular imaging through MRS and MRSI for radiation oncology. The main limitations are noise corruption of encoded time signals and their processing by exclusive reliance on the fast Fourier transform (FFT). We then present certain novel possibilities for optimization through advanced signal processing methods, notably quantum-mechanical spectral analysis for metabolite quantifications via the fast Padé transform (FPT) [1,2].

25.2 Achievements of MRSI in radiation oncology

Molecular imaging has been vital to radiation oncology. By helping to define complex target geometries and surrounding healthy tissue, volumetric imaging was the pivotal spur for advances such as Intensity Modulated Radiation Therapy (IMRT) [3]. While MRS and MRSI have been used more widely in cancer diagnostics [4], their main applications within radiation oncology have been for brain tumors and prostate cancer. The achievements of MRS and MRSI in these two areas of radiation oncology are now briefly summarized.

25.2.1 Brain tumors

Clearly, in radiation neuro-oncology, the most delicate clinical decisions are made, requiring maximal information of the highest possible reliability. In no other domain of oncology have MRS and MRSI become so widely incorporated into clinical practice. Currently, ratios of certain metabolites have mainly been used. These include: choline (Cho) a marker of membrane damage and cellular proliferation whose resonant frequency is at $\sim$3.2 ppm (parts per million) in relation to Nitrogen-Acetyl Aspartate (NAA) an indicator of viable neurons, resonating at $\sim$2.0 ppm, or to creatine (Cr) at $\sim$3.0 ppm, a marker of cerebral energy metabolism.

Incorporating MRSI into RT planning for primary brain tumors can improve control, while reducing complications. Traditionally, the clinical target volume for

RT of gliomas was generated by adding uniform margins of 2-3 cm to the area of T_2 hyper-intensity. This is done because the so-called peri-tumoral uncertain zone, which appears normal on MRI, is frequently infiltrated by tumor when examined histopathologically [4]. By using MRSI to determine areas of high Cho/NAA or Cho/Cr, the shape and size of the clinical target volume can often be better identified, with more confident sparing of uninvolved brain tissue. Ratios of Cho/Cr have also been used to tailor radiation dose levels to the glioma grade within a map of clinical target volumes. A phase I dose intensification trial for gliomas using IMRT is being planned using the information provided by Cho/Cr ratios from MRSI together with functional MRI [5].

A major challenge in neuro-oncology is the detection of tumor which has recurred after RT. Subsequent to RT for glioma, the appearance of a new contrast-enhancing lesion on MRI could be the result of radiation necrosis, but could also be due to recurrent tumor. Distinction between recurrent glioma and radiation necrosis with the appearance of new contrast-enhancing lesions has been facilitated by MRSI [2]. We have performed a meta-analysis of all available published data comparing radiation necrosis and recurrent primary brain tumors in new contrast enhancing lesions post-RT. The ratios of Cho/Cr and Cho/NAA were greater in the latter compared to the former ($p = 0.000001$, $p = 0.000007$, respectively). However, as per our study to be published, there was no ideal cutpoint which unequivocally identified recurrent tumor, such that up to 50% of the brain tumors could be mistaken for radiation necrosis.

25.2.2 Prostate cancer

Nearly 20% of patients with newly diagnosed prostate cancer are treated with RT. A major dilemma is that harm to nearby organs (bladder and rectum) is dose-related, occurring with increasing frequency at precisely the radiation total dose levels for which the chance of cure is greatest (70 Gy). Accurate target definition is therefore of utmost importance to spare the surrounding healthy tissue. While MRI has been widely used for diagnosing, staging and managing prostate cancer, it poorly distinguishes benign from cancerous prostatic lesions. In comparison to MRI alone, MRSI can enhance the accuracy of detecting prostate cancer and differentiating it from benign prostatic hypertrophy (BPH). Most applications of *in vivo* MRSI in prostate cancer diagnostics rely upon the ratio of choline to citrate. Citrate, with resonant frequencies in the spectral region between 2.5 ppm and 2.7 ppm, is considered to reflect healthy secretory activity of prostate epithelium. However, low citrate is also seen in normal stromal prostate, as well as in metabolically atrophic prostate secondary to therapeutic interventions, including RT [2].

Among 67 patients with biopsy-proven prostate cancer who were followed up after external beam radiation therapy (EBRT), MRI and MRSI findings prior to therapy were found to be stronger predictors of outcome compared to clinical variables [6]. The criteria for considering a volume of tissue having clearly malignant metabolism on MRSI was that the choline peak was unequivocally larger than the citrate peak [6].

Detection of so-called "dominant intra-prostatic lesions" that should receive a boost dose of RT has also been facilitated via MRSI. Advantages of this approach include better sparing of surrounding healthy tissue and fewer complications, together with a more effective treatment of the tumor [2].

Also, MRSI has been used to compare the effectiveness of three-dimensional conformal EBRT relative to permanent prostate implantation through brachytherapy among 50 patients with low risk prostate cancer [7]. Via MRSI, it could be concluded that permanent implants were more effective than EBRT in generating metabolic atrophy. However, on the basis of MRS and MRSI, it was not possible to determine with certainty which RT method was more successful in curing the prostate cancer [7].

It has also been suggested that MRSI could improve assessment of recurrence after RT, which causes fibrotic and atrophic changes that distort the glandular anatomy. These are manifested by low T_2-weighted signal intensity on MRI, which is difficult to differentiate from prostate cancer. The appearance of choline suggests that there has been a local recurrence. Salvage therapy could then be guided by MRSI. It has been reported that MRSI detected recurrence of prostate cancer accurately after RT in over 80% of cases [8].

25.3 Limitations in MRS and MRSI due to the FFT

Notwithstanding the above-described achievements, there are important shortcomings of the current applications of MRS and MRSI in radiation oncology. Many of these limitations are due to reliance on the FFT which converts the encoded time signal, or free induction decay (FID), into its spectral representation in the frequency domain. The FFT is a non-parametric, low-resolution spectral estimator. Post-processing through fitting is required to obtain estimates of metabolite concentrations [1]. In other words, since it is non-parametric, the FFT supplies only the total shape of spectral structures, i.e., the overall lineshape envelope, but does not provide quantification. The peak parameters are extracted afterwards in a post-processing stage, by fitting the obtained structures to a sum of Gaussians or Lorentzians, or both. Thus, much of the intermediate information, typical of quantum-mechanical phase interference effects inherent in the complex-valued time signals is totally overlooked in the process of fitting the lineshape envelopes. As a consequence, fitting cannot provide the needed accurate values for the actual position, width, height and phase of each metabolite, nor the true number of physical resonances [1]. As in quantum theory of resonance scattering, these phase effects are at the very core of every resonance phenomenon and without their proper account, no valid spectral analysis is possible. Over-modeling (spurious peaks) as well as under-modeling (undetected genuine metabolites) regularly occurs in all the known fitting recipes.

Although unsubstantiated claims were repeatedly put forward for objectivity, in fact, the inherent subjectivity of fitting has conclusively been demonstrated. This becomes particularly troublesome for overlapping resonances [1,2].

Using the FFT, a shape spectrum is obtained from pre-assigned frequencies whose minimal separation is determined by the total acquisition time, T. The FFT spectrum is defined only on the Fourier grid points $k/T(k = 0, 1, 2, ...)$. Attempts to improve resolution by increasing T and thereby decreasing the distance $1/T$ between the grid points, lead to another problem, because clinical FIDs become heavily corrupted with background noise for longer T. Since FIDs fall off exponentially, the larger signal intensities are observed early in the encoding. It is thus advantageous to encode all FIDs as rapidly as possible, i.e., to avoid long T at which mainly noise will be measured. Because of these two mutually exclusive requirements within the FFT, attempts to improve resolution invariably lead to a worsening of the signal to noise ratio (SNR). The FFT also lacks extrapolation capabilities. Yet another reason for its low resolution is that the FFT is a linear mapping, since its transformation coefficients or weights are independent of the FID points [1,2].

25.3.1 Limitations of the FFT in radiation neuro-oncology

Several of the problems with current applications of MRSI in radiation neuro-oncology are related to resolution and SNR. Attempts to improve the SNR have usually entailed either increasing the acquisition time, or the volume of brain tissue from which data is acquired. The latter leads to recording heterogeneous voxels with a mixture of tissue types. Because of the importance of achieving volumetric coverage of brain tumors, which themselves are often heterogeneous, the SNR issues regarding MRSI are of major concern for radiation neuro-oncology. Insufficient resolution and SNR limit the capability of MRSI to detect small foci of brain tumors. Resolution and SNR for brain tumor diagnostics via MRSI have been improved by using higher magnetic field strengths. However, detection of residual brain tumor is troublesome even with 3T scanners. Another strategy has been to use short echo times (TE) in attempts to capture clinically important metabolites that decay rapidly [9]. However, at short TE the pitfalls of relying upon fitting can be exacerbated. Moreover, since T_2 relaxation times of various metabolites differ, peak ratios can be affected by changes in TE [2]. Consequently, reliance upon metabolite ratios becomes even more problematic.

In our systematic review [2], none of the metabolite ratios estimated via the FFT unequivocally distinguished brain tumors from non-malignant brain pathology. Low NAA can reflect loss of neurons with infiltration of brain tissue by tumor. However, since it is a marker of neuronal viability, NAA can also be decreased in almost any brain abnormality. On the other hand, choline can be low in very small or necrotic brain tumors or tumors otherwise containing mixed tissues. The cutpoints for metabolite ratios used to indicate the presence of malignant brain tissue are arbitrary, and, as such, vary substantially in the literature. As noted, alterations in the ratios of Cho/NAA and Cho/Cr may occur with radiation necrosis and other treatment effects. There are no cutpoints of these ratios that provide unequivocal distinction from recurrent brain tumors.

There is a much richer store of potentially informative metabolites for radiation neuro-oncology than is currently extracted using *in vivo* one-dimensional (1D) MRS and MRSI. These metabolites have been identified using two-dimensional (2D) MRS (which, though applicable *in vivo*, has not been amenable to volumetric coverage and quantification [1, 2]) and *in vitro* MRS (which requires biopsy and analysis of excised specimens at high magnetic fields). Many of the most diagnostically important components overlap. In the lipid lactate region, e.g. highly esterified cholesteryl esters, triglycerides as well as lactate may help identify high grade tumors [2]. The closely-lying components of total choline (phosphocholine, glycerophosphocholine and free choline) between 3.2 ppm to 3.3 ppm may also be informative regarding brain tumor characteristics [2]. *In vitro* MRS together with histopathology may provide further insights. For example, taurine at around 3.3 ppm to 3.4 ppm showed a highly significant correlation with apoptotic cell density in astrocytomas [10]. Moreover, unlike lipids at 0.9 ppm and 1.3 ppm, this association is independent of necrotic tissue. Thus, taurine may represent an MR-detectable biomarker of apoptosis that is unrelated to necrosis. Assessment of taurine levels *in vivo* has been difficult due to overlap with myoinositol, choline and glucose [10].

As discussed, the FFT relies upon post-processing via fitting in attempts to quantify, but this is non-unique. In this way, the number of resonances can only be surmised. For example, most authors fit two peaks (glutamine and creatine) in the region between 3.8 ppm and 4.0 ppm. However, Opstad et al. [11] included a third peak at 2.9 ppm (glutathione) and reported a better fit. Another relevant example of the problem of fitting when overlapping resonances are present concerns the possibility that the peak at around 2 ppm contains other metabolites besides NAA, namely lipids at 2.05 ppm, and glutamine-glutamate at 2.1 ppm. The data concerning myoinositol are also contradictory for distinguishing brain tumors from non-neoplastic processes. The peak at 3.56 ppm has been most frequently attributed to myoinositol alone. However, some authors have viewed this as a combined myoinositol-glycine peak [12]. Fitting procedures are especially problematic in the presence of large amounts of mobile lipids. The prominent broad resonances at 0.9 ppm and 1.3 ppm are not fully modeled by the baseline spline functions of post-processing fitting models. This can lead to incorrect estimates of lactate as well as alanine. Inappropriate fit of the whole spectrum with phasing errors can lead to an entirely wrong result [13]. Thus, the non-uniqueness of fitting could well have important consequences for radiation neuro-oncology.

25.3.2 Limitations of the FFT in radiation oncology of the prostate

There are also a number of problems with the current reliance upon ratios of Cho to citrate for prostate cancer assessment. On the one hand, as mentioned, low citrate is seen in normal stromal prostate, and in metabolic atrophy. Moreover, high citrate is usually observed in BPH, even with coexistent malignancy [14]. Other limitations of current applications of MRS and MRSI in prostate cancer diagnostics include poor resolution and lack of consistent added diagnostic value when used with MRI [2]. Notwithstanding progress in coil design and other technological advances, resolution remains an important drawback to wider application of MRSI for prostate cancer diagnostics and management. Attempts to improve resolution and SNR by increasing the static magnetic field strength are noted to affect the spectral shape of citrate and its ratio to choline, and thus are considered of questionable benefit [15]. Another drawback of Fourier-based MRSI is low sensitivity for detecting smaller prostate cancers [2].

25.4 Optimal solutions via advanced signal processing by the fast Padé transform

The FPT is an advanced signal processor, particularly appropriate for *in vivo* MRS and MRSI [1, 2, 4, 16]. The FPT is a high-resolution, parametric estimator, which unequivocally determines the true number K of metabolites. It exactly reconstructs the spectral parameters from which metabolite concentrations, including those from very tightly overlapping resonances, can be reliably computed [2, 16]. Once the spectral parameters, such as the fundamental frequencies and the associated amplitudes $\{\omega_k, d_k\}\,(1 \le k \le K)$ of the given time signal $\{c_n\}\,(0 \le n \le N-1)$ of length N and sampling time $\tau = T/N$ have been retrieved, the corresponding complex-valued total shape spectrum is automatically generated via $\sum_{k=1}^{K} d_k/(z^{-1} - z_k^{-1})$:

$$c_n = \sum_{k=1}^{K} d_k z_k^n \qquad \Longrightarrow \qquad \sum_{k=1}^{K} \frac{d_k}{z^{-1} - z_k^{-1}} = \frac{P_{K-1}(z^{-1})}{Q_K(z^{-1})}. \tag{25.1}$$

Here, $z = \mathrm{e}^{i\tau\omega}, z_k = \mathrm{e}^{i\tau\omega_k}, \mathrm{Im}(\omega_k) > 0, \omega = 2\pi\nu, \omega_k = 2\pi\nu_k$ where ω and ν are angular and linear frequency. In (25.1), the spectrum $\sum_{k=1}^{K} d_k/(z^{-1} - z_k^{-1})$ is explicitly summed up to give the polynomial quotient P_{K-1}/Q_K which is the para-diagonal Padé approximant. Also frequently employed is the diagonal Padé approximant, P_K/Q_K.

The Padé approximant, or the fast Padé transform, as alternatively called in signal processing, is long known as the work-horse of theoretical physics, including quantum mechanics. Therefore, the FPT is the method of choice for spectral analysis of general FIDs as well as those in MRS. Time-frequency duality implies that the inverse fast Padé transform (IFPT) computed from the Padé spectrum from (25.1) will yield the $\{c_n\}$ as a sum of K damped complex exponentials. Similar to the inverse FFT, the ability of the IFPT to exactly retrieve the input FID irrespective of the level of noise corruption is precisely the feature which justifies the use of the term "transform" in the FPT. This can be seen by casting the Padé spectrum from (25.1) into its equivalent form of continued fractions (CF) [1]. Namely, every signal point $\{c_n\}$ $(0 \le n \le N-1)$ can be exactly reconstructed for any noise level from the general analytical expression for the expansion coefficients $\{a_n\}$ in the CF [1]. This determines that the optimal mathematical model for the frequency spectrum of these time signals is prescribed quantum-mechanically as the finite-rank response Green function in the form of the unique ratio of two polynomials, i.e., the FPT. Similarly to the time domain, where the Schrödinger time evolution operator predicts the FID as the sum of damped exponentials, the same quantum physics automatically prescribes that the frequency spectrum is given by the Green function via the Padé quotient of two polynomials. This is the true origin of the unprecedented algorithmic success of the FPT, via its demonstrable, exact reconstructions [1,2].

To cross-validate its finding, the FPT uses two equivalent, but conceptually different versions denoted by $\text{FPT}^{(+)}$ and $\text{FPT}^{(-)}$. Their diagonal forms have the following representations for the exact infinite-rank Green function, which is defined as the Maclaurin series with the time signal points $\{c_n\}$ as the expansion coefficients:

$$\text{Exact}: \ G(z^{-1}) \equiv \sum_{n=0}^{\infty} c_n z^{-n-1} \tag{25.2}$$

$$\approx G_K^{(+)}(z) \equiv \frac{P_K^+(z)}{Q_K^+(z)} = \frac{\sum_{r=1}^{K} p_r^+ z^r}{\sum_{s=0}^{K} q_r^+ z^s} \quad : \ \text{FPT}^{(+)}, \tag{25.3}$$

$$\approx G_K^{(-)}(z^{-1}) \equiv \frac{P_K^-(z^{-1})}{Q_K^-(z^{-1})} = \frac{\sum_{r=0}^{K} p_r^- z^{-r}}{\sum_{s=0}^{K} q_r^- z^{-s}} \quad : \ \text{FPT}^{(-)}. \tag{25.4}$$

The expansion coefficients $\{p_r^\pm, q_s^\pm\}$ of the numerator $P_K^\pm(z^{\pm 1})$ and denominator $Q_K^\pm(z^{\pm 1})$ polynomials can be extracted exactly and uniquely from the given signal points $\{c_n\}$ by solving only one system of linear equations from the definitions (25.3) and (25.4), after truncation of the Maclaurin series for $G(z^{-1})$ at $n = N-1$. Spectra $P_K^\pm(z^{\pm 1})/Q_K^\pm(z^{\pm 1})$ can equivalently be written in their canonical forms:

$$\frac{P_K^\pm(z^{\pm 1})}{Q_K^\pm(z^{\pm 1})} = \frac{p_K^\pm}{q_K^\pm} \prod_{k=1}^{K} \frac{z^{\pm 1} - z_{k,P}^\pm}{z^{\pm 1} - z_{k,Q}^\pm}. \tag{25.5}$$

Roots of the characteristic equations, $Q_K^\pm(z^{\pm 1}) = 0$, have the solutions $z_k^{\pm 1} \equiv z_{k,Q}^\pm$ $(1 \le k \le K)$ that represent one constituent part of the fundamental harmonics via $\{d_k^{\pm}, z_{k,Q}^{\pm}\}$. By definition, the corresponding amplitudes $d_k^{\pm}$ are the Cauchy residues of the spectra from (25.5). For non-degenerate roots $(z_{k',Q}^{\pm}) \neq z_{k,Q}^{\pm}\, k' \neq k$, we have, $d_k^{\pm} = \lim_{z^{\pm 1} \to z_{k,Q}^{\pm}} \{(z^{\pm 1} - z_{k,Q}^{\pm})[P_K^{\pm}(z^{\pm 1})/Q_K^{\pm}(z^{\pm 1})]\}$, so that:

$$d_k^{\pm} = \frac{P_K^{\pm}(z_{k,Q}^{\pm})}{(\mathrm{d}/\mathrm{d}z_{k,Q}^{\pm})Q_K^{\pm}(z_{k,Q}^{\pm})} = \frac{p_K^{\pm}}{q_K^{\pm}} \prod_{k'=1}^{K} \frac{z_{k,Q}^{\pm} - z_{k',P}^{\pm}}{(z_{k,Q}^{\pm} - z_{k',Q}^{\pm})_{k' \neq k}} \tag{25.6}$$

$$\implies \quad d_k^{\pm} \propto (z_{k,Q}^{\pm} - z_{k,P}^{\pm}), \tag{25.7}$$

where $(\mathrm{d}/\mathrm{d}z_{k,Q}^{\pm})Q_K^{\pm}(z_{k,Q}^{\pm})$ is the first derivative of the denominator polynomials. Thus, each amplitude has the meaning of a metric in the sense of the distance given by the separation between the pole and zero, $d_k^{\pm} \propto z_{k,Q}^{\pm} - z_{k,P}^{\pm}$. This is consistent with the mathematical complex analysis according to which the Cauchy residue describes the behavior of line integrals of a meromorphic function around the given pole. This completes the reconstruction of the $2K$ complex fundamental parameters $\{\omega_k^{\pm}, d_k^{\pm}\}$ in the FPT$^{(\pm)}$. Both $G_K^{(+)}(z)$ and $G_K^{(-)}(z^{-1})$ from (25.3) and (25.4) approximate the same Green function $G(z^{-1})$. The Maclaurin series $G(z^{-1})$ is convergent for $|z| > 1$ and divergent for $|z| < 1$, i.e., outside and inside the unit circle, respectively. The FPT$^{(-)}$ via $G_K^{(-)}(z^{-1})$ is defined in terms of the same variable z^{-1} as $G_K^{(-)}(z^{-1})$. Therefore, $G_K^{(-)}(z^{-1})$ converges outside the unit circle, $|z| > 1$, but does so faster than the original Maclaurin series. Hence, convergence acceleration of $G(z^{-1})$ by the FPT$^{(-)}$. On the other hand, the FPT$^{(+)}$ through $G_K^{(+)}(z)$ employs the variable z and, and as such, converges inside the unit circle ($|z| < 1$), where the exact Green function $G(z^{-1})$ diverges. Hence, analytical continuation $G(z^{-1})$ by the FPT$^{(-)}$. Overall, the FPT$^{(+)}$ and FPT$^{(-)}$ are optimally suited to work inside and outside the unit circle, respectively. Nevertheless, by the Cauchy concept of analytical continuation, they are both well defined everywhere in the complex plane with the exception of the pole positions that are located at $z^{\pm 1} = z_{k,Q}^{\pm}$. However, physical spectra $G_K^{(\pm)}(z^{\pm 1}) = P_K^{\pm}(z^{-1})/Q_K^{\pm}(z^{-1})$ in the FPT$^{(\pm)}$ are perfectly well defined even at the poles, since $z^{\pm 1} - z_{k,Q}^{\pm} \neq 0$ for real frequencies that are of interest in practice. The internal cross-validation in the fast Padé transform is achieved upon full convergence in the FPT$^{(\pm)}$ leading to the agreements $\omega_k^{+} \approx \omega_k^{-} \equiv \omega_k$ and $d_k^{+} \approx d_k^{-} \equiv d_k$ where $\{\omega_k, d_k\}$ are the genuine amplitude and frequencies from the time signal (25.1). It is in this straightforward way that the FPT$^{(\pm)}$ is able to solve exactly the harmonic inversion problem (quantification) by using only the sampled time signal $\{c_n\}$ to reconstruct all its constituent fundamental frequencies and amplitudes $\{\omega_k, d_k\}$ according to (25.1).

The spectra $P_K^{\pm}(z^{\pm 1})/Q_K^{\pm}(z^{\pm 1})$ are meromorphic functions, since poles are their only singularities. Poles $\{z_{k,Q}^{\pm}\}$ and zeros $\{z_{k,P}^{\pm}\}$ of these spectra are the roots of $Q_K^{\pm}(z^{\pm 1}) = 0$ and $P_K^{\pm}(z^{\pm 1}) = 0$, respectively. Here, as elsewhere, the harmonic variables $z_k^{\pm 1}$ are denoted by $z_{k,Q}^{\pm}$ and $z_{k,P}^{\pm}$, where the subscripts Q and P refer to the polynomials denominator $Q_K^{\pm}(z^{\pm 1})$ and numerator $P_K^{\pm}(z^{\pm 1})$ polynomials. The spectral poles and zeros provide a physical parametrization of the studied general system which produced the time signals as a response to an external excitation. All the other physical quantities from quantum physics (state density, transition probabilities, oscillator strengths, etc) can be successfully generated from the spectral poles and zeros. The spectral amplitudes, as the intensities of the given time signal, are also directly connected to the spectral poles and zeros via $d_k^{\pm} \propto z_{k,Q}^{\pm} - z_{k,P}^{\pm}$ as per (25.7). The FPT$^{(\pm)}$ can be conceived as the union of two representations, one being pFPT$^{(\pm)}$ (poles of FPT$^{(\pm)}$) and the other zFPT$^{(\pm)}$ (zeros of FPT$^{(\pm)}$). All these features establish the fundamental role of poles and zeros of the response function or the Green function in description of any physical system.

Stability of systems is obtained if stable poles prevail. Stable poles are physical or genuine, since they are robust against external perturbations. Unstable poles oscillate widely even under the smallest perturbation and they never converge with the increased degree of the Padé polynomials. This pattern is reminiscent of random fluctuations as typical of noise or noise-like corruption of the system. Stable and unstable poles can be unequivocally detected in the FPT$^{(\pm)}$ by comparing the spectral poles and zeros. Physical poles and zeros do not coincide with each other, $z_{k,Q}^{\pm} \neq z_{k,P}^{\pm}$. By contrast, "noisy" features are characterized by pole-zero confluences $z_{k,Q}^{\pm} \approx z_{k,P}^{\pm}$. These characteristics have direct implications for the spectral amplitudes. Specifically, it follows from $d_k^{\pm} \propto z_{k,Q}^{\pm} - z_{k,P}^{\pm}$ via (25.7) that the genuine (spurious) resonances have non-zero (zero) amplitudes, $d_k^{\pm} \neq 0\,(z_{k,Q}^{\pm} \neq z_{k,P}^{\pm})$ and $d_k^{\pm} \approx 0\,(z_{k,Q}^{\pm} \approx z_{k,P}^{\pm})$, respectively. Overall, the spurious resonances have coincident or near-coincident poles and zeros. These confluent pairs are called Froissart doublets and they correspond to zero or near-zero amplitudes. Feeble, vanishingly small amplitudes of spurious, Froissart resonances are the cause for their marked instability against the weakest perturbations. Adding even an infinitesimally small amount of Gaussian random noise to time signals would completely change the distribution of spurious frequencies and amplitudes in the complex plane. Suppose that K is the stabilized, i.e., exact number of resonances, and that we still continue to compute the Padé spectra for higher polynomial degrees $K+m$ $(m = 1, 2, 3, ...)$. Then every newly found resonance for a positive integer m will be spurious with $z_{k,Q}^{\pm} = z_{k,P}^{\pm}$ $(k = K+m)$. This would lead to cancellations of all the denominator and numerator terms with spurious poles $(z^{\pm 1} - z_{k,Q}^{\pm})$ and spurious zeros $(z^{\pm 1} - z_{k,P}^{\pm})$ in the canonical forms (25.5) of the spectra. Therefore, pole-zero cancellation with the ensuing stabilization of the computed spectra:

$$\frac{P_{K+m}^{\pm}(z^{\pm 1})}{Q_{K+m}^{\pm}(z^{\pm 1})} = \frac{P_K^{\pm}(z^{\pm 1})}{Q_K^{\pm}(z^{\pm 1})} \qquad (m = 1, 2, 3, ...). \tag{25.8}$$

In Padé-based reconstruction, the true number K of fundamental harmonics is treated as yet another unknown parameter. Its exact value is unequivocally determined upon convergence of the reconstructed frequencies and amplitudes as well as spectra, by implication. We gradually increase the degree of the Padé polynomials and the exact K is detected when all the genuine frequencies and amplitudes have reached their stable values. This ensures reconstruction of all the physical resonances. With any further increase of the polynomials' degree beyond this latter value K, new resonances emerge, but they are all exceedingly unstable and, hence, spurious. Through the concept of Froissart doublets or pole-zero cancellations, a powerful means is thus provided through the FPT of determining whether a given reconstructed resonance is true or spurious [1, 2]. This "Signal-Noise Separation" (SNS) has been demonstrated to be successful not only for noise-free, but also noise-corrupted MR FIDs [2]. Note that the concept of SNS and Froissart doublets appear even if noiseless FIDs are used, in which case computational round-off errors act as random noise. Remarkably, the largest noise suppression in the FPT occurs near the reconstructed genuine fundamental frequencies, in accordance with the variational nature of the FPT [1].

One of the main reasons for the popularity of the FFT in signal analysis for clinical purposes is its steady convergence with increasing signal length N, such that there are no major troublesome surprises as the values of N are systematically augmented. In contrast, nearly all parametric estimators are unstable as a function of N. This is manifested in dramatic oscillations (spikes, Gibbs ringing and other artificial spectral structures) that appear prior to convergence. Where to stop the computations under wild convergence patterns is a matter of concern. Needless to say, such spurious findings are totally anathema to the clinical demands for reliable spectral information. The FPT shares and improves upon the favorable qualities of the FFT. Unlike most other parametric estimators, the FPT demonstrates stable convergence with increasing signal length [1, 2]. A key advantage of the FPT relative to the FFT is that convergence is also rapid. This means that even at short signal lengths, the FPT can assess concentrations of the main metabolites that remain undetected by the FFT in its total shape spectra (envelopes) that are the only output available. The spectrum in the FPT does not use the fixed Fourier mesh k/T ($k = 0, 1, 2, ...$) in the frequency domain, and can be computed at any frequency $\omega = 2\pi\nu$. Consequently, resolution is not pre-determined by T. The Fourier conundrum between augmenting T for improved resolution and increasing SNR is thereby obviated by the FPT. This is especially important for detection of short-lived metabolites, such as lipids, glutamine, glutamate and myoinositol that are informative for brain tumor diagnostics.

Besides interpolation, the FPT has extrapolation capabilities. The FFT places a sharp cut-off to the FID beyond T, with added zeros via zero-filling (zero-padding) or using the FID's periodic extension, with no new information in either case. In the FPT, it is the very form of the polynomial quotient from (25.1) which achieves extrapolation beyond T, thereby enhancing resolution [1, 2]. Moreover, the FPT provides non-linear mapping, as the transformation coefficients are dependent upon the FID points. Thus, as opposed to the FFT, whose linearity preserves noise from the FID, the non-linearity of the FPT permits noise suppression [1].

25.5 Applications of the FPT for MRS data from the brain and prostate

25.5.1 Quantification benchmarking with synthesized time signals

Benchmark studies [2,14,16] show that the FPT can yield highly accurate metabolite concentrations from MRS signals from normal and malignant tissue. These include studies from FIDs that closely match those encoded via MRS from the brain of a healthy volunteer [17].

Figure 25.1 illustrates reconstruction by the $\text{FPT}^{(\pm)}$ using the noise-free FID from (25.1) comprised of 25 complex attenuated exponentials. Froissart doublets as spurious resonances are seen on panels (a) and (b) through the coincidence of spectral poles and zeros. Only a small fraction of Froissart doublets are shown in the displayed window of the entire Nyquist range. In the $\text{FPT}^{(+)}$ from panel (a), genuine and spurious resonances are totally separated by the opposite signs $\text{Im}(\nu_k^+) > 0$ and $\text{Im}(\nu_k^+) > 0$ of their imaginary frequencies. Thus the $\text{FPT}^{(+)}$ provides the exact separation of the physical from the noise-like content of the FID. By contrast, in the $\text{FPT}^{(-)}$ from panel (b), physical and unphysical resonances are mixed together, as they all have the same sign $\text{Im}(\nu_k^+) > 0$ of their imaginary frequencies. Nevertheless, the emergence of Froissart doublets remains clear also in the $\text{FPT}^{(-)}$, with the ensuing unambiguous identification of spurious resonances. Due to the pole-zero cancellation, each Froissart doublet has precisely the zero value of their amplitudes, as seen on panel (c). This serves as a further check for the consistency and fidelity of separation of genuine from spurious resonances. Note that the full auxiliary lines on the left column in Fig. 25.1 are drawn merely to transparently delineate the areas with Froissart doublets.

The component and total shape spectra in the $\text{FPT}^{(-)}$ are shown on panels (d)–(f) in Fig. 25.1. A component shape spectrum is generated for each resonance separately from the reconstructed fundamental frequencies and the corresponding amplitudes. The total shape spectrum is the sum of all the component shape spectra. It is seen on panel (e) that the component shape spectra converge at $N_P = 220$. This is due to the stability of the estimation of spectral parameters at $N_P = 220$. The same convergence also occurs for the total shape spectrum at $N_P = 220$ on panel (f). However, an intriguing situation occurs prior to convergence of the component shape spectra. Namely, on panel (d) at $N_P = 180$, the 11th peak is missing, and the 12th peak is overestimated. Yet, the corresponding total shape spectrum at $N_P = 180$ has fully converged on panel (f). Here, an apparent indication of

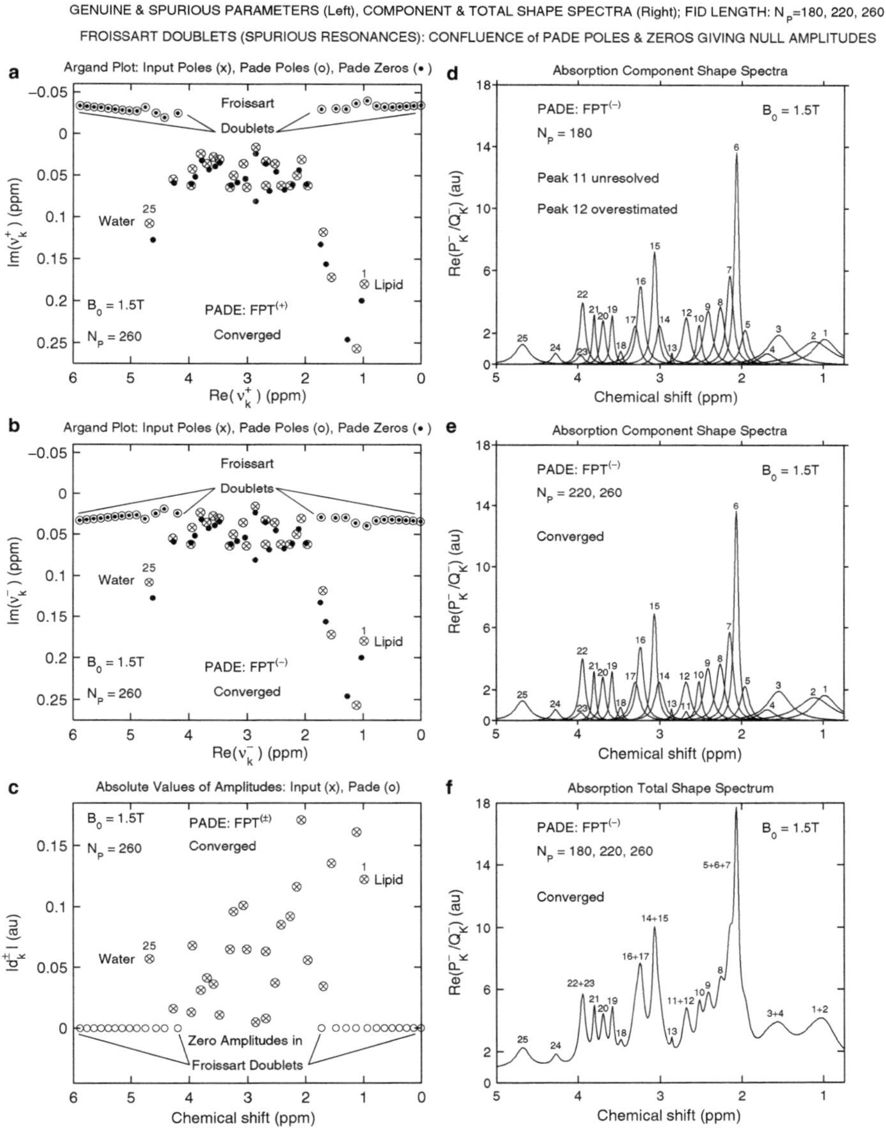

Fig. 25.1 Reconstruction of parameters and absorption shape spectra by the FPT$^{(\pm)}$ for the noise-free synthesized FID with $K = 25$ complex damped harmonics. The spectra correspond to time signals encoded clinically at the magnetic field strength $B_0 = 1.5$T from healthy human brain [17]. Panels (**a**) and (**b**) display genuine and spurious frequencies, whereas the corresponding amplitudes are shown on panel (**c**). The component shape spectra in the FPT$^{(-)}$ for each reconstructed physical resonance are shown on panels (**d**) and (**e**) at the partial signal lengths $N_P = 180$, 220 and 260 for which the total shape spectra or envelopes given on panel (**f**) are indistinguishable from each other

convergence of the envelope spectrum at $N_P = 180$ is the fact that no difference exists between any two envelope spectra computed at $N_P = 180, 220, 260$. This is so in particular for $N_P = 180$ because the area of the 12th peak is overestimated precisely by the amount of the corresponding area of the missing 11th peak on panels (d). The two identical total shape spectra at $N_P = 180$ and $N_P = 220$ on panels (f) contain 24 and 25 resonances, respectively. Such a discrepancy in the number of reconstructed resonances is not detected by the residual or error spectra (not shown). This implies that it is not reliable to use the converged total shape spectra and the related residual or error spectra as the only criterion for determining the number of reconstructed resonances. Precisely this latter criterion is used in all fittings, within MRS and beyond, that rely heavily upon the residual spectrum defined as the difference between the spectrum from the FFT and a modeled spectrum. The right column of Fig. 25.1 clearly shows how precarious it is to surmise which components are hidden underneath a spectral structure. Thus, rather than reconstructing the mobile lipids under the two broad structures in the range 1-2 ppm, as done unambiguously by the FPT, equally acceptable (in the least-square sense) results of fitting by the usual methods from MRS could "reconstruct" two, three, four or more peaks that would all give the same absorption total shape spectrum from 1 ppm to 2 ppm. This is reminiscent of the Lanczos paradox [1] of fitting the same experimental data with 3 identical curves with widely different set of parameters. Even more serious problems with clinically unacceptable uncertainties stemming from fittings are found in other parts of the spectrum from panel (c) in Fig. 25.1. In particular, any attempts to use fitting to ascertain that the peaks close to 2.7 ppm are, in fact, almost degenerate would be practically impossible.

We also consider a noise-corrupted time signal $\{c_n + r_n\}$, where $\{c_n\}$ is the same noiseless FID employed in Fig. 25.1. Here, $\{r_n\}$ is zero-mean complex-valued random Gaussian white noise (orthogonal in its real and imaginary parts). The standard deviation σ of $\{r_n\}$ is set to be 0.00289 RMS, where RMS is the root-mean-square of the noise-free FID, $\{c_n\}$. Reconstructions by the FPT$^{(+)}$ are illustrated in Fig. 25.2 for genuine and spurious resonances that appear in the entire Nyquist range. The only difference between Froissart doublets for the noise-free $\{c_n\}$ and noise-corrupted $\{c_n + r_n\}$ time signals from Figs. 25.1 and 25.2 is that the latter are more irregularly distributed. This is expected due to the presence of the random perturbation $\{r_n\}$ in the noisy FID. However, this difference is irrelevant, since the only concern to SNS is that noise-like or noisy information is readily identifiable by pole-zero coincidences. Once the Froissart doublets are identified and discarded from the whole set of results, only the reconstructed parameters of the genuine resonances will remain in the output data. Crucially, the latter set of Padé-retrieved spectral parameters also contains the exact number K_G of genuine resonances as the difference between the total number $K \equiv K_T$ of all the found resonances and the number K_F of Froissart doublets, $K_G = K_T - K_F$. In Fig. 25.2, we used a quarter of the full signal length $N_P = N/4 = 1024/256$, which corresponds to the Padé polynomial degree $K = 128$ in spectra $P_K^{\pm}(z^{\pm 1})/Q_K^{\pm}(z^{\pm 1})$. In the whole Nyquist range, the FPT$^{(+)}$ and FPT$^{(-)}$ find 103 Froissart doublets, $K_F = 103$. Therefore, the number K_G of genuine resonances reconstructed by the FPT$^{(+)}$ and FPT$^{(-)}$ is given

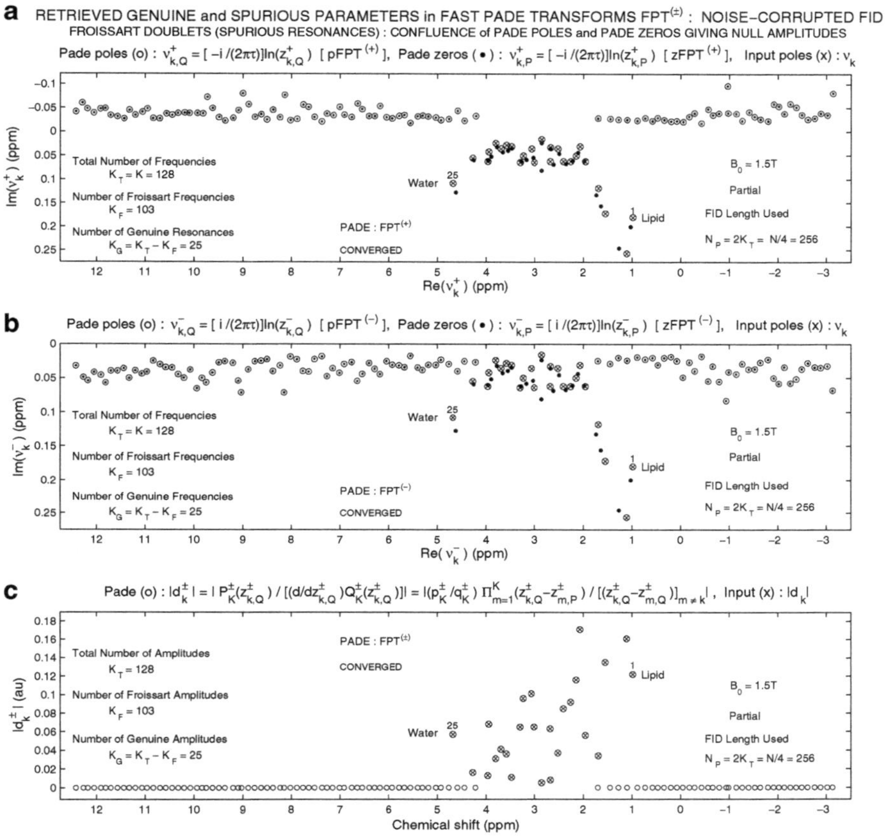

Fig. 25.2 Reconstruction of frequencies and amplitudes in the whole Nyquist interval by the $\text{FPT}^{(\pm)}$ at the signal length $N/4 = 256$ ($N = 1024$) for the noise-corrupted synthesized FID with $K = 25$ complex damped harmonics. Panels (**a**) and (**b**) show genuine and spurious frequencies, whereas the corresponding amplitudes are given on panel (**c**). In panel (**a**), the $\text{FPT}^{(+)}$ completely separates the genuine from spurious frequencies into two disjoint regions $\text{Im}(\nu_k^+) > 0$ and $\text{Im}(\nu_k^+) < 0$, respectively. In the $\text{FPT}^{(-)}$ from panel (**b**), the imaginary parts of the genuine and spurious frequencies have the same sign, $\text{Im}(\nu_k^+) > 0$. The joint feature of panels (**a**) and (**b**) is a clear and large set of pole-zero coincidences that define spurious resonances. In both panels (**a**) and (**b**), the immediate neighborhood of the interval with the genuine resonances is the least infiltrated with Froissart doublets. The reconstructed converged amplitudes associated with the genuine resonances are identical in the $\text{FPT}^{(+)}$ and $\text{FPT}^{(-)}$. Panel (**c**) shows that all the Froissart amplitudes are zero-valued

by $K_G = 128 - 103 = 25$, in exact agreement with the corresponding value of the input data. Overall, Froissart doublets simultaneously achieve three important goals: (i) noise reduction, (ii) dimensionality reduction as per (25.8) and (iii) stability enhancement. Stability against perturbations of the physical time signal under study is critical to the reliability of spectral analysis. The main contributor to instability of systems is its spurious information. Being inherently unstable, spuriousness is unambiguously identified by the twofold signature of Froissart doublets (pole-zero coincidences and zero-valued amplitudes) and, as such, discarded from the output data in the FPT. What is left is genuine information which is stable.

25.5.2 Application of the FPT to in vivo MRS time signals from the brain

We next compare the performance of the FPT with the FFT for a clinical MRS time signal of length $N = 2048$ encoded from the brain of a healthy volunteer at the static magnetic field strength of 4T [18]. In Fig. 25.3 we present absorption total shape spectra at three partial signal lengths for the FFT (left column) and the FPT (right column). At the top and middle panels of Fig. 25.3 the most dramatic differences between the FFT and FPT are seen at the shortest signal lengths ($N/32 = 64$) and ($N/16 = 128$), respectively. Here, the FFT presents no meaningful information, whereas nearly 90% of the NAA concentration is predicted by the peak at around 2.0 ppm by the FPT at ($N/16 = 128$). At half signal length ($N/2 = 1024$) at the bottom panel, the FFT has not demonstrated the accurate ratio between Cr and Cho at 3.0 and 3.3 ppm; these two metabolites still incorrectly appear as almost equal. The triplet of glutamine and glutamate near 2.3 ppm can be discerned at half signal length only by the FPT, and not by the FFT. By contrast, at half signal length ($N/2 = 1024$) the FPT resolves all the physical metabolites for which every peak parameter is accurately extracted, including the case of overlapping resonances. Furthermore, while the FFT demands the total signal length ($N = 2048$) to fully resolve all the metabolites, the difference between the two FPT spectra at $N = 1024$ and $N = 2048$ is buried entirely in the background noise, as shown in [2]. In other words, in this illustration, the FPT spectra at half-signal length can be treated as fully converged. As seen in Fig. 25.3, the FPT produces no spurious metabolites or other spectral artefacts in the process of converging steadily as a function of increased signal length. Clearly, the FPT exhibits a much faster convergence rate than that in the FFT [1, 2].

25.5.3 Application of the FPT to time signals from normal and malignant prostate

In our analysis of MR time signals as encoded *in vitro* from normal and cancerous prostate, the FPT reconstructed all the physical resonances, including multiplets and closely overlapping peaks of different metabolites [2, 14]. The FPT exactly reconstructed all the input spectral parameters for the data corresponding to two types

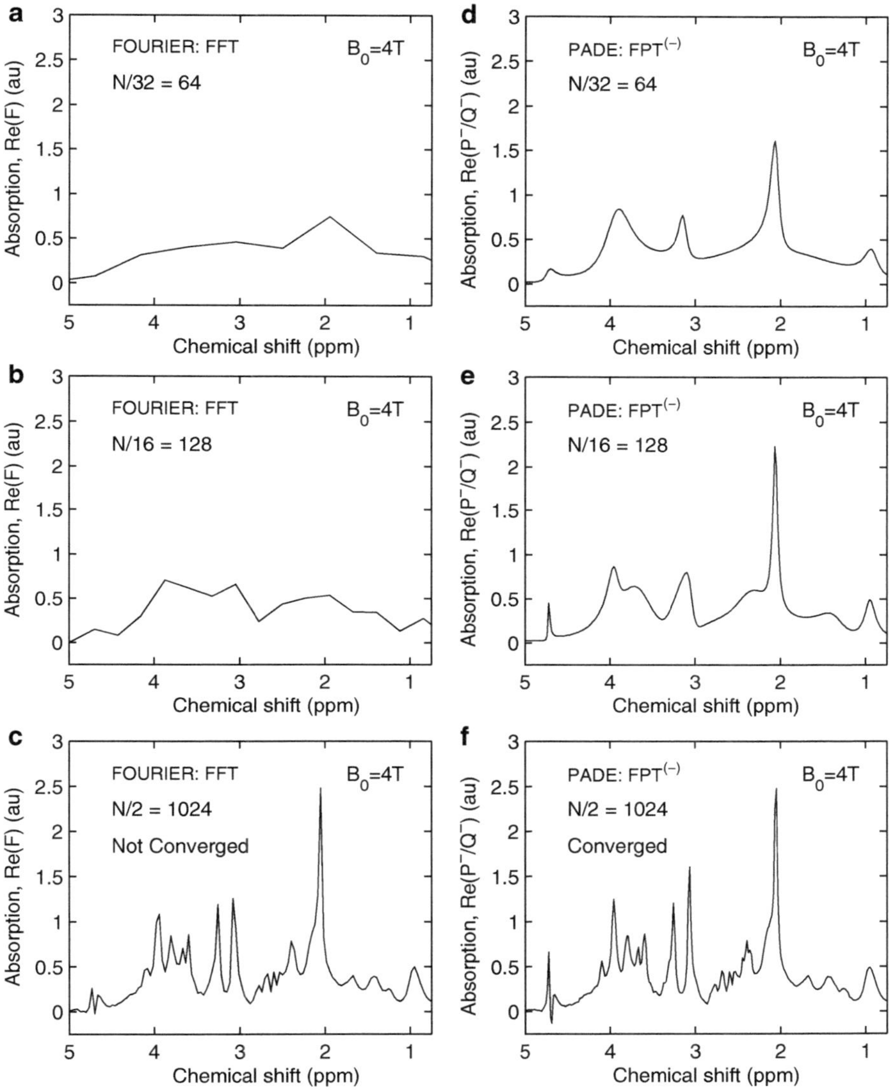

Fig. 25.3 Fourier and Padé absorption spectra computed using the time signal (divided by 10000) at three partial signal lengths ($N/32 = 64, N/16 = 128, N/2 = 1024$), where the full signal length is $N = 2048$, as encoded in [18] at 4T from occipital grey matter of a healthy volunteer

of normal prostate tissue and malignant prostate. Padé-based reconstruction yielded the exact spectral frequencies and amplitudes of all the resonances and provided certainty about their true number. The "spectral crowding" problem does not hinder the FPT, which via parametric analysis, without fitting or numerical integration of

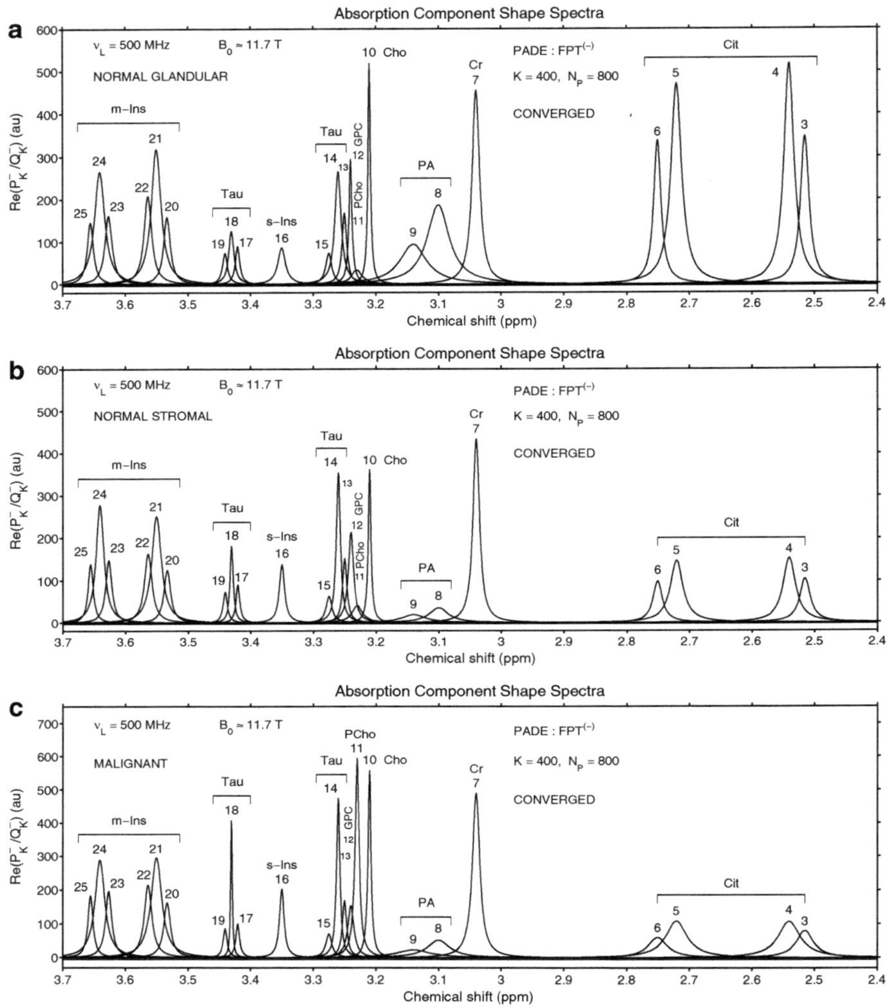

Fig. 25.4 Converged absorption component spectra reconstructed by the fast Padé transform at partial signal length $N_P = 800$, for normal glandular prostate (top panel (**a**)), normal stromal prostate (middle panel (**b**)) and malignant prostate (bottom panel (**c**)) within the region between 2.4 ppm to 3.7 ppm derived from *in vitro* data of [19]

peak areas, retrieved all the multiplets and closely overlying resonances of different metabolites for normal glandular and stromal prostate, as well as for prostate cancer. This was achieved at short signal lengths, which implies that the discussed problems due to poor resolution could be circumvented [2, 14]. Figure 25.4 shows the Padé-reconstructed component spectra for data from normal glandular and normal stromal prostate, as well as for the malignant prostate data.

25.6 Conclusions and future perspectives

We analyze the potential and limitations of magnetic resonance spectroscopy and spectroscopic imaging particularly for the brain and prostate cancer diagnostics with special relevance to radiation therapy. There is an increasing awareness in clinical oncology that these two non-invasive modalities could revolutionize not only tumor diagnostics, but also image-guided surgery, post-operative followup and surveillance screening. These assessments of experts are based on the extracted information on a handful metabolites or their concentration ratios. Such information is not accessible directly from encoded time signals, but rather it becomes available through mathematical reconstructions via quantifiable data analysis. As presently illustrated, far more information about the metabolic content of cancerous versus surrounding normal tissue could be obtained by reliance upon the unequivocal Padé-based high-resolution quantitative signal processing instead as of the qualitative Fourier estimation accompanied with ambiguous fitting. This information as the added value is also vital for target definition and dose planning. Moreover, richer metabolic information could provide better distinction between recurrent tumor and non-malignant changes due to radiation therapy. Advances in signal processing based on the fundamental theory and concepts of quantum-mechanical spectral analysis will be invaluable for further progress in this field whose data analysis and interpretation needs to go beyond the already exhausted phenomenological approaches.

Acknowledgements This work was supported by Cancerfonden, the King Gustav the 5th Jubilee Fund, the Karolinska Institute Fund and by the Signe and Olof Wallenius Stiftelse, to which the authors are grateful.

References

1. Belkić, Dž.: Quantum Mechanical Signal Processing and Spectral Analysis, Institute of Physics Publishing, Bristol (2005)
2. Belkić, Dž., Belkić, K.: Signal Processing in Magnetic Resonance Spectroscopy with Biomedical Applications, Taylor & Francis Group, London (2010)
3. Bortfeld, T., Phys. Med. Biol. **51**, R363-R379 (2006)
4. Belkić, K.: Molecular Imaging through Magnetic Resonance for Clinical Oncology, Cambridge International Science Publishing, Cambridge (2004)
5. Narayana, A., Chang, J., Thakur, S., Huang, W., Karimi, S., Hou, B., Kowalski, A., Perera, G., Holodny, A., Gutin, P., Br. J. Radiol. **80**, 347-354 (2007)
6. Joseph, T., McKenna, D., Westphalen, A., Coakley, F., Zhao, S., Lu, Y., Hsu, I., Roach, M., Kurhanewicz, J., Int. J. Radiat. Oncol. Biol. Phys. **73**, 665-671 (2009)
7. Pickett, B., Kurhanewicz, J., Pouliot, J., Weinberg, V., Shinohara, K., Coakley, F., Roach, M., Int. J. Radiat. Oncol. Biol. Phys. **65**, 65-72 (2006)
8. Westphalen, A., McKenna, D., Kurhanewicz, J., Coakley, F., J. Endourol. **22**, 789-794 (2008)
9. Hattingen, E., Pilatus, U., Franz, K., Zanella, F., Lanfermann, H., J. Magn. Reson. Imag. **26**, 427-431 (2007)

10. Opstad, K., Bell, B., Griffiths, J., Howe, F., Br. J. Cancer **100**, 789-794 (2009)
11. Opstad, K., Provencher, S., Bell, B., Griffiths, J., Howe, F., Magn. Reson. Med. **49**, 632-637 (2003)
12. Novotny, E., Fulbright, R., Pearl, P., Gibson, K., Rothman, D., Ann. Neurol. **54** (Suppl.), 25-31 (2003)
13. Auer, D., Gössl, C., Schirmer, T., Czisch, M., Magn. Reson. Med. **46**, 615-618 (2001)
14. Belkić, Dž., Belkić, K., J. Math. Chem. **45**, 819-858 (2009)
15. Kim, C., Park, B., J. Comp. Assist. Tomogr. **32**, 163-172, (2008)
16. Belkić, Dž., Phys. Med. Biol. 51, 2633-2670 (2006)
17. Frahm, J., Bruhn, H., Gyngell, M., Merboldt, K., Hänicke, W., Sauter, R., Magn. Reson. Med. **9**, 79-93 (1989)
18. Tkáč, I., Andersen, P., Adriany, G., Merkle, H., Uğurbil, K., Gruetter, R., Magn. Reson. Med. **46**, 451-456 (2001)
19. Swanson, M., Zektzer, A., Tabatabai, Z., Simko, J., Jarso, S., Keshari, K., Schmitt, L., Carroll, P., Shinohara, K., Vigneron, D., Kurhanewicz, J., Magn. Reson. Med. **55**, 1257-1264 (2006)

Part IV
Future Trends in Radiation Research and its Applications

Chapter 26
Medical Applications of Synchrotron Radiation

Exploring New Paths in Radiotherapy

Yolanda Prezado, Immaculada Martínez-Rovira, and the ID17 Biomedical Beamline (ESRF)

Abstract This chapter describes the state-of-art of synchrotron radiation therapies in the treatment of radioresistant tumors. The tolerance of the surrounding healthy tissue severely limits the achievement of a curative treatment for some brain tumors, like gliomas. This restriction is especially important in children, due to the high risk of complications in the development of the central nervous system. In addition, the treatment of tumors close to an organ at risk, like the spinal cord, is also restrained. One possible solution is the development of new radiotherapy techniques would exploit radically different irradiation modes, as it is the case of synchrotron radiotherapies. Their distinct features allow to modify the biological equivalent doses. In this chapter the three new approaches under development at the European Synchrotron Radiation Facility (ESRF), in Grenoble (France), will be described, namely: stereotactic synchrotron radiation therapy, microbeam radiation therapy and minibeam radiation therapy. The promising results obtained in the treatment of high grade brain tumors in preclinical studies have paved the way to the forthcoming clinical trials, currently in preparation.

26.1 Introduction

The use of X-rays in medicine started almost immediately after being discovered by Wilhelm Conrad Röntgen in 1895. Leopold Freund treated the first five patients

---

Y. Prezado (✉)
ID17 Biomedical Beamline, European Synchrotron Radiation Facility, 38043 Grenoble, France
e-mail: prezado@gmail.com

I. Martínez-Rovira
ID17 Biomedical Beamline, European Synchrotron Radiation Facility, 38043 Grenoble, France

Institut de Técniques Energètiques, Universitat Politècnica de Catalunya, E-08028 Barcelona, Spain
e-mail: immamartinez@gmail.com

G.G. Gómez-Tejedor and M.C. Fuss (eds.), *Radiation Damage in Biomolecular Systems*, Biological and Medical Physics, Biomedical Engineering,
DOI 10.1007/978-94-007-2564-5_26,

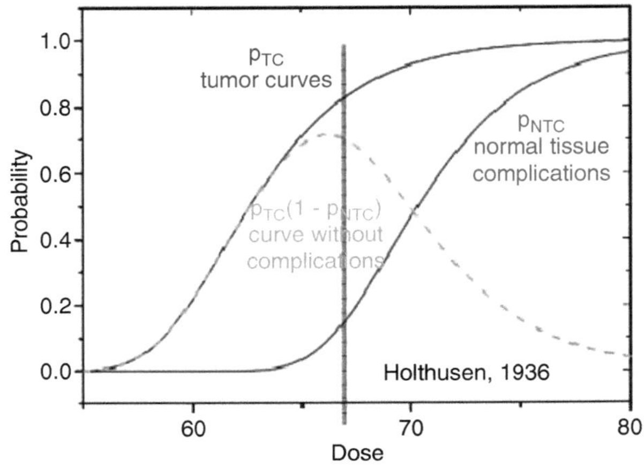

Fig. 26.1 Example of dose-response relationships in radiotherapy. Taken from [7]. The therapeutic window corresponds to the range of doses for which the probability of tumor control is significantly higher than the normal tissue complication probability

a year later, in 1896 [1, 2]. Despite intensive research and development work in conformal radiotherapy, there are still some radioresistant tumors, like gliomas, for which a radical radiotherapy treatment is usually not feasible at hospitals. They are one of the most frequent brain tumors, with a incidence of 5 to 10 per 100000 in general population [3]. The gold standard treatment, chemo-radiotherapy with temozolomide [4–6], provides only a slight increase of survival.

Discussion of the possible benefit of a radiotherapy treatment must always consider simultaneously the effects on tumor response and on normal-tissue damage. If the first one is measured by determining the proportion of tumors that are controlled, then a sigmoid relationship to dose is expected. If normal-tissue damage is quantified according to certain end-points (radionecrosis, for example), there will also be a rising curve of toxicity. See Fig. 26.1. As radiation dose is increased, there will be a tendency for tumor response to augment. The same is true for normal-tissue damage. The term therapeutic window describes the (possible) difference between tumor control dose and the normal tissue tolerance dose. It corresponds to the range of doses for which the probability of tumor control is much higher than the probability of producing deleterious side effects in the normal-tissue.

For particularly radioresistant tumours, like gliomas, the dose-response curves for tumor control and normal tissue complications lay in close proximity, resulting in only palliative treatments in conventional radiotherapy. To reach a curative radiation dose, the risk of serious damage to normal tissues would be unacceptable.

The beam type (photons, electrons, protons, etc), the beam quality and the dose delivery methods (fractionation scheme, dose rate, spatial distribution, etc) have a direct impact on the biological effect of the radiation [8]. The modification of any of the aforementioned parameters implies a different biological response. This might lead to a shift of the dose-response curve of healthy tissue complications to higher doses, opening the therapeutic window for gliomas.

The quest for a curative treatment with ionising radiation is in the origin of the development of three new radiotherapy techniques at the Biomedical Beamline of the European Synchrotron Radiation Facility (ESRF): Microbeam Radiation Therapy (MRT), the most recently implemented, Minibeam Radiation Therapy (MBRT), and Stereotactic Synchrotron Radiation Therapy (SSRT). A brief overview on the Biomedical Beamline at the ESRF will be presented. Then, the state-of-art of the three techniques will be described.

26.2 The Biomedical Beamline at the ESRF

The ESRF is one the three synchrotrons of highest energy and brilliance in the world [9]. It has 49 beamlines, with enegies varying from infrared to several hundreds of keV. Among them, the ID17 Beamline is devoted to biomedical research [10], with a great concentration of efforts nowadays in the development of radiotherapy techniques with synchrotron radiation.

The irradiation conditions differ from conventional radiotherapy in source type, beam energy, intensity, geometry, etc. At ID17 the X-ray source consists in two wigglers with 15 cm (wiggler 1) and 12.5 cm (wiggler 2) periods, respectively, and a maximum magnetic field of 1.6 T [11, 12].

ID17 is one of the two long ESRF beamlines. Two experimental stations are available. In the first one, located around 40 meters from the source, the white beam is mainly used for the MRT and MBRT programs. A second experimental hutch is hosted at a satellite building, outside the ESRF main experimental hall, at about 150 m from the source. In this station a monochromator (two bent Si (111) crystals, in Laue geometry) is used to tune the X-ray beam at the desired energy in the 20-100 keV range, with a narrow bandwidth of a few tens of eV [11, 13]. The monochromatic beam is then collimated by means of tungsten leaves (slits) located 5.2 meters before the target. A more detailed technical description of the radiotherapy setup can be found elsewhere [11, 12, 14].

26.3 Exploring the limits of dose-volume effects: MRT and MBRT

As explained in Sect. 30.1, the dose delivery methods (fractionation scheme, dose rate, spatial distribution, etc.) influence the biological outcome of the radiotherapy treatments. This idea is the basis for the development of two new radiotherapy techniques at the ESRF: microbeam radiation therapy and, more recently, minibeam radiation therapy. They present several distinct features with respect to conventional methods:

1. Submillimetric field sizes are used. The beam widths range from 25 to 100 μm in the case of MRT and from 500 to 700 μm in MBRT.

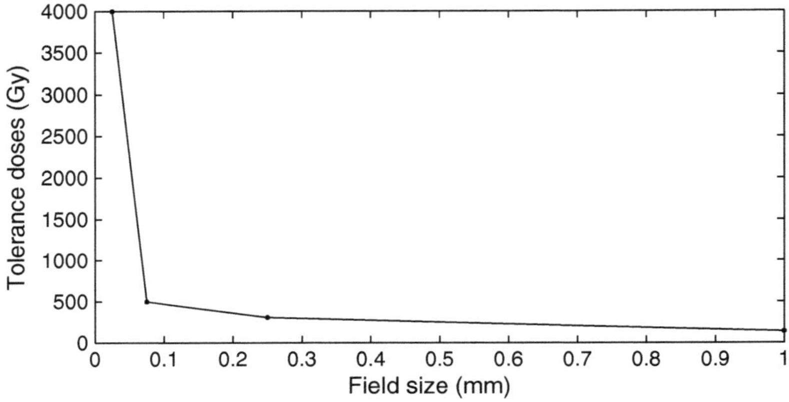

Fig. 26.2 Reconstruction of the results obtained by Zeman et al. [18]. The mice brain tolerance doses (radionecrosis) after irradiation with deuteron beams remain almost constant for field sizes larger than 0.1 mm. Below that threshold, a rapid increase is observed

2. The dose is spatially fractionated: very high doses (≥50 Gy) are delivered in one fraction by using arrays of intense parallel beams. The interbeam separation is 200 μm or 400 μm in the case of MRT and 600 μm in MBRT.
3. The X-ray spectrum ranges from 50 to 500 keV, with a mean energy at around 99 keV [15].
4. Extremely high dose rates (≥5000 Gy/s) have been used up to date in order to provide a fast irradiation.

The use of submillimetric field sizes allows to explore the limits of what is called dose-volume effects: the smaller the field size is, the higher the tolerances of the healthy tissues are [16, 17]. This is a phenomenon known from the 50s. Mice brain were irradiated with deuteron beams of different sizes in a series of experiments [18] in order to study the possible biological hazards of the cosmic rays in the astronauts. The reconstruction of those results is depicted in Fig. 26.2. The tolerance doses (radio-necrosis) remain almost constant for the different field sizes until a certain threshold is reached (0.1 mm in this case), below which the tolerances dramatically increase. The same phenomenon was observed in experiments with high energy photons [19]: the tolerances grew exponentially when field sizes smaller than a certain threshold were used. This effect might be explained by the stem cell depletion hypothesis: for each organ it exists a critical volume that can be repopulated by a single survival stem cell that migrates from the nearby tissue to recover the tissue damaged by the radiation [20]. Following that hypothesis, the spatial fractionation of the dose would provide a further gain in tissue sparing due to the biological repair of the microscopic lesions by the minimally irradiated contiguous cells [21, 22].

Therefore the combination of submillimetric field sizes and a spatial fractionation of the dose would lead to the shift of the normal tissue complication probability curve towards higher doses widening the therapeutic window for gliomas. This has been confirmed by in vivo experiments as it will be presented hereafter.

Potential advantages of MRT and MBRT over the existing clinical radiotherapy and radiosurgery methods might include the following: (i) a higher normal tissue tolerance, allowing the use of higher and potentially curative doses in those clinical cases in which cure is not possible today; (ii) due to the very small penumbras (10-50 μm in comparison with several millimetres in radiosurgery) those techniques could be employed to treat tumours close to organs at risk like the spinal cord or the brain stem, or for the treatment of illnesses like epilepsy, Parkinson, etc, with negligible secondary effects; (iii) a potentially more effective combination with tumour-dose-enhancement agents based on high Z-contrast elements, because of the larger photoelectric cross section of the MRT/MBRT lower beam energies; (iv) MRT/MBRT might also produce a temporary disruption of the blood brain barrier in very small brain regions (10–20 mm^3) for selective delivery of a chemotherapy drug to small areas in the brain, improving the therapeutic index in an exponential way.

26.3.1 Microbeam Radiation Therapy (MRT)

MRT explores the limits of dose-volume effects. The irradiation is carried out by means of an array of parallel X-ray microbeams (from 25 to 50 μm width), with a centre-to-centre (c-t-c) distance between them of 200 or 400 μm. The microbeams are produced by a multi-slit collimator that spatially fractionates in the horizontal direction the beam coming from the synchrotron source [23]. Targets are then vertically scanned through the microfractionated beam to deliver microplanes of X-rays. The MRT irradiation scheme results in dose profiles consisting of a pattern of peaks and valleys, i.e., with high doses in the microbeam paths and low doses in the spaces between them. See Fig. 26.3. Synchrotron microbeams possess two essential features: negligible divergence (allowing the production of sharply defined beam edges in tissue) and high flux (enabling a fast irradiation process that prevents motion artifacts of the subject caused by cardiosynchronous pulsation [24]).

MRT has several interesting properties, which appear to challenge many of the current paradigms in conventional radiation therapy. The preclinical studies in MRT have confirmed a remarkable healthy tissue sparing [21, 25–30]. In addition, malignant tissue mass appears to respond to MRT by significant growth delay and, in some cases, complete tumour ablation [31–39], despite the small fraction of the tumour mass irradiated with the high dose microbeams. This might indicate that MRT involves other biological mechanisms different from a direct damage by ionising radiation, which are not yet fully understood. The preferential effect on malignant tissues has been mainly attributed to selective effects of microbeams on immature tumor vessels versus lack of microbeam effects on the differentiated normal vasculature [22]. No clear experimental proof of that hypothesis has been given in the literature and the latest studies failed to reveal important damage to tumor vessels after MRT [34]. In parallel, Dilmanian et al. [22] have shown that the sparing effect of MRT seems to depend mostly on the valley dose. The brain-sparing

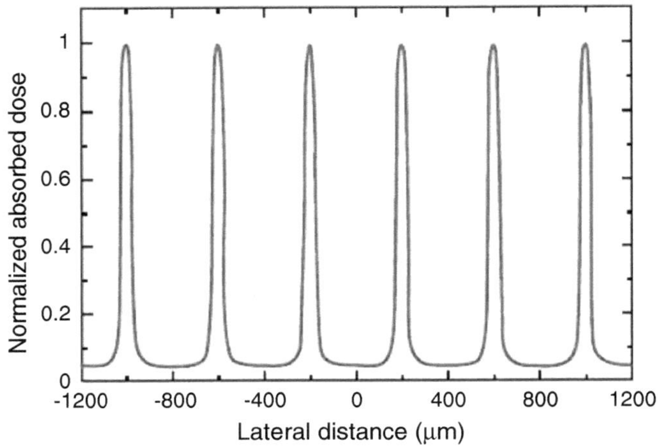

Fig. 26.3 Monte Carlo simulation of the MRT dose profile in the centre of an anthropomorphic head phantom. The dose distribution consists of a pattern of peaks (areas of high dose in the microbeam paths) and valleys (zones of low dose in the areas between the microbeams)

effect (measured by the onset of the appearance of white matter necrosis) vanishes only when the valley dose approaches the tissue tolerance to broad beams [22]. Consequently, the beam quality must be kept below 200 keV [40]. Otherwise, the lateral scattering filling the valleys would be too high, and the sparing effect in the healthy tissues would be lost [40].

Due to the success of the preclinical studies, clinical trials in MRT are being prepared at the ESRF. The progression from studies on implanted tumors in rodents to those on spontaneous tumors in humans includes changes in many parameters such as size of the host and the target volume, location, biological characteristics of the tumor and the tumor bed. Because of this complexity, it has been decided to break down the process into smaller steps. The first phase will consist in the treatment of spontaneous tumors in larger pet animals, cats and dogs. The expected results will not allow the determination of tumoricidal doses or safe normal tissue exposures for humans, but the extrapolation from similarly sized animals with spontaneous tumors to human patients will be safer than from rodents with implanted tumors. In addition these treatments can be considered as an early warning system for delayed radiation reactions because of the shorter life time of the animals.

For the preparation of these clinical trials, intensive work in different Medical Physics aspects has been performed during the last years. The main subjects will be discussed hereafter.

One important step in the path from preclinical studies with small animals to clinical trials with human beings is the assessment of the beam energy providing the best balance between tumor treatment and healthy tissue sparing. In spatially fractionated techniques like MRT, a relevant dosimetric quantity is the peak-to-valley dose ratio (PVDR). It gives a measurement of the peak dose in relation to the valley dose between two peaks and it is likely to be an important parameter in tissue sparing [22, 36]. The PVDR should be low in the tumor in order to inhibit any possible repair mechanisms and high in the healthy tissues, always keeping the valley dose below the tolerances for the healthy tissues [22]. Therefore, the optimum energy is the one maximizing the ratio between the PVDR in the healthy tissue (HT) and the one in the tumor ($PVDR_{HT/Tumor}$). By using Monte Carlo simulations it was shown that the ratio $PVDR_{HT/Tumor}$ increases up to energies around 175 keV, where the highest ratio is achieved [40]. This energy is higher than the mean energy of the spectrum currently used at the ESRF [41]. The hardening of the MRT spectrum remains an open question for clinical trials since it is needed to assess the possible biological effects due to the increase of the microbeams penumbras with the beam energy.

The definition of safe dose limits for the clinical trials has been performed by evaluating the maximum peak and valley doses achievable in the tumour while keeping the valley doses in the healthy tissue under tolerances. By using Monte Carlo simulations and the linear-quadratic model [42], it has been determined that for the most conservative case considering an unidirectional irradiation and a centrally located tumour in an adult, the largest peak and valley doses achievable in the tumour are 55 Gy and 2.6 Gy, respectively, with a maximum entrance dose of 200 Gy [43].

In addition, to treat patients, dose calculations in computer tomography data of the actual patient anatomy (voxelized structures) are required. This is done by what is called treatment-planning system (TPS). None of the commercial TPS were valid for synchrotron radiation therapies because of the difference in energy, machine geometry and parameters respect to conventional radiotherapy. That is why a TPS for MRT, based on Monte Carlo simulations (PENELOPE code), is under development at the ESRF. It will provide a 3-D dose distribution of the peak, valley and PVDR values. The main challenges of this development are: i) the reduction of the long computation times required to achieve the needed statistics in the small voxels employed in the calculations: ii) the experimental measurements of those small fields sizes to be able to benchmark the TPS against the measured dose distributions, following international recommendations [44].

No model that could be adapted for the development of a dosimetry protocol for MRT existed. It was, however, desirable to develop dosimetry protocols following international recommendations, such as the widely used IAEA TRS 398 [45], based on dose absorbed to water. For clinical trials the tolerances of the different dosimetric quantities and distributions must also fulfill the criteria imposed by French legislation [46]. The dosimetry of MRT is challenging due to the extremely high spatial resolution needed and the high dose rates used that induce saturation problems in the ionization chambers. The absolute dosimetry is currently performed in two steps. First the dose for "broad" beam (2×2 cm^2) configuration is measured with a semiflex chamber (PTW 31010). The conversion of the dose deposited with that "broad" beam is then translated to peak dose by means of some Monte Carlo calculated factors that take into account the field size dependence of the dose deposition. Those theoretical factors have been experimentally verified by using gafchromic films.

26.3.2 Minibeam Radiation Therapy (MBRT)

As explained in Sect. 26.3.1, the thin microbeams (and their associated small beam spacing) need high dose rates, only available at synchrotrons nowadays. This limits their widespread clinical implementation. In addition, the high lateral scattering produced by beam energies higher than 200 keV would lead to healthy tissue sparing loss [40]. The requirement of low-energy beams limits the dose penetration to the tissue. To overcome those difficulties, A. Dilmanian et al. [29] proposed the so called Minibeam Radiation Therapy (MBRT). They have hypothesized that beams as thick as 0.68 mm keep (part) of the sparing effect observed in MRT [29]. Moreover, from MRT preclinical studies there are indications that a wider beam results in a higher tumoricidal effect [39]. In addition, the use of higher beam energies is feasible in MBRT [47], resulting in a lower entrance dose to deposit the same integral dose in the tumor. The dose profiles of minibeams are not as vulnerable as the ones of microbeams to beam smearing from cardiac pulsations, therefore high dose rates are not needed and it is conceptually possible to extend this technique by using modified X-ray equipment.

An original method was developed and tested at the ESRF ID17 biomedical beamline to produce the minibeam patterns [48]. It utilizes a specially developed high-energy white-beam chopper whose rotation is synchronized with the vertical motion of the target moving at constant speed. Each opening of the chopper generates a horizontal beam print.

In parallel, a dosimetry characterization of MBRT was performed [48]. The good agreement between Monte Carlo simulations and the experimental measurements opened the door to the biological studies in MBRT. Recently, interlaced minibeams were produced by slightly modifying the duty cycle of the chopper (53%). In this configuration, two orthogonal arrays interlace at the target. A quasi-homogeneous dose distribution in the tumor is achieved while the healthy tissue still benefit from the spatial fractionation of those submillimetric beams.

Preclinical trials in MBRT have already started at the ESRF and they are ongoing. Several radiobiology studies (in vitro and in vivo) showed that MBRT widens the therapeutic window for gliomas: extremely high dose tolerances of healthy rat brains accompanied by a factor three gain in mean survival time of treated tumor bearing rats was observed [49]. This indicates that MBRT might allow the use of higher and potentially curative doses in clinical cases where the tolerance doses of healthy tissues impose a limit on the dose delivered to the tumor if conventional therapy is used. Improvement of the outcome is expected by using image-guidance, chemo-radiotherapy, etc. in future studies.

Other preclinical research opportunities for MBRT are the creation of small lesions in the submillimeter range to mitigate Parkinson disease or epilepsy using minibeams [29].

26.4 Stereotactic Synchrotron Radiation Therapy (SSRT)

SSRT consists in loading the tumor with a high atomic number (Z) element and irradiating it with monochromatic X-rays from a synchrotron source (tuned at an optimal energy) in stereotactic conditions. The high Z element injected in the patient selectively accumulates in the brain tumor as a consequence of the permeability of the blood-brain barrier due to invasive growing of the tumor. At energies of few tens of keV, the high photoelectric cross section of high Z materials like Iodine, results in a great number of photoelectric interactions. Due to the short range of the products of those interactions there is an enhancement of the dose deposited locally in the tumor. This leads to improved dose distributions when compared to conventional high energy treatment [50, 51]. The use of monochromatic X-rays optimises the dose distributions with respect to a spectrum [52]. Hence, the synchrotrons are ideal for this therapeutic modality since they provide very intense monochromatic X-rays.

Several preclinical studies were carried out and are still ongoing using monochromatic X-rays at the ESRF biomedical beamline. Two different approaches were simultaneously developed. The first one is contrast enhanced Synchrotron Radiation Therapy (SRT) with extracellular agents like Iodine [51, 53]. A enhancement of life span of around 200 % was achieved. The second one uses some chemotherapy drugs containing platinum [54, 55] or iodine compounds [56, 57]. The studies using platinum compounds reached a mean survival increase close to 700 %. Those drugs penetrate the cell and bound to DNA. Due to the intrinsic toxicity of those drugs only small concentrations (ppm) can be brought to the DNA and therefore it is not clear whether there is a physical dose enhancement in this chemo-radiotherapy modality. For the aforementioned reasons, the ESRF is planning the clinical trials following the first avenue. With this aim different dosimetric aspects have been assessed and will be described hereafter.

First, the beam energy providing the best balance between tumor treatment and healthy tissue sparing in a human head had to be assessed. In SSRT Monte Carlo simulations in anthropomorphic head phantoms showed that even if energies around 50 keV are the ones providing the highest dose enhancement factor in the tumor, 80 keV renders a good compromise between a high dose deposition in the target (up to 82 Gy) with healthy tissue doses within tolerable levels [52, 58, 59]. It is expected that the net gain in deposited dose in the tumor (32 Gy) with respect to conventional radiotherapy (50 Gy) will result in an increase of tumor control probability.

To be able to treat patients, a Monte Carlo based TPS has been developed and benchmarked against experimental measurements [60]. The calculation engine will be integrated in the ISOgray^{TM} system sold by the French firm DOSIsoft (Cachan, France). A three-step Monte Carlo simulation has been implemented in order to compute the dose in the patient from the TPS, considering the features of SSRT: particles of medium energies; beamline geometry; contrast media in the target. Simulations were compared to measurements conducted under clinical irradiation conditions. Measurements were performed using ionization chambers and radiosensitive films. The agreement between the TPS calculations and the experimental data fits within the limits recommended by the IAEA [44].

In parallel a protocol for synchrotron radiation dose measurements was developed [61]. It is based on absorbed dose to water [45] and inspired by the IAEA recommendations: the reference depth for absolute dosimetry is 2 cm depth in water and a semiflex chamber (PTW 31010) is used. However, standard dosimetry protocols and reference calibration protocols of ionization chambers assume a uniform exposure of the chamber. This condition cannot be met at the synchrotron since at least one of the transverse dimension of the beam is smaller than the dimensions of any ionization chamber. It has been shown that to integrate the dose rate while scanning the ionization chamber with a constant and well known speed is equivalent to measure the dose deposited with an uniform irradiation [61]. The protocol allows to measure the absolute dose with an accuracy of 2%, within the recommended limits reported in international protocols and French laws for patients treatments. A detailed description is beyond the scope of this chapter and it can be found in [61].

26.5 Summary

Synchrotron radiation is a innovative tool in the biomedical research field and, in particular, for the treatment of brain tumors. It is an example of multidisciplinary research in which Medical Physics plays a major role, from the design of new techniques to their posterior clinical implementation: from dose calculations, definition of safe dose escalation schemes, establishment of experimental dosimetry protocols, modelling of new detector systems, etc. In this chapter the three new radiotherapy techniques under development at the ESRF have been described. Thanks to the promising results obtained in the preclinical phase, the ESRF is walking towards the clinical trials in SSRT and MRT. The first patients are expected in 2011. MBRT is the most recently implemented technique at the ESRF. The first biological studies suggest that this technique could be successfully applied in clinical cases where tissue tolerances are a limit for conventional methods. An asset of MBRT is the possibility to be extended outside synchrotron sources with a cost-effective equipment.

References

1. S. Birkenhake, R. Sauer, Cellular and Molecular Life Sciences 681, 51–7 (1995)
2. L. Freund, "Grundriß der gesamten Radiotherapie für praktische Ärzte", Berlin, Urban & Schwarzenberg, 1903
3. J.M. Legler, L.A. Ries, M.A. Smith, J.L. Warren, E.F. Heineman, R.S. Kaplan et al., J. Natl. Cancer. Inst. 91, 1382–90 (1999)
4. R. Stupp, P. Dietrich, S.O. Kraljevic, A. Pica, I. Maillard, P. Maeder, R. Meuli et al., J. Clin. Oncol. 20, 1375–1382 (2002)

5. R. Stupp, W.P. Mason, M.J. van den Bent, M. Weller, B. Fisher, J.B. Martin et al., N. Engl. J. Med. 352, 987–996 (2005)
6. R. Stupp, M.E. Hegi, W.P. Mason, M.J. van den Bent, M.J. Taphoorn, R.C. Janzer, et al., Lancet Oncol. 10, 45966 (2009)
7. H. Holthusen, Strahlentherapie 57, 254–69 (1936)
8. G.G. Steel, "Basic Clinical Radiobiology", 3rd Ed., Edward Arnold, 2002
9. http://www.esrf.fr
10. W. Thomlinson, P. Suortti, D. Chapman, NIM A 543, 288–96 (2008)
11. H. Elleaume, A. Charvet, P. Berkvens, G. Berruyer, T. Brochard, Y. Dabin, et al., NIM A 428, 513–527 (1999)
12. H. Requardt, A. Bravin, Y. Prezado, E. Bräuer-Krisch, M. Renier, Th. Brochard et al., AIP conference proceedings 1234, 161–164 (2010)
13. P. Suortti, S. Fiedler, A. Bravin, T. Brochard, M. Mattenet, M. Renier et al., J. Synchrotron Radiat. 7, 340–347 (2000)
14. M. Renier, Th. Brochard, C. Nemoz, H. Requardt, E. Bräuer, F. Estève, et al., Eur. J. Radiol. 68S, 147–150 (2008)
15. I. Martínez-Rovira, J. Sempau and Y. Prezado, submitted to Med. Phys.
16. H.J. Curtis, Rad. Res. Suppl. 7, 250–7 (1967)
17. Y.R. Lawrence, X.Q. Li, I. El Naqa, C.A. Hahn, L.B. Marks, T.E. Merchant, A.P. Dicker, Int. J. Radiat. Oncol. Biol. Phys. 76, S20–27 (2010)
18. W. Zeman, H.J. Curtis, El. Gebhard, W. Haymaker, Science 130, 1760–1 (1959)
19. J.W. Hopewell, K.R. Trott, Radiother. Oncol. 56, 283–8 (2000)
20. R.J. Yaes, A. Kalend, Int. J. Radiat. Oncol. Biol. Phys. 14, 1247–59 (1988)
21. D. Slatkin, P. Spanne, F. A. Dilmanian, J.O. Gebbers, J. A. Laissue, Proc. Natl. Acad. Sci. U.S.A. 92, 8783–8787 (1995)
22. F.A. Dilmanian, T. Button, G. Le Duc, et al., Neuro. Oncol. 4, 26–38 (2002)
23. E. Bräuer-Krisch, H. Requardt, T. Brochard, G. Berruyer, M. Renier, J.A. Laissue and A. Bravin, Rev. Sci. Instrum. 80, 074301 (2009)
24. B.P. Poncelet, V.J. Wedeen, R.M. Weisskopf, M.S. Cohen, Radiol. 185, 645–651 (1992)
25. J.A. Laissue, N. Lyubimova, H.P. Wagner, et al., SPIE 3770, 38–45 (1999)
26. F.A. Dilmanian, G. Morris, G. Le Duc et al., Cell Mol Biol (Noisy-le-grand) 47, 485–493 (2001)
27. J.A Laissue, H. Blattmann, M. Di Michiel, D.N. Slatkin, et al., SPIE 4508, 65–73 (2001)
28. N. Zhong, G. M. Morris, T. Bacarian, E. M. Rosen, F. A. Dilmanian, Radiat. Res. 160, 133–142 (2003)
29. F. A. Dilmanian, Z. Zhong, T. Bacarian, H. Benveniste, P. Romanelli, R. Wang, J. Welwart, T. Yuasa, E. M. Rosen, D. J. Anschel, Proc. Natl. Acad. Sci. U.S.A. 103, 9709–9714 (2006)
30. B. Van der Sanden, E. bräuer-Krisch, E. A. Siegbahn, C. Ricard, J.C. Vial, J. Laissue, Int. J. Radiat. Oncol. Biol. Phys. 77, 1545–1552 (2010)
31. J.A. Laissue, G. Geiser, P.O. Spanne, et al., Int. J. of Cancer 78, 654–660 (1998)
32. R. Serduc, T. Christen, J. Laissue, R. Farion, A. Bouchet, B. van der Sanden et al., Phys. Med. Biol. 53, 3609–3622 (2008)
33. R. Serduc, Y. van de Looij, G. Francony, O. Verdonck, B. Boudewijn van der Sanden, J. Laissue et al., Phys. Med. Biol. 53, 1153–1166 (2008)
34. A. Bouchet, B. Lemasson, G. Le Duc, C. Maisin, E. bräuer, E.A. Siegbahn et al., Int. J. Radiat. Oncol. Biol. Phys. 78, 1503–1512 (2010)
35. H.M. Smilowitz, H. Blattmann, E. Bräuer-Krisch, et al., J. Neurooncol. 78, 135–43 (2006)
36. P. Régnard, G. Le Duc, E. Bräuer-Krisch, et al., Phys. Med. Biol. 53, 861–878 (2008)
37. F. A. Dilmanian, G. Morris, N. Zhong, T. Bacarian, J. F. Hainfeld, J. Kalef-Ezra, L. J. Brewington, J. Tammam, E M Rosen, Radiat. Res. 159, 632–641 (2003)
38. M. Miura, H. Blattmann, E. Bräuer-Krisch, A. Bravin, A.L. Hanson, M. M. Nawrocky, P. L. Micca, D. N. Slatkin, J. A. Laissue, "Radiosurgical palliation of aggressive murine SCCVII squamous cell carcinomas using synchrotron-generated X-ray microbeams", Br. J. Radiol. 79, 71–75 (2006)

39. R. Serduc, E. bräuer-Krisch, A. Bouchet, L. Renaud, T. Brochard, A. Bravin, J. A. Laissue, G. Le Duc, J. Synchrotron Radiat. 16, 587–590 (2009)
40. Y. Prezado, G. Fois, G. Le Duc, A. Bravin, "Gadolinium dose enhancement studies in microbeam radiation therapy", Med. Phys. 36, 3568–74 (2009)
41. E.A. Siegbahn, J. Stepanek, E. Bräuer-Krisch, A. Bravin, Med. Phys. 33, 3248–59 (2006)
42. J.F. Fowler, Br. J. Radiol. 62, 679–694 (1989)
43. I. Martínez-Rovira, J. Sempau, J.M. Fernández-Varea, A. Bravin, Y. Prezado, Phys. Med. Biol. 55, 4375–4388 (2010)
44. International Atomic Energy Agency (IAEA), "Commissioning and Quality Assurance of Computerized Planning Systems for Radiation Treatment of Cancer", AEA Technical Reports Series TRS430 (Vienna: IAEA) (2004)
45. International Atomic Energy Agency (IAEA), "Absorbed dose determination in external beam radiotherapy: An International Code of Practice for Dosimetry based on Standards of Absorbed Dose to Water", IAEA Technical Reports Series 398 (Vienna: IAEA) (2005)
46. Journal Officiel de la Republique Francaise (JORF), "Décision du 7 juillet 2007 fixant les modalités du controle de qualité interne des installations de radiothérapie externe", n. 209 du 9 septembre 2007
47. Y. Prezado, S. Thengumpallil, M. Renier, A. Bravin, Med. Phys. 36, 4897–4902 (2009)
48. Y. Prezado, M. Renier, A. Bravin, J. Synchr. Radiat. 16, 582–586 (2009)
49. Y. Prezado, S. Sarun, S. Gil, P. Deman, A. Bouchet, G. Le Duc, submitted to J. Synchr. Radiat.
50. A.V. Mesa, A. Norman, T.D. Solberg, J.J. Demarco, J.B. Smathers, Phys. Med. Biol. 44, 1955–1968 (1999)
51. J.F. Adam, A. Joubert, M.C. Biston, A.M. Charvet, M. Peoc'h, J.F. Le Bas, J. Balosso, F. Estève, H. Elleaume, Int. J. Radiat. Oncol. Biol. Phys. 64, 603–611 (2006)
52. C. Boudou, J. Balosso, F. Estève, H. Elleaume, Phys. Med. Biol. 50, 4845–4851 (2005)
53. J.F. Adam, H. Elleaume, A. Joubert, M.C. Biston, A.M. Charvet, J. Balosso, J.F. Le Bas, F. Estève, Int. J. Radiat. Oncol. Biol. Phys. 57, 1413–1426 (2003)
54. M. C. Biston, A. Joubert, J. F. Adam, H. Elleaume, S. Bohic, A. M. Charvet, F. Estève, N. Foray, J. Balosso, Cancer Res. 64 2317–2323 (2004)
55. J. Rousseau, C. Boudou, R. F. Barth, J. Balosso, F. Estève, H. Elleaume, Clin. Cancer Res. 13, 575 5195–5201 (2007)
56. J. F. Adam, M. C. Biston, J. Rousseau, C. Boudou, A. M. Charvet, J. Balosso, F. Esteve, H. Elleaume, Phys. Med. 24, 92–97 (2008)
57. J. Rousseau, J. F. Adam, P. Deman, T. D. Wu, J. L. Guerquin-Kern, B. Gouget, R. F. Barth, F. Estève, H. Elleaume, J. Synchrotron Radiat. 16, 573–581 (2009)
58. M. Edouard, D. Broggio, Y. Prezado, F. Estève, H. Elleaume, J.F. Adam, Med. Phys. 37, 2445–2456 (2010)
59. Y. Prezado, G. Fois, M. Edouard, C. Nemoz, M. Renier, H. Requardt, F. Estève, J. F. Adam, H. Elleaume, A. Bravin, Med. Phys. 36, 725–732 (2009)
60. M. Vautrin, Y. Prezado, I. Martínez-Rovira, P. Deman, M. Edouard, H. Elleaume et al., to be submitted to Med. Phys.
61. Y. Prezado, M. Vautrin, I. Martínez-Rovira, A. Bravin, F. Estéve, H. Elleaume, P. Berkvens, J.F. Adam, Med. Phys. (2011)

Chapter 27
Photodynamic Therapy

Slávka Kaščáková, Alexandre Giuliani, Frédéric Jamme, and Matthieu Refregiers

Abstract Treatments based on absorption of electromagnetic radiation may be categorized according to the photon wavelength range. On the one hand, radiotherapy is based on X-rays delivery to tissues and is widely spread and recognized for cancer treatment. On the other hand, photodynamic therapy (PDT) involves low energy radiation in the visible and near infrared range in combination with a drug referred to as the photosensitizer. A short overview of conventional radiotherapy and accelerator-based therapy is first presented. Then PDT is introduced and its mechanisms are reviewed along with the factors affecting its outcome. The domains of application of this therapy are presented through a discussion of the most used photosentizers. Finally we present new developments in the field that would permit the combination of potentialized radiotherapy and photodynamic therapy.

27.1 Introduction

The use of an electromagnetic radiation for treatment of diseases is a vast subject. Indeed, the particular illnesses to be cured may be as diverse as age-related macular degeneration, psoriasis or brain cancer. Moreover, the wavelength range of the incoming radiation extends from near infrared down to hard X-rays, and such a large photon energy domain may be achieved using a variety of light sources.

S. Kaščáková • M. Refregiers
Synchrotron SOLEIL, L'Orme des Merisiers, 911 92 Gif sur Yvette, France

A. Giuliani (✉) • F. Jamme
Synchrotron SOLEIL, L'Orme des Merisiers, 911 92 Gif sur Yvette, France
Cepia, Institut National de la
e-mail: alexandre.giuliani@synchrotron-soleil.fr

Recherche Agronomique, BP 716 27, 443 16 Nantes, France

G.G. Gómez-Tejedor and M.C. Fuss (eds.), *Radiation Damage in Biomolecular Systems*, Biological and Medical Physics, Biomedical Engineering,
DOI 10.1007/978-94-007-2564-5_27, © Springer Science+Business Media B.V. 2012

Although this work is devoted to photodynamic therapy (PDT), it begins with a brief overview of radiotherapy. Indeed, radiotherapy is the most widely spread therapy using photons. Our overview is also partial as it is centred on accelerator-based therapy and especially on synchrotron radiation facilities. Synchrotron radiation is emitted over a continuous spectrum (from terahertz to hard X-rays) with high intensity and with a small divergence. Considering medical applications, wavelength tunability allows monochromatic beams to be generated at virtually any photon energy. This allows enhancement of diagnostic images and therapeutic doses upon selection of the most effective photon energy or the best contrast for an image or a tomogram. It means a more effective dose delivery in therapeutic modalities and a lower dose to reach greater image quality in diagnosis. A large activity at synchrotron facilities is devoted to analysis. Although this subject is clearly beyond the scope of this contribution, it is worth mentioning that X-Rays have been reported to be a very useful tool for the localisation of metallic compounds [1] and metal associated drugs [2].

However, considering synchrotrons radiation facilities for patient's treatment, only five medical beamlines operate in the world. These particular beamlines, defined as those being able to receive animals or patients for clinical studies and treatments, are: X17 at the National Synchrotron Light Source (NSLS) (Brookhaven National Laboratory, USA), ID17 at European Synchrotron Radiation Facility (ESRF) (Grenoble, France), a bending magnet based beamline at Canadian Light Source (CLS) (University of Saskatchewan, Canada), BMIT at Australian Synchrotron (Melbourne, Australia) and BL20 located at SPring-8 (Riken, Japan).

At a first glance, this offer may seem somewhat restricted. However, it must be emphasis on the fact that the number of medical beamline has doubled within ten years.

Photodynamic therapy employs visible and near infrared light as the activation source in combination with a photosensitizer. Although synchrotron radiation contains the visible part of the electromagnetic spectrum, it has never been used to trigger PDT. The contribution of synchrotron radiation (SR) to the study of PDT is found in the spectroscopic and imaging field. Indeed, SR based infrared microspectroscopy has been shown to be a powerful analytical tool to assess individual cells changes without any labelling. In comparison to conventional instruments, thanks to the brightness of the source, high signal-to-noise ratio and high spatial resolution are routinely achieved [3]. This technique is able to reach down to 5 μm spatial resolution, and allows monitoring global proteome changes and phenotypic heterogeneity at the individual cell level [4, 5]. Interestingly, the observed infrared spectral features could be related to consequences of particular types of cell death [6].

The mechanisms of action of PDT are reviewed below along with its advantages and limitations. The field of PDT is active and has seen interesting development in the last decade. Especially, combinations of PDT with radiotherapy sound very promising.

27.2 Radiotherapy

27.2.1 Conventional and spatially resolved radiotherapy

Radiation therapy with X-rays beams is one of the most common and efficient technique for cancer treatment. It is the golden standard for many types of cancers. Currently, radiation delivery techniques are based on external radiotherapy or internal radiotherapy also known as brachytherapy. Technical innovations in these fields always aim at improving the dose deposition and distribution to decrease normal tissue toxicity. The precision of external radiation therapy has been markedly improved by innovative technical development of various imaging modalities and irradiation devices. Recently, intensity modulated radiotherapy (IMRT) and volumetric modulated arc therapy (VMAT) have shown benefits especially in decreasing acute treatment-related toxicity in either definitive or palliative re-irradiated cases [7]. As a result, since 1994 when the first clinical IMRT with modern delivery technology was used for head and neck cancers at Baylor College of Medicine [8], the tumour control rate by radiation therapy improved significantly. IMRT became an excellent alternative to surgery for early stage lung cancer, low-risk prostate cancer and for asymptomatic or mildly symptomatic brain tumours [9].

However, brain tumours radiotherapy may cause considerable long time scale healthy tissue damage such as pituitary diseases, hormone depletion, demyelinization and white matter radionecrosis [10–13]. The latter two may lead to severe cognitive dysfunctions and dementia. Complications of irradiation may also arise as a consequence of vasogenic oedema following disruption of the blood-brain barrier [14, 15].

Microbeam radiation therapy (MRT) is a spatially fractionated radiotherapy that uses an array of microscopically thin (25 to 100 μm width) and nearly parallel synchrotron-generated X-ray beams separated by 100 to 200 μm centre-to-centre distances [16]. The high flux of synchrotron light allows very high rates of dose deposition (several hundreds Gy within less than 1 s). MRT was initiated at the National Synchrotron Light Source (Brookhaven National Laboratory, USA) [16, 17] and then later developed at the European Synchrotron Radiation Facility (ESRF, Grenoble, France) [18–20]. The properties of microbeams that make them good candidates for tumour therapy are (a) their sparing effect on normal tissues, including the central nervous system [17–22], and (b) their preferential damage to tumours, even when administered from a single direction [22, 23]. MRT is currently considered one of the most exciting applications of synchrotron X-rays in medical research. Despite MRT potential, such high-intensity microbeams can only be produced by synchrotron radiation sources, which is a practical limitation for clinical implementation. Therefore, following the principle of spatial fractionation, an extension of the MRT method has been proposed by Dilmanian et al. [24] from the National Synchrotron Light Source (Brookhaven National Laboratory, USA) and termed minibeam radiation therapy (MBRT). In MBRT, the beam thickness ranges from 500 to 700 μm with a separation between two adjacent minibeams of the same magnitude. These aspects are discussed in greater details in Chapter 26.

27.2.2 X-Ray potentialized radiation treatments

Norman and co-workers first proposed the concept of radiation dose enhancement in the nineties [25,26] using a simple irradiation scheme and a conventional scanner x-ray source. Iodinated contrast media were injected to patients and accumulated preferentially in the tumour interstitium owing to the increased permeability of the blood-brain barrier caused by an invasive tumour growth [25, 26]. At kilo electron-volt photon energies, the high photoelectric cross sections of iodine result in substantial interactions with the incident radiation. The high linear energy transfer and short range of photoelectric interaction products lead to a localized dose enhancement.

Several agents have been evaluated as potential radiosensitizers. One of the oldest one is platinum; a metal present in chemotherapeutic agents and studied for its radiosensitizing activity [27, 28]. The efficiency of the treatment with platinum-complexes has been assessed by synchrotron radiation *in vitro* and *in vivo* conditions [29–33].

The use of monochromatic X-rays tuned at the optimal energy could significantly improve the dose delivery [34]. Synchrotron sources, providing flux and tuneable monochromatic X-rays are therefore ideal for this potentialized radiation treatments. Gadolinium, a high-atomic number (Z) element has been recently investigated in combination with synchrotron microbeam radiation therapy [35–37].

While stereotactic radiation therapy is a type of external radiation delivery involving tumours treatments with focused beam, the synchrotron stereotactic radiation therapy (SSRT) is defined by the additional presence of high-atomic number (Z) element to reach a radiation dose enhancement specific to the tumour when irradiated in stereotactic conditions [38]. SSRT is developed in details in chapter 26.

27.3 Photodynamic therapy

Aside from all the improvements in radiotherapy, such as the above mentioned approaches of X- ray potentialized radiation treatments (stereotactic and microbeam radiation therapy), there is a real need for safer therapies, to limit and reduce the toxic effect of X-ray radiation on healthy tissues [39,40]. Another approach would use radiation in a particular wavelength range that would not be toxic. Such a method is photodynamic therapy (PDT). PDT shares several of particularities with radiotherapy and its newest developments. First at all, it shares with the potentialized therapy the use of a drug in combination with the radiation. Moreover, the radiation used in PDT is administrated very locally. But PDT has particular and unique features.

Indeed, PDT is attractive, because it combines elements, which are harmless separately, namely a drug referred to as the photosensitizer, oxygen and an electromagnetic radiation [41]. In PDT, visible light is used to electronically excite the photosensitizer, which in the presence of oxygen leads to the production of oxygen-related cytotoxic intermediates (such as singlet oxygen or free radicals). Those are responsible for cell death and tissue response [41–43]. The history of PDT has been reviewed in details by MacDonald and Dougherty [44] and more recently by Hamblin and Mroz [45], and is not presented here. However, it is interesting to note, that although PDT has been approved for clinical use only quite recently (randomized clinical trials were initiated with photosensitizer Photofrin® in 1987 [46]), the first observations of photodynamic effect are known since more than hundreds of years [47].

Since 1987, significant progresses have been made in the development of photosensitizers leading to approbation for varieties of indications, such as treatment of (pre-) cancerous conditions like superficial gastric cancer, Barett's oesophagus, palliative treatment of head and neck cancers, and skin malignancies [41–43, 48]. Apart from cancer diseases, PDT is successfully applied to the treatment of age-related macular degeneration, psoriasis and scleroderma. In rheumatology, PDT is being tested against arthritis [42–44, 46, 48, 49]. Finally, the application of PDT in microbiology to target microorganisms should not be disregarded [43]. Excellent articles and reviews on the clinical status of PDT may be found and the reader is referred to them [41–43, 48, 49].

27.3.1 Mechanism of photodynamic effect

There are two generally accepted mechanisms describing the photosensitizing effect, usually referred to as type I and type II. As seen in Fig. 27.1, the initiating step of the photosensitizing mechanism is the absorption of photons by the sensitizer, leading to population of an unstable excited state. From this excited state, the drug molecule can either relax back to the ground state by fluorescence (which may be used for diagnostic purposes) or undergo intersystem crossing to a relatively long-lived excited state such as the lowest energy triplet state. From the triplet state, the photosensitizer can follow two pathways. In the so-called type I, the photosensitizer may directly react with an organic substrate and transfer an hydrogen atom or an electron to form a radical. These intermediates are then prone to reactions with oxygen leading to peroxides, superoxide ions or hydroxyl radicals, which initiate a free chain reaction. The photosensitizer in its triplet state may also transfer its energy directly to ground-state molecular oxygen ($^3O_2$), to form singlet oxygen ($^1O_2$), a highly reactive oxygen species [50], which is the basis of the type II pathway.

Both type I and type II reactions may occur simultaneously, and the ratio between these processes depends on the type of photosensitizer used, the concentrations of oxygen, the substrate nature, as well as its binding affinity to the sensitizer [50, 51]. Because the effects of PDT drugs are oxygen dependent, photosensitization typically does not occur in anoxic areas. However, the oxygen dependency of all photosensitizers has been challenged by the observation of Park et al. [52], who presented a study of the light-induced antiviral activity of photosensitizer hypericin

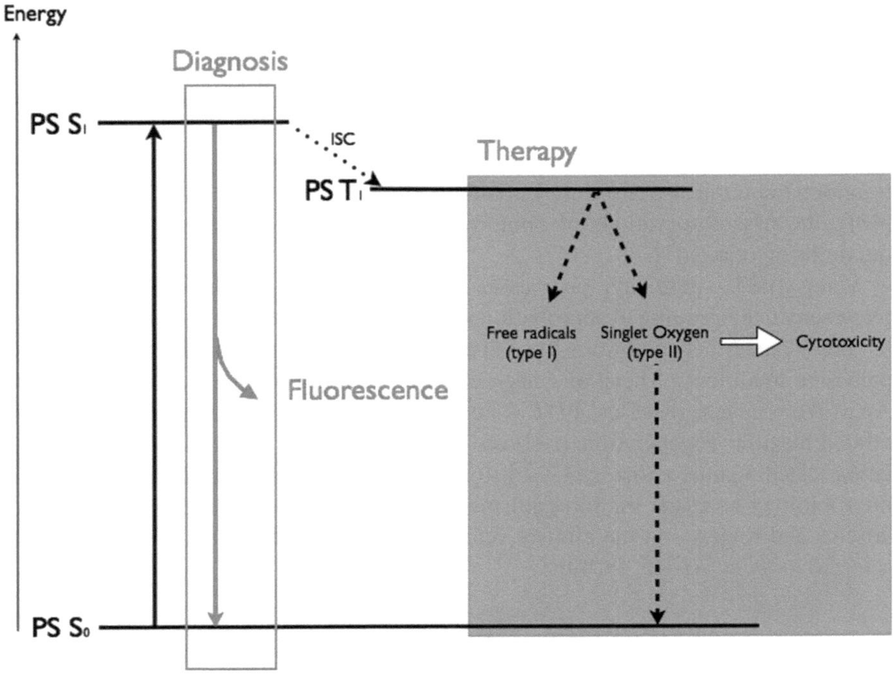

Fig. 27.1 Simplified representation of Jablonski diagram to illustrate the mechanisms of photodynamic therapy

and its analog hypocrellin A as a function of oxygen concentration. A significant reduction of the light-induced antiviral activity for both photosensitizers has been observed when oxygen levels were lowered. In contrast to hypocrellin A, hypericin under hypoxic conditions still showed virucidal activity. Considering these results, mechanisms involving only oxygen are not sufficient to completely account for the entire activity of this photosensitizer. Additional mechanisms are likely to contribute to its photobiological activity. The photogenerated pH-drop proposed by group of Petrich [52, 53] has been supported by the group of Miskovsky et al. [54, 55]. The photo-induced biological activity of hypericin on enzymes has also been extensively investigated [56, 57]. However, it is important to stress out that, although hypericin presented an antiviral activity in hypoxic environment, its activity was reduced by two orders of magnitude with respect to aerobic conditions.

27.3.2 Advantages of PDT

Reactive oxygen species (ROS) formed by PDT are extremely reactive and have short half-life. The half-life of singlet oxygen in biological systems is below 0.04 μs giving an action radius lower than 0.02 μm [58, 59]. Therefore PDT only affects biomolecules in close proximity to regions where ROS are generated. However, it means that the localization of photosensitizer during irradiation is of crucial importance for the treatment efficiency.

The mechanisms of action of PDT together with the importance of the photosensitizer localization during irradiation provide the following advantages. First, PDT is spatially specific: only cells near the photosensitizer may be affected. The use of a localized and spatially well defined energy source leads to a confined ROS production, thus making PDT ideal to target tumour cells without harming the surrounding tissue. Second, the photosensitizer is not cytotoxic until illuminated, thus allowing excess of unbound reagent to be cleared from the body. Third, the choice of photosensitizer and the treatment parameters, such as the drug-light interval, the total PDT dose and the light fluence can help selecting the primary biological targets. For example treatment of age-related macular degeneration deliberately exploits the vascular response [60].

In contrast to radiotherapy, the DNA is not the major target in PDT. Most of the photosensitizers localize in the cell membranes and induce a somatic cell death [61]. As a result, tissue responses are very rapid, and owing to the lack of mutagenicity of visible light and photosensitizer alone, PDT can be repeated multiple times, without apparent induction of resistance.

27.3.3 *Factors that affect PDT efficacy*

It is commonly accepted, that singlet oxygen is the predominant cytotoxic agent produced during PDT. Thus, the effectiveness of this treatment modality is largely determined by the efficiency of its production [41–44, 46]. As mentioned above, many factors influence how efficiently singlet oxygen is generated in a PDT process, including the type of photosensitizer used, light intensity and wavelength, and oxygen concentration. Much of the improvements in PDT have concerned developments in photosentizers chemistry and light sources.

27.3.3.1 Photosensitizers

The ideal photosensitizer for *in vivo* PDT would have several specific properties. First, it would exhibit no dark toxicity, carcinogenicity or mutagenicity. Second, it would have a poor tendency to aggregate in an aqueous medium. Third, this photosentizer would show high molar extinction coefficient in the near infrared where tissue penetration is maximum and a high quantum yield for singlet oxygen generation. It would also be easy to target and to deliver to tumours. Clearance from the body should be so that the patient would not suffer from prolonged skin-photosensitized toxicity. So far, no photosensitizer satisfies all of these requirements [41, 42, 44].

Table 27.1 Photosensitizers for malignant and other disease [62]

Photosensitizer	Trade name	Potential indication
BPD-MA	Verteporfin	Basal-cell carcinoma; Age-related macular degeneration*
HPD (partially purified) porfimer sodium	Photofrin	Cervical*, endobronchial*, oesophageal*, bladder and gastric cancers*, and brain tumours
mTHPC	Foscan	Head and neck tumours*, prostate and pancreatic tumours
5-ALA	Levulan	Basal-cell carcinoma, head and neck, gynaecological tumours
		Diagnosis of brain, head and neck, and bladder tumours
5-ALA-methylesther	Metvix	Basal-cell carcinoma*
5-ALA benzylesther	Benzvix	Gastrointestinal cancer
5-ALA hexylesther	Hexvix	Diagnosis of bladder tumours
SnET2	Purlytin	Cutaneous metastatic breas cancer, basal-cell carcinoma, Kaposi's sarcoma, prostate cancer
Boronated protoporphyrin	BOPP	Brain tumours
HPPH	Photochlor	Basal-cell carcinoma
Lutetium texaphyrin	Lutex	Cervical, prostate and brain tumours
Phthalocyanine-4	Pc 4	Cutaneous/subcutaneous lesions from diverse solid tumours origins
Taporfin sodium	Talaporfin	Solid tumours from diverse origins

*Indications that are registered in one or more countries (all other indications are in development). 5-ALA, 5-aminolevulinic acid; BPD-MA, benzoporphyrin derivative-monoacid ring A; HPD, haematoporphyrin derivative; HPPH, 2-(1-hexyloxyethyl)-2-devinyl pyropheophorbide-alpha; mTHPC, meta-tetrahydroxyphenylchlorin; SnET2, tin ethyl etiopurpurin.

Table 27.1 presents the most commonly used photosensitizers and precursors in clinical applications of PDT. One of the first systematically studied for clinical PDT is an haematoporfyrin derivate, Photofrin®. Although Photofrin® is the most commonly used photosensitizer, it has several limitations. It is plagued by prolonged cutaneous phototoxicity, which can last up to 4–6 weeks [41, 63]. Near infrared radiation penetrates the deepest in superficial tissues. In this wavelength range most tissue chromophores absorb weakly [49, 64]. Typically, the effective penetration depth is about 2 to 3 mm at 630 nm and increases to 5 to 6 mm at longer wavelength (700 to 800 nm) [64]. Since blue light does not penetrate very deeply into tissues, the excitation of the Soret band is not useful. Consequently, a weaker absorption bands at 630 nm, known as the Q-band, is used for treatment [63]. Unfortunately, due to low extinction coefficient at this wavelength, high concentrations of Photofrin® are required to achieve adequate photodynamic effect. The limitations of Photofrin® have stimulated research in the development of new photosensitizers.

The synthesis of improved (second generation) photosensitizers moved towards modified tetrapyrolic compounds, such as benzoporphyrin (Visudyne®) and chlorine derivatives (Foscan®) type.

Visudyne® is a liposomal formulation benzoporphyrin. This photosensitizer is active at 690 nm allowing a deeper penetration. Studies revealed that most of the clinical response from Visudyne® sensitization is based on vascular disruption and shutdown due to its localization in these compartments [65]. However, because of its effectiveness in the obliteration of neovessels, this compound is being developed as a first-line treatment for age-related macular degeneration (AMD) [60, 65]. AMD is a leading cause of blindness and its pathophysiology involves the abnormal growth of blood vessels in the choriocapillaris. These leaky vessels cause loss of central vision. The use of Visudyne® in PDT of AMD has been marketed. In this treatment, the irradiation takes place few minutes after intravenous injection of Visudyne®, while the drug is still in the vascular compartment. The PDT affects vascular endothelial cells, resulting in thrombosis and vessel closure, followed by curtailed of the visual loss. Visudyne® is now the standard treatment of this disease and the most popular application of PDT.

Foscan®, a formulation of meta-tetrahydroxyphenylchlorin in water-free ethanol and propylene glycol, is a second-generation photosensitizer which has been granted European approval as a PDT agent for prostate and head and neck cancer [48,63,66]. This chlorine derivative, has shown to be more potent that Photofrin®, since it needs very low drug dose (0.15 mg kg^{-1} compared to 2.0 mg kg^{-1} for Photofrin®), as well as lower light doses (20 J cm^{-2} rather than 150 J cm^{-2}) to produce similar PDT results [66]. The reasons for this effectiveness have been attributed to more suitable photophysical properties, such as increased molar absorption coefficient and infrared shift of the maximum absorption wavelength [67]. However, significant complications have been observed [68]. Given the highly efficient nature of this photosensitizer and the short time needed to create PDT, great care must be taken to protect all the body regions that are not to be submitted to therapy. If used with care, excellent clinical and cosmetic outcomes have been obtained for cutaneous squamous cell and basal cell lesions, head and neck lesions [62]. Concerning gastrointestinal tract treatment PDT can lead to fistulas and circumferential fibrosis in the oesophagus. The consequences of the photosensitizer potency modification using a shorter excitation wavelength (514 nm) with reduced penetration depth, made the treatment of early oesophageal lesions possible with minimal side effects [68]. This particular example highlights the advantage of PDT and especially the possible modulation of the treatment efficiency.

Unfortunately one of the limitations of Foscan®, is the pain experienced by the patients at the site of drug injection. This inconvenience is due to the highly hydrophobic nature of the drug and its practical insolubility in water [69], leading to a commercial formulation based on water-free ethanol and propylene glycol solution for intravenous injection. Finally, this drug exhibits poor tumour selectivity [70, 71].

A need for increased selectivity of photosensitizers for tumour over healthy tissue clearly appears. A targeted approach employs the utilization of ligands, which can bind specifically to neovascular endothelium or cellular markers to target tumour. While antibody-conjugates have received most of the attention [72–75], cellular transformations offer other potential targets to exploit. Growth factor receptors, hormonal, insulin, transferrin, glucose, folic acid, and low-density lipoprotein receptors have been investigated as cellular markers for targeting [76–79]. The latter, based on the assembly of the photosensitizer with low-density-lipoproteins (LDL), make LDL as molecular carriers [80]. This approach is similar to the encapsulation of photosensitizers within liposomal formulations decorated with targeting moiety on the surface of liposomes. In this way, the photosensitizer is not only addressed to the tumour cells, but owing to the large size of carrier, important amount of photosensitizer molecules can be delivered to a single target [79, 80]. Additionally, the advantage of such carriers is that it can favor the pharmacokinetics of hydrophobic photosensitizers, which otherwise suffer from aggregation in the bloodstream, followed by fast clearance from the body. The rationale for use of molecular delivery systems for photosensitizers is therefore similar to the delivery of chemotherapeutic drugs. Carrier-mediated delivery allows increased accumulation of molecule at the target site. Drug delivery approaches broaden the clinical repertoire of photosensitizers and reduce the precision that is needed in light delivery.

27.3.3.2 Light sources

With the wide range of clinical applications of PDT, the improvements in PDT light sources have occurred during the last decades. Light delivery to most anatomic sites is now achievable.

The basic requirements for light source used in PDT are (i) to cover the wavelength region of optimal absorption of a given photosensitizer and (ii) to generate adequate power at this wavelength. Typically, 1–5 W of usable power is required in the wavelength range 630–850 nm at irradiances of up to several hundred mW cm^{-2} so that treatment itself can be less time consuming, i.e. tens of minutes. Lasers, light emitting diodes (LEDs) and filtered lamps are the three main classes of PDT light sources used in clinic. The choice of a light source and delivery mode is usually based on the nature and location of disease. The main issue is, to deliver enough light from the source to the target tissue.

Depending on accessibility of treated surface, the light can be used directly from the source with no delivery system; via a lens system; or via a single fibre-optic (e.g. placed through the instrument channel of an endoscope) with or without a microlens tip. For intracavitary treatments, the goal is to disperse the light isotropically from the fibre. In the case of spherical cavities such as resection cavity after surgical debulking of brain tumours this can be achieved for example by using an inflatable balloon applicator in conjunction with laser and optical fibre. The applicator can be also shaped so that more irregular body cavities can be matched. The balloon material itself can scatter light instead of being filled with light-scattering liquid. However, for interstitial approaches, cylindrically diffusing fibres is preferred, so that large volume of tissue can be irradiated. In addition, depending on the size of treated area, if necessary, multiple diffusing fibres can be used simultaneously.

27.3.3.3 Tissue oxygenation

Along with drug and light, for most photosensitizer, tissue oxygenation is a crucial parameter, which influences the outcome of PDT. As mentioned above, the efficiency of PDT relies on singlet oxygen and other ROS formation. Under anoxic conditions, PDT is strongly reduced, although for some photosensitizers, *in vitro* results indicate oxygen-independent cytotoxic pathways [52, 81]. This remains to be proven *in vivo*.

Two types of hypoxia can be distinguished: the first is related to the tumour physiological development and in the second case oxygen depletion is induced by PDT itself. In PDT, two mechanisms can account for such a limitation: 1) the photochemical consumption of oxygen during the process and 2) the perturbation of the microvasculature,

An interesting consequence of the dependency of PDT on tissue oxygenation is the observed effect of fluence. In clinical situation, higher fluences have been thought to be favourable because of shorter irradiation time. However, there is now considerable evidence, that PDT is less effective under such conditions [82–84]. This effect has been attributed to oxygen depletion due to an important consumption during the photochemical reactions. Obviously, the photochemical consumption is exceeding the ability of microvasculature to deliver oxygen to the irradiated tissue areas. *In vitro* measurements and photochemical calculations confirmed that a limitation to singlet oxygen production should exist *in vivo*, when the oxygen supply is limited and when the photosensitizer tissue concentration and fluence are both high [82, 85–88]. The efficiency of PDT may be increased with prolonged treatment and reduced fluence [89]. Therefore, the choice of the proper treatment parameters (fluence or additionally, fractionation of light delivery) may help to limit the oxygen depletion.

It may be concluded, that despite all the developments in light source technology and photosensitizers, PDT dosimetry remains complex. Many factors may influence the treatment response, such as (i) inhomogeneous localization of photosensitizer within the treated site and its concentration time dependence, (ii) differences in optical properties of the organs, (iii) photosensitizer destruction during the therapy and (iv) oxygen depletion. Owing to their dynamic changes during PDT, real-time monitoring of all previous parameters is mandatory.

27.3.4 New developments in PDT

Although PDT offers numerous of advantages over other forms of cancer treatments, its main disadvantage is the poor accessibility of light to more deeply lying malignancies. While spatial control of the illumination provides specificity in tissue destruction, ironically, it seems to be also its limitation. For an efficient treatment response, whole target must be accessible to the radiation. The effective penetration depth is around 5 to 6 mm in the wavelength range from 700 to 800 nm [44, 49, 64, 89]. This means that photosensitizers with longer absorbing wavelengths and higher efficiency to absorb light at these wavelengths would be potentially more effective PDT agents. Unfortunately, this is not the case for most photosensitizers.

To overcome these limitations, external light sources are applied in an invasive fashion (interstitial treatments) in which optical fibres are placed intratumourally using needles. If larger area needs to be illuminated, multiple fibres can be inserted [90, 91]. However, without improved resources for dosimetry and light delivery, the efficiency of interstitial treatment can be poor and non-identified metastatic disease can be left untreated. Therefore, along with interstitial PDT, other approaches (two-photon excitation, chemiluminescence-based activation and/or the PDT activation using the X-ray excitation), have been suggested and some of them will be further discussed.

27.3.4.1 Chemiluminiscent approach

Molecular flashlight [92] or intracellular activation of PDT [93] are two different alternatives to external illumination sources, which are based on same noninvasive strategy, i.e. chemiluminescent activation. In 1994, the group of J. Petrich [92] proposed exploiting the existence of light mediated by the enzymatic oxidation of D-luciferin to oxyluciferin [94] for the excitation of a photosensitizer. In this study, hypericin was used as photosensitizer and its antiviral activity has been investigated. They demonstrated that co-incubating hypericin with all the necessary components induced a 10-fold decrease in the viral infectivity. Later on Theodossiou et al. [95] studied the PDT effect of Rose Bengal activated based in luciferase-transfected NIH 3T3 murine fibroblasts. Although the control groups exhibited 100% survival rate, the cells treated with Rose Bengal and D-luciferin exhibited 10% survival rate. The observed toxicity resulted mainly in singlet oxygen production, since lycopene presence (a singlet oxygen quencher) reversed to 90 % survival rate. This approach may lead to exploration of other chemiluminescent-systems.

27.3.4.2 Combinations of PDT with radiotherapy

Combination of nanoparticles absorbing X-rays and PDT is a novel approach to deep cancer treatment, proposed recently by Chen et al. [96]. In this concept, molecules of photosensitizer are coated onto or attached to X-ray luminescent nanoparticles. When the nanoparticle-photosensitizer conjugates are targeted to the tumour and stimulated by X-rays, the particles emits light that triggers PDT.

Owing to the deep penetration of X-rays, PDT may now become efficient for deep-seated malignancies. The advantages of this combination of two effective therapies are the activation by a single energy source and a lowered risk of radiation damage to healthy tissue. However, the success of this approach depends on the overlap of the emission spectrum of nanoparticle and the photosensitizer's absorption. Implicitly, an efficient energy transfer between the nanoparticle and photosenstizer is required.

To prove the concept, Chen has reported [97] synthesis of water-soluble scintillation nanoparticles using Tb and Ce doped LaF3 nanostructures. The X-ray luminescence, from LaF3:Ce3+, Tb3+ and LaF3:Tb3+ and generation of singlet oxygen by a photosensitizer in presence of X-Ray excited nanoparticles have been demonstrated [98].

The first results of this novel approach are promising and it is therefore of further interest to work on the development of a nanoparticle photosensitizer conjugates suitable for *in vivo* applications. Synchrotron radiation has its role in this approach, not only from the point of view of conjugate development, but more important, for the application of this conjugate for the treatment. We believe that combination of this new PDT system with one of the most exciting applications of synchrotron X-rays in medical research, namely the microbeam radiation therapy, can definitely lead to more efficient treatment responses.

References

1. Ortega R, Bohic S, Tucoulou R, Somogyi A, Deves G, Anal Chem **76**(2): 309–314 (2004)
2. Modok S, Scott R, Alderden RA, Hall MD, Mellor HR, Bohic S, Roose T, Hambley TW, Callaghan R, Br J Cancer **97**(2): 194–200 (2007)
3. Fisher SE, Chalmers JM, Marcelli A, Byrne HJ, Lyng F, Lasch P, Miller LM, Dumas P, Gardner P, Moss D, Biomedical applications of synchrotron infrared microspectroscopy: A practical approach. In: Moss D (ed.) RSC Analytical Spectroscopy Series. (2010)
4. Chio-Srichan S, Refregiers M, Jamme F, Kascakova S, Rouam V, Dumas P, Biochim Biophys Acta **1780**(5): 854–860 (2008)
5. Mamoon AM, Gamal-Eldeen AM, Ruppel ME, Smith RJ, Tsang T, Miller LM, Photodiagnosis Photodyn Ther **6**(2): 105–116 (2009)
6. Buriankova L, Nadova Z, Jancura D, Refregiers M, Yousef I, Mikes J, Miskovsky P, Laser Physics Letters **7**(8): 613–620 (2010)
7. Dahele M, Senan S, J R Coll Physicians Edinb **40**(2): 136–143 (2010)
8. Webb S, Br J Radiol **78**: S64–S72 (2005)
9. Bhide SA, Nutting CM, BMC Med **8**: 25 (2010)
10. Skowronska-Gardas A, Pedziwiatr K, Chojnacka M, Radiother Oncol **70**: 269–273 (2004)
11. Moretti R, Torre P, Antonello RM, Cattaruzza T, Cazzato G, Bava A, Ukmar M, Korczyn AD, J Neurol Sci **229–230**: 195–200 (2005)
12. Toogood AA, Growth Horm IGF Res (Suppl A) **14**: 118–124 (2004)
13. Ng WK, Curr Diagn Pathol **9**: 124–136 (2003)

14. St George EJ, Kudhail J, Perks J, Plowman PN, J Neurosurg **97**: 631–634 (2002)
15. Levegrun S, Hof H, Essig M, Schlegel W, Debus J, Int J Radiat Oncol Biol Phys **59**(3): 796–808 (2004)
16. Slatkin DN, Spanne P, Dilmanian FA, Sandborg M, Med Phys **19**: 1395–1400 (1992)
17. Slatkin DN, Spanne P, Dilmanian FA, Gebbers JO, Laissue JA, Proc Nat Acad Sci USA **92**(19): 8783–8787 (1995)
18. Thomlinson W, Berkvens P, Berruyer B, Blattmann H, Brauer-Krisch E, Brochard T, Charvet AM et al, Cell Mol Biol **46**(6): 1053–1063 (2000)
19. Brauer-Krisch E, Requardt H, Regnard P, Corde S, Siegbahn E, Leduc G, Brochard T, Blattmann H, Laiissue J, Bravin A, Phys Med Biol **50**:3103–3111 (2005)
20. Serduc R, Verant P, Vial JC, farion R, Rocas L, Remy C, Fadlallah T, Brauer E, Bravin A, Laissue J, Blattmann H, van der Sanden B, Int J Radiat Oncol Biol Phys **64**: 1519–1527 (2006)
21. Serduc R, van de Looij Y, Francony G, Verdonck O, van der Sanden B, Laissue J, Farion R, Brauer-Krisch E et al, Phys Med Biol **53**: 1153–1166 (2008)
22. Dilmanian FA, Qu Y, Liu S, Cool CD, Gilbert J, Hainfeld JF, Kruse CA, Laterra J, Lenihan D, Nawrocky MM, Pappas G et al, Nucl Instrum Methods Phys Res A **548**(1–2): 30–37 (2005)
23. Laissue JA, Geiser G, Spanne PO, Dilmanian FA, Gebbers JO, Geiser M, Wu XY, Makar MS, Micca PL, Nawrocky MM, Joel DD, Slatkin DN, Int J Cancer **78**(5): 654–660 (1998)
24. Dilmanian FA, Zhong Z, Bacarian T, Benveniste H, Romanelli P, Wang R, Welwart J, Yuasa T, Rosen EM, Anschel DJ, Proc Natl Acad Sci USA **103**(25): 9709–9714 (2006)
25. Norman A, Iwamoto KS, Cochran ST, Invest Radiol **26**: 120–121 (1991)
26. Mesa AV, Norman A, Solberg TD, Demarco JJ, Smathers JB, Phys Med Biol **44**: 1955–1968 (1999)
27. Douple EB, Richmond RC, Br J Cancer Suppl **3**: 98–102 (1978)
28. Techer BA, Holden SA, Radiat Res **109**(1): 58–67 (1987)
29. Hall MD, Dillon CT, Zhang M, Beale P, Cai Z, Lai B, Stampfl AP, Hambley TW, J Biol Inorg Chem **8**(7): 726–732 (2003)
30. Ferrer S, Med Clin (Barc) **124**(7): 271–273 (2005)
31. Bencokova Z, Balosso J, Foray N, J Synchrotron Radiat **15**(Pt1): 74–85 (2008)
32. Biston MC, Joubert A, Adam JF, Elleaume H, Bohic S, Charvet AM, Esteve F, Foray N, Balosso J, Cancer res **64**(7): 2317–2323 (2004)
33. Rousseau J, Barth RF, Fernandez M, Adam JF, Balosso J, Esteve F, Elleaume H, J Neurooncol **98**(3): 287–295 (2010)
34. Boudou C, Balosso J, Esteve F, Elleaume, Phys Med Biol **50**: 4841–4851 (2005)
35. De Stasio G, Rajesh D, Ford JM, Daniels MJ, Erhardt RJ, Frazer BH, Tyliszczak T, Gilles MK, Conhaim RL, Howard SP, Fowler JF, Esteve F, Mehta MP, Clin Cancer Res **12**(1): 206–213 (2006)
36. Regnard P, Brauer-Krisch E, Tropes I, Keyrilainen J, Bravin A, Le Duc G, Eur J Radiol **68** (3 Suppl): S151–S155 (2008)
37. Prezado Y, Fois G, Le Duc G, Bravin A, Med Phys **36**(8): 3568–3574 (2009)
38. Deman P, Edouard M, Besse S, Vautrin M, Elleaume H, Adam JF, Esteve F, La Revue de Medecine Interne **31**(8): 586–589 (2010)
39. Wagner HP, Support Care Cancer 7: 293–294 (1999)
40. Ribi K, Relly C, Landolt MA, Alber FD, Boltshauser E, Grotzer MA, Neuropediatrics **36**: 357–365 (2005)
41. Hasan T., Ortel B, Solban N, Pogue B, Photodynamic therapy of cancer. In: Kufe, Bast, Hait, Hong, Pollock, Weichselbaum et al (ed.) Cancer medicine, 7th edn. Hamilton (Ontario): B. C. Decker, Inc. (2006)
42. Dougherty TJ, J Clin Laser Med Surg **20**: 3–7 (2002)
43. Hamblin MR, Mroz P, Advances in photodynamic therapy: basic, translational and clinical, Artech House, Norwood (2008)
44. MacDonald IJ, Dougherty TJ, J Porph Pthalocyn **5**(2): 105–129 (2001)
45. Hamblin MR, Mroz P, History of PDT: The first hundred years. In: Hamblin MR and Mroz P (ed.) Advances in photodynamic therapy: basic, translational and clinical, Artech House, Norwood (2008)

46. Doughery TJ, Henderson BW, Gomer CJ, Jori G, Kessel D, Korbelik M, Moan J, Peng Q, J Natl Cancer Inst **90** (12): 889–905 (1998)
47. Raab C (1900) Uber die wirkung fluoreszierender stoffe auf infusoria. Z Biol **39**: 524–526
48. Brown SB, Brown EA, Walker I, Lancet Oncol **5**: 497–508 (2004)
49. Wilson BC, Patterson MS, Phys Med Biol **53**: R61–R109 (2008)
50. Redmond RW, Photophysics and photochemistry in photodynamic therapy. In: Hamblin MR and Mroz P (ed.) Advances in photodynamic therapy: basic, translational and clinical, Artech House, Norwood (2008)
51. Foote CS, Mechanisms of photo-oxydation. In: Doiron DR, Gomer CJ (ed.) Porphyrin localization and treatment of tumours, Alain R. Liss, New York (1984)
52. Park J, English DS, Wannemuehler S, Carpenter S, Petrich JW, Photochem Photobiol **68**: 593–597 (1998)
53. Gai F, Fehr MJ, Petrich JW, J Am Chem Soc **115**: 3384–3385 (1993)
54. Sureau F, Miskovsky P, Chinsky L, Turpin PY, J Am Chem Soc **118**: 9484–9487 (1996)
55. Mirossay L, Mirossay A, Kocisova E, Radvakova I, Miskovsky P, Mojzis J, Physiol Res **48**(2): 135–141 (1999)
56. Kocanova S, Miskovsky P, Hornakova T, Hritz J, Jancura D, Chorvat D, Mateasik A, Ulicny J, Refregiers M, Maurizot JC, Photochem Photobiol **82**(3): 720–728 (2006)
57. Miskovsky P, Current Drug Targets **3**: 55–84 (2002)
58. Moan J, J Photochem Photobiol B: Biol **6**: 343–344 (1990)
59. Moan J, Berg K, Photochem Photobiol **53**: 549–553 (1991)
60. Bressler NM, Arch Opthalomology **119**: 198–207 (2001)
61. Gomer CJ, Luna M, Ferrario A, Wong S, Fisher AM, Rucker N, J Clin Laser Med Surg **14**: 315–321 (1996)
62. Dolmans DEJGJ, Fukumura D, Jain RK, Nat Rev Cancer **3**: 380–387 (2003)
63. Ethirajan M, Saenz C, Gupta A, Dobhal MP, Pandey RK, Photosensitizers for photodynamic therapy and imaging In: Hamblin MR and Mroz P (ed.) Advances in photodynamic therapy: basic, translational and clinical, Artech House, Norwood (2008)
64. Star WM, Phys Med Biol **42**(5): 763–787 (1997)
65. Schmidt-Erfurth U, Miller JW, Sickenberg M et al, Arch Ophthalmology **117**: 1177–1187 (1999)
66. Allison RR, Downie GH, Cuenca R, Hu XH, Childs CJH, Sibata C, Photodiagnosis and Photodynamic Therapy **1**: 27–42 (2004)
67. Jones HJ, Vernon DI, Brown SB, Br J Cancer **89**: 398–404 (2003)
68. Grosjean P, Savary J, Wagniers G, Lasers Med Sci **8**: 235–243 (1993)
69. Kascakova S, Kruijt B, de Bruijn HS, van der Ploeg-van den Heuvel A, Robinson DJ, Sterenborg HJCM, Amelink A, J Photochem Photobiol B: Biol **91**: 99–107 (2008)
70. Kascakova S, de Visscher S, Kruijt B, de Bruijn HS, van der Ploeg-van den Heuvel A, Sterenborg HJCM, Witjes MJH, Amelink A, Robinson DJ, Lasers Med Sci, in press
71. Triesscheijn M, Ruevekamp M, Out R, van Berkel TJC, Schellens J, Baas P, Stewart FA, Cancer Chemother Pharmacol **60**: 113–122 (2007)
72. Del Governatore M, Hamblin MR, Shea ChR, Rizvi I, Molpus KG, Tanabe KK, Hasan T, Cancer Res **60**: 4200–4205 (2000)
73. Del Governatore M, Hamblin MR, Piccinini EE, Ugolini G, Hasan T, British Journal of Cancer **82**(1): 56–64 (2000)
74. Oseroff AR, Ohuoha D, Hasan T, Bommer JC, Yarmush ML, Proc Natl Acad Sci USA **83**: 8744–8748 (1986)
75. Soukos NS, Hamblin MR, Keel S, Fabian RL, Deutsch TF, Hasan T, Cancer Res **61**: 4490–4496 (2001)
76. Chen B, Pogue BW, Hoopes PJ, Hasan T, Crit Rev Eukaryot Gene Expr **16**(4): 279–305 (2006)
77. Sharman WM, van Lier JE, Allen CM, Advanced Drug Delivery Reviews **56**: 53–76 (2004)
78. Rosenkranz AA, Jans DA, Sobolev AS, Immunol Cell Biol **78**: 452–464 (2000)
79. Kascakova S, Nadova Z, Mateasik A, Mikes J, Huntosova V, Refregiers M, Sureau F, Maurizot JC, Miskovsky P, Jancura D, Photochem Photobiol **84**: 120–127 (2008)

80. Kascakova S, Refregiers M, Jancura D, Sureau F, Maurizot JC, Miskovsky P, Photochem Photobiol **81**: 1395–1403 (2005)
81. Foote CS, Photochem Photobiol **54**(5): 659 (1991)
82. Foster TH, Murant RS, Bryant RG, Knox RS, Gibson SL, Hilf R, Radiat Res **126**: 296–303 (1991)
83. Robinson DJ, de Bruijn HS, van der Veen N, Stringer MR, Brown SB, Star WM, Photochem Photobiol **67**: 140–149 (1998)
84. Busch TM, Lasers Surg Med **38**: 494–499 (2006)
85. Foster TH, Gibson SL, Gao L, Hilf R, Proc SPIE **1645**: 104–114 (1992)
86. Foster TH, Gao L, Radiat Res **130**: 379–383 (1992)
87. Foster TH, Hartley DF, Nichols MG, Hilf R, Cancer Res **53**: 1249–1254 (1993)
88. Nichols MG, Foster TH, Phys Med Biol **39**: 2161–2181 (1994)
89. Grossweiner LI, Hill JH, Lobraico RV, Photochem Photobiol **46**: 911–917 (1987)
90. Kruijt B, van der Ploeg-van den Heuvel A, de Bruijn HS, Sterenborg HJ, Amelink A, Robinson DJ, Lasers Surg Med **41**(9): 653–664 (2009)
91. Zhu TC, Finlay JC, Hahn SM, J Photochem Photobiol B: Biol **79**(3): 231–241 (2005)
92. Carpenter S, Fehr MJ, Kraus GA, Petrich JW, Proc Natl Acad Sci USA **91**: 12273–12277 (1994)
93. Laptev R, Nisnevitch M, Siboni G, Malik Z, Firer MA, British Journal of Cancer **95**: 189–196 (2006)
94. McElroy WD, DeLuca M, Bioluminescence and chemiluminescence. Academic Press: New-York (1981)
95. Theodossiou T, Hothersall JS, Woods EA, Okkenhaug K, Jacobson J, MacRobert AJ, Can Res **63**: 1818–1821 (2003)
96. Chen W, Zhang J, Journal of Nanoscience and Nanotechnology **6**: 1159–1166. (2006)
97. Liu Y, Chen W, Wang S, Joly AG, Westcott S, Woo BK, X-ray luminescence of $LaF_3 : Tb^{3+}$ and $LaF_3 : Ce^{3+}$, J Appl Phys **103** (063105):1–7 (2008)
98. Liu Y, Chen W, Wang S, Joly AG, Appl Phys Lett **92** (043901): 1–3 (2008)

Chapter 28
Auger Emitting Radiopharmaceuticals for Cancer Therapy

Nadia Falzone, Bart Cornelissen, and Katherine A. Vallis

Abstract Radionuclides that emit Auger electrons have been of particular interest as therapeutic agents. This is primarily due to the short range in tissue, controlled linear paths and high linear energy transfer of these particles. Taking into consideration that ionizations are clustered within several cubic nanometers around the point of decay the possibility of incorporating an Auger emitter in close proximity to the cancer cell DNA has immense therapeutic potential thus making nuclear targeted Auger-electron emitters ideal for precise targeting of cancer cells. Furthermore, many Auger-electron emitters also emit γ-radiation, this property makes Auger emitting radionuclides a very attractive option as therapeutic and diagnostic agents in the molecular imaging and management of tumors. The first requirement for the delivery of Auger emitting nuclides is the definition of suitable tumor-selective delivery vehicles to avoid normal tissue toxicity. One of the main challenges of targeted radionuclide therapy remains in matching the physical and chemical characteristics of the radionuclide and targeting moiety with the clinical character of the tumor. Molecules and molecular targets that have been used in the past can be classified according to the carrier molecule used to deliver the Auger-electron-emitting radionuclide. These include (1) antibodies, (2) peptides, (3) small molecules, (4) oligonucleotides and peptide nucleic acids (PNAs), (5) proteins, and (6) nanoparticles. The efficacy of targeted radionuclide therapy depends greatly on the ability to increase intranuclear incorporation of the radiopharmaceutical without compromising toxicity. Several strategies to achieve this goal have been proposed in literature. The possibility of transferring tumor therapy based on the emission of Auger electrons from experimental models to patients has vast therapeutic potential, and remains a field of intense research.

N. Falzone (✉) • B. Cornelissen • K.A. Vallis
Gray Institute For Radiation Oncology and Biology, Department of Oncology, Radiobiology Research Institute, Churchill Hospital, Headington, Oxford OX3 7LJ, United Kingdom
e-mail: nadia.falzone@oncology.ox.ac.uk

G.G. Gómez-Tejedor and M.C. Fuss (eds.), *Radiation Damage in Biomolecular Systems*, Biological and Medical Physics, Biomedical Engineering,
DOI 10.1007/978-94-007-2564-5_28,

28.1 Introduction and present status

Many localized primary cancers can be treated effectively by surgery, radiotherapy or a combination of both. Once a tumor has metastasized or there is a possibility that it has done so with the dissemination of tumor cells beyond the primary cancer site, then chemotherapy is usually the treatment modality of choice [1, 2]. The administration of anticancer drugs has several drawbacks, such as the lack of selectivity, toxicity to normal cells, fast elimination from the blood circulation, and the acquired or intrinsic multi-drug resistance of cancer cells [3]. Therefore, in the last two decades, cancer research has turned to a more selective, targeted approach, focused on the development of anticancer therapies with improved efficacy and reduced peripheral toxicity. Present-day treatment modalities include immunotherapy, anti-angiogenesis therapy, molecularly-targeted agents, gene-therapy, radionuclide therapy or combinations of these therapies [4].

The notion of selectively delivering radionuclides that emit charged particles such as Auger electrons, β or α-particles, to cancer cells via a targeting moiety whilst sparing normal tissue began at the turn of the twentieth century when Ehrlich *et al.* [5] postulated the idea of the "Magic Bullet". This idea only gained clinical application when Pressman and Keighley introduced the concept of radiolabelled anti-tumor antibodies for cancer detection [6]. Pressman *et al.* showed that monoclonal antibodies (mAbs) labeled with radioactive iodine could be used in animals as a diagnostic agent [7]. The concept was further expanded by Bale *et al.* who demonstrated that tumor localizing radio-labeled antibodies could be used as a therapeutic agent for treating experimental neoplasms. The modern era of targeted cancer therapy commenced in 1975 with the seminal publication of Kohler and Milstein [8] describing the technology for mass production of mAbs with high specificity towards tumor-associated antigens, also referred to as the "hybridoma technology". This opened the door to preclinical and clinical research in targeted radiotherapy using mAbs, a field now known as radio-immunotherapy (RIT). Since the 1980s radionuclide therapy gained renewed interest and has since become an area of intense research [9, 10], with other targeting strategies apart from mAb being investigated including peptides, small molecules and nano-materials [11–14].

Radionuclides that emit Auger electrons have been of particular interest as therapeutic agents (Table 28.1). This is primarily due to the short range in tissue (fraction of a nanometer to $\sim 1\,\mu m$), controlled linear paths and high linear energy transfer (LET) (~ 4 to $\sim 25\,keV\ \mu m-^{1}$) of these particles [15]. Taking into consideration that ionizations are clustered within several cubic nanometers around the point of decay [16] the possibility of incorporating an Auger emitter in close proximity to the cancer cell DNA has immense therapeutic potential thus making nuclear targeted Auger-electron emitters ideal for precise targeting of cancer cells [17]. In spite of this studies incorporating Auger radiation therapy remain mostly pre-clinical [18–23]. This can in part be ascribed to obstacles such as identifying suitable tumor selective delivery vehicles and increasing tumor retention [17, 24].

Table 28.1 Characteristics of some Auger emitting radionuclides adapted from Cornelissen and Vallis [139]. The Auger yield is the mean number of Auger and Coster-Kronig electrons emitted per decay. The Auger energy is the average kinetic energy of Auger and Coster-Kronig electrons emitted per decay

Radionuclide	Half-life (days)	Auger yield	Auger energy (keV)
^{67}Ga	3.26	4.7	6.26
^{99m}Tc	0.25	4.0	0.89
^{111}In	2.8	14.7	6.75
^{123}I	0.55	14.9	7.42
^{125}I	59.4	24.9	12.24
^{195m}Pt	4.02	33.0	22.53
^{201}Tl	3.04	36.9	15.27

However, one of the main challenges of targeted radionuclide therapy remains in matching the physical and chemical characteristics of the radionuclide and targeting moiety with the clinical character of the tumor.

28.2 Selection of an auger electron-emitting radionuclide for therapy

Although most clinical targeted radionuclide therapies are based on β^- emitters, Auger electron emitters could provide a therapeutic advantage [25]. Auger electrons that decay close to DNA, have relative biological efficacy similar to that of α particles [26]. Compared to α radiation, however, Auger electrons have shorter path lengths making them less toxic when they decay outside the target cell [17]. Therefore Auger-emitters have higher anti-tumor efficacy associated with lower toxicity. There are several factors that should be taken into consideration when selecting an Auger-electron emitting radionuclide for targeted radiotherapy. The efficacy of an ideal radionuclide depends largely on (a) the number of electrons emitted per decay, (b) the ratio of penetrating (X- and γ-rays) to non-penetrating (electron or β-particle) forms of radiation, (c) the physical half-live vs. effective half-life (d) suitable chemistry for the radiolabelling process.

28.2.1 Electron yield

The electron yield of an Auger emitter includes both Auger and internal conversion (IC) electrons as well as the total energy per decay. Auger electron emitters can be divided in two major groups (Table 28.1), namely halogens (^{125}I, ^{123}I, ^{77}Br, ^{88m}Br) and metals (^{201}Tl, ^{195m}Pt, ^{193m}Pt, ^{111}In, ^{114m}In, ^{99m}Tc, ^{67}Ga, ^{55}Fe and ^{51}Cr). However not all the Auger electron emitters are suitable for therapeutic use owing to their very long half-lives (e.g. $T_{1/2}(^{55}\text{Fe}) = 2.7\,\text{y}$). Among the suitable radioisotopes, ^{201}Tl has the greatest yield of electrons with 36.9 electrons/decay and average energy of 15.27 keV. Other suitable Auger emitting radio-isotopes with high yields include, ^{195m}Pt and ^{125}I with 33 electrons/decay (average energy 22.53 keV) and 25.8 electrons/decay (average energy 19.4 keV) respectively. ^{111}In and ^{123}I have intermediate electron yield (14.7 and 14.9 electrons/decay, respectively) associated with average energies of 32.7 and 27.6 keV, respectively. Furthermore, the amount of energy deposited per decay in a five nanometer sphere is much greater for ^{195m}Pt (2000 eV) than for either ^{125}I (1000 eV), ^{123}I (550 eV), or ^{111}I (450 eV) [27].

The total energy emitted by ^{67}Ga (34.4 keV) is similar to that of ^{111}In or ^{123}I, but is carried by fewer electrons (5.0 electrons/decay) thus providing a greater overall average electron energy. In particular, ^{67}Ga emits an abundant number of electrons with energy of 8.4 keV that have a range in tissue of 2.0 μm, whereas the range of 99% of the electrons emitted by ^{111}In or ^{125}I is much less than 1 μm [28]. It has been suggested that ^{67}Ga is potentially a more useful Auger electron emitter than ^{111}In or ^{125}I for targeted radiotherapy, as nuclear incorporation is not a prerequisite due to its higher range as is the case with ^{111}In and ^{125}I [29]. However, crossfire to non-targeted cells due to its longer range could to lead to undesired toxicity when targeting small metastasis or disseminated cancer cells. ^{99m}Tc provides a similar electron yield (5.1 electrons/decay) as ^{67}Ga but two-fold lower total energy (16.3 keV) [30].

28.2.2 Ratio of penetrating to non-penetrating radiation

Many radionuclides do not exclusively emit low energy electrons during decay, but also emit γ-radiation. The γ-radiation from ^{111}In, ^{123}I, or ^{99m}Tc has long since been used in nuclear medicine to visualize tumor and normal tissue distribution of the radiotherapeutic agent in patients and in preclinical animal models by single photon emission computed tomography (SPECT) [30]. However, the moderate-high energy but low LET γ-emissions from ^{111}In (171 and 245 keV) can irradiate and potentially kill non-targeted normal cells. Therefore it is important to consider the ratio of penetrating (X- and γ-) to non-penetrating (electron or β-particle) forms of radiation (p/e), emitted from Auger electron emitters [31]. The p/e ratios are also reflected in the percentage of total energy that is emitted as Auger or IC electrons for the different radionuclides.

28.2.3 Physical and effective half live

The physical half-live of the radionuclide should closely match the effective half-life [1, 15]. A too short physical half-live would necessitate repeated administration to ensure adequate dose delivered to the tumor before biological clearance. On the other hand a very long physical half-live could lead to under dosing of the tumor if an inadequate number of decays occur before excretion. Analogues to this argument are that of the dose rate effect [32]. Low dose rates are less damaging than high dose rates, as with fractionated external beam therapy, the total dose from continues low dose radionuclide therapy is less biologically effective than a single dose of the same magnitude. Therefore, the most suitable physical half-live should vary between a few hours to that of a few days when targeting disseminated cancer cells. On the other hand, longer physical half-lives might be desirable in the treatment of solid tumors where high uptake is required. Furthermore, stable non-toxic decay products are desirable to minimize normal tissue toxicity.

28.2.4 Suitable chemistry for the radiolabelling process

Production of Auger electron emitters for therapy should be economically viable and allow preparation to high specific activity and purity. Efficient incorporation into a selective carrier molecule is also a prerequisite. Once inside the target tissue, the selective carrier molecule should be able to associate with the DNA complex for a time corresponding to the radionuclide half-life [33]. Prolonged intracellular retention can be achieved by using various residualizing agents for indirect halogen labeling [34]. In addition, cellular excretion can be limited if the radionuclides are of metal type, e.g. indium or platinum, this is due to the intracellular retention of metal containing catabolic products [35, 36].

28.3 Targeting strategies - the clinical experience

The first requirement for the delivery of Auger emitting nuclides is the definition of suitable tumor-selective delivery vehicles to avoid normal tissue toxicity [17]. The explosive growth of antibody targeted cancer therapy has expanded the development of novel pharmaceuticals for targeted radionuclide therapy. After the FDA approval of the first monoclonal antibodies for clinical use [37] many other targeting strategies have been explored for targeted radionuclide therapy. Molecules and molecular targets that have been used in the past can be classified according to the carrier molecule used to deliver the Auger-electron-emitting radionuclide. These include (1) antibodies, (2) peptides, (3) small molecules, (4) oligonucleotides and peptide nucleic acids (PNAs), (5) proteins, and (6) nanoparticles. A schematic overview of the trafficking and nuclear localization of Auger-electron-emitters is shown in Fig. 28.1. Auger electron emitters can only be considered as an effective compliment to other treatment modalities or a possible alternative to chemotherapy if targeted delivery of the radionuclide complex could accomplish total eradication of disseminated tumor cells and micro-metastasis. From the vast number of cell culture studies performed it has been shown that nuclear localization of the radionuclide is a key factor for the induction of high LET type cytotoxicity in mammalian cells [38]. Therefore, only targets on cancer cells that

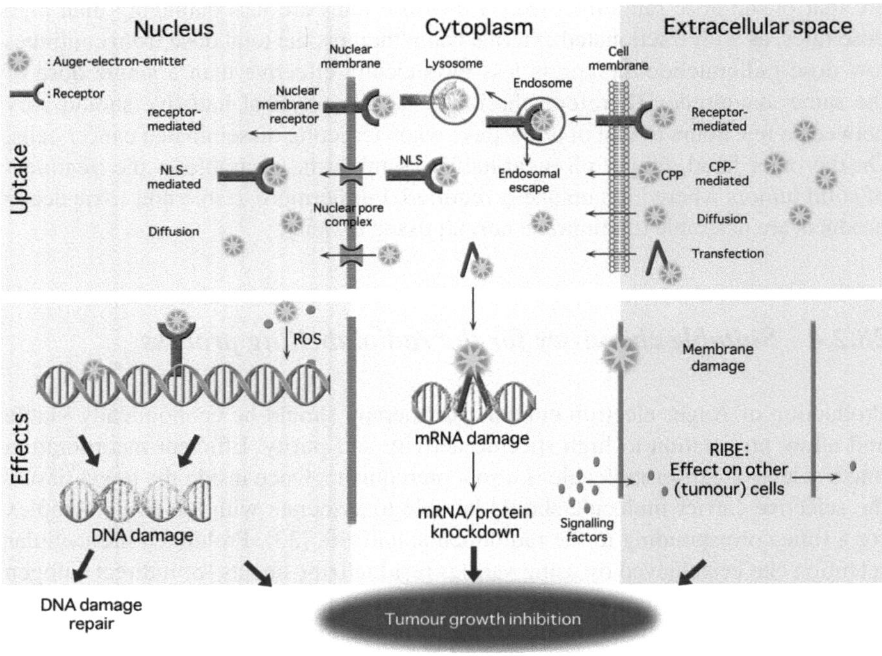

Fig. 28.1 A schematic overview of the trafficking and nuclear localization of Auger-electron-emitting radiopharmaceuticals (AER). AER can damage and kill cells by inducing membrane damage, or by internalizing into the cell. Internalization can occur via receptor- or cell-penetrating-peptide- mediated endocytosis, or happen via diffusion or transfection. Once in the cell, AER internalized by endocytosis escape from the endosome, or are degraded by lysosomes. Nuclear localization takes place via receptor-mediated nuclear transport, or via the nuclear pore complex, which can happen via diffusion for smaller molecules, or via nuclear localization sequence (NLS)-mediated active transport. AER can bind covalently to DNA, intercalate in the DNA helix, interact with chromatin via receptors and scaffolding proteins, or cause ROS species. All of the above results in DNA damage, which is either repaired, or leads to cytotoxicity. Even in cells which have not been affected directly by AER, the radiation-induced bystander effect (RIBE) can induce cytotoxicity via excreted signaling factors [139]

have shown promise in laboratory models have been exploited for Auger electron radiotherapy. We restrict the discussion to those agents that have been tested in clinical trials.

28.3.1 Radioimmunotherapy

One of the first agents to be evaluated against tumor antigens was monoclonal antibodies. Auger electron emitters can be used to label mAbs that target either the cell surface or the cytoplasm through a receptor-mediated internalizing process.

Several approaches have been reported exploiting the overexpression of antigens and or receptors in specific cancers, these include targeting the epidermal growth factor receptor (EGFR) [39, 40], the human epidermal growth factor receptor (HER2) [41–43] or the glucagon-like receptor (GLP-1R) [44] as well as numerous cell surface epitopes [45,46] to name but a few. Furthermore, Auger emitter labeled mAbs may be better suited to treat hematological malignancies compared to solid tumors due to their limited range, preliminary studies have shown promising results [47, 48]. A major disadvantage of radioimmunotherapy (RIT) is that the antigen being targeted is not always expressed homogeneously in the cancer cell resulting in heterogeneous dose distribution in the target tissue [49]. Another limitation, highlighted by the results of clinical studies using murine mAbs, was the production of human anti-murine immunoglobulin antibodies (HAMA), after repeated administration [37]. This limitation can be addressed by chemical modification of the mAbs, through production of chimeric mAbs, or complete humanization of the protein [50]. Another consideration, in the case of ^{125}I, it is not possible to deliver the nuclide to the nucleus using directly radiolabelled antibodies which bind to cell surface antigens as ^{125}I-labeled monoclonal antibodies are catabolized in lysosomes ultimately yielding free ^{125}I-iodide which is rapidly excreted from the cells [49]. Therefore a molecular target that not only internalizes but is targeted to the nucleus, or more specifically to the DNA is a requirement for inducing lethal lesions.

28.3.1.1 Epidermal growth factor receptor targeting

The EGFR is a cell surface signaling glycoprotein and the first member of the Type 1 family of transmembrane peptide growth factor receptors which also includes the HER2, HER3, and HER4 receptors [51,52]. Overexpression of EGFR is common in many malignancies including cancers of the breast, ovary, head and neck, lung, bladder, and colon as well as in glioblastomas [53]. Apart from overexpression (up to 100 fold in some tumor cells compared to normal cells), EGFR also plays a role in proliferation and is associated with poor prognosis, which taken together makes EGFR an attractive therapeutic target [54]. Therapeutic interventions exploiting the overexpression of EGFR include monoclonal antibodies (mAbs) that block ligand binding and tyrosine kinase inhibitors that interfere with receptor autophosphorylation and propagation of mitogenic signaling [55].

An extensively studied example of an Auger emitter using this targeting strategy is that of ^{125}I-labelled mAb-425 [56–60]. The internalizing anti-EGFR antibody mAb-425 was originally studied in the 90's as a new potential therapy for glioblastoma multiforme (GBM), an EGFR-positive aggressive brain tumor with limited therapeutic options and high morbidity. mAb-425, proved successful in *in vitro* assays, as it reduced clonogenic survival of EGFR-overexpressing cells, in contrast to cells with lower EGFR membrane expression [61]. ^{125}I-mAb425 as well as ^{131}I-mAb-425 reduced tumor growth in mouse xenograft models. In a large phase II clinical trial, a total of 192 patients with GBM were treated intravenously with ^{125}I-mAb-425 following surgery and radiation therapy [62]. Treatment with

^{125}I-mAb-425 alone resulted in a median survival of 14.5 months, compared to the control arm of patients receiving standard care, but not ^{125}I-mAb-425, which had a median survival of just 10.2 months. Combination of ^{125}I-mAb-425 and temozolomide (a chemotherapeutic) provided the greatest survival benefit with a median survival of 20.4 months. This combination was safe and well tolerated with little added toxicity. Nuclear localization of the antibody was not reported.

28.3.1.2 Targeting cell surface epitopes

The A33 antigen, has shown some promise as a target for RIT in the treatment of colorectal cancer. Not only is the antigen homogeneously expressed by >95% of colon cancers [63], but phase I/II clinical trials with ^{125}I murine mAb in colon carcinoma showed favorable biodistribution and evidence of tumor response [64]. Of the 20 patients with advanced chemotherapy-resistant colon cancer treated with a single ^{125}I-mAb A33 dose, all showed localization of ^{125}I to sites of disease and sufficient retention even after a period of 4–6 weeks to make external imaging possible. Furthermore, no dose limiting toxicity was observed, however immunogenicity precluded repeat dosing. Studies to evaluate the efficacy of humanized mAb A33 in patients with colorectal carcinoma continue [65–67]. In a separate study, ^{125}I-labeled CO17-1A mAbs, recognizing a tumor associated epitope on colorectal cancer was administered in escalating single doses to 28 patients with metastatic colorectal cancer, although no severe toxicity was noted no objective responses were observed [68].

All the targeting strategies mentioned so far make use of an unmodified antibody, whether internalizing or not. However, introduction of a nuclear localizing sequence (NLS) on the antibody would result in greater nuclear localization, closer to the DNA, which could increase cytotoxicity, and at the same time protect the radioimmunoconstruct from lysosomal degradation. This strategy was explored in a few studies by the Reilly group [43,69–74]. In a paper by Chen *et al.* conjugation of the SV-40 large T antigen NLS to an ^{111}In-labelled anti-CD33 antibody was reported for the first time [75]. Conjugation of eight NLS peptides to the antibody increased nuclear localisation 8-fold and decreased clonogenic survival of CD33 positive cells. In 7 out of 9 cases, ^{111}In-anti-CD33-NLS reduced clonogenic survival of cells obtained from plasma of myeloid leukemia (AML) patients. Kersemans *et al.* demonstrated that ^{111}In-anti-CD33-NLS could reduce clonogenic survival of multidrug resistant AML cell lines [76]. Similarly, Costantini *et al.* showed that addition of up to six NLS peptides to the ^{111}In-labelled anti-HER2 antibody trastuzumab (Herceptin) increased internalisation and nuclear localisation of the antibody and decreased clonogenic survival of HER-2 positive cells [77]. ^{111}In-trastuzumab-NLS was shown to overcome IGF-receptor-induced trastuzumab resistance in breast cancer cells *in vitro* and in a mouse xenograft model [78]. A major shortcoming of conventional RIT however, remains the slow extravasation and clearance of intact antibodies from the blood [79], often leading to myelo suppression [30]. Strategies to overcome this include radiolabelled antibody fragments and analogs, such as Fab fragments and affibody molecules, in place of whole antibodies [38]. Alternatively, overexpression of specific cell-surface receptors by certain tumors can be exploited using radiolabelled peptides.

28.3.2 Peptide-receptor radionuclide therapy

Radiolabelled peptides have been studied extensively for use in radionuclide therapy. This is due to their excellent binding efficiencies, selectivity, and favorable pharmacokinetic characteristics Tumor targeting with peptides has found both diagnostic and therapeutic applications. In the case of cancer diagnosis and radiotherapy, a radioligand is usually attached to the regulatory peptide carrier by the aid of a chelator. A wide variety of chelating agents (e.g., diethylenetriamine-pentaacetic acid [DTPA] and 1,4,7,10-tetraazacyclododecane- 1,4,7,10-tetraacetic acid [DOTA]) have been developed for convenient radiolabeling of peptides. The most studied peptides for peptide-receptor radionuclide therapy (PPRT) are the radiolabeled somatostatin analogs. Apart from somatostatin analogs, many other peptides have been developed for PRRT, including cholecystokinin-2/gastrin receptors (CCK-2r), gastrin-releasing peptide receptors (GRP-r), vasoactive intestinal peptide receptors-1 (VPAC1-r), melanocortin-1 receptors (MCR-1r), neurotensin receptors-1 (NTR-1), neuropeptide Y-Y1 receptors (NP–Y Y1r), $\alpha\nu\beta3$ integrins, gonadotropin-releasing hormone receptors (GnRHr-I), and glucagon-like peptide-1 receptors (GLP-1r). These receptors are overexpressed on various tumor types and can be targeted with peptide analogs with high affinity [80].

28.3.2.1 Somatostatin Targeting

Somatostatin (SMS), is a naturally occurring cyclic 14- or 28-amino acid peptide [81, 82] which binds to the somatostatin receptors-SSTRs (sst_1, ss_{t2}, sst_3, sst_4, and sst_5). SSTRs are expressed in most gastroenteropancreatic neuroendocrine tumors (GEPNETs) as well as in some other malignancies including breast cancer, neuroblastoma, and lymphomas [14, 83]. This renders SSTRs as ideal targets for peptide-receptor radionuclide therapy (PRRT). However, few Auger emitter complexes for PRRT have been tested in clinical trials.

^{111}In- DTPA-derivatised octreotide (Octreoscan®, Covidien, Hazelwood, MO) was developed in the late 1980s for the diagnostic imaging of somatostatin (sst_2) overexpressing tumors taking advantage of the two γ-photon emissions of ^{111}In; 171 keV (90%) and 245 keV (94%) [84] Since then it has been used for PRRT exploiting the Auger electron emissions, rather than the γ-emissions used for tumor imaging [85–89] Although in vitro experiments with ^{111}In-DTPA-octreotide noted a therapeutic effect dependent on the internalization ability of the complex [90], limited clinical success has been noted. The lower efficacy reported in larger tumors is most likely related to the limited range of Auger electrons [91], as the decay of

^{111}In has to occur close to the cell nucleus to be tumoricidal. It has been suggested that ^{111}In-DTPA-octreotide is not ideal for PRRT [83]. Nonetheless ^{111}In-DTPA-octreotide treatment could potentially be achieved by selecting patients with lower tumor volumes, or by increasing its uptake into the nucleus of SSTR-positive tumor cells.

A possible strategy was explored by Ginj *et al.* who reported the use of an octreotide analogue (TOC) conjugated to the SV40 large-T antigen nuclear localization sequence (NLS), labeled with ^{111}In-DOTA [92]. ^{111}In-NLS-DOTA-TOC showed enhanced cellular uptake and a 6-fold increase in the cellular retention in SSTR-positive rat AR4-2J pancreatic tumor cells, compared to its parent compound lacking the NLS sequence. Moreover, nuclear uptake was 45-fold higher when NLS was incorporated in ^{111}In-DOTA-TOC. A drawback to the use of peptides for PRRT, is that they are mainly excreted via the kidneys, where renal retention could lead to nephrotoxicity and eventual kidney failure [14]. Several methods have been developed to overcome this, including structural modification of the radiolabeled peptides, and the use of renal protectors [79, 93–96]. Adequate protection may enable the administration of higher activity doses, thereby improving the efficacy of therapy.

28.3.2.2 Receptor targeting

Another approach is to take advantage of a peptide ligand which accumulates in the nucleus. Several peptide ligands and their receptors have been shown to localize in the nucleus and this may be mediated by NLS present in the ligands or receptors [66]. The receptor-ligand complex, through the action of a NLS in the juxtamembrane region of EGFR, translocates to the nucleus, coming into close proximity with DNA [39,97–99]. This nuclear translocation was exploited by Reilly et al. in the design of ^{111}In-DTPA-hEGF, which binds to the EGFR, forming a ligand-receptor complex which is internalized and is selectively radiotoxic to EGFR-overexpressing tumor cells [39, 97]. Nuclear localization of ^{111}In-DTPA-hEGF was measured to be more than 10% of the internalized protein. A fraction of the ^{111}In-DTPA-hEGF in the nucleus became associated with chromatin. Furthermore, DNA double strand break formation in EGFR-overexpressing cells was increased 7-fold. Proliferation, as well as clonogenic survival and *in vivo* tumor growth were inhibited by ^{111}In-DTPA-hEGF in EGFR overexpressing cells and tumors [100, 101]. Because ^{111}In-DTPA-hEGF uses the EGFR as a carrier to transport it into the nucleus of the target cell, it was possible to increase the cytotoxicity of ^{111}In-DTPA-hEGF by increasing the nuclear translocation of the EGFR by interfering with the intracellular trafficking by using a selective tyrosine kinase inhibitor [97]. A phase I clinical trial has been undertaken to test the safety of ^{111}In-DTPA-hEGF in patients with chemotherapy-resistant metastatic breast cancer. Tumor uptake of ^{111}In-DTPA-hEGF was noted in several patients although the initial administered dose was very modest ($<$2290 Mbq), and no significant toxicity was reported [30]. A multi-dose Phase I trial is now planned by the same group.

28.3.3 Small molecules

Iododeoxyuridine (IUdR), a thymidine analog, is incorporated into DNA in proliferating cells during DNA synthesis. Therefore, radioiodonated IUdR could be used to target the DNA and exploit the different cell proliferating kinetics between tumors and normal tissue. ^{125}I-IUdR (5-[^{125}I]iodo-2′-deoxyuridine) is the most extensively studied Auger-electron emitting radiopharmaceutical in this class [102, 103]. Since the first pre-clinical animal studies in the 1970s [104–106], some preliminary results on clinical applications have been published [107–109]. ^{125}I-IUdR or indeed, ^{123}I-IUdR is taken up by tumor cells, where it is phosphorylated by thymidine kinase, trapped in the cell, and incorporated into DNA during the S-phase of the cell cycle instead of thymidine. Ex vivo, nuclear localization in tumor tissue was shown using micro-autoradiography [103, 107]. Although ^{125}I-IUdR has high intranuclear uptake after systemic administration [108,110], the extremely fast metabolic degradation however, makes it unsuitable for systemic tumor targeting [111]. Therefore, Marinari *et al.* investigated the metabolism and selectivity of ^{125}I-IUdR incorporation after direct or locoregional (intracavitary) administration in patients with breast, colon and gastrointestinal cancer [107, 108]. Tumor to non-tumor ratios were excellent, however the inhomogeneous distribution within the tumor mass and need for repeated administration, makes it unsuitable for intratumoral administration [11]. The main disadvantages of labeled deoxyribonucleotides are their short biological half-life and the restriction of their cytotoxic effects to cells during the S-phase of the mitotic cycle. Although the percentage of cells in S-phase in the tumor is higher than in normal tissues, only subsets of cells can be targeted. Radiation-induced bystander effects, prevalent at low radiation dose and low dose rate (a characteristic of Auger emitter targeted radiotherapy) are thought to be responsible for the remaining cytotoxicity [111–113]. Interestingly, when the radiopharmaceutical was co-administered with methotrexate (MTX), an antimetabolite that enhances IUdR uptake by DNA-synthesizing cells, a therapeutic effect was observed [114]. As a result of these findings, a combination of MTX and ^{125}IUdR have been used as treatment for neoplastic meningitis [109]. This preliminary study was conducted on a single patient only, the combination of repeated doses of MTX and a single dose of ^{125}IUdR was well tolerated with no evidence of central nervous system toxicity. Although there was a biological relapse after initial improvement this study demonstrated the efficacy of combining different treatment modalities.

Meta-iodobenzylguanidine (mIBG) is a guanethidine analog that is actively concentrated and stored in neuroblastoma cells via the norepinephrine transporter. There has been considerable interest in using ^{131}I-mIBG for cancer therapy. However, adverse hematological toxicity has limited its clinical use. Therefore an Auger emitting agent may be more beneficial than the β-emitting analog. Initial clinical studies on the use ^{125}I-mIBG in the treatment of neuroblastoma in children showed no clear therapeutic benefit [115, 116]. Recent studies by He et al. noted the ability of ^{123}I-mIBG to kill cultured neuroblastoma cells with its short-range

Auger and conversion electrons while sparing cells of haematopoietic lineage [117]. Their findings suggest that ^{123}I-MIBG may be useful for residual small volume and micrometastatic neuroblastoma.

28.4 Future perspectives

Although radionuclide therapy has been available for many years, few methods involving Auger-electron emitters are routinely used in the clinical setting. The exceptions are ^{125}I labeled MIBG (meta-iodobenzylguanidine) for treatment of pheochromocytoma and neuroblastoma [118, 119] and radiolabelled somatostatin analogues for treatment of neuroendocrine tumors [120]. Many Auger-electron emitters also emit γ-radiation, this property has been exploited for decades in nuclear medicine for imaging [121]. Thus the use of Auger emitting radionuclides as therapeutic and diagnostic agents is a very attractive option in the molecular imaging and management of tumors. The challenge remains however to increase intranuclear incorporation of the radiopharmaceutical without compromising toxicity. Several strategies to achieve this goal have been proposed and can be broadly classified as;

1. Strategies addressing the expansion of the family of potential agents that target the cell nucleus. Candidates include DNA intercalating agents, DNA binding (minor/major) groove molecules, single-strand antisense oligonucleotides or peptide nucleic acids capable of forming triplexes with specific double-stranded DNA sequences [122–125], nucleotides targeting single genes [126], internalizing antibodies [127] and biocompatible nanotechnologies [13, 21, 128, 129]. Additionally, strategies that increase receptor binding affinity and/or increases the number of receptors on tumor cells by, e.g., gene therapy [130, 131].
2. Strategies that combine targeted radiotherapy with other treatment modalities, such as radiosensitizer pre-treatment or chemotherapy. It is well established that combinations of immunotherapy and chemotherapy shown significant efficacy in disease management [132–134]. The combination of chemotherapy and radioimmunotherapy (RIT) could be exploited to gain therapeutic advantage [37].
3. Strategies aimed at reducing the dose to normal organs. A major obstacle in the effectiveness of the radio-labeled antibodies, especially against solid tumors, are extended residence in the bloodstream, resulting in a high marrow dose, and poor penetration into some parts of the tumor, which leads to non-uniform distribution of absorbed radiation dose [38]. To resolve these drawbacks, antibody fragments (Fab, $F(ab')_2$) or genetically engineered entities (Fv) have been used in an attempt to shorten the residence time, improve penetration and address dose inhomogeneity [135]. Alternatively, the use of pre-targeting with unlabeled antibody or antibody fragments containing an additional affinity site (e.g., streptavidin) followed by a chase of a small, rapidly excreted radiolabeled

molecular conjugate (e.g., radiobiotin) can be used [136,137]. Often, this strategy includes the use of an unlabeled conjugate to clear the blood of any residual circulating antibody.

4. The choice of radionuclide. Using Auger emitters with high electron yields and varying half-lives would enable exploitation of the strong influence of dose rate in determining the therapeutic efficacy of molecules directly or not directly bound to DNA [103]. ^{195m}Pt is a much more efficient Auger emitter with 33 electrons/decay compared to 24.9 for ^{125}I and 14.7 of ^{111}In. In addition, ^{119}Sb has recently been identified as a potent Auger emitter for therapy [138]. However availability and cost related considerations would affect routine use of the more exotic Auger emitters.

28.5 Conclusion

There is now ample evidence that a major determinant of biological damage caused by Auger emitters is highly dependent on the intranuclear localization and specifically the proximity of Auger-electron emitters to the double-stranded nuclear DNA. Furthermore, due to the short ranges of these electrons, Auger emitters are extremely radiotoxic when they decay close to DNA. However decay outside the cell is associated with little toxicity. This quality is not true of either α- or β-emitting isotopes, and can therefore not be exploited in targeted radiotherapy with these isotopes. This knowledge can be used to design and develop novel radiopharmaceuticals for targeting the nuclei of cancer cells. The possibility of transferring tumor therapy based on the emission of Auger electrons from experimental models to patients has vast therapeutic potential, and remains a field of intense research.

References

1. Carlsson, J., Forssell Aronsson, E., Hietala, S.-O., Stigbrand, T., Tennvall, J., Radiother. Oncol. **66**(2), 107–117 (2003)
2. Sgouros, G., Roeske, J.C., McDevitt, M.R., Palm, S., Allen, B.J., Fisher, D.R., Brill, A.B., Song, H., Howell, R.W., Akabani, G., In collaboration with the SNM MIRD Committee: Wesley E. Bolch, A.B.B., Darrell R. Fisher, Roger W. Howell, Ruby F. Meredith, George Sgouros, Barry W. Wessels,, Pat B. Zanzonico: MIRD Pamphlet No. 22 (Abridged): Radiobiology and Dosimetry of {alpha}-Particle Emitters for Targeted Radionuclide Therapy. J. Nucl. Med. **51**(2), 311–328 (2010)
3. Schally, A., Nagy, A., Trends Endocrinol Metab **15**, 300–310 (2004)
4. Carlsson, J., Stigbrand, T., Adams, G.P., In: Stigbrand, T., Carlsson, J., Adams, G.P. (eds.) Targeted Radionuclide Tumor Therapy. pp. 1–11. Springer Netherlands, (2008)
5. Ehrlich, P., Herta, C.A., Shigas, K.Z., Physiol Chem **61**, 379–392 (1904)
6. Pressman, D., Keighley, G., Journal of Immunology **59**, 141–146 (1948)
7. Pressman, D., Day, E., Blau, M., Cancer Res **17**, 845 (1957)
8. Kohler, G., Milstein, C., Nature **256**(5517), 495–497 (1975)

9. Meredith, R.F., International Journal of Radiation Oncology Biology Physics **66**(2, Supplement 1), S7–S7 (2006)
10. De Nardo, S.J., De Nardo, G.L., International Journal of Radiation Oncology, Biology and Physics **66**, S89–S95 (2006)
11. Adelstein, S., Kassis, A., Bodei, L., Mariani, G., Cancer Biother Radiopharm **18**, 301–316 (2003)
12. Mező, G., Manea, M., Expert Opinion on Drug Delivery **7**(1), 79–96 (2010)
13. Scheinberg, D.A., Villa, C.H., Escorcia, F.E., McDevitt, M.R., Nat Rev Clin Oncol **7**(5), 266–276 (2010)
14. van Essen, M., Krenning, E.P., Kam, B.L.R., de Jong, M., Valkema, R., Kwekkeboom, D.J., Nat Rev Endocrinol **5**(7), 382–393 (2009)
15. Kassis, A.I., Adelstein, S.J., J Nucl Med **46**, 4S–12S (2005)
16. Kassis, A.I., International Journal of Radiation Biology **80**(11–12), 789–803 (2004)
17. Buchegger, F., Perillo-Adamer, F., Dupertuis, Y., Bischof Delaloye, A., European Journal of Nuclear Medicine and Molecular Imaging **33**(11), 1352–1363 (2006)
18. Barendswaard, E.C., Humm, J.L., O'Donoghue, J.A., Sgouros, G., Finn, R.D., Scott, A.M., Larson, S.M., Welt, S., J Nucl Med **42**(8), 1251–1256 (2001)
19. Fischer, T., Schomäcker, K., Schicha, H., Int J Radiat Biol **84**(12), 1112–1122 (2008)
20. Fondell, A., Edwards, K., Ickenstein, L.M., Sjöberg, S., Carlsson, J., Gedda, L., Eur J Nucl Med Mol Imaging **37**, 114–123 (2010)
21. Fonge, H., Lee, H., Reilly, R.M., Allen, C., Molecular Pharmaceutics **7**(1), 177–186 (2009)
22. Ickenstein, L.M., Edwards, K., Sjfberg, S., Carlsson, J., Gedda, L., Nuclear Medicine and Biology **33**, 773–783 (2006)
23. Zou, Y., Xiao, X., Li, Y., Zhou, T., Oncology Reports **21**, 379–386 (2009)
24. Kassis, A.I., Semin Nucl Med **38**(5), 358–366 (2008)
25. Behr, T.M., Béhé, M., Löhr, M., Sgouros, G., Angerstein, C., Wehrmann, E., Nebendahl, K., Becker, W., European Journal of Nuclear Medicine and Molecular Imaging **27**(7), 753–765 (2000)
26. Nikjoo, H., Lindborg, L., Phys Med Biol **55** R65–R109 (2010)
27. Sastry, K.S.R., Howel, l.R.W., Rao, D.V., et al.: Dosimetry of Auger emitters: Physical and phenomenological approaches., vol. 27. DNA Damage by Auger Emitters. Taylor and Francis, London (1988)
28. Howell, R.W., Med Phys **19**, 1371–1383 (1992)
29. Govindan, S.V., Goldenberg, D.M., Elsamra, S.E., Griffiths, G.L., Ong, G.L., Brechbiel, M.W., et al: Radionuclides linked to a CD74 antibody as therapeutic agents for B-cell lymphoma: comparison of Auger electron emitters with β-particle emitters. J Nucl Med **41**, :2089–2097 (2000)
30. Reilly, R.M.: Monoclonal Antibody and Peptide-Targeted Radiotherapy of Cancer. John Wiley & Sons, Inc., Hoboken, New Jersey (2010)
31. Bernhardt, P., Forssell-Aronsson E, Jacobsson, L., Skamemark, G., Acta Oncologica **40**, 602–608 (2001)
32. Barendswaard, E.C., O'Donoghue, J.A., Larson, S.M., Tschmelitsch, J., Welt, S., Finn, R.D., Humm, J.L., J Nucl Med **40**(10), 1764–1768 (1999)
33. Unak, P., Braz Arch Biol Technol **45**, 97–100 (2002)
34. Foulon, C.F., Reist, C.J., Bigner, D.D., Zalutsky, M.R., et al., Cancer Res **60**, 4453–4460 (2000)
35. Shih, L.B., Thorpe, S.R., Griffiths, G.L., et al., J Nucl Med **35**, 899–908 (1994)
36. Press, O.W., Shan, D., Howell-Clark, J., et al., Cancer Res **56**, 2123–2129 (1996)
37. Milenic, D.E., Brady, E.D., Brechbiel, M.W., Nat Rev Drug Discov **3**(6), 488–499 (2004)
38. Britz-Cunningham, S.H., Adelstein, S.J., J Nucl Med **44**(12), 1945–1961 (2003)
39. Reilly, R.M., Kiarash, R., Cameron, R.G., Porlier, N., Sandhu, J., Hill, R.P., Vallis, K., Hendler, A., Gariepy, J., J Nucl Med **41**(3), 429–438 (2000)
40. Reilly, R.M., Kiarash, R., Sandhu, J., Lee, Y.W., Cameron, R., Hendler, A., et al:, J Nucl Med **41**, 903–911 (2000)

41. Mattes, M.J., Goldenberg, D.M., Eur J Nucl Med Mol Imaging **35**, 1249–1258 (2008)
42. Michel, R.B., Castillo, M.E., Andrews, P.M., Mattes, M.J., Clin Cancer Res **10**, 5957–5966 (2004)
43. Costantini, D.L., Chan, C., Cai, Z., Vallis, K.A., Reilly, R.M., J Nucl Med **48**, 1357–1368 (2007)
44. Wild, D., Behe, M., Wicki, A., Storch, D., Waser, B., et al., J Nucl Med **47**, 2025–2033 (2006)
45. Woo, D.V., Li, D., Mattis, J.A., Steplewski, Z., . Cancer Res **49**, 2952–2958 (1996)
46. Daghighian, F., Barendswaard, E., Welt, S., Humm, J., Scott, A., Willingham, M.C., McGuffie, E., Old, L.J., Larson, S.M., J Nucl Med **37**(6), 1052–1057 (1996)
47. Michel, R.B., Rosario, A.V., Andrews, P.M., Goldenberg, D.M., Mattes, M.J., Clin Cancer Res **11**, 777–786 (2005)
48. Mattes. M. J,. Int J Radiat Biol **84**, 389–399 (2008)
49. Karagiannis, T.C., Hell J Nucl Med **10**(2), 82–88 (2007)
50. Milenic, D.E., Seminars in Radiation Oncology **10**(2), 139–155 (2000)
51. Boonstra, J., Rijken, P., Humbel, B., Cremers, F., Verkleij, A., van Bergen, E.N., P, H., Cell Biol International **19**, 413–430 (1995)
52. Rubin, I., Yarden, Y., Annals Oncol **12** (Suppl 1), S3–S8 (2001)
53. Ciardello, F., Tortora, G., Clin Cancer Res **7**, 2958–2970 (2001)
54. Hynes, N.E., MacDonald, G., Curr Opin Cell Biol **21**, 177–184 (2009)
55. Klein, S., Levitzki, A., . Curr Opin Cell Biol **21**, 185–193 (2009)
56. Bender, H., Takahashi, H., Adachi, K., al., e., Cancer Res **52**(1), 121–126 (1992)
57. Derui, L., Woo, D.V., Emrich, J., al., e., Am J Clin Oncol **15**(4), 288–294 (1992)
58. Li, L., Quang, T.S., Gracely, E.J., et al., J Neurosurg **113**(2), 192–128 (2010)
59. Quang, T.S., Brady, L.W., Int J Radiat Oncol Biol Phys **58**(3), 972–975. (2004)
60. Woo, D.V., Li, D.R., Brady, L.W., Emrich, J., Mattis, J., Steplewski, Z., Front Radiat Ther Oncol **24**, 47–63 (1990)
61. Bender, H., Takahashi, H., Adachi, K., Belser, P., Liang, S.H., Prewett, M., Schrappe, M., Sutter, A., Rodeck, U., Herlyn, D., Cancer Res **52**(1), 121–126 (1992)
62. Li, L., Quang, T.S., Gracely, E.J., Kim, J.H., Emrich, J.G., Yaeger, T.E., Jenrette, J.M., Cohen, S.C., Black, P., Brady, L.W., J Neurosurg (2010), doi:10.3171/2010.2.JNS091211
63. Garin-Chesa, P., Sakamoto, J., Welt, S., Real, F.X., Rettig, W.J., Lloyd J. O: Organ-specific expression of the colon cancer antigen A33, a cell surface target for antibodybased therapy. Int J Oncol **9**, 465–471 (1996)
64. Welt, S., Scott, A., Divgi, C., Kemeny, N., Finn, R., Daghighian, F., Germain, J., Richards, E., Larson, S., Old, L., Journal of Clinical Oncology **14**(6), 1787–1797 (1996)
65. Scott, A.M., Lee, F.-T., Jones, R., Hopkins, W., MacGregor, D., Cebon, J.S., Hannah, A., Chong, G., U, P., Papenfuss, A., Rigopoulos, A., Sturrock, S., Murphy, R., Wirth, V., Murone, C., Smyth, F.E., Knight, S., Welt, S., Ritter, G., Richards, E., Nice, E.C., Burgess, A.W., Old, L.J., Clinical Cancer Research **11**(13), 4810–4817 (2005)
66. Welt, S., Ritter, G., Williams, C., Cohen, L.S., John, M., Jungbluth, A., Richards, E.A., Old, L.J., Kemeny, N.E., Clinical Cancer Research **9**(4), 1338–1346 (2003)
67. Welt, S., Ritter, G., Williams, C., Cohen, L.S., Jungbluth, A., Richards, E.A., Old, L.J., Kemeny, N.E., Clinical Cancer Research **9**(4), 1347–1353 (2003)
68. Meredith, R.F., Khazaeli, M.B., Plott, W.E., Spencer, S.A., Wheeler, R.H., Brady, L.W., et al. : Initial clinical evaluation of iodine-^{125}-labeled chimeric 17-1A for metastatic colon cancer. J Nucl Med **36**, 2229–2233 (1995)
69. Chen, P., Wang, J., Hope, K., et.al., J Nucl Med **47**(5), 827–836. (2006)
70. Cornelissen, B., McLarty, K., Kersemans, V., Scollard, D., Reilly, R.M., Nucl Med Biol **35**(1), 101–110 (2008)
71. Costantini, D.L., Bateman, K., McLarty, K., Vallis, K.A., Reilly, R.M., J Nucl Med **49**(9), 1498–1505 (2008)
72. Costantini, D.L., McLarty, K., Lee, H., Done, S.J., Vallis, K.A., M., R.R., J Nucl Med **51**(7), 1084–1091 (2010)
73. Costantini, D.L., Villani, D.F., Vallis, K.A., Reilly, R.M., J Nucl Med **51**(3), 477–488 (2010)

74. Kersemans, V., Cornelissen, B., Minden, M.D., Brandwein, J., Reilly, R.M., J Nucl Med **49**(9), 1546–1554. (2008)
75. Chen, P., Wang, J., Hope, K., Jin, L., Dick, J., Cameron, R., Brandwein, J., Minden, M., Reilly, R.M., J Nucl Med **47**(5), 827–836 (2006), doi:47/5/827 [pii]
76. Kersemans, V., Cornelissen, B., Minden, M.D., Brandwein, J., Reilly, R.M., J Nucl Med **49**(9), 1546–1554 (2008), doi:jnumed.107.047399 [pii] 10.2967/jnumed.107.047399
77. Costantini, D.L., Chan, C., Cai, Z., Vallis, K.A., Reilly, R.M., J Nucl Med **48**(8), 1357–1368 (2007). doi:jnumed.106.037937 [pii] 10.2967/jnumed.106.037937
78. Costantini, D.L., Villani, D.F., Vallis, K.A., Reilly, R.M., J Nucl Med **51**(3), 477–483. doi:jnumed.109.069716 [pii] 10.2967/jnumed.109.069716
79. Vegt, E., de Jong, M., Wetzels, J.F.M., Masereeuw, R., Melis, M., Oyen, W.J.G., Gotthardt, M., Boerman, O.C., J Nucl Med **51**(7), 1049–1058 (2010)
80. Schottelius, M., Wester, H.J., Methods **48**, 161–177 (2009)
81. Guillermmet-Guibert, J., Lahlou, H., Cordelier, P., Bousquet, C., Pyronnet, S., Susini, C., J Endocrinol Invest **28** ((11 Suppl)), 5–9 (2005)
82. Csaba, Z., Dournand, P., Neuropeptides **35**, 1–23 (2001)
83. Pool, S.E., Krenning, E.P., Koning, G.A., van Eijck, C.H.J., Teunissen, J.J.M., Kam, B., Valkema, R., Kwekkeboom, D.J., de Jong, M., Seminars in Nuclear Medicine **40**(3), 209–218 (2010)
84. de Jong, M., Breeman, W.A.P., Kwekkeboom, D.J., Valkema, R., Krenning, E.P., Accounts of Chemical Research **42**(7), 873–880 (2009)
85. Valkema, R., De Jong, M., Bakker, W.H., et al: : Phase I study of peptide receptor radionuclide therapy with [In-DTPA]octreotide: The Rotterdam experience. . Semin Nucl Med **32**, 110–122 (2002)
86. Anthony, L.B., Woltering, E.A., Espenan, G.D., et al:, Semin Nucl Med **32**, 123–132 (2002)
87. Buscombe J. R, Caplin M. E, J, H.A., J Nucl Med **44**, 1–6 (2003)
88. Delpassand, E.S., Sims-Mourtada, J., Saso, H., et al, Cancer Biother Radiopharm **23**, 292–300 (2008)
89. Limouris, G.S., Chatziioannou, A., Kontogeorgakos, D., et al., Eur J Nucl Med Mol Imaging **35**, 1827–1837 (2008)
90. Capello, A., Krenning, E.P., Breeman, W.A., J Nucl Med **44**, 98–104 (2003)
91. Capello, A., Krenning, E., Bernard, B., Reubi, J.C., Breeman, W., de Jong, M., Eur J Nucl Med Mol Imaging **32**(11), 1288–1295 (2005)
92. Ginj, M., Hinni, K., Tschumi, S., Schulz, S., Maecke, H.R., J Nucl Med **46**(12), 2097–2103. (2005)
93. Gotthardt, M., van Eerd-Vismale, J., Oyen, W.J., et al., J Nucl Med **48**(4), 596–601 (2007)
94. Lambert, B., Cybulla, M., Weiner, S.M., et al., Radiat Res **161**(5), 607–611 (2004)
95. Melis, M., Vegt, E., Konijnenberg, M.W., et al., J Nucl Med **51**(6), 973–977 (2010)
96. Vegt, E., Eek, A., Oyen, W.J., de Jong, M., Gotthardt, M., Boerman, O.C., Eur J Nucl Med Mol Imaging **37**(2), 226–234 (2010)
97. Bailey, K.E., Costantini, D.L., Cai, Z., Scollard, D.A., Chen, Z., Reilly, R.M., Vallis, K.A., J Nucl Med **48**(9), 1562–1570 (2007)
98. Lo, H.W., Hsu, S.C., Hung, M.C., Breast Cancer Res Treat **95**(3), 211–218 (2006)
99. Lin SY, Makino K, Xia W, Matin A, Wen Y, Kwong KY, Bourguignon L, Hung MC: Nuclear localization of EGF receptor and its potential new role as a transcription factor. Nat Cell Biol **3**(9), 802–808 (2001)
100. Chen, P., Cameron, R., Wang, J., Vallis, K.A., Reilly, R.M., J Nucl Med **44**(9), 1469–1478 (2003)
101. Cai, Z., Chen, Z., Bailey, K.E., Scollard, D.A., Reilly, R.M., Vallis, K.A., J Nucl Med **49**(8), 1353–1361 (2008)
102. Bloomer, W.D., Adelstein, S.J., Nature **265**(5595), 620–621 (1977)
103. Bodei, L., Kassis, A.I., Adelstein, S.J., Mariani, G., Cancer Biother Radiopharm **18**, 861–877 (2003)
104. Burki, H.J., Koch, C., Wolff, S., Curr Top Radiat Res Q **12**, 408–425 (1978)

105. Feinendegen, L.E., Radiat Environ Biophys **12**, 85–99 (1975)
106. Krisch, R.E., Krasin, F., Sauri, C.J., Int J Radiat Biol Relat Stud Phys Chem Med **29**, 37–50 (1976)
107. Mariani, G., Cei, A., Collecchi, P., Baranowska-Kortylewicz, J., Van den Abbeele, A.D., Di Luca, L., Di Stefano, R., Viacava, P., Ferdeghini, E.M., Di Sacco, S., Salvadori, P.A., Bevilacqua, G., Adelstein, S.J., Mosca, F., Kassis, A.I., J Nucl Med **34**(7), 1175–1183 (1993)
108. Mariani, G., Collecchi, P., Giuliani, L., Baranowska-Koktylewicz, J., Di Luca, L., Meucci, G., Viacava, P., Van Den Abbeele, A.D., Salvadori, P.A., Di Sacco, S., Adelstein, S.J., Bevilacqua, G., Kassis, A.I., Annals of the New York Academy of Sciences **698**(1), 204–211 (1993)
109. Rebischung, C., Hoffmann, D., Stéfani, L., Desruet, M.D., Wang, K., Adelstein, S.J., Artignan, X., Vincent, F., Gauchez, A.S., Zhang, H., Fagret, D., Vuillez, J., Kassis, A.I., Balosso, J., International Journal of Radiation Biology **84**(12), 1123–1129 (2008)
110. Kassis, A.I., J Nucl Med Allied Sci **34**, 315–319 (1990)
111. Kassis, A.I., Adelstein, S.J., Mariani, G., Q J Nucl Med **40**(301–319) (1996)
112. Boyd M, Ross SC, Dorrens J, Fullerton NE, Tan KW, Zalutsky MR, et al., J Nucl Med **47**, 1007–1015 (2006)
113. Xue LY, Butler NJ, Makrigiorgos GM, Adelstein SJ, Kassis AI: Bystander effect produced by radiolabeled tumor cells in vivo. Proc Natl Acad Sci U S A **99**, 13765–13770 (2002)
114. Kassis, A.I., Dahman, B.A., Adelstein, J.S., Acta Oncologica **39**(6), 731–737 (2000)
115. Hoefnagel, C.A., Smets, L.A., Voute, P.A., de Kraku, J., J Nucl Med **32**, 361–362 (1992)
116. Sisson JC, Shapiro B, Hutchinson RJ, Shulkin BL, Zempel S: Survival of patients with neuroblastoma treated with ^{125}I-MIBG. Am J Clin Oncol **19**, 144–148 (1996;)
117. He, Y., Das, B., Baruchel, S., Kumar, P., Wiebe, L., Reilly, R.M., Nuclear Medicine Communications **25**, 1125–1130 (2004)
118. Hoefnagel, C.A., Q J Nucl Med **43**, 336–343 (1999)
119. Sisson, J.C., Shapiro, B., Hutchinson, R.J., Zasadny, K.R., Mallette, S., Mudgett, E.E., Wieland, D.M., J Nucl Biol Med **35**(4), 255–259 (1991)
120. Forssell-Aronsson, E.B., Nilsson, O., Benjegard, S.A., Kolby, L., Bernhardt, P., Molne, J., Hashemi, S.H., Wangberg, B., Tisell, L.-E., Ahlman, H., J Nucl Med **41**(4), 636–642 (2000)
121. Monsieurs, M., Brans, B., Bacher, K., Dierckx, R., Thierens, H., European Journal of Nuclear Medicine and Molecular Imaging **29**(12), 1581–1587 (2002)
122. Martin, R.F., D'Cunha, G.: ^{125}I decay in synthetic oligodeoxynucleotides. In: Howell RW, Narra VR, Sastry KSR, Rao DV (eds.) Biophysical Aspects of Auger Processes, American Association of Physicists in Medicine Symposium Series No 8., vol. 153. American Institute of Physics, Woodbury, NY (1992)
123. Panyutin, I.G., Neumann, R.D., Acta Oncol 35:817 (1996)
124. Karamychev, V.N., Panyutin, I.G., Kim M-K, et al., J Nucl Med **41**, 1093 (2000)
125. Panyutin, I.V., Luu, A.N., Panyutin, I.G., Neumann, R.D., Radiat Res **156**(158) (2001)
126. Sedelnikova, O.A., Luu, A.N., Karamychev, V.N., et al., Int J Radiat Oncol Biol Phys **49**, 391 (2001)
127. Behr, T.M., Béhé, M., Sgouros, G., Cancer Biother Radiopharm **17**, 445 (2002)
128. Ashwin, A.B., Vyomesh, P., Julie, G., Guofeng, Z., Alioscka, A.S., Andrius, M., Richard, D.L., Roberto, W., J. S. G,, James, F.R., ACS Nano **3** (2), 307–316 (2009)
129. Muthu, M.S., Wilson, B., Nanomedicine **5**(2), 169–171 (2010)
130. Dwyer, R.M., Bergert, E.R., O'Connor, M.K., Gendler, S.J., Morris, J.C., Clin Cancer Res **11**, 1483–1489 (2005)
131. Buchsbaum, D.J., Chaudhuri, T.R., Zinn, K.R., J Nucl Med **46**(Suppl 1), 179S–186S (2005)
132. Pegram, M.D., Konecny, G.E., O'Callaghan, C., Beryt, M., Pietras, R., Slamon, D.J., Journal of the National Cancer Institute **96**(10), 739–749 (2004)
133. Slamon, D.J., Leyland-Jones, B., Shak, S., Fuchs, H., Paton, V., Bajamonde, A., Fleming, T., Eiermann, W., Wolter, J., Pegram, M., Baselga, J., Norton, L., New England Journal of Medicine **344**(11), 783–792 (2001)

134. Wilson, W.H., Dunleavy, K., Pittaluga, S., Hegde, U., Grant, N., Steinberg, S.M., Raffeld, M., Gutierrez, M., Chabner, B.A., Staudt, L., Jaffe, E.S., Janik, J.E., Journal of Clinical Oncology **26**(16), 2717–2724 (2008)
135. Imam, S.K., Cancer Biother Radiopharm **16**, 237–256 (2001)
136. Weiden, P.L., Breitz, H.B., Crit Rev Oncol Hematol **40**, 37–51 (2001)
137. Press, O.W., Corcoran, M., Subbiah, K., et al., Blood **98**, 2535–2543 (2001)
138. Thisgaard, H., Jensen, M., Applied Radiation and Isotopes **67**(1), 34–38 (2009)
139. Cornelissen, B., Vallis, K.A., Current Drug Discovery Technologies **7** (2010)

Chapter 29
Using a Matrix Approach in Nonlinear Beam Dynamics for Optimizing Beam Spot Size

Alexander Dymnikov and Gary Glass

Abstract Beam focusing is understood as the result of non-linear motion of a set of particles. As a result of this motion, we have the beam spot on the target. The set has a volume (the phase volume, or emittance). For a given brightness, the phase volume is proportional to the beam current and vice versa. The beam has an envelope surface. All particles of the beam are located inside of this surface, inside of this beam envelope. For the same phase volume (or beam current) the shape of the beam envelope can be different. We say the beam envelope is optimal if the spot size on the target has a minimum value for a given emittance. The essential feature of our optimization is the matrix approach for non-linear beam motion. In this approach we obtain and use analytical expressions for the matrizant (or transfer matrix) and for the envelope matrix. This matrix technique is known as the Matrizant method.

29.1 The phase space and the phase moment space

The nonlinear differential equations of charge particles motion along an optical rectilinear axis z are written in the phase space y, where we use the following notations for 6-dimensional and 3-dimensional vectors:

$$y \equiv \begin{pmatrix} x_1 \\ x_2 \\ x_3 \\ x_1' \\ x_2' \\ x_3' \end{pmatrix} = \begin{pmatrix} x \\ x' \end{pmatrix}$$

A. Dymnikov (✉) • G. Glass
Louisiana Accelerator Center, The University of Louisiana at Lafayette, P.O. Lafayette, LA 70504-2410, USA
e-mail: dymnikov@louisiana.edu; glass@louisiana.edu

G.G. Gómez-Tejedor and M.C. Fuss (eds.), *Radiation Damage in Biomolecular Systems*, Biological and Medical Physics, Biomedical Engineering,
DOI 10.1007/978-94-007-2564-5_29, © Springer Science+Business Media B.V. 2012

$$x \equiv \begin{pmatrix} x_1 \\ x_2 \\ x_3 \end{pmatrix}, \quad x' \equiv \frac{dx}{dz} = \begin{pmatrix} x_1' \\ x_2' \\ x_3' \end{pmatrix}$$

Here y is a 6-dimensional phase vector, vectors x_1 and x_2 are transverse coordinates of a charged particle, x_3 is the absolute time, which is proportional to the arrival time of a charged particle at the plane (x_1, x_2), defined by the expression

$$x_3 \equiv ct.$$

In the last expression c is the speed of light, t – time.

29.2 The m-power vector

In the general case we define recursively the m-power of the n-vector y as the vector y^m, where

$$y^m \equiv \begin{pmatrix} y_1 y^{m-1}(1) \\ \dots \\ y_n y^{m-1}(n) \end{pmatrix}.$$

This vector is called the m-moment of the vector y or in short, the m-moment, where the auxiliary vector $y^l(j)$ is defined by the recursive relation:

$$y^l(j) \equiv \begin{pmatrix} y_j y^{l-1}(j) \\ \dots \\ y_n y^{l-1}(n) \end{pmatrix}, \quad y^m(1) \equiv y^m, \quad y^1 \equiv y, \quad y^0(j) = 1, \quad j = 1, \dots, n,$$
$$l = 1, \dots m.$$

The m-moment vector y^m has C_{n-1+m}^{n-1} scalar elements, where

$$C_{n-1+m}^{n-1} = \frac{(n-1+m)!}{(n-1)!m!}.$$

As an example, for $n = 3$ we obtain three m-moment vectors: y^1, y^2 and y^3.

$$y^1 = y^1(1) = y = \begin{pmatrix} y_1 \\ y_2 \\ y_3 \end{pmatrix}, \quad y^1(1) = y^1, \quad y^1(2) = \begin{pmatrix} y_2 \\ y_3 \end{pmatrix}, \quad y^1(3) = y_3,$$

$$y^2 = y^2(1) = \begin{pmatrix} y_1\, y^1(1) \\ y_2\, y^1(2) \\ y_3\, y^1(3) \end{pmatrix} = \begin{pmatrix} y_1^2 \\ y_1\, y_2 \\ y_1\, y_3 \\ y_2^2 \\ y_2\, y_3 \\ y_3^2 \end{pmatrix},$$

$$y^2(2) = \begin{pmatrix} y_2\, y_1(2) \\ y_3\, y_1(3) \end{pmatrix} = \begin{pmatrix} y_2^2 \\ y_2\, y_3 \\ y_3^2 \end{pmatrix}, \qquad y^2(3) = (y_3\, y_1(3)) = y_3^2,$$

$$y^3 = y^3(1) = \begin{pmatrix} y_1\, y^2(1) \\ y_2 y^2(2) \\ y_3\, y^2(3) \end{pmatrix} = \begin{pmatrix} y_1^3 \\ y_1^2\, y_2 \\ y_1^2\, y_3 \\ y_1\, y_2^2 \\ y_1\, y_2\, y_3 \\ y_1\, y_3^2 \\ y_2^3 \\ y_2^2\, y_3 \\ y_2\, y_3^2 \\ y_3^3 \end{pmatrix}.$$

Similarly, the m-power of the differential n-vector operator $\partial(y)$ is a differential m-moment operator $\partial^m(y)$, where

$$\partial(y) \equiv \begin{pmatrix} \partial_1(y) \\ \partial_2(y) \\ \dots \\ \partial_n(y) \end{pmatrix}, \qquad \partial_\alpha(y) \equiv \frac{\partial}{\partial y_\alpha}, \qquad \alpha = 1, 2, \dots, n.$$

29.3 The m-moment phase vector and the phase moment space

The m-moment phase vector of the n-vector y is denoted by $y\langle m\rangle$ and is defined by the following expression:

$$y\langle m\rangle \equiv \begin{pmatrix} y^1 \\ y^2 \\ \dots \\ y^m \end{pmatrix}.$$

The phase vector $y\langle m\rangle$ is a vector having α component:

$$\alpha = \sum_{\lambda=1}^{m} \alpha_\lambda, \quad \alpha_\lambda = C_{5+\lambda}^{5} = \frac{(5+\lambda)\,!}{5\,!\lambda\,!}$$

29.4 The Taylor expansion of the equations of motion

The equations of particle beam motion usually are written as:

$$x'' = f(x,\, x') = f\,(y)\,,$$

where

$$x'' \equiv \frac{d^2x}{dz^2}, \quad f\,(y) \equiv \begin{pmatrix} f_1\,(y) \\ f_2\,(y) \\ f_3\,(y) \end{pmatrix}.$$

Using the Taylor expansion of the vector function $f(y)$ to transform it into a finite series, the equations of motion can be written as

$$x_i'' = \sum_{\lambda=1}^{m} \tilde{f}^{(i,\lambda)} y^\lambda, \quad i = 1,\, 2,\, 3,$$

where $\tilde{f}^{(i,\lambda)}$ is a row vector of dimension equal to the number α_λ of combinations of 6 elements λ at a time:

$$\alpha_\lambda = C_{5+\lambda}^{5} = \frac{(5+\lambda)\,!}{5\,!\lambda\,!}$$

The quantity y^λ is the λ-moment of the vector y. Using the unit 6-vectors $i\,(\lambda),\ \lambda = 4, 5, 6$, where

$$i\,(4) \equiv \begin{pmatrix} 0 \\ 0 \\ 0 \\ 1 \\ 0 \\ 0 \end{pmatrix}, \quad i\,(5) \equiv \begin{pmatrix} 0 \\ 0 \\ 0 \\ 0 \\ 1 \\ 0 \end{pmatrix}, \quad i\,(6) \equiv \begin{pmatrix} 0 \\ 0 \\ 0 \\ 0 \\ 0 \\ 1 \end{pmatrix},$$

we can write the equations of motion in the matrix form in the following way:

$$\frac{dy}{dz} = \begin{pmatrix} \tilde{i}\,(4) & 0 & \dots & 0 \\ \tilde{i}\,(5) & 0 & \dots & 0 \\ \tilde{i}\,(6) & 0 & \dots & 0 \\ \tilde{f}^{(1,1)} & \tilde{f}^{(1,2)} & \dots & \tilde{f}^{(1,m)} \\ \tilde{f}^{(2,1)} & \tilde{f}^{(2,2)} & \dots & \tilde{f}^{(2,m)} \\ \tilde{f}^{(3,1)} & \tilde{f}^{(3,2)} & \dots & \tilde{f}^{(3,m)} \end{pmatrix} \times y\,\langle m \rangle .$$

29.5 The equation for the phase moments

The equations $\frac{dy^s}{dz}$ for the phase moments can be obtained with the same accuracy as equations for $\frac{dy}{dz}$. This is possible by substituting equations

$$x_i'' = \sum_{\lambda=1}^{m} \tilde{f}^{(i,\lambda)} y^{\lambda}, \quad i = 1, 2, 3,$$

into the expression for $\frac{dy^s}{dz}$. The result will be:

$$\frac{dy^s}{dz} = \sum_{\lambda=s}^{m} P^{\,s\lambda}\, y^{\lambda},$$

where $P^{s\lambda}$ is an $\alpha_s \times \alpha_\lambda$ block matrix.

With the nonlinear equations given by equations for x_i'' it is possible to associate a linear equation for the phase moments given by

$$\frac{dy\,\langle m \rangle}{dz} = P^{(m)} y\,\langle m \rangle\,.$$

Here the matrix $P^{(m)}$ has the form of an upper triangular block:

$$P^{(m)} \equiv \begin{pmatrix} P^{11} & P^{12} & \dots & P^{1m} \\ 0 & P^{22} & \dots & P^{2m} \\ \dots & \dots & \dots & \dots \\ 0 & 0 & \dots & P^{mm} \end{pmatrix},$$

where

$$P^{11} = \begin{pmatrix} \tilde{i}\,(4) \\ \tilde{i}\,(5) \\ \tilde{i}\,(6) \\ \tilde{f}^{(1,1)} \\ \tilde{f}^{(2,1)} \\ \tilde{f}^{(3,1)} \end{pmatrix}, \quad P^{1\lambda} = \begin{pmatrix} 0 \\ 0 \\ 0 \\ \tilde{f}^{(1,\lambda)} \\ \tilde{f}^{(2,\lambda)} \\ \tilde{f}^{(3,\lambda)} \end{pmatrix}, \quad \lambda = 2, \ldots, m.$$

We call the matrix $P^{(m)}$ the coefficient matrix. The solution of the linear equation for y,

$$\frac{dy\,\langle m\rangle}{dz} = P^{(m)} y\,\langle m\rangle\,,$$

coincides with the solution of the equation

$$x_i'' = \sum_{\lambda=1}^{m} \tilde{f}^{(i,\lambda)} y^{\lambda}, \quad i = 1, 2, 3,$$

which has been obtained by the successive-approximaton method. The method of reducing the equation

$$x_i'' = \sum_{\lambda=1}^{m} \tilde{f}^{(i,\lambda)} y^{\lambda}, \quad i = 1, 2, 3,$$

to the form of the equation

$$\frac{dy\,\langle m\rangle}{dz} = P^{(m)} y\,\langle m\rangle$$

is referred to as the method of embedding in phase moment space. In this method the ideas that were originally presented in ref. [1] have been developed further in ref. [2].

Writing a non-linear equation in a linearized form makes it possible to construct the solution using the matrizant. This solution will be independent of the initial vector y_0, whereas the solution of the nonlinear equation is sought for each value of y_0. The use of matrices for solving nonlinear problems was first proposed by Brown [2].

The solution of the equation

$$\frac{dy\,\langle m\rangle}{dz} = P^{(m)} y\,\langle m\rangle$$

is written in terms of matrizant $Y(P^{(m)}, z/z_0)$ in the form:

$$y\,\langle m\rangle = Y(P^{(m)}, z/z_0)\, y_0\,\langle m\rangle\,, \quad Y(P^{(m)}, z_0/z_0) = I\,\langle\alpha\rangle,$$

where $I\,\langle\alpha\rangle$ is the unit $\alpha \times \alpha$ matrix.

The matrizant has, in the same way as the coefficient matrix $P^{(m)}$, the form of an upper triangular block:

$$Y(P^{(m)}, z/z_0) = \begin{pmatrix} Y^{11} & Y^{12} & \dots & Y^{1m} \\ 0 & Y_{22} & \dots & Y^{2m} \\ \dots & \dots & \dots & \dots \\ 0 & 0 & \dots & Y^{mm} \end{pmatrix}.$$

The matrizant satisfies the differential equation:

$$\frac{dY(P^{(m)}, z/z_0)}{dz} = P^{(m)}\, Y(P^{(m)}, z/z_0).$$

The block matrix Y^{11} is the matrizant of the linear equation in the phase space y:

$$\frac{dy}{dz} = P^{11}\, y, \quad y = Y^{11}\, y_0.$$

The matrix function Y^{ik} as well as P^{ik}, where $k \geq i$, are rectangular matrices with C_{n-1+i}^{n-1} rows and C_{n-1+k}^{n-1} columns.

The matrix function $Y^{\lambda\lambda}$ is determined by the equation:

$$\left(Y^{11} y_0\right)^{\lambda} = Y^{\lambda\lambda} y_0^{\lambda}.$$

The off-diagonal block matrices Y^{ik}, $(k \geq i)$ have the form:

$$Y^{ik}\,(z/0) = \sum_{l=i+1}^{k} \int_0^z Y^{ii}\,(z/\tau)\, P^{il}\,(\tau)\, Y^{lk}\,(\tau/0)\, d\tau.$$

These matrices are calculated in the following way: first $Y^{i,i+1}$, then $Y^{i,i+2}$, and so on up to $Y^{i,k} (i = 1, \dots, m, k = i + 1, \dots, m)$.

29.6 Nonlinear beam dynamics in the quadrupole lens

The quadrupole lenses are widely used for forming and transportation of charged particle beams themselves as well as a part of more complicated systems. Focusing systems from the quadrupole lenses are often used to reduce the lateral cross section of a beam. The purpose of these systems is to obtain the minimum of a beam spot size ρ for the initially diverging beam. These systems are used in ion microprobe, where the beam spot size ρ defines the microprobe resolution. To find the best microprobe focusing system means to find the system which produces for a given beam current (or a given emittance) the smallest beam spot size on the target. What is the lower limit $\rho = \rho_m$ for a given emittance which it is possible to obtain in quadrupole systems? To solve this problem it is not enough to consider the linear approximation of the equation of a beam motion. We must take into consideration the nonlinear terms.

As an example, we consider the charge particle motion in the magnetostatic quadrupole lens. The equation of motion in x-plane of the third approximation for the monochromatic beam is written in the form:

$$x'' = -\beta^2 f(z)x - \frac{3}{2}\beta^2 f(z)x\,(x')^2 - \frac{1}{2}\beta^2 f(z)xyy' + \beta^2 f'(z)xyy' + \\ +\beta^2 f(z)yx'y' + \frac{1}{12}\beta^2 f''(z)x^3 + \frac{1}{4}\beta^2 f''(z)xy^2$$

For a magnetic lens

$$\beta \equiv \sqrt{\frac{B_{12}}{p}},$$

where

$$B_{12} = \frac{\partial B_x}{\partial y} = \frac{\partial B_y}{\partial x}, \quad B_x = \frac{B_x^*}{(B\rho)_0^*}, \quad B_y = \frac{B_y^*}{(B\rho)_0^*}, (B\rho)_0^* = \frac{m_0^* c}{q}, \quad p = \frac{p_m^*}{p_{0m}^*}.$$

The quantities B_x^*, B_y^*, p^* and q denote magnetic field induction in x and y planes, momentum and charge of an arbitrary particle, while m_0^* and p_0^* denote rest mass and rest momentum. The quantities with index "m" denote the same quantity for the reference particle.

The function $f(z)$, normalized to unit, represents the distribution of the magnetic gradient of the quadrupole lens along the optical axis z.

The equations for y-plane are obtained from the equations in x-plane by the replacements

$$x \to y, \quad y \to x, \quad \beta^2 \to -\beta^2.$$

In the paraxial case, where the equation of motion is the equation of the first approximation

$$x'' = -\beta^2 f(z)x,$$

we have one four dimensional phase moment of the first order,

$$\tilde{x}[1] \equiv \{x_1, x_2, x_3, x_4\}, \quad x_1 \equiv x, \quad x_2 \equiv x', \quad x_3 \equiv y, \quad x_4 \equiv y',$$

and two two dimensional phase moments of the first order,

$$\tilde{x}_x[1] \equiv \{x_1, x_2\}, \quad x_1 \equiv x, \quad x_2 \equiv x'$$

and

$$\tilde{y}_x[1] \equiv \{y_1, y_2\}, \quad y_1 \equiv y, \quad y_2 \equiv y'.$$

The appropriate moments of the third order are written in the following forms:

$$x^3(1) = \begin{pmatrix} x_1\, x^2(1) \\ x_2\, x^2(2) \\ x_3\, x^2(3) \\ x_4\, x^2(4) \end{pmatrix}, \quad x^2(1) = \begin{pmatrix} x_1^2 \\ x_1\, x_2 \\ x_1\, x_3 \\ x_1\, x_4 \\ x_2^2 \\ x_2\, x_3 \\ x_2\, x_4 \\ x_3^2 \\ x_3\, x_4 \\ x_4^2 \end{pmatrix}, \quad x^2(2) = \begin{pmatrix} x_2^2 \\ x_2\, x_3 \\ x_2\, x_4 \\ x_3^2 \\ x_2\, x_3 \\ x_4^2 \end{pmatrix},$$

$$x^2(3) = \begin{pmatrix} x_3^2 \\ x_3\, x_4 \\ x_4^2 \end{pmatrix}, \quad x^2(4) = x_4$$

$$x_x^3(1) = \begin{pmatrix} x_1\, x_x^2(1) \\ x_2\, x_x^2(2) \end{pmatrix}, \quad x_x^2(1) = \begin{pmatrix} x_1^2 \\ x_1\, x_2 \\ x_2^2 \end{pmatrix}, \quad x_x^2(2) = x_2^2, \quad x_x^3(1) = \begin{pmatrix} x_1^3 \\ x_1^2 x_2 \\ x_1 x_2^2 \\ x_2^3 \end{pmatrix},$$

$$x_y^3(1) = \begin{pmatrix} y_1^3 \\ y_1^2 y_2 \\ y_1 y_2^2 \\ y_2^3 \end{pmatrix}.$$

If the motion of the monochromatic beam is described by the equation of the third order approximation, we have two phase moments of the third order:

$$\tilde{x}_x[3] \equiv \{x_1,\, x_2,\, x_1^3,\, x_1^2 x_2, x_1 x_2^2, x_2^3, x_1 y_1^2, x_1 y_1 y_2, x_1 y_2^2, x_2 y_1^2, x_2 y_1 y_2, x_2 y_2^2\}$$

and

$$\tilde{y}_x[3] \equiv \{y_1,\, y_2,\, y_1^3,\, y_1^2 y_2, y_1 y_2^2, y_2^3, y_1 x_1^2, y_1 x_1 x_2, y_1 x_2^2, y_2 x_1^2, y_2 x_1 x_2, y_2 x_2^2\}.$$

To the linear equation of the first approximation we associate a linear equation for the phase moments of the first order:

$$dx[1]/dz = P_x^{11}(z) x[1],$$

where

$$P_x^{11}(z) = \begin{pmatrix} 0 & 1 \\ -\beta^2 f(z) & 0 \end{pmatrix}.$$

To the nonlinear equation of the third approximation we can associate a linear equation for the phase moments of the third order:

$$dx[3]/dz = P_x(z)x[3], \quad P_x(z) = \begin{pmatrix} P_x^{11}(z) & P_x^{12}(z) \\ 0 & P_x^{22}(z) \end{pmatrix},$$

where

$$P_x^{12}(z) = \beta^2 \begin{pmatrix} 0 & 0 & 0 & 0 & 0 & 0 & 0 & 0 & 0 & 0 \\ \frac{f''(z)}{12} & 0 & -\frac{3f(z)}{2} & 0 & \frac{f''(z)}{4} & f'(z) & -\frac{f(z)}{2} & 0 & f(z) & 0 \end{pmatrix},$$

$$P_x^{22}(z) = \begin{pmatrix} 0&3&0&0&0&0&0&0&0&0 \\ 0&0&2&0&0&0&0&0&0&0 \\ 0&0&0&1&0&0&0&0&0&0 \\ 0&0&0&0&0&0&0&0&0&0 \\ 0&0&0&0&0&2&0&1&0&0 \\ 0&0&0&0&0&0&1&0&1&0 \\ 0&0&0&0&0&0&0&0&0&1 \\ 0&0&0&0&0&0&0&0&2&0 \\ 0&0&0&0&0&0&0&0&0&1 \\ 0&0&0&0&0&0&0&0&0&0 \end{pmatrix} + \beta^2 f(z) \begin{pmatrix} 0&0&0&0&0&0&0&0&0&0 \\ -1&0&0&0&0&0&0&0&0&0 \\ 0&-2&0&0&0&0&0&0&0&0 \\ 0&0&-3&0&0&0&0&0&0&0 \\ 0&0&0&0&0&0&0&0&0&0 \\ 0&0&0&0&1&0&0&0&0&0 \\ 0&0&0&0&0&1&0&0&0&0 \\ 0&0&0&0&-1&0&0&0&0&0 \\ 0&0&0&0&0&-1&0&1&0&0 \\ 0&0&0&0&0&0&-1&0&1&0 \end{pmatrix}$$

29.6.1 Optimization

Beam focusing is understood as the result of non-linear motion of a set of particles. As a result of this motion, we have the beam spot on the target. The set has a volume (the phase volume, or emittance). For a given brightness, the phase volume is proportional to the beam current and *vice versa*. The beam has an envelope surface. All particles of the beam are located inside of this surface, inside of this beam envelope. For the same phase volume (or beam current) the shape of the beam envelope can be different. We say the beam envelope is *optimal* if the spot size on the target has a *minimum* value for a given emittance. The beam of a given emittance is defined by a set of two matching slits: objective and divergence slits. For a given emittance *em*, the shape of the beam envelope is the function of the half-width (or radius) r_1 of the objective slit and of the distance l_{12} between two slits. The size r_2 of the second (divergence) slit is determined by the expression: $r_2 = em \times l_{12}/r_1$. The optimal parameters r_1, r_2 and l_{12} determine the *optimal beam envelope* or the *optimal matching slits*.

29.6.2 Matrix approach

The essential feature of our optimization is the matrix approach for non-linear beam motion. In this approach we obtain and use analytical expressions for the matrizant (or transfer matrix) and for the envelope matrix. This matrix technique is known as the Matrizant method. [3] This technique is used for solving the equations of motion, the non-linear differential equations in 4-dimensional phase space accurate to terms of the third order. These equations are replaced by two vector linear equations (for x- and y- planes) in the 12-dimensional phase moment space or by one equation in the 24-dimensional phase moment space. Writing the non-linear equations in a linearized form allows us to construct the solution using a 12×12 (or 24×24) third order matrizant.

As a result of this linearization it becomes possible to use all the advantages of linear differential equations over non-linear ones, including the independence of the matrizant of the choice of the initial point of the phase space.

Assuming an uniform density of particles in the plane of the two slits, knowing the third order matrizant and choosing by a random method N particles we can obtain the position of these particles in the image plane or in the target (or specimen) plane.

From the matrizant we can also find the spherical aberration coefficients in the object space *csx, csy*, and *csxy*. For the fields with the quadrupole symmetry (with the third order matrizant) these coefficients in the image space are written in the form: $csx_{im} = csx/d_1^3$, $csy_{im} = csy/d_2^3$, $csxy_{im} = csxy/d_1d_2^2 = csyx/d_2d_1^2$.

We study the evolution of the phase moment vector, which contains the elements of the phase moments of first and third order. The envelope matrix is taken as the matrix of the second moments of the distribution of this vector over the totality of the phase coordinates. We consider the case of a small density beam; then, beam self-field as well as particle collisions can be neglected and the distribution function satisfies the Liouville's equation. The integration is done over the object and aperture slits. We find the analytical form of the 12×12 (or 24×24) initial envelope matrix for the fields with quadrupole symmetry as a function of *em*, r_1 and l_{12}. This matrix is normalised by equating the first diagonal element to r_1^2. Thus the average radius of the beam is determined by the first diagonal element of the envelope matrix, which is a function of the position along the axis.

References

1. K.L. Brown, R. Belbeoch, P. Bounin, Rev. Sci. Instr. **35** 481 (1964)
2. A.D. Dymnikov, G.M.Osetinskij, Fiz. Elem. Chastits At. Yadra **20** 694 (1989); Sov. J. Part Nucl. **20** 293 (1989)
3. A. D. Dymnikov, R. Hellborg, Nucl. Instr. Meth. **A 330** 323 (1993)

Chapter 30
Future Particle Accelerator Developments for Radiation Therapy

Michael H. Holzscheiter and Niels Bassler

Abstract During the last decade particle beam cancer therapy has seen a rapid increase in interest, and several new centers have been built, are currently under construction, or are in an advanced stage of planning. Typical treatment centers today consist of an accelerator capable of producing proton or ion beams in an energy range of interest for medical treatment, i.e. providing a penetration depth in water of about 30 cm, a beam delivery system to transport the produced beam to the patient treatment rooms, and several patient stations, allowing for an optimal usage of the continuously produced beam. This makes these centers rather large and consequently expensive. Only major hospital centers situated in an area where they can draw on a population of several million can afford such an installation. In order to spread the use of particle beam cancer therapy to a broader population base it will be necessary to scale down the facility size and cost. This can in principle be done by reducing the number of treatment rooms to one, eliminating the need of an elaborate beam delivery system, and thereby reducing the building size and cost. Such a change should be going in parallel with a reduction of the accelerator itself, and a number of approaches to this are currently being pursued. If successful, such developments could eventually lead to a compact system where all components would fit into a single shielded room, not much different in size from a typical radiation vault for radiotherapy with X-rays.

M.H Holzscheiter (✉)
University of New Mexico, Albuquerque, NM 87131, USA
e-mail: michael.holzscheiter@cern.ch

N. Bassler
University of Aarhus, 8000-C Aarhus, Denmark
e-mail: bassler@phys.au.dk

G.G. Gómez-Tejedor and M.C. Fuss (eds.), *Radiation Damage in Biomolecular Systems*, Biological and Medical Physics, Biomedical Engineering,
DOI 10.1007/978-94-007-2564-5_30, © Springer Science+Business Media B.V. 2012

30.1 Motivation

While a sharp worldwide increase in the number of hospital based particle beam cancer therapy centers can be observed for the last decade, one of the main reasons for proton and ion therapy centers not to be even more widely spread and accessible to a larger population base is the very large physical size of such facilities, and the consequently very high initial investment cost. The accelerator system together with the subsequent components needed to transport the high energy charges particles to the various patient treatment stations, the actual treatment rooms, and the office and workshop space to support such a multi-room facility, typically occupies an area the size of a soccer field. The overall building cost is driven up due to the demands for massive amounts of shielding material, needed to protect patients, staff, and the environment from unwanted radiation. The total cost of a proton center will run around 100 million Euros, depending on the individual circumstances, with a heavy ion treatment center being still more expensive. Several studies [1] have shown that a potential patient base of 10 million population is necessary to economically justify such an investment. Of this cost roughly one third goes towards the accelerator and beam delivery systems.

A significant reduction of building size can be achieved when concentrating on a single treatment room only, but then the cost of the accelerator would become a larger portion of the overall investment, and in addition may not be as fully utilized, as unavoidable down times between patients will cause the accelerator to idle for a significant fraction of the time. Nevertheless, a single room system at a reduced cost may allow hospitals with a smaller pool of potential patients to offer particle beam therapy. While the immediate goal may seem a significant reduction of the overall facility cost, both in the construction phase and in the continuing maintenance of the system to enable a wider spread of particle therapy, one must also consider other consequences. Less staff will be needed to operate such a facility, but less staff will also be available for continuing research in the still developing field of particle therapy. We strongly believe that large multi-room installations with a significant staffing of physicists, biologists, and medical researchers are needed to drive the development of particle beam cancer therapy forward. These centers should be relieved to some extend from economic considerations by funding from government agencies and public and private foundations, to allow staff and beam time to be used for continuing research. For these larger centers a reduction in the accelerator size may not be a dominating goal, as its cost only represents a relatively small portion of the overall budget, and the flexibility of a larger accelerator, capable of feeding multiple beam lines, may be beneficial for the much needed research activity. But increased accessibility of particle beam cancer therapy to patients worldwide, based on the technological and clinical development achieved in the large research centers, will require a significant reduction of the price of entry into the field.

The largest gain would be achieved if the actual acceleration system could be made small enough to fit directly into the treatment room, especially into a size room comparable to standard X-ray therapy rooms. Then hospitals could consider upgrading part of their radiation oncology installation to proton therapy. Several proposals have been put forward to achieve such a size and cost reduction, and we will discuss a selection of these in this report.

Introducing new accelerator concepts to the medical community is especially difficult due to the stringent requirements on an accelerator used for treatments of human patients. Above all, the beam obtained from the accelerator must meet all specifications set forth by the clinical treatment plans, without any exceptions. This not only refers to beam energy, but also to beam shape and delivery options (spot scanning, passive scattering, etc.). If the time structure significantly deviates form current standards, one needs to be assured that the biological effects are not altered. This will require a number of additional test experiments, both in-vitro and in-vivo, before a system otherwise meeting the basic requirements of energy and intensity can be fully certified for medical use.

Additionally one typically requires a system availability of more than 95% to avoid extended interruptions of patient treatments due to technical failures. Considering that the actual accelerator is only a small part of a complex chain between ion source and patient, it may be necessary to call for an up-time of the actual accelerator of 98% or more. As a back-up solution one may consider setting up agreements with other particle therapy centers in the same geographical region to transfer patients when technical problems endanger the completion of an already started treatment course. Alternatively one may consider using more than one accelerator per facility [2] and allowing all beam lines to be served by all accelerators, or even to duplicate installations of a facility to achieve maximum accessibility to treatment for patients already enrolled. Continued, possibly in-beam, quality assurance will need to be developed for new systems, and a fail safe operation needs to be achieved through complex control systems adapted to the specific acceleration scheme. Background radiation from the acceleration column in terms of physical dose and particle spectrum must be understood and be bound by the ALARA (As Low As Reasonably Achievable) concept.

Considering the stringent requirements on performance and quality assurance placed on medical installations as well as the long lead time to introduce new technology and to obtain approval by the relevant authorities in different countries, a logical approach obviously is to start from proven standard technology and modify it such that it can meet new requirements with a minimum of deviations from standard approved systems.

30.2 Reducing the facility footprint

The following paragraphs will show various examples from simple modifications of facility lay-out to using proven modifications to the two widely used standard accelerator schemes, cyclotrons and synchrotrons, to achieve a smaller foot print, thereby making the technology of hadron therapy available to more clinics. Most of the development ongoing at this time is focusing on protons only, even though the technologies can be extended to heavier ions in most cases at the expense of higher cost and space requirements. We also briefly describe possible variations to the accelerator scheme, including non scaling Fixed Field Alternating Gradient (FFAG's) accelerators and various combinations of cyclotrons and synchrotrons which may have important advantages over existing systems beyond just reduction in size.

30.2.1 Single room options

Two suppliers of commercial proton therapy systems, IBA and Varian Medical Systems, offer relatively small, single room proton therapy systems based upon approved technology. Both systems use compact cyclotrons for acceleration of protons to the 230 MeV energy required for 30 cm penetration depth into a human target. The accelerators are adaptations of the standard systems offered by the two companies, optimized for size and weight. Both systems follow the standard lay-out of separating the accelerator and energy selection from the patient treatment room, allowing full shielding of stray radiation produced in the process of beam delivery. These are the smallest possible fully approved systems available today.

One step further away from the standard lay-out is the system offered by Still River Systems. The essential philosophy behind this system is to integrate a compact accelerator system and the energy selection process directly into the gantry rotating around the patient. In the Monarch^{250} the company advertises a small synchro-cyclotron based on superconducting technology mounted on a rotatable arch, aiming the beam at the patient from arbitrary directions. While at first this seems to be only a small modification to standard systems, it actually represents a significant step away from known centers. The massive shielding which is normally found between the accelerator and the patient has been removed, the beam travels in a straight line from the accelerator output to the patient. With passive energy selection, necessary for cyclotrons, and beam shaping (it is not known to the authors if this shall be a passive or active system) done this close to the patient position, shielding the patient from out-of-field background radiation must be a major concern. No information is available to the authors at the time of writing about shielding requirements and potential radiation background, nor about beam monitoring and quality assurance options foreseen for this short beam line.

The Monarch^{250} is offered at a price of approximately 25 M$, which compared to a typical price tag of 125 M$ price tag of a 4 room treatment installation does not seem to offer a significant financial savings. But smaller size and overall cost of this integrated system would make proton therapy available to smaller clinics or specialized private installations which may not have access to the patient base of 10 million inhabitants in the geographic capture area necessary to justify a multi-room system.

To our knowledge the Monarch^{250} has at the time of writing not yet received final approval by the Food and Drug Administration, but the company expects this approval to be forthcoming within this year. According to information in various press organs, several hospitals have placed an option on this system and the first installation is expected to take place in 2011 at the Siteman Cancer Center at Barnes-Jewish Hospital and Washington University School of Medicine in St. Louis, MO. [3]

30.2.2 Miniaturization of Standard Accelerators

Cyclotrons are not the only system which can benefit from modern technology. Using superconducting materials cyclotrons today are much more compact than systems of the same output energy 20 or 30 years ago. Similar developments are foreseeable for synchrotrons, and attempts to compactify medical synchrotrons exist, mostly in Russia.

The patents for one such system has been purchased by Texas based ProTom International, Inc. Offered under the name Radiance 330 it is a compact proton synchrotron with fully variable beam energy ranging from 30-330 MeV, has an external ring diameter of approximately 5 meter and weighs only 15 tons. The system is offered in a standard configuration of separate accelerator room and treatment room(s), but has not yet been FDA approved.

But this must by no means be the smallest size. At the Budker Institute for Nuclear Physics in Novosibirsk a table top synchrotron has been designed with a footprint of only 1.6 x 1.6 m^2 for an energy range from 12 to 200 MeV and an RF sweep from 7.4 to 26.5 MHz in 3.5 milliseconds. Multi-turn injection had been demonstrated but the stored beam was 2 orders lower in intensity than expected. Also, at the time no RF was installed and no extraction was provided. The project was abandoned for unknown reasons, but a system like this could easily rival the compact cyclotrons, and could even be mounted on a gantry, similar to the Still River system.

30.2.3 Fixed Field Alternating Gradient (FFAG) accelerators

Aside from the standard cyclotrons and synchrotrons, used nowadays for virtually all hospital based particle therapy installations, the fixed-field alternating gradient accelerator (FFAG) promises to combine the advantages of continuous operation found in cyclotrons with the flexibility of active energy change available in a synchrotron. In a FFAG variable energy extraction at kHz rate is possible.

The FFAG closely resembles a synchrotron, using a specific combination of focussing and deflecting magnets to steer the particles onto a closed orbit, intersecting acceleration cavities on each loop. The main difference consists of the fact that instead of varying the magnetic field strength during the acceleration process in order to keep the particles on the prescribed orbit, the FFAG uses a temporally fixed field with a strong radial gradient. Particles with larger energies move to slightly

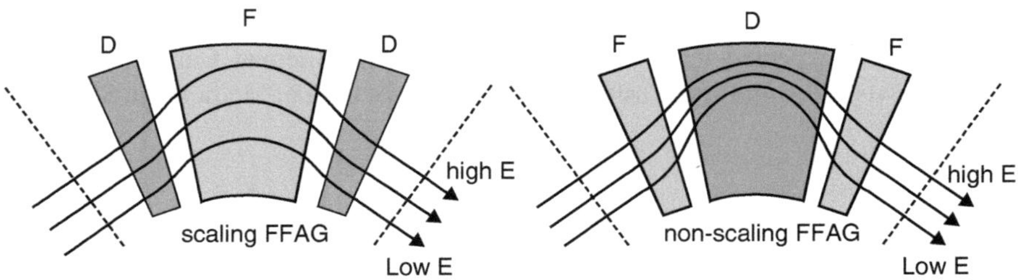

Fig. 30.1 Principle of magnetic system of Fixed Field Alternating Gradient accelerator. Scaling FFAGs (left) produce constant orbit shapes, avoiding beta-tune resonances. If the acceleration is sufficiently rapid, non-scaling FFAG's (right) which are much easier to design can be employed

larger orbits, where the bending field is larger, thus assuring that all particles remain confined to a narrow ring. This resembles the operation of a synchrotron but does not require the machine to be operated in pulsed acceleration cycles.

FFAG's are not exactly a new invention. The idea of fixed-field alternating-gradient synchrotrons was developed independently in Japan, the United States, and Russia by Tihiro Ohkawa [4], Keith Symon [5], and Andrei Kolomensky [6], and the first prototype was already operational in early 1956.

But as the magnets needed for FFAGs are quite complex, only through improved computer modeling and magnet technology available during the last decade has the technical realization of such a machine become possible. An important step was the realization that some of the complexity of the magnet design can be relaxed if the acceleration process is sufficiently fast. In the standard FFAG design the bending field increases at a high power of the radius in such a way that the higher energy orbits move outwards without changing shape see Fig. 30.1. The advantage of this design, called *scaling* FFAG is in avoiding betatron oscillations, and guaranteeing beam stability throughout the acceleration process. But if the acceleration process is fast enough, the particles can pass through any possible betatron resonances without building up a damaging oscillation amplitude. This permits the dipole field to increase only linear with radius, making the magnets smaller and simpler to construct.

These newer, *non-scaling* FFAGs are prime candidates for medical physics type accelerators and are the basis of two projects in the United Kingdom, EMMA (Electron Model for Many Applications) and PAMELA (Particle Accelerator for MEdical Applications).

EMMA is located at the Daresbury Laboratory in the UK and is intended as a proof of principle for stable operation of *non-scaling* FFAGs. The ring of approximately 5 m diameter comprises 42 cells with 19 acceleration cavities [7]. EMMA, being intended as a proof of principle experiment, aims at electron acceleration from 10 to 20 MeV and is flexible in design, allowing rapid repair and changes, but also equipped with a large variety of diagnostic equipment. The project achieved a major milestone in 2010 circulating the first beam in the ring.

In the same laboratory, PAMELA (Proton Accelerator for MEdicaL Applications) has been initiated as a design study for Hadron Therapy. It has been found that the lattice design of EMMA could not be transferred directly to a proton/carbon machine [8] as it was difficult to achieve the packing density with a realistic magnet design, making the space in the straight sections insufficient for the RF acceleration cavities. A new lattice design has been proposed [9] to overcome these difficulties. This design offers extraction energies up to 250 MeV for protons and 68 MeV/amu for carbon ions and consists of 12 cells on a radius of 6.25 meters.

30.2.4 Linear Accelerators

Another interesting concept merging the continuous beam production of a cyclotron and the flexibility in energy selection offered by RF acceleration systems has been proposed and developed by U. Amaldi from the TERA (TErapia con Radiazioni Adroniche) Foundation. A beam extracted from a cyclotron at an energy of 30 - 60 MeV is chopped at 200 - 400 Hz and injected into a 3 GHz side coupled linac (SLC) for acceleration from 30 to 210 MeV. One of the advantages of the cyclotron/linac combination, named Cyclinac, is the availability of direct beams from the cyclotron for the production of radio pharmaceuticals, as well as for eye tumor treatments. Hospitals which already own a radio pharmacy could add a linear accelerator to provide an energy boost to the output of the cyclotron and add proton therapy to their program.

The ensuing beam consists of a sequence of 5 μsec pulses with a delay time between pulses of 2.5 - 5 ms. The number of particles per pulse can be varied from pulse to pulse within milliseconds by changing the power delivered to the accelerating modules of the linear accelerator and by changing the charge injected into the acceleration chain from a computer controlled ion source. The output energy of the system is variable as it is in a synchrotron, the time structure of the beam is ideal for spot scanning, and the transverse emittance is about 10 times smaller than found in cyclotrons and synchrotrons.

Latter allows for smaller gaps in focussing and deflection magnets, reducing the cost of beam delivery systems and gantries. The original design for the SLC called for a 11 m long structure with 24 tanks and permanent quadrupole magnets, an RF peak power of 30 MW, and a beam current of 33 μA [10]. A four tank prototype named LIBO (LInear BOoster) was constructed and tested at CERN. Each of the tank consisted of 15-16 accelerating cells and between two successive tanks a FODO structure of permanent magnetic quadruploe magnets was installed for beam focussing. This system achieved an acceleration of protons from 62 to 74 MeV in 1 meter distance [11]. Based on this success a complete design has been finished. The first fast cycling accelerator is planed for the Institute for Diagnostics and RAdiotherapy (IDRA) in Biella, Italy. The same concept can be applied to the production of carbon ion beams for therapy, and a design of a Carbon BOster for Therapy in Oncology (CABOTO) has been developed. Using 18 modules with 3 tanks consisting of 17-21 acceleration modules each, carbon ions and hydrogen molecules can be accelerated from 120 MeV/amu to 400 MeV/amu.

Not only is the combination of a low energy cyclotron and a linac comparable in size and carries only 1 quarter of the weight of a super conducting cyclotron [12] for carbon ions of 700 tons, but the transverse emittance of the linac beam is about 5 times smaller than the one of standard accelerators. This makes it possible to use smaller and lighter magnets for the beam delivery system. A high-frequency proton linac rotating around the patient has been patented by TERA [13]. A 6 GHz radiofrequency has been chosen for TULIP (TUrning LInac for Proton therapy) to reduce the length of the linac sections. The beam is deflected after the cyclotron by a first bending magnet, accelerated to an intermediate energy using part of the SLC structure, bent to the horizontal direction again, accelerated to the full energy, and finally bent to a vertical down direction. Such a rotating linac can provide a proton beam cycling at hundreds of Hertz, ideal for treatment schemes like distal edge tracking [14].

30.3 Novel Technology Approaches

Modern particle accelerators have become known to grow ever bigger with each increase of energy range being explored. Synchrotrons increase in radius to allow higher and higher energies to be confined on a closed orbit with existing magnet technology. The currently largest circular accelerator, the LHC at CERN, is occupying an underground tunnel of 27 km circumference. This has led to the understanding that the design of the next generation accelerators will be based on linear geometries. But even these machines are getting larger and larger, and can easily occupy 100's of meters or more. The main reason for this is that all accelerator schemes currently used rely on accelerating particles in a vacuum waveguide with a relatively modest upper limit in break down voltage of a few tens of MeV per meter. To achieve drastic progress in reducing the size of future accelerators an entirely new approach is needed, and medical applications may possibly benefit from these developments.

During the last decade new technologies for the production and linear acceleration of heavy charged particles have been emerging and are frequently discussed, amongst other applications, in the context of hadron therapy. The direct drive accelerator (DDA) [15], the dielectric wall accelerator (DWA), and the acceleration using laser produced plasmas are the most important examples and the two latter ones will be discussed here.

30.3.1 Dielectric Wall Accelerator

The dielectric wall accelerator (DWA) system, employing a variety of advanced concepts to achieve extreme high electric field gradients, is being developed at the Lawrence Livermore National Laboratory by the group of G. J. Caporaso

[16]. Originally developed as a possible replacement for high current electron accelerators for x-ray production it uses high voltage transmission lines and fast switching devices to generate a pulsed electric field on he inside of a high gradient insulating tube. A traveling wave is set up on the axis of the tube which pushes the bunch of charged particles injected from an appropriate ion source forward. Alternating insulators and conductors together with short pulse durations for the applied voltage pulses have proven to provide electric field strength of 100 MV/m and beyond. The design is based on the concept of a coreless induction accelerator first introduced in 1970 [17].

The three critical technical components of the system are the high gradient insulators (HGI) [18], dielectric materials with high bulk breakdown strength [19] for the pulse forming lines, and very fast closing switches compatible with operation at high voltage gradients [20].

Failure of an insulator in vacuum, i.e. surface flashover, occurs at a field strength inversely proportional to the width of the applied electrical pulse. This observation is the key motivation for the operational mode of the accelerator proposed by Caporaso. A high gradient insulator (HGI) consists of alternating layers of conductor and insulator, typically on a scale of less than 1 mm. The HGI still exhibits the same inverse dependence of flashover field strength on pulse width, but has an improved overall performance by a factor of 4 [19]. The reason for this performance enhancement can be found by studying the trajectories of electrons generated by field emission near the surface of the HGI. While in standard monolithic insulators these electrons frequently bombard the surface, leading to an amplification of the charge density, the periodic microstructure along the surface of an HGI deflects the electrons away from the insulator surface, provided the ratio of conductor to insulator thickness lies in a certain range [21].

While original coreless induction accelerators used liquid dielectrics it is highly desirable to replace these with solid materials. Castable dielectric materials with high bulk breakdown strength and variable permittivity have been developed for some time and a particular solution is a blend of nanoparticles of $BaSrTiO_2$ and various epoxy bases. Dielectric constants between about 3 and 45 have been achieved by varying the concentration of the nanoparticles in the epoxy. A sample transmission line of dimensions 4 cm $\times$ 56 cm with embedded electrodes at 0.8 mm separation was constructed and charged repeatedly with 400 ns wide pulses at increasing voltage amplitude. Failure occurred at 141 kV, representing a field stress of 170 MV/m.

In order to generate the traveling pulse wave on the inside of the HGI tube in Fig. 30.2 "Blumlein" structures coupled to fast closing switches need to be employed. These switches should exhibit fast rise times and low resistance in the "on" position. One possible candidate under investigation is the use of wide band gap materials illuminated below bandgap width by lasers. As the below bandgap light is able to penetrate the material several centimeters one can place electrodes on opposite sides of thin wavers to take advantage of the full bulk breakdown strength of the material [20]. Initial tests performed by the group of G. Caporaso failed at a field strength across the gap of only 30 MV/m, but it could be concluded that this was due to edge

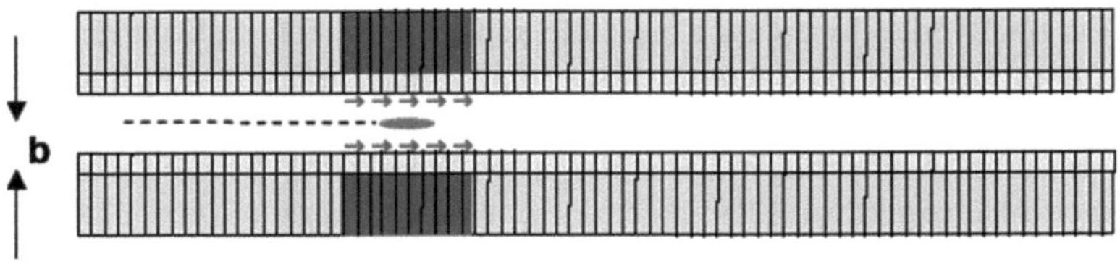

Fig. 30.2 Architecture of a short pulse DWA for proton therapy. The beam tube with diameter b is a high gradient insulator. The vertical lines indicate the conductors of the transmission lines that supply pulsed voltages across the HGI. The arrows denote the tangential electric field along the insulator surface that provides the acceleration of the particle bunch

effects enhancing the field by a factor of about 10 to 300 MV/m, which is in excess of the bulk breakdown strength of the material. The group is currently working on developing an enhancement free, integrated switch package [16].

A test system named F.A.S.T. for First Article System Test [16] comprising a small number (7) of Blumleins, photoconductive switches, and a high gradient insulator tube has been constructed and serves as testbed for the fully integrated system. The number of Blumleins, and thereby the maximum acceleration achievable, has been kept small to allow easy disassembly and repair in case of failure on some components. In order to accelerate protons through F.A.S.T. an initial velocity sufficient to cross the HGI gap during the 3 ns pulse duration, requiring at least 200 keV at the entrance of F.A.S.T. is required. This was achieved using 5 20 ns induction cells.

The technology, showing a clear promise towards a lightweight, compact (2 meters or less) source of 200 MeV protons, could potentially revolutionize proton beam cancer therapy, making it available to small hospitals and private clinics, as the foreseen size of the overall system would allow to retrofit existing IMRT installations to provide intensity modulated proton therapy (IMPT) instead. But before patients could benefit from protons accelerated by a Dielectric Wall Accelerator many steps beyond achieving the required energy will be necessary. Exact dosimetric information for the beam from the DWA will need to be established. Beam delivery concepts and treatment planning studies need to be performed and verified by in-vitro and in-vivo experiments. As with all projects relying on new technology, it is not possible to predict when it will become available commercially, but strong commercial interest is exemplified by Tomotherapy, Inc. joining the effort and providing partial funding of the continuing development. An artist's conception of such a future installation is shown in Fig. 30.3.

30.3.2 Laser Plasma Accelerators

If a high-intensity laser [22] or a particle beam [23] passes through a plasma, a fast, large amplitude plasma wave, also known as a wakefield, is generated. Field

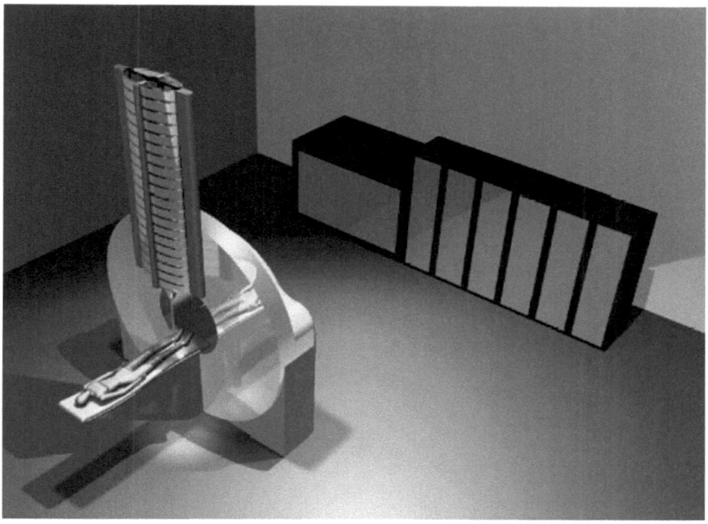

Fig. 30.3 Artist's rendering of a Dielectric Wall Accelerator based Particle Therapy System

strengths in such a wakefield can exceed a hundred gigavolt per meter. With the development of ultra intense lasers in the last decade the possibility is suggested that compact accelerating systems could be build at a fraction of the size and cost of standard systems available today.

During the last decade the field has evolved rapidly. Ultrahigh intensity lasers have been shown to achieve accelerating fields in the range of 10 TV/meter, surpassing the field strength of standard accelerators by sixth orders of magnitude, and the generation of mono energetic electrons of GeV energies have been reported [24–26]. Laser-based acceleration of protons and ions also has seen significant progress in recent years and many authors have suggested this technology as an alternative for particle beam therapy [27–29].

Laser-driven ions in the MeV/u range also have been reported to exhibit short pulse length, high intensity, and low transverse emittance. Unfortunately their exponential energy spectrum had an almost 100% energy spread [30, 31].

The underlying physics of laser plasma acceleration of ions is the process of target normal sheath acceleration (TSNA) [32]. Initially the ultra-high intensity laser pulse incident on a target accelerates a significant number of electrons to energies of several MeV. These electrons can traverse thin target foils, generating electric fields in excess of 1 TV/m. This field ionizes material present on the back surface of the foil and accelerates the ions. Predominantly one finds protons from hydrocarbon contamination on the the target surface, but heavier ions can be generated after careful cleaning of the surface. Experiments have reported acceleration of protons to more than 60 MeV [30], fluor ions to above 100 MeV [31], and palladium to 225 MeV [33].

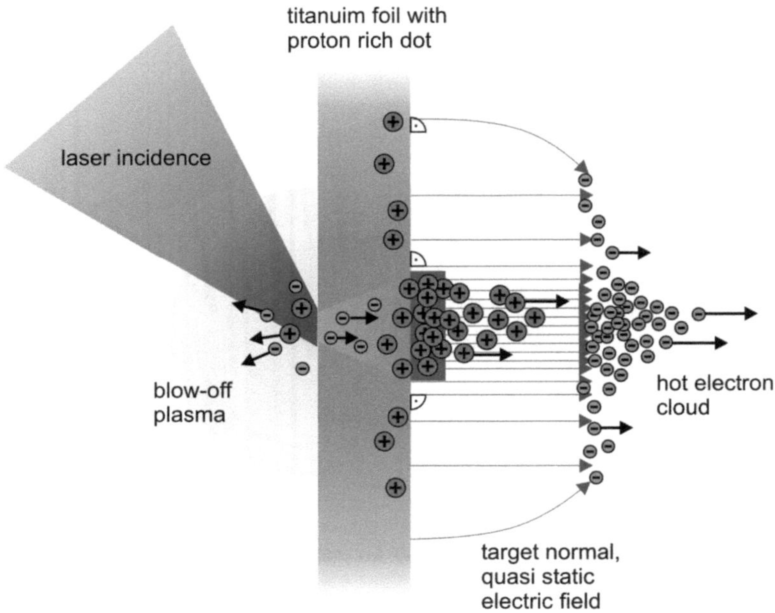

Fig. 30.4 A terawatt laser pulse is focused onto the front side of the target foil, where it generates a blow-off plasma and subsequently accelerates electrons. The electrons penetrate the foil, ionize hydrogen and other atoms at the back surface and set up a Debye sheath. The distribution of the hot electron cloud causes a transversely inhomogeneous accelerating field

The problem of the very broad, exponential energy spectrum can be explained by the inhomogeneous distribution of electrons in the sheath causing a transverse inhomogeneous accelerating field. Ions originating from different locations on the back surface of the target foil experience different accelerating fields. This problem has been overcome by using micro-structured targets consisting of a thin high Z metal foil and one or several small proton rich dots on the back surface [34]. If the transverse dimensions of such dots is smaller than the acceleration sheath the produced protons are only subjected to the central part of the field and thus experience a homogeneous potential. Following a proposal by Esirkepov et al. [35] the group of H. Schwoerer [34] used an experimental set-up where the target consisted of a 5 μm titanium foil coated with 0.5 μm of PMMA on the back surface. The PMMA layer of the sample was microstructured, leaving PMMA dots of $(20 \times 20)\ \mu\text{m}^2$. The laser was then aligned to hit the target foil directly opposite of one of these dots. The laser used in this experiment was a 10 TW Ti:Sapphire laser with 80 fs pulse duration, 10 Hz repetition rate, and a pulse energy on target of 600 mJ. The generated protons were energy analyzed and the spectrum exhibited a narrow feature peaked around 1.2 MeV on top of a broad, exponential background. This feature contained typically 10^8 protons per 24 msr, and had a full-width half-maximum of 300 keV (or 25% of the mean energy).

Hegelich et al. report production of a quasi-monoenergetic C^{5+} beam using a 20 TW/0.8 ps laser pulse from the Los Alamos National Laboratory TRIDENT laser facility [36]. In typical vacuum conditions of 10^{-6} mbar surface contaminations containing protons are always present. As these protons have the largest charge-to-mass ratio, they are accelerated predominantly. Pre-treatment of the target foils at temperatures of about 1,100K using a current passing through the target can reduce adsorbed protons to an unobservable level, allowing higher-Z ions to be the dominant species [31].

The single shot nature of the TRIDENT laser and the shot-to-shot fluctuations of the laser parameters complicate the analysis, but the group reports monoenergetic C^{5+} and C^{6+} formed from a an ultra-thin layer of graphitic carbon on a 20 μm palladium foil. C^{5+} with a charge-to-mass ratio of 0.42 is accelerated dominantly, and due to the small spatial extent of the carbon layer, all of the carbon ions are accelerated at once at the peak of the accelerating field. Following the acceleration of the carbon ions the field is still very strong and the next highest charge-to-mass ratio ions, Pd^{22+}, gain a large fraction of the energy before the field decays. For the leading bunch of C^{5+} ions a mean energy of 36 MeV (3 MeV/amu) with a full-width half-maximum of 0.5 MeV/amu was observed. Computer simulations using the code BILBO (Backside Ion Lagrangian Blow-Off) point to a decrease of attainable energy and an increase of the width of the energy spectrum with increasing thickness of the carbon layer.

Achieved particle energy of these experiments is already in the right range for fusion applications, but the number of ions will have to be drastically increased. For medical applications this is just the other way around, the particle number is sufficient, but the energy will have to be increased dramatically. With the rapid development of ultra-high intensity lasers it is expected that this progress will come rapidly.

The group at the Research Center Dresden-Rossendorf (FZD) using the 150 TW laser DRACO (Dresden Laser Acceleration Source) reports proton energies of 17 MeV from unstructured targets and describes a linear scaling of proton energy with laser power. This scaling indicates that clinically relevant energies of laser-accelerated protons should be achievable at the level of 1 petawatt [37].

At the Institute of Optics and Quantum Electronics at the University of Jena the fully diode pumped ultrahigh peak power laser system POLARIS (Petawatt Optical Laser Amplifier for Radiation Intensive Experiments) is being developed. Presently, this laser system reaches a peak power of some ten terawatt [38]. The last amplifier, which will boost the output energy to the 100 J level, is nearly completed and will soon be commissioned. The expected performance of the final system is 1 Petawatt power, 100 J energy on target, 120 fs pulse duration, and a repetition rate of 0.03 Hz. Such a system will provide the required high repetition rate of the laser system driving the initial acceleration process to allow varying as many experimental parameters as possible to study the physics underlying the acceleration processes.

The fact that these developing laser systems are true table-top systems and that the light beams can be delivered under any angle in space to the production targets using a simple set of mirrors could allow for a very compact and light weight gantry system [39]. The primary laser beam could also be shared amongst several production targets, allowing a multiple room facility on a small foot print.

But before such systems can be realized a long line of physical and biological research will be necessary to understand the real-time dosimetric properties of the beam and any deviations in relative biological efficiency of laser-accelerated protons from standard sources which may be due to the extremely different time structure of the beam delivery. The high pulse dose, the broad energy spectrum, and the short pulse duration present formidable challenges to the dose planning. First in-vitro irradiations have been performed [42]. To facilitate studies of biological outcome in terms of a dose response relation ship it is necessary to develop precise dosimetry, allowing the determination of the energy spectrum and pulse-to-pulse fluctuations of the beam profile and intensity. The group of J. Pawelke at the University of Dresden has designed, constructed, and tested such an integrated Dosimetry and Cell Irradiation system (IDOCIS), incorporating a Faraday cup insert for precise absolute dosimetry and a cell holder insert for the positioning of cell samples or dosimetric detectors like EBT radiochromic films and CR-39 solid track detectors in the incoming proton beam [43].

Efforts have started to sharpen the energy spectrum of the ion beam using magnetic spectrometers and moveable apertures [40]. With the help of high magnetic fields protons with different energies can be separated in space allowing the selection of only parts of the spectrum by blocking the unwanted particles. Afterwards, with further magnetic fields, the desired particles can be merged again into one beam. One straightforward way of delivering radiation therapy that is similar to the conventional procedure is to use a small collimator within such a device to get monoenergetic beams. These can then be used to produce spread out Bragg peaks (SOBPs) by varying the selected energy and the beam intensity between several laser shots. As this would potentially lead to a drastic reduction in available beam intensity, Wilkens and Schell [41] have proposed a method based on the above energy selection, which, instead of selecting only one beam energy modifies the relative intensity of the different energy bins such that a SOBP can be produced in a single shot.

30.4 Conclusions

The sharp increase in particle beam therapy has initiated an active discussion on a number of approaches on how to minimize the size and cost of particle therapy facilities in order to allow a higher accessibility to this promising treatment modality. From simple modifications of standard designs to novel acceleration schemes many ideas have been brought into this discussion. None of these systems has reached the level of approval for medical use yet. The next years will show if reduced systems with co-located accelerator and patient treatment table can meet the safety standards set for medical applications. Initial tests of the dielectric wall accelerators show promising results, but a full energy system still needs to be build, the beam characteristics studied, and a delivery system including online monitoring and quality assurance developed and certified. Laser plasma accelerators are progressing rapidly and first in vitro experiments have already been performed. It will be interesting to see how this technology can be scaled up to clinically relevant energies. While all proposals discussed here could eventually lead to smaller, and thereby cheaper, treatment centers, one should not forget the fact, that the utmost important component in all this is the patient. The requirement on medical equipment are very stringent and new ideas have to undergo numerous tests before being allowed for medical use. The development of particle therapy began in 1946 with Robert Wilson's seminal paper, and more than 60 years later, cancer therapy with particle beams from cyclotrons and synchrotrons is still improving through careful studies of biological effects, treatment delivery options, online monitoring and quality assurance. Any new technology entering the field will have to measure up to these standards, for the safety of the patients and for the continuing growth of this field. But we can be convinced that particle beam cancer therapy will conquer a growing position in cancer care.

Acknowledgements This work was partially supported by the DFG under contract WE3565-3 and the NSF under grant # CBET 0853157. MHH acknowledges support by the EU through a Marie Curie Fellowship under contract # PIIF-GA-2009-234814.

References

1. U. Mock et al., Radioth. Oncol. **73** Sup. 2 S29-S34 (2004); B. Glimelius et al., Acta Oncologica **44** 836- 849 (2005)
2. J.B. Flanz, Nucl. Inst. Meth. B **261** 768-772 (2007)
3. www.stillriversystems.com
4. T. Ohkawa, Phys. Soc. Japan Symp. Nucl. Phys. Tokyo, (1953)
5. K. Symon, D. Kerst, L. Jones L. Laslett, Terwiliger K., Phys. Rev. **103** 1837 (1956)
6. A. A. Kolomensky, Sov. Phys. JETP **6** 231 (1957)
7. R. Edgecok et al., Proc. of EPAC 2008, Genoa, Italy (2009)
8. K. Peach, et al., Proc of PAC 2009, Vancouver, BC, Canada
9. S. Sheehy, Proceedings pf PAC 2009, Vancouver, BC, Canada
10. U. Amaldi, et al., Nucl. Inst. Meth. A **521** 512-529 (2004)
11. U. Amaldi, et al., Nucl. Inst. Meth. A **620** 563-577 (2010)
12. http://archade.fr/english/
13. U. Amaldi, S. Braccini, G. Magrin, P. Pearce, R. Zennaro, Ion acceleration system for medical and/or other applications, Patent WO 2008/081480 A1
14. U. Oelfke, T. Bortfeld, Technol. Cancer Res. Treat. **2-5** 401 (2003)
15. O. Heid, T. Hughes, Proceedings of HB2010, Morschach, Switzerland (2010)
16. G.J. Caporaso, et al. Proceedings PAC 09, Vancouver, BC, Canada (2009)
17. A. Pavloski, et al., Sov. At. En. **28** 549 (1970)
18. S.E. Sampayan, et al., IEEE Trans. Diel. and Elec. Ins. **7** (3) pg. 334 (2000)

19. J. Harris, et. al., Appl. Phys. Lett. **93**, 241502 (2008)
20. J. Sullivan, J. Stanley, IEEE International Power Modulator Symposium (27th) and High-Voltage Workshop, Washington, DC, May 14-18, p. 215 (2006)
21. J. Leopold, et. al., IEEE Trans. Diel. And Elec. Ins. **12** (3) 530 (2005)
22. T. Tajima, J.M. Dawson, Phys. Rev. Lett. **43** 267 (1979)
23. P. Chen, J.M. Dawson, R.W. Huff, T. Katsouleas, Phys. Rev. Lett. **54** 693-696 (1985)
24. J. Faure, et al., Nature **431** 541-544 (2004)
25. S. Mangles, et al., Nature **431** 535-538 (2004)
26. C. Geddes, et al., Nature **431** 538-541 (2004)
27. M.J. Martin, Natl. Cancer Institute **101** 450-451 (2009)
28. G. Kraft, S. Kraft, New J. Phys. **11** 025001
29. P.R. Bolton, et al., Nucl. Inst. Methods A**620** 1392-1400 (2010)
30. R.A. Snavely, et al., Phys. rev. Lett. **85** 2945-2948 (2000)
31. M. Hegelich, et al., Phys. Rev. Lett. **89** 085002-1-4 (2002)
32. S. Hatchett, et al., Phys. Plasmas **5** 2076-2082 (2000)
33. M. Hegelich, et al., Phys. Plasmas **12** 056314 (2005)
34. H. Schwoerer, et al., Nature **439** 445-448 (2006)
35. T. Esirkepov, et al., Phys. Rev. Lett. **89** 175003 (2002)
36. B.M. Hegelich, et al., Nature **439** 441-444 ((2006)
37. K. Zeil, et al., NJP **12** 045015 (2010)
38. J. Hein, et al., Conference Proceedings LEI 2009, ed. Dan Dumitras, AIP Conf. Proc. **1228** 159-174
39. C.-M. Ma, et al., Laser Physics **16** 639-646 (2006)
40. W. Luo, E. Fourkal, J. Li, C.-M. Ma, Med. Phys. **32** 794806 (2005)
41. S. Schell, J.J. Wilkens, Phys. Med. Biol. **54** N459-N466 (2009)
42. S.D. Kraft, et al., **12** 085003 (2010)
43. C. Richter, et al., Phys. Med. Biol. **56** 1529-1543 (2011)

Index

G.G. Gómez-Tejedor and M.C. Fuss (eds.), *Radiation Damage in Biomolecular Systems*, Biological and Medical Physics, Biomedical Engineering,
DOI 10.1007/978-94-007-2564-5, © Springer Science+Business Media B.V. 2012

MIX
Papier aus verantwortungsvollen Quellen
Paper from responsible sources
FSC® C105338
FSC
www.fsc.org

If you have any concerns about our products,
you can contact us on
ProductSafety@springernature.com

In case Publisher is established outside the EU,
the EU authorized representative is:
Springer Nature Customer Service Center GmbH
Europaplatz 3, 69115 Heidelberg, Germany

Printed by Libri Plureos GmbH
in Hamburg, Germany